Chemistry

FOR CHRISTIAN SCHOOLS®

SECOND EDITION

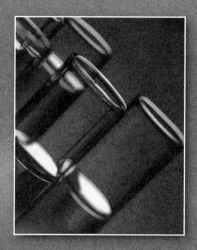

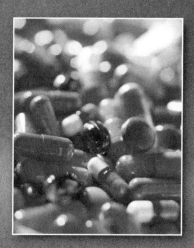

Chemistry
FOR CHRISTIAN SCHOOLS®
SECOND EDITION

Heather E. Cox
Thomas E. Porch, D.M.D.
John S. Wetzel, M.S.

 Bob Jones University Press
Greenville, South Carolina 29614

Consultants and Contributors

George Matzko, Ph.D.
Professor and
  Department Head,
  Department of Chemistry
Chairman,
  Division of Natural Science
Bob Jones University

Verne Biddle, Ph.D.
Professor,
  Department of Chemistry
Bob Jones University

Bill Harmon
Instructor
Bob Jones Academy

Contributors to the First Edition

George L. Mulfinger, M.S.
Brian S. Vogt, Ph.D.
John R. Wolsieffer, Ph.D.
John E. Jenkins

Editors

Doug Rumminger
Robert Grass

Composition and Design

US Color

The fact that materials produced by other publishers may be referred to in this volume does not constitute an endorsement of the content or theological position of materials produced by such publishers. Any references and ancillary materials are listed as an aid to the student or the teacher and in an attempt to maintain the accepted academic standards of the publishing industry.

"The Ant" from *Verses from 1929 On* by Ogden Nash. Copyright© 1935 by Ogden Nash. First appeared in *The Saturday Evening Post*. Reprinted by permission of Little, Brown and Company, Inc.

CHEMISTRY for Christian Schools®
Second Edition

Heather E. Cox
Thomas E. Porch, D.M.D.
John S. Wetzel, M.S.

Produced in cooperation with the Bob Jones University Division of Natural Science of the College of Arts and Science and Bob Jones Academy.

for Christian Schools is a registered trademark of Bob Jones University

©2000 Bob Jones University Press, Greenville, South Carolina 29614

All rights reserved. Printed in the United States of America.

ISBN 1-57924-421-1

EAN 978-1-57924-421-7

15 14 13 12 11 10 9 8 7 6 5 4 3

CONTENTS

Introduction vii
Chapter 1: Science, Chemistry, & You 1
   1A Science and the Scientific Method 1
   1B The History of Chemistry 6
   1C Chemistry and You 13

Chapter 2: Matter 21
   2A The Classification of Matter 21
   2B Energy in Matter 29
   2C The States of Matter 36

Chapter 3: Math: The Central Language of Science 43
   3A The Measurement of Matter 43
   3B Reporting Measurements 48
   3C Organized Problem Solving 58
   3D Density 60

Chapter 4: Atomic Structure 67
   4A The Development of Atomic Models 67
   4B The Quantum Model 80
   4C Numbers of Atomic Particles 88

Chapter 5: Elements 101
   5A The Periodic Table 101
   5B Periodic Trends 110
   5C Descriptive Chemistry 117

Chapter 6: Chemical Bonds 135
   6A How and Why Atoms Bond 135
   6B The Quantum Model and Bonding 148

Chapter 7: Describing Chemical Composition 163
   7A Oxidation Numbers 163
   7B Nomenclature 170
   7C The Mole 181

Chapter 8: Describing Chemical Reactions 199
   8A Writing Equations 200
   8B Types of Reactions 207
   8C Stoichiometry 212

Chapter 9: Gases 227
   9A The Nature of Gases 227
   9B Gas Laws 234
   9C Gases and the Mole 242

Chapter 10: Solids & Liquids 259
   10A Intermolecular Forces 259
   10B Solids 262
   10C Liquids 272

Chapter 11: Water ... 289
- 11A The Water Molecule ... 289
- 11B The Reactions of Water ... 298
- 11C Water in Compounds ... 304

Chapter 12: Solutions ... 309
- 12A The Dissolving Process ... 309
- 12B Measures of Concentration ... 317
- 12C Colligative Properties ... 322
- 12D Colloids ... 328

Chapter 13: Thermodynamics & Kinetics ... 337
- 13A Thermodynamics ... 337
- 13B Kinetics ... 351

Chapter 14: Chemical Equilibrium ... 365
- 14A Theories of Chemical Equilibrium ... 365
- 14B Applications of Equilibrium Chemistry ... 377

Chapter 15: Acids, Bases, & Salts ... 389
- 15A Definitions and Descriptions ... 389
- 15B Equilibria, Acids, and Bases ... 393
- 15C Neutralization ... 406

Chapter 16: Oxidation–Reduction ... 417
- 16A Redox Reactions ... 417
- 16B Electrochemical Reactions ... 431

Chapter 17: Organic Chemistry ... 443
- 17A Building an Organic Compound ... 443
- 17B Hydrocarbons ... 446
- 17C Substituted Hydrocarbons ... 456
- 17D Organic Reactions ... 466

Chapter 18: Biochemistry ... 475
- 18A Carbohydrates ... 476
- 18B Proteins ... 481
- 18C Lipids ... 484
- 18D Cellular Processes ... 487

Chapter 19: Nuclear Chemistry ... 501
- 19A Natural Radioactivity ... 501
- 19B Induced Reactions ... 513

Appendixes ... 524

Glossary ... 528

Index ... 542

Photograph Credits ... 550

To the Student

Why are you studying chemistry this year? Is it just another course to fulfill a graduation requirement, or are you "science-oriented" and cannot wait to learn more about the world around you? In either case, keep two verses in mind: I Corinthians 10:31b, ". . .whatsoever ye do, do all to the glory of God," and Psalm 19:1, "The heavens declare the glory of God; and the firmament sheweth his handywork." Studying God's creation at the atomic and molecular level will reinforce your faith and enable you to realize the orderliness of all that surrounds you.

As you study chemistry this year, you will study about the makeup of matter, the design of atoms, the periodic table, bonding, formulas, reactions, solutions, acids, bases, and several specialized branches of chemistry—topics that make up a comprehensive high school chemistry course. Sounds overwhelming, doesn't it? This course does cover a lot of ground, but fundamental ideas are used to explain each concept. Words, not mathematical equations, are used to describe the content. For example, descriptions of the forces between positive and negative electrical charges help to explain where electrons exist in atoms, why molecules have unique three-dimensional shapes, why some substances dissolve in water while others do not, and why salt has a very high melting point.

You will find that this text has been designed to help you learn. Important terms have been boldfaced in the text and defined in the glossary. Sample problems, with their solutions, are included in the text. Review questions at the end of each chapter section and at the end of each chapter give you an opportunity to practice and solidify what you have learned. These tools can help you learn and understand the facts of chemistry better.

What Is Christian About Chemistry?

Some people have developed the idea that higher mathematics and science have little to do with the Bible or the Christian life. They may think that because chemistry deals with scientific facts, or because it is not pervaded with evolutionary ideas, there is no need to study it from a Christian perspective. Those with ideas such as these fail to see the many ways that Christianity and chemistry are related. Listed below are several reasons that Christian students should study chemistry.

1. Your knowledge of and faith in God can be increased by a detailed study of Creation. When God wished to show Job His wisdom, power, and greatness, He gave Job a tour of the universe that He had created (Job 38-41). After seeing the intricacies, the grandeur, and the

splendor of the universe, Job declared, "I know that thou canst do every thing, and that no thought can be withholden from thee. . . . I have heard of thee by the hearing of the ear: but now mine eye seeth thee" (Job 42:2, 5). Like Job, you can learn new things about God through a study of His creation.

2. Christians can be more effective for the Lord when they combine strong academics with genuine faith in God. Daniel, Moses, Paul, and Luke are examples of men who used their good education to serve the Lord. No matter what career the Lord has planned for you, you should prepare as well as you can. Chemistry is a necessary part of the preparation for many vocations.

3. Chemistry offers unique opportunities for the development of Christian character in your life. Discipline, diligence, accuracy, organization, inquisitiveness, and thoroughness are just a few of the characteristics that you can develop from your study of chemistry.

4. Christian students must be aware of scientific evidences for Creation. Often chemistry is involved in the debate on the origin of the universe and life. You should be informed about the laws of thermodynamics, the formation of chemical bonds, and the structure of atoms. While impressive scientific evidence will not win unsaved men to Christ, it is still important. Christians should be able to show that their faith is sound, reasonable, and superior to any alternative.

5. Spiritual truths must serve as moral guidelines when chemistry is applied in our society. Today's chemists can do many things. They can perform genetic engineering, build many new chemical factories, and make advancements in nuclear power. Yet sometimes the things scientists *could* do are not the things they *should* do. Wisdom, discernment, and sound values—things not learned from academic textbooks—must guide these decisions. The world needs dedicated Christians who are qualified to speak out about these important topics.

6. Christians have been instructed to subdue and care for God's creation. A knowledge of chemistry will help you know how to use, yet not abuse, what God has provided. Even if you are not a scientist, you are a member of a society that is grappling with the issues of acid rain, nuclear waste, clean air, noise pollution, energy sources, and water purification. Being a good steward means becoming informed of the issues and taking an active part in the use of God's creation.

A Plan of Attack

Tough nuts can be cracked open by a combination of force and strategy—one without the other will not work. Succeeding in chemistry is much like cracking open a nut: it takes a combination of diligent study and an intelligent approach. Nothing takes the place of study, but this plan of attack can make your work more efficient. If chemistry seems like a tough nut to crack, follow these suggestions:

1. Relate new material to previously learned concepts. By doing this, you will avoid being overcome with many isolated facts. Only a few concepts will be totally new to you. You will have seen most of the material in your physical science and biology courses.
2. Look for basic explanations of complicated ideas. What seems complicated will become more easily comprehended when you understand "why."
3. Develop an appreciation for God's creation. Seeing how marvelous it is will whet your appetite for more.
4. Expose yourself to the practical applications of chemistry. Many of the review questions illustrate how the theory in the text is applied in the real world.
5. Use the mathematical skills you have learned in the past. Most problems in chemistry rely on the math you learned in algebra. Organize your work, and do it neatly.
6. Keep an open mind about enjoying what you are learning. Small children often miss out on enjoying many delicious foods because they make up their minds ahead of time that they will not like them. Do not deny yourself the fascination of chemistry because you think you will not like it.

We hope that your study of chemistry will bring you an increased understanding of the world around you, a solid preparation for future studies, and a heightened interest in science. More importantly, may your studies this year help you to honor Christ and to do His will.

That in all things he might have the preeminence.
Colossians 1:18

1A Science and the Scientific Method   page 1
1B The History of Chemistry   page 6
1C Chemistry and You   page 13
   FACETS: The Scientific Method of Antoine Lavoisier   page 11

Science, Chemistry, & You

What's the Connection?

What is science? For centuries science was a branch of philosophy, and scientists were more like debaters than researchers. Scholars reasoned and argued about the nature of truth and the universe. In the late 1500s science changed drastically. Pure reasoning and argumentation gave way to observation and experimentation. This new way of examining the universe developed into modern science. As scientists studied the creation through observation and experimentation, chemistry emerged as a distinct discipline. With the rich legacy provided by Old Testament craftsmen, Greek philosophers, and medieval alchemists, the science of chemistry soon flourished. Chemistry now touches so many areas of our lives that every person should have some knowledge of this subject.

Science and the Scientific Method

Inductive Reasoning: Starting with the Facts

Science had remained a stagnant body of knowledge for nearly two thousand years before a certain man received a professorship at the University of Pisa. Within a few months, Galileo asked questions that revolutionized the scientific community. Instead of relying on prevailing secular and religious philosophies,

History of Chemistry

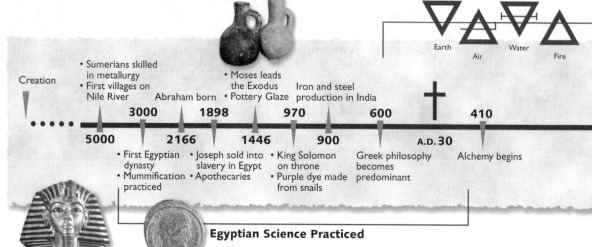

Egyptian Science Practiced

King Tutankhamen's headdress

Galileo and his telescope

he made observations and collected data on his own. He argued that inductive reasoning, not deductive reasoning, was the proper and logical method of approaching a problem. **Deductive reasoning** intends to prove that the conclusion must be true, whereas **inductive reasoning** intends to show that the conclusion is probable or most likely. In sound, valid deductive reasoning the argument is logical and always true. Inductive reasoning can present solid evidence, but still produce a faulty conclusion. Inductive reasoning could then be summarized as proceeding from the known fact to the unknown conclusion. This idea of inductive reasoning was vastly different from the prevailing belief that logic always leads to truth. In the past, observations that went against accepted philosophies were rejected. Many thought that although the senses could be deceived, rational thought could not. Galileo's experiments brought the conflict between logic and physical evidence to a climax. Just how strong was this controversy? Galileo was pronounced a heretic, imprisoned, and exiled by the Roman Catholic Church because he viewed the sun as the center of the solar system. At this time in history the Roman Catholic church firmly believed that the earth was the center of the solar system.

Timeline: Age of Greek Thought through Future of chemistry

Above the line:
- 1500 — Geber begins study of chemistry
- 1600 — Greek philosophy openly challenged
- 1774 — Lavoisier's Table of 31 elements
- 1792 — Dalton's atomic theory
- 1804 — Organic chemistry begins; Döbereiner introduces triad concept
- 1864 — (Dimitri Ivanovich Mendeleev)
- 1898 — Mendeleev publishes periodic table
- 1913 — Henry Moseley counts protons in a nucleus
- 2000 — 100 known elements

Below the line:
- 722 — Chemistry first used to treat disease
- 1600 — Priestley discovers oxygen
- 1790 — World's first chemical society in Pennsylvania
- 1802 — Wollaston discovers palladium & platinum
- 1829 — Newlands arranges elements by atomic mass
- 1869 — Curie discovers radium & polonium
- 1912 — Neils Bohr's atomic structure
- 1954 — 115 known elements

Marie Curie

Neils Bohr

Sample Problem Determine whether the following arguments use deductive or inductive reasoning.
   a. The rainfall in Seattle has been more than 20 in. every year for the past 10 years. Therefore, the rainfall next year will be more than 10 in.
   b. Mount McKinley is higher than Mount Whitney, and Mount Whitney is higher than Mount Washington. Therefore, Mount McKinley is higher than Mount Washington.

Solution
   a. This argument intends to show that the conclusion is probable or most likely, but cannot prove that the rainfall will occur. Therefore, the argument uses inductive reasoning.
   b. This argument compares statements and arrives at a conclusion that must be true if the statements are true. Therefore, the argument uses deductive reasoning.

The Scientific Method: Not a Formula, But a Process

Inductive reasoning starts with observations. After making observations, scientists often see a problem worth investigating. From their observations they propose general, tentative statements called **hypotheses** (hi PAH thuh seez). These hypotheses serve as a basis for further experimentation. To determine whether their hypotheses are true, scientists must gather more information, usually from carefully controlled experiments and surveys. Hypotheses, then, provide a way of organizing the data, and thereby suggest specific questions to be answered through the design of the controlled experiment. If the research supports or refutes the hypotheses, scientists can then frame theories. A theory that is backed by many additional observations and studies is sometimes called a scientific law. A scientific law holds true under normal circumstances.

In the Bible there are recorded exceptions to scientific laws. The laws defined by man do not confine God. Many of Christ's miracles are exceptions to known scientific laws. For instance, a law of science states that energy can neither be created nor destroyed, only changed in form. However, when Christ fed the 5,000 he created *ex nihilo* (from nothing) enough fish and bread to feed the crowd. Creation is also an exception to this law.

Scientists have modified and expanded the inductive approach to fit their needs. The result is called the **scientific method.** The scientific method of inquiry is a general, inductive approach to discovering information about our universe. This general process can be broken down into several steps: recognizing a problem, making observations, proposing a hypothesis, organizing and analyzing data, framing a theory, and verifying the theory. Is this list different from one you have seen before? If so, do not be surprised. There are many versions of the scientific method. Most of these versions can be considered correct, since they merely describe the same process with different words.

The inductive approach is fundamental to the study of the universe. Researchers use it to investigate the mysteries of God's creation and to learn how to use natural laws for man's benefit. **Science*** is the systematic study of nature based on observations. Although science is a large number of previously established facts, it is also an ongoing activity, a process.

Theories: Just a Beginning, Not an End

When hypotheses seem workable and are supported by observations, they are called **theories.** Theories play a double role in scientific research. They usually serve to solve scientific problems, but they often introduce new ones! Theories help scientists organize a body of data as well as suggest new avenues of research. A theory proposed by William Wollaston (WOOL uh stuhn) in the early 1800s served these two purposes. It helped

---

*science: *scio* (L. verb – to know) *Scientia* (L. noun – knowledge)

Wollaston solve a short-range problem, but it also led him to an unexpected, but major, scientific discovery.

Wollaston was searching for some material that he could use as a durable container while producing sulfuric acid. After carefully studying platinum, he concluded that it was inert. **Inert** means that it will not readily react with another chemical; therefore platinum would be an ideal material for his container. Unfortunately, the platinum metal produced in the 1800s was too brittle to be shaped into desired forms. Wollaston hypothesized that impurities caused the brittleness. Wollaston analyzed the metal and found some impurities that had been overlooked. He carefully isolated two substances and identified their properties. When he found no record of substances with these properties, he concluded that he had discovered two new elements. These elements, now called palladium and rhodium, have become important ingredients in high-temperature alloys. Wollaston's fortune resulted from his production of malleable platinum, but his fame came from his discovery of two elements. His theory about impurities in platinum solved one problem, but it also opened a new avenue of research.

William Wollaston

When you examine chemical theories, keep in mind that they are not above question. Theories should be examined, not blindly accepted. Thinking through the theories used in this text will provide a more thorough understanding of the concepts. Is it possible that you as a student could make a scientific discovery that would improve people's lives? Yes, but science requires both the mental exercise of logic and a knowledge of scientific theories. These two ingredients, logic and knowledge, will allow you to recognize a good idea when it passes by.

Scientific Laws: Descriptions, Not Rules

The scientific method relies on observations, but observations have limitations. Observations can be biased and in error; thus, conclusions resulting from the scientific method can also be in error. Consequently, the scientific method cannot determine truth with absolute certainty. This is an important point, for many students confuse scientific laws with "truth." Simply labeling an idea as a scientific "law" does not make that idea infallible. Only the Word of God is infallible. Many "laws of science" have met a hasty end when confronted with new evidence. Nonetheless, tested, workable **laws** do correctly describe the behavior of matter in God's universe and form the foundations of true science. Scientific discovery also relies heavily on the integrity of the scientist doing the experiment. Because of the large amount of published scientific research, only a small fraction of experiments are ever repeated by others. The scientist who is a

Christian must rely on the power of God to keep him from any form of scientific misconduct. Pressures are often placed on scientists, tempting them to cut corners in order to be the first to publish a new theory or discovery. Scientists who are successful have much prosperity in terms of financial grants, promotions, honor, and recognition in the scientific community.

If this discussion makes you think that scientific ideas are a bit unstable, you are right! But this instability is both a limitation and a strength of science. The constant examination of laws and theories allows scientists to upgrade their ideas with each new discovery. Laws and theories can and should be continually verified by new observations. While man's descriptions of creation may change, the God who established the foundational truths of the universe will not change. As Psalm 102:25-27 states, "Of old hast thou laid the foundation of the earth: and the heavens are the work of thy hands. They shall perish, but thou shalt endure: yea, all of them shall wax old like a garment; as a vesture shalt thou change them, and they shall be changed: But thou art the same, and thy years shall have no end."

Section Review Questions 1A

1. Determine whether the following statements use inductive or deductive reasoning.
   a. Dr. Blackford has performed 100 cardiac bypass surgeries, and all were successful. When Dr. Blackford performs this surgery on my grandpa Harvey it will be successful.
   b. Everything that the Bible teaches is true. The Bible teaches that man is sinful by nature. Therefore, it must be true that man is sinful by nature.
   c. If humanism is correct, then man is inherently good. Inherent goodness requires loving others; and yet war, deceit, and treachery have been continuously present in the world throughout human history. Thus, humanism must not be correct.
2. Formulate your own hypothesis and list one way in which you could test that hypothesis.
3. Discuss the importance of integrity in science.

1B   The History of Chemistry

Everything in the material universe is composed of chemicals. You are sitting on chemicals, wearing chemicals, and eating chemicals with every meal. **Chemistry,** then, is the study of chemicals—that is, all **matter** (anything that takes up space and has mass)—and the changes that the chemicals undergo. This definition summarizes the broad scope of chemistry. In the past, however, the scope of chemistry was not so broad. During ancient times, for example, chemistry had primarily a practical application. As time passed and theory and reasoning became important, chemistry became more philosophical. The emphasis eventually changed again as experimentation replaced reason. Today theory, experimentation, and practical application are all important parts of chemistry.

The Age of Practical Skills: Old Testament Times

Chemistry developed as a practical skill during Old Testament times. As the descendants of Adam gained the necessary skills to make weapons, tools, and other utensils, they laid the early foundations of chemistry. By the time of Abraham, man had already reached a surprising degree of chemical technology. The Sumerians, for example, were well skilled in **metallurgy*** (MET tal lur jee), the science of obtaining metals from their ores. Some of the first metals to be smelted were gold, silver, and copper. These soft metals were easily shaped into jewelry and coins. Old Testament metallurgists also manufactured bronze, an alloy made from molten tin and molten copper. The production of bronze provided the right material for making strong metal weapons.

Early apothecary shops sold many goods, but they focused on the chemicals used for medicinal purposes.

The Sumerians were not the only civilization developing their chemical technology at this time. The Egyptians were also making great progress, especially in medicine. By the time Joseph was sold into slavery, **apothecaries** (ah POTH ih care eez) were an important part of Egyptian culture. These early pharmacists prepared and sold a wide variety of chemicals and herbs. Several of these substances were even effective medicines. An apothecary's stockroom would reveal such "prescriptions" as copper salts, used as an antiseptic; magnesia, used as a laxative; opium, used as a sedative; and various herbs, used to treat diseases.

A clay tablet that dates from about the same period as the Exodus gives detailed recipes for ceramic finishes on pottery and tiles. The Phoenicians, who are credited with being the first to make soap, used snails to make a dye that served as the basis of a thriving textile business. Their purple cloth was a prized commodity at the time King Solomon was rebuilding the temple.

---

*metallurgy: metalurg (Gk. – metalworking)

The development of processed iron had a great military influence on the ancient world. Iron swords were used throughout the Middle East and Greek peninsula over one thousand years before Christ. Until then, the best armament had been bronze swords, but those splintered into pieces under a heavy blow from the stronger iron weapons. When the Philistines occupied Canaan during the days of Saul, they did not allow the Israelites to make iron weapons (I Sam. 13:19-22). Ironically, the Lord used young David's stone instead of an iron sword to defeat the champion of the Philistine army.

The Age of Critical Thought: Greek Influences

From 600 B.C. through the time of the early church, Greek philosophers debated their ideas about matter. Unlike their predecessors, the Greeks desired knowledge rather than practical skills. Luke noted this fact in Acts 17:21: "For all the Athenians and strangers which were there spent their time in nothing else, but either to tell, or to hear some new thing." The Greeks were the first to introduce an organized approach to chemistry. This use of critical thinking was important to the Greeks, for they prided themselves on their disciplined minds and bodies. Just as the Olympic competitions tested physical strength, logical arguments tested the mind.

The theories developed by the Greek philosophers were so logical that they dominated science for the next two thousand years. In some ways, however, this domination hindered the development of chemistry. The Greeks' high regard for logic and their lack of technology prevented them from making observations and doing experiments. Because the Greeks did not test their ideas, their theories contained several major errors. For instance, they thought that the universe was made of four elements: earth, air, fire, and water. Their critical thought and philosophy could carry science only so far. A new direction was needed for science to advance, and that direction was experimentation.

The Age of Applied Experimentation: The Alchemists

During the time of Christ, a large scientific community formed in Alexandria, Egypt. Greeks who traveled to this center of learning blended their deductive logic with the ancient skills of the apothecaries. It was in ancient writings from this Greco-Egyptian city that the word *chemia*, from which the word *chemistry* is derived, first appeared. Although some claim that *chemia* comes from the name of a legendary apothecary, the word probably

While alchemists are usually thought of as frauds or deluded fools, they made important contributions to the field of chemistry.

comes from the Egyptian word for "black." Either way, the label "black" fits well, for the strange skills practiced in Alexandria were much like black arts.

As the influence of the Roman Empire declined, the Arabs took control of Alexandria. Fascinated by the black arts, Arab scholars learned all they could. As "chemia" became a part of their culture, they added the Arabic prefix al (the) and called it **alchemy** (AHL keh mee). Alchemy spread into Europe with the Arab conquest of Spain. One of the first major alchemists in Europe was Jabir ibn-Hayyan. Geber [(JAE bare) the English version of his name] wrote many books about chemical techniques. He distilled acetic acid from vinegar, prepared nitric acid, and studied the transmutation of metals.

The alchemists gradually forsook practical studies to pursue the secret of converting ordinary metals into gold. Many royal courts financed such searches for an unlimited source of gold. Alchemists also believed that magical powders could cure diseases and extend lifespans. These false hopes fueled the legends of "elixirs of life" and "fountains of youth" that lured Ponce de León to the New World.

Despite their failures, alchemists did provide modern chemistry with a rich legacy of techniques and laboratory equipment. Alchemists developed techniques such as distillation, sublimation, precipitation, and crystallization. These techniques, in a

An early pharmacy

more refined form, are still used today. Alchemists were also master designers of glassware and porcelain. But the most important contribution of the alchemists was their experimental approach. Francis Bacon noted this contribution when he wrote:

> *Alchemy was like the man who told his sons he had left them gold buried somewhere in his vineyard; where they by digging found no gold, but by turning up the [dirt] about the roots procured a plentiful vintage.*

Alchemy made its first steps toward modern chemistry under the guidance of the Swiss alchemist Philippus Paracelsus. In the early 1500s Paracelsus promoted the use of chemicals to treat disease. This application of chemistry to medicine became the forerunner of modern pharmacology. **Pharmacology** is the science of making, using, and studying the effects of medicinal drugs. The relationship between chemistry and medicine is even more important today. As a result, all students studying to be medical doctors, nurses, or pharmacists are required to study chemistry.

The Rise of Modern Chemistry: The Transition

Not until the middle of the 1600s did chemists dispute the Greek idea of the four basic elements. Robert Boyle proposed a completely new definition of elements. He said that elements are substances that cannot be chemically decomposed into simpler substances. Earth, air, fire, and water could not be called elements by this new definition. Chemists were on the verge of discovering the basic units of matter.

One of the most significant elements to be discovered was oxygen. In 1774, Joseph Priestley heated mercuric calx (mercury (II) oxide) and obtained a gas in which substances easily burned. Shortly thereafter, the French chemist Antoine Lavoisier (lah VWAH zee aye) observed that several substances *gained* rather than lost weight after combustion. Lavoisier concluded that the gas Priestly had discovered was a common substance in the air that combined with other substances in combustion. Lavoisier named Priestley's discovery oxygen. Lavoisier's use of the scientific method and his reliance on careful measurements served as a model for many other chemists. His work earned him a spot among the founders of modern chemistry.

Chemists of the 1700s used glassware like this as they identified the elements.

FACETS of CHEMISTRY: The Scientific Method of Antoine Lavoisier

Substances lose weight when they burn, right? After all, a hefty log turns into a small pile of ashes after a night in the campfire! Observations like this led many scientists to assume that burning always decreases the mass of a substance. Early scientists thought that burning allowed a mysterious substance called phlogiston to escape. The loss of phlogiston supposedly accounted for the decrease in weight during burning. Yet one observant scientist saw something that threatened the widely accepted phlogiston theory. On November 1, 1772, the French chemist Antoine Lavoisier delivered a sealed note to the secretary of the French Academy of Sciences.

> "About eight days ago I discovered that sulfur in burning, far from losing weight, on the contrary, gains it; it is the same with phosphorus. This increase in weight arises from a prodigious quantity of air that is consumed during combustion. The discovery, which I have established by experiment which I regard as decisive, has led me to think that what is observed in the combustion of sulfur and phosphorus may well take place in the case of all substances that gain weight by combustion."

Lavoisier had observed a significant gain in weight when phosphorus and sulfur burned. This contradicted the prevailing idea that all substances lost weight when they burned. Lavoisier recognized that this contradiction posed a problem well worth investigating. Were sulfur and phosphorus isolated exceptions to the phlogiston theory, or was the entire theory faulty? After careful study of the matter, Lavoisier proposed a daring hypothesis: Substances gain something from the atmosphere when they burn. This idea went against the accepted theory and would not be listened to without experimental proof. To prove that his hypothesis was correct, Lavoisier knew he must first identify the "something" from the atmosphere that substances gain when they burn.

At first Lavoisier suspected that carbon dioxide caused the increase in mass. Several experiments soon proved that carbon dioxide would not support combustion of any type. The discovery of oxygen by fellow chemist Joseph Priestly gave several valuable clues. When heated, mercuric calx (a compound of oxygen and mercury) releases large quantities of oxygen gas and leaves silvery, elemental mercury behind. Lavoisier duplicated Priestley's procedure with this red-orange compound. The results of this experiment gave Lavoisier necessary insight into his problem. Perhaps oxygen was the component of air that combined with burning substances and increased their masses. To test this idea, Lavoisier developed a controlled experimental procedure to produce mercuric calx from mercury and ordinary air and then from mercury and pure oxygen. Both procedures produced the same compound. Oxygen was the substance in the air that combined with substances as they burned!

Lavoisier began with a problem, made observations, researched and defined the problem, and then developed a hypothesis. He then conducted experiments in which he gathered more observations, chose his solution, and then verified it. His consistent use of the scientific method in his chemical investigation did not guarantee success, but it did keep him on the right path. With the help of other researchers and a keen mind, Lavoisier made a discovery that changed the theoretical framework of chemistry.

Modern chemistry has many practical applications.

Chemistry Today

By 1800 chemistry had become an academic discipline. In the New World several colleges made chemistry a part of their curriculum. Benjamin Rush was the first professor of chemistry in the United States. His lectures at the Philadelphia Medical School set the foundation stone of chemical education in the new nation. At Princeton, professors attacked the old Greek ideas and used demonstrations and experiments in their teaching. These early professors foreshadowed the importance the United States would achieve in chemistry within the next two centuries.

As chemical knowledge expanded throughout the world, specialized branches of chemistry developed. Scientists were soon forced to concentrate in one area. The branch of **organic chemistry** developed as a result of investigations by Friedrich Wöhler. Until 1826 it was commonly believed that organic compounds, those that contain carbon, could be produced only by living organisms. Wöhler amazed the scientific community by synthesizing urea (a waste product from animals) from two "inorganic" compounds. This discovery provided the foundation for a new branch of chemistry: the study and synthesis of organic compounds. Urea is still used today in fertilizers, manufacturing of plastics, and pharmaceutical preparations.

Table 1A-3 The Major Branches of Chemistry

Branch	Description
Inorganic chemistry	The study of all elements other than carbon and their components
Organic chemistry	The study of compounds containing carbon
Biochemistry	The study of the chemical processes in living things—plants, animals, and man
Nuclear chemistry	The study of radioactivity, the nucleus, and the changes that the nucleus undergoes
Physical chemistry	The foundational theories that allow detailed examinations of interactions between substances and the accompanying energy changes
Analytical chemistry	The techniques used in all branches of chemistry in which chemists devise equipment and methods to (1) discover what substances are in a sample of material (qualitative) and (2) determine how much of each component it contains (quantitative)

Soon other branches of chemistry developed. **Inorganic chemistry, analytical chemistry, physical chemistry, nuclear chemistry,** and **biochemistry** all expanded into separate fields. Today these branches overlap considerably. For example, a chemist who studies the rate at which aspirin (an organic compound) forms could be called a physical organic chemist. Each field will be covered to some extent in this textbook.

Just as the branches of chemistry overlap with each other, the entire subject of chemistry overlaps with other subjects. For example, the production of new elements and the study of subatomic particles blend physics with nuclear chemistry. Likewise, studies of protein structure combine biology and biochemistry. Today chemistry contains so many important topics that it is now the largest scientific discipline in the world. Therefore, chemistry can be considered the fundamental or central science.

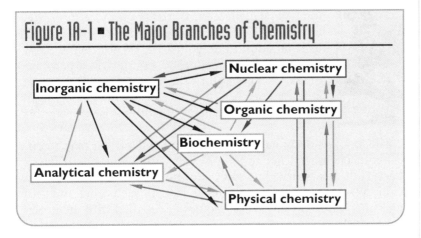

Section Review Questions 1B

1. Can you think of any material object that is not influenced by science?
2. Construct a time line on the history of chemistry. Include important terms and people.
3. Evaluate the statement: "Chemisty can be considered the fundamental or central science."

1C Chemistry and You

At this moment you have a set of preconceived ideas about what chemistry is, what this course holds in store for you, and how chemistry will affect your life. Take a minute to focus on your attitude toward chemistry. You may already be interested in the marvels of God's creation, and you may be looking forward to learning more about its composition. You may be thinking about chemistry's reputation for being a difficult course. You may see it as just another requirement for graduation and admission to some college program. You might be asking, "How can chemistry benefit me in the future?" That is a fair question, and it deserves a straightforward answer.

Character Development

A chemistry course provides a unique opportunity for you to develop academic and personal character. Learning about electrons, periodic tables, and acids is important because of the scientific ideas involved. But the study of chemistry also promotes self-discipline by requiring diligence and organization. As challenging material comes your way, you have the opportunity to respond with effort and a determination to do your best. I Corinthians 10:31 says, "Whether therefore ye eat, or drink, or whatsoever ye do, do all to the glory of God." The goal of doing your best to please God will enable you to keep trying even when difficulties arise during this course.

Successful laboratory work requires organization, skill, attention to details, and accurate observations.

This book is designed to help you understand the major concepts in chemistry, not to supply you with thousands of facts. Facts are necessary, but they do not stand independently. Concepts and principles must relate facts to each other and provide an organized framework. Do not be satisfied with memorizing facts. Instead, strive to see relationships between facts and to understand the concepts and principles. Throughout the course, note how many times the main concepts and principles are repeated in the different sections of chemistry.

The process of investigating chemistry in the laboratory also allows you to develop your character. Before you begin an experiment, organize and schedule your work. While carrying out an experiment, use precise physical skills, such as measuring and filtering. Report observations accurately and honestly. At the end of each experiment, analyze your work, identify errors, and develop logical conclusions based on experimental results. Habits of carefulness and industriousness developed in the laboratory will serve you well in the future.

Vocational Preparation

Why study chemistry? It will help prepare you for your life's work. A knowledge of chemistry helps those who will work in

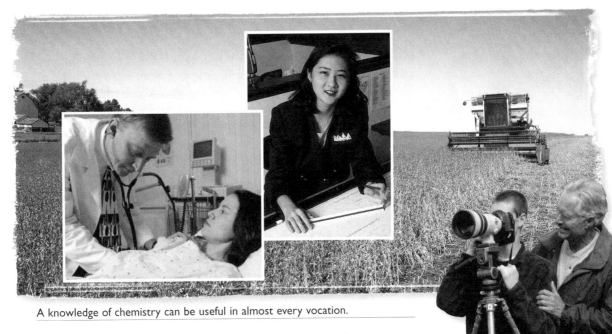

A knowledge of chemistry can be useful in almost every vocation.

both technical and nontechnical jobs. No matter what occupation you foresee for yourself, a chemistry course can help you reach your potential in that field.

Obviously, chemical researchers need a strong background in chemistry. Some research chemists work in the area of pure science. They seek to extend scientific knowledge for the sake of finding out new facts about matter. Although the main intent of these researchers is not to develop new commercial products, practical applications often result from their work. Most chemists, however, work in the area of applied science. They seek to use scientific knowledge to create useful products. Chemists in many industries are constantly developing new fabrics, cosmetics, plastics, fuels, paints, foods, pesticides, drugs, and fertilizers. All chemists, whether they are working with pure or applied science, need a strong foundation of chemical knowledge. High-school chemistry serves as the cornerstone of this foundation and as a steppingstone to further study.

Photographers, architects, farmers, cosmetologists, culinary artists, doctors, nurses, dentists, dental hygienists, meteorologists, geologists, pharmacists, and X-ray technicians are not practicing chemists. Nevertheless, they all rely on chemistry to do their jobs. A knowledge of chemical principles is recommended, if not required, for jobs such as these.

A course in chemistry can benefit you even if you do not foresee a career in chemistry or some other technical field. What help is chemistry to homemakers, mechanics, and businessmen? More than you might think. Homemakers who provide nutritious meals want to substitute polyunsaturated fats for saturated fats. Chemistry can help these homemakers choose the best foods, as well as understand the reasons for their choices. Mechanics should know about octane numbers, material properties of tools and metals, and chemical emission standards to be knowledgeable in their field. Future businessmen can also benefit from chemistry. Today's society is becoming more safety conscious every day. A business owner must be thoroughly familiar with hazardous material in the workplace and environmental pollutant management. To be a knowledgeable consumer, you need an understanding of the materials that you buy for your own use and to sell in your business. No matter what your future vocation may be, a knowledge of the basic principles of chemistry can help you function better in our society.

A close look at a list of ingredients reveals many additives.

Garbage continues to be buried in our landfills.

Participation in Public Policy—Making Decisions

Most science-based public policy issues are chemistry based. In our democratic society it is important to have informed citizens who are able to evaluate these issues. For example:

1. Should our country build more nuclear power plants?
2. Are incinerators a good way to handle our garbage, or should we continue to bury it?
3. Is biotechnology, such as genetic testing, a benefit or a danger?
4. Should we continue to depend on petroleum, or should we develop alternative energy sources?
5. Which food additives should the government regulate?
6. Is global warming a real danger?
7. Is it safe to irradiate food to kill bacteria?

The list is endless. The answers to these questions will affect not only your quality of life but the life of generations to come, until the return of the Lord. Unfortunately, just at a time when more and more chemistry-based issues need to be decided, the educational system is producing fewer and fewer chemists. In addition, the availability of chemistry teachers is also declining in both public and private schools.

Expanded Christian Witness

Those who have accepted Jesus Christ as their personal Savior have an additional, compelling reason to learn about chemistry. Christians are in this world to tell others about Christ and His redeeming work. In many cases they can witness for Christ more effectively if they have prepared themselves in a wide variety of fields, including chemistry.

As diplomats for Christ and His kingdom, Christians must do all they can to present the gospel clearly to as many people as possible. Good political diplomats not only refrain from bad behavior, but they also try to relate to the people of their host country. They follow local traditions, dress well, and do their best to learn the local language. Diplomats for Christ also have this twofold task of refraining from bad behavior and relating to the people of this world. They must develop their personalities, educate themselves, meet people, and express the gospel well. It would be a tragedy if a Christian neglected to develop himself and became a hindrance to the spread of the gospel.

God does not want His people to be ignorant. Christians can be more effective for God if they seize opportunities to learn. Although God can use sincere but unschooled people, He does not approve of wasting an opportunity to develop the mind. Scripture tells of many men who were used by God to do great things after they had diligently prepared themselves. The apostle Paul commanded respect across the world because of his formal schooling, his fluency in several languages, and his knowledge of literature. These abilities opened many doors to Paul and gave him an expanded witness. Luke, the author of Acts, was a medical doctor and an eloquent writer. He was privileged to use both of these skills during the days of the early church. Moses, most noted for his spiritual leadership, was also learned in the medicinal and chemical knowledge of the Egyptians. This knowledge and the respect it commanded very likely helped him to gain access to Pharaoh's court and equipped him to lead the Israelites. Daniel had a tremendous influence on the Babylonian and Medo-Persian empires because of his great character and wisdom. He gained these qualities by studying diligently as a teenager. Each of these men had a great impact for the Lord because of the way he prepared himself for his service. None of these spiritual heroes could have been as effective as they were had they neglected their opportunities to learn.

Enhanced Appreciation of God

"By the word of the Lord were the heavens made; and all the host of them by the breath of his mouth. He gathereth the waters of the sea together as an heap: he layeth up the depth in storehouses. Let all the earth fear the Lord: let all the inhabitants of the world stand in awe of him. For he spake, and it was done; he commanded, and it stood fast" (Ps. 33:6-9).

God's voice spoke into existence a universe that is wondrous to behold. Stars, some so large that they would reach out to the orbit of Mars if they replaced the sun, are scattered across galaxies. Our earth hangs in space with nothing but carefully balanced forces holding it in place. The beauty of a sunset, a forest, and the Grand Canyon deserve a reverent respect. Even the casual

The closer we study Creation, the more we learn about the Creator.

observer can see proofs of God's existence and His awesome power. Romans 1:20 says that "the invisible things of him from the creation of the world are clearly seen, being understood by the things that are made, even his eternal power and Godhead."

Like everyone else, chemistry students see evidences of God's power and existence. They also have the opportunity to observe many additional attributes of God in His creation. Orderliness and logic can be seen in the structure of crystals. A concern for details is evident in the architecture of atoms. Natural laws such as the laws of thermodynamics reveal God's sovereignty. God's goodness is obvious when we consider the abundant provisions He has given to sustain life. The intricate structure of a DNA molecule generates a sense of beauty and creativity. Yes, nature shows God's omnipotence, but it also displays many of His other attributes. A careful study of the universe will enhance your appreciation of its Creator.

Section Review Questions 1c

1. Evaluate why it is important to study chemistry.
2. Think of a career that interests you and tell how you should use this chemistry course to help prepare for that career.
3. How can you individually expand your Christian witness by studying chemistry?

Chapter Review

Coming to Terms

deductive reasoning
law
inductive reasoning
chemistry
hypothesis

matter
scientific method
metallurgy
science
apothecary

theory
alchemy
inert
pharmacology

Review Questions

1. What is the difference between inductive and deductive reasoning? Which type of reasoning do scientists use to develop theories?
2. By what criteria did the Greeks judge their scientific ideas? What criteria do modern scientists use?
3. What crucial difference separates the Greek philosophers from modern chemists?
4. Can scientific laws be proved to be wrong? Why or why not?
5. Are there any substances in the material universe not made of chemicals?
6. Describe how men in Old Testament times practiced chemistry. Why did the Greeks pursue science? The alchemists?
7. Although alchemists are sometimes criticized, they developed several things that benefit modern chemists. List these benefits.
8. What is the difference between organic chemistry and inorganic chemistry?
9. Name four ways in which knowledge of chemistry can benefit Christians.
10. Classify each of the following activities as pure chemistry, applied chemistry, or both.
    a. Determining how much energy is released when iron forms rust
    b. Determining how the size of silver grains on photographic film affects the resolution of the finished print
    c. Isolating a chemical compound from the leaves of a newly discovered tropical plant
    d. Developing a method of purifying the polluted well water in a small town
    e. Determining what elements exist on a distant star
11. The text mentioned how knowledge and training enabled Moses, Daniel, King Solomon, Paul, and Luke to serve the Lord more effectively. Match each of these men above with the one Scripture verse that describes him.
    a. I Kings 3:5-10
    b. II Timothy 2:15
    c. Philippians 3:4-8
    d. Daniel 1:3-6
    e. Colossians 4:14
    f. Proverbs 3:13-18
    g. Hebrews 11:24-27
12. Investigation—Choose a career involving chemistry that interests you and write a one-paragraph summary about that career.

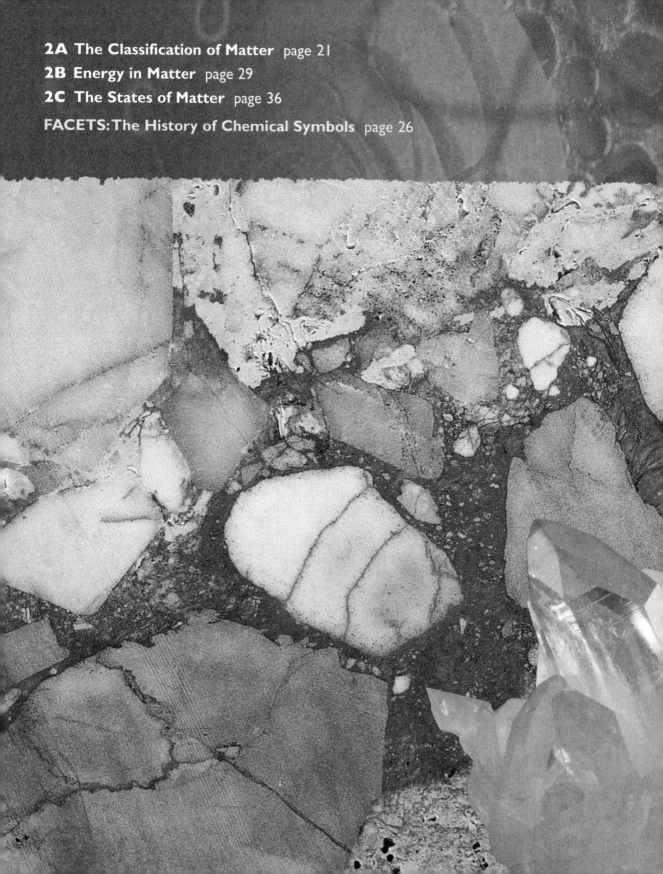

2A The Classification of Matter   page 21
2B Energy in Matter   page 29
2C The States of Matter   page 36
FACETS: The History of Chemical Symbols   page 26

Matter

What Chemistry Is All About

The materials that make up the physical universe were created out of nothing (*ex nihilo*) by God. There is no conclusive proof for or against this statement. It is a statement of faith and the foundational truth in a Christian perspective of chemistry.

"Through faith we understand that the worlds were framed by the word of God, so that things which are seen were not made of things which do appear" (Heb. 11:3).

What is matter? *Matter* is difficult to define because it encompasses practically everything. To establish a meaning for this basic term, scientists describe matter by the way it is measured. Since the measurable properties common to all matter are volume and mass, **matter** is operationally described as "anything that takes up space and has mass."

2A The Classification of Matter

The world that God spoke into existence contains a variety of materials. Cars, electrons, people, sand, uranium, trees, and sulfuric acid are just a few of the many forms that matter takes. A study of all this matter needs the organization that a good classification scheme can provide. Although there are many possible classification systems, this text will use one that divides matter into two major categories: pure substances and mixtures. Distinctions will be made between the two categories based on the physical and chemical properties of matter.

Chemical and Physical Properties

Properties are the distinguishing characteristics of matter. Scientists divide these characteristics into two classes: physical and chemical. The **physical properties** of a material are related to the physical relationships among the particles in that material. What is the physical appearance of the material? How many particles are there? How closely are they packed together? Physical properties of matter are often defined as properties that a person can measure without changing the actual composition of the material. Color, shape, texture, odor, taste, electrical conductivity, and density can be measured without changing the material.

Chemists use several physical properties to describe matter. **Density** describes how the particles are "packed" into a material. Dense objects have many particles packed into a relatively small space. Less dense objects have fewer particles in the same space. We know that wood is less dense than water because a piece of wood will float on water. Materials that are **malleable** (MAL ee a bul) can be easily hammered into shapes. Some materials can be stretched into thin wires. This property is called **ductility** (duk TIL i tee). Gold is the most malleable and ductile metal. One ounce of gold (28.35 grams) can be hammered out to cover 300 square feet! (The average bedroom is only 180 square feet (12×15 feet).) Since it is such a soft metal, gold is usually mixed with another metal such as copper or silver to give it strength. **Conductivity** measures the ability of a material to transfer heat or electricity between its particles. Silver is the most electrically conductive metal, but because of silver's high cost, copper is usually used for the wiring in our homes.

A second class of characteristics describes how matter acts in the presence of other materials. Scientists call these characteristics **chemical properties.** In order to determine the chemical

Table 2A-1

	Physical Change	Chemical Change
Definition	A change in state or shape that does not alter the identity of the material	A change in the composition of the particles of a material that alters the identity of the material
Particle changes	The positions of the particles may change; association of particles unchanged	The positions and associations of the particles will change.
Results	Chemical properties are not altered.	New substances are formed that have different chemical and physical properties.
Examples	Ice melting, flour ground from wheat, sugar dissolved in water	Metal rusting, oil burning, wood rotting, food being digested

properties of material, scientists must know the kinds of changes that the material can undergo. Gasoline reacts with oxygen. This reaction is called *combustion*, or burning. The fact that gasoline will react with oxygen is one of its chemical properties. Each material has an individual set of chemical properties.

The terms *physical change* and *chemical change* are closely associated with the two sets of properties. Changes that occur in a material without changing the identity of the material are **physical changes.** Physical changes cause changes in both state and shape. By "state," we mean whether a chemical substance is found as a solid, liquid, or gas. Boiling is a physical change in which a material changes from its liquid state to its gaseous state, but the identity of the material is not altered.

Chemical changes, or **chemical reactions,** are changes in the identity of a material, that is, changes that result in a different material with a different composition and different properties. When iron rusts, it undergoes a chemical change. The iron particles combine with oxygen particles to form rust. Not only is this new substance totally different from oxygen or iron, but it also has a different physical appearance. The chart below contains a comparison of physical and chemical changes.

Processing iron involves both chemical changes (purifying the ore) and physical changes (molten to solid).

A tin can lid undergoes a chemical change when it corrodes. Water undergoes physical change when it changes between ice, liquid, and steam.

Chapter 2A

Oil-and-water salad dressing is a heterogeneous mixture.

The Division of Matter

Matter can be divided into two categories: pure substances and mixtures. A pure substance consists of only one type of matter, which cannot be separated into other kinds of matter by any physical processes. Pure substances can be further subdivided into elements and compounds, both of which will be discussed in detail in later sections. A **mixture** is a material that can be separated by physical means into two or more pure substances. There are two types of mixtures—heterogeneous and homogeneous. All **heterogeneous** mixtures (physical combinations of pure substances), if examined carefully, appear to have two or more distinct regions called *phases*. Consider oil and vinegar salad dressing. When the dressing is undisturbed, there are two distinct phases, the oil (the top layer) and the vinegar (the bottom layer). No matter how vigorously it is shaken, there will always be two separate phases—oil and vinegar, and if left undisturbed, it will again form the two distinct layers. The individual glittering particles of quartz and mica can be clearly seen as separate phases in a piece of granite—another example of a heterogeneous mixture. On the other hand, matter that shows only a single phase is said to be a **homogeneous** mixture (also known as a **solution**). The physical properties appear to be the same throughout. If you mix sugar with water and stir it until it is completely dissolved, it appears as a single liquid phase—

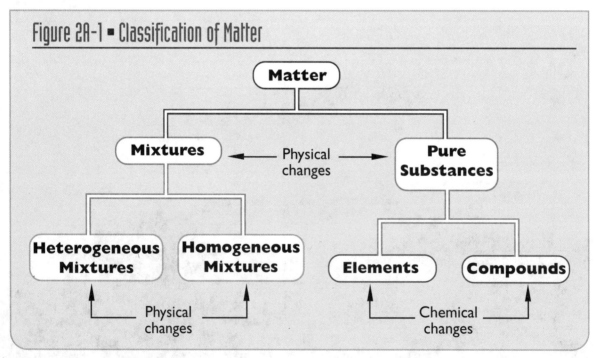

Figure 2A-1 • Classification of Matter

sugar water. If you melt gold with silver, you get a single phase of white gold. Air is a good example of a homogeneous mixture of gases.

Since both solutions and pure substances are considered homogenous matter, how can we distinguish between them? The major identifying characteristic is that solutions are of variable composition, whereas pure substances are of a fixed composition. For example, a solution such as sugar water does not need to be of any particular proportions of sugar or water to be considered sugar water. It could vary from just barely sweet to syrupy. But a pure substance of baking soda (sodium bicarbonate) has specific proportions of the elements that compose it—sodium, carbon, oxygen, and hydrogen. Change the proportion of any one of these elements, and you no longer have baking soda.

Elements and Their Symbols

A pure substance that cannot be broken down into a simpler substance by ordinary chemical means is an **element.** Men have known about several elements, such as gold, silver, sulfur, tin, and lead, since biblical times. Scientists discovered many others shortly after the rise of modern science in the 1800s. In addition, other elements have recently been synthesized in the laboratories of nuclear scientists. You can see these elements represented on any periodic table. (See inside back cover of text.)

Atoms are the particles in elements. They are the smallest particles that maintain the physical and chemical characteristics unique to the element they compose. For example, a lump of sulfur is made up of sulfur atoms.

Elements such as neon, helium, and argon, whose atoms do not naturally combine or bond together are called **monatomic** (mon uh TOM ik) **elements.** Monatomic elements are rare, for most atoms bond with other atoms to form molecules. Elements whose atoms bond into two-atom units are called **diatomic** (die uh TOM ik) **elements.** Oxygen, O_2, and hydrogen, H_2, are examples of diatomic elements. By the same line of reasoning, elements composed of multi-atom units are called **polyatomic elements.** For instance, sulfur often exists in the form of eight atoms bonded into an S_8 unit.

Each element has a special **symbol** that represents its name. The first letter of the name of the element often serves as its symbol (H for hydrogen, N for nitrogen, and O for oxygen). Frequently the names of more than one element have the same first letter. To avoid confusion in these cases, a second lower-case letter is used in the symbol. The second lower-case letter is usually related to the sound of the element's name. The single letter C is the symbol for carbon, Ca stands for calcium, and Cd represents cadmium.

The first letter of an element's symbol is *always* capitalized, but a second letter is *always* given in the lower case. Careless writing of symbols can result in serious errors. Consider the symbol for the element cobalt: Co. If written carelessly as CO, this symbol

FACETS OF CHEMISTRY: The History of Chemical Symbols

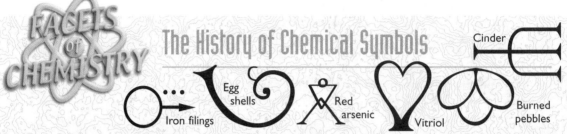

A student in the early 1800s sat down to study his chemistry text. Chemistry was his favorite subject, but reading the text was one task he always dreaded. He quickly became mired in a bog of strange symbols that made no sense at all. The student recognized one symbol but then found that another symbol meant the same thing. One book in the university's library used twenty different symbols for mercury. Another book used fourteen different symbols for lead. "This mess of symbols must be cleaned up and replaced with something better," he muttered.

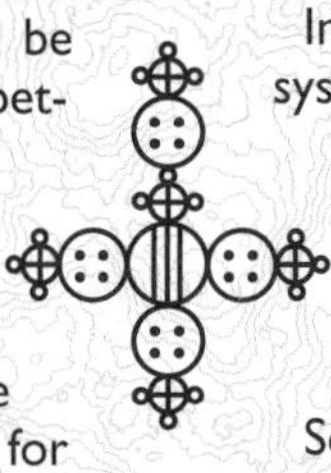

Metallurgists developed the first symbols for elements. The metals they worked with eventually became associated with gods and planets that were known in ancient times. Ancient symbols for the metals tell of the Egyptian, Persian, Greek, and Roman associations.

- ☉ **Gold** – The "perfect" metal received the symbol of perfection and divinity.
- ☽ **Silver** – The metal with the luster of moonlight received the shape of the moon for its symbol.
- ♀ **Copper** – The goddess Venus supposedly rose out of the sea off the coast of Cyprus, which was known for its copper mines. The symbol for copper pictures the looking glass of the goddess of beauty.
- ♄ **Lead** – This metal was associated with Saturn, the god of harvest, and his scythe.
- ♂ **Iron** – The metal used in weapons received a symbol that showed the lance and shield of Mars, the god of war.
- ♃ **Tin** – The thunderbolt of Jupiter served as tin's symbol.
- ☿ **Mercury** – This flowing liquid metal became associated with Mercury, the messenger of the gods. Mercury's wand was used as the symbol.

The alchemists produced many new symbols. Drawing their symbols often required artistic skill and time. The alchemists probably used their symbols to create a scholarly atmosphere and to hide their ignorance of chemistry.

In the early 1800s John Dalton designed a system of symbols to illustrate the different kinds of atoms. He drew various symbols inside circles to designate the different elements. Dalton represented compounds by combining the symbols of the elements that were in the compounds.

Soon after Dalton introduced his system, a Swedish chemist named Jons Berzelius developed a system of abbreviations. He included several revolutionary ideas in his system: "It is easier to write an abbreviated word than to draw a figure which has little analogy with words. The chemical signs ought to be letters for the greater facility of writing and not disfigure a printed book. I shall therefore take for the chemical sign the initial letter of the Latin name of each chemical element. If the first two letters be common to two metals, I shall use both the initial letter and the first letter they have not in common."

Soon Berzelius's abbreviations became accepted and understood all over the world. Today the conglomeration of confusing symbols has yielded to a unified, reasonable, and understandable system of chemical symbols.

would represent the compound carbon monoxide, a poisonous gas found in automobile exhaust.

For some elements that were known in ancient times, the Latin names serve as the basis of the symbols. For example, *cuprum*, the name for copper, comes from the Latin for "from the island of Cyprus." Today, it is generally accepted that the discoverer of the element has the honor of naming the newly discovered element, subject to the approval of the International Union of Pure and Applied Chemistry (IUPAC). Some elements were named for colors (iridium, from the Latin for rainbow), some for people (curium, for Pierre and Marie Curie, early researchers of radioactivity), some for places (californium, for California), some for heavenly bodies (helium, from the Greek word for sun), and others for miscellaneous Greek and Latin words (bromine, from the Greek word for stench).

Compounds and Their Formulas

Compounds are made up of atoms from two or more different elements that have been chemically bonded together. As you may have guessed, there are many more compounds than elements. Just as symbols represent elements, formulas represent compounds. **Formulas** tell the type and number of atoms that are present in compounds. The formula CO represents carbon monoxide, a molecule that can bind with the oxygen-carrying portions of red blood cells and destroy their effectiveness. This formula tells chemists that the individual units of this compound contain one carbon atom that has been bonded to one oxygen atom. Another compound of carbon and oxygen is carbon dioxide (CO_2). This compound consists of one carbon atom bonded to two oxygen atoms. Other common compounds, their formulas, and the atoms they contain are listed in Table 2A-2.

Table 2A-2 Common Compounds and Their Formulas

Compound	Formula	Atoms
Ammonia	NH_3	1 nitrogen, 3 hydrogen
Rust	Fe_2O_3	2 iron, 3 oxygen
Sucrose (table sugar)	$C_{12}H_{22}O_{11}$	12 carbon, 22 hydrogen, 11 oxygen
Slaked lime	$Ca(OH)_2$	1 calcium, 2 oxygen, 2 hydrogen
Salt	NaCl	1 sodium, 1 chlorine
Water	H_2O	2 hydrogen, 1 oxygen

Numbers written at the lower right of a chemical symbol are called **subscripts***. They indicate the number of atoms or groups of atoms in a formula. A unit of water (H_2O) contains two hydrogen atoms and a single oxygen atom (the 1 is assumed). When a subscript follows a group of symbols that are surrounded by parentheses, it refers to the entire group. $Ca(OH)_2$ contains two OH groups for a total of two oxygen atoms and two hydrogen atoms. A number in front of a formula is called a **coefficient** and refers to the entire unit. Thus 5 Fe_2O_3 (rust) refers to five Fe_2O_3 groups, for a total of ten iron atoms and fifteen oxygen atoms.

Sample Problem How many atoms of each element are present in each of the following groups?
   a. $Na_2S_2O_3$ **b.** $Mg(NO_3)_2$ **c.** 3 $CaBr_2$

Solution
- **a.** The subscripts show that two Na atoms, two S atoms, and three O atoms are present.
- **b.** No subscript after the Mg implies that only one atom is present. The subscript of 2 after the (NO_3) means that two NO_3 groups are present for a total of two N atoms and six O atoms.
- **c.** A single $CaBr_2$ group would have one Ca atom and two Br atoms. The coefficient 3 refers to three of these groups. Tripling the number of atoms in one group gives a total of three Ca atoms and six Br atoms.

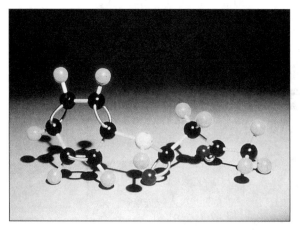

Molecules are the smallest independent units in many compounds. A **molecule** consists of two or more atoms that are chemically bonded together. H_2O is a molecule, as is NH_3 and $C_{12}H_{22}O_{11}$. Diatomic and polyatomic elements are also made up of molecules; O_2 and S_8 qualify because they contain two or more atoms.

---

*subscript: *sub* (L. – under, below) + *scriptus* (L. – written)

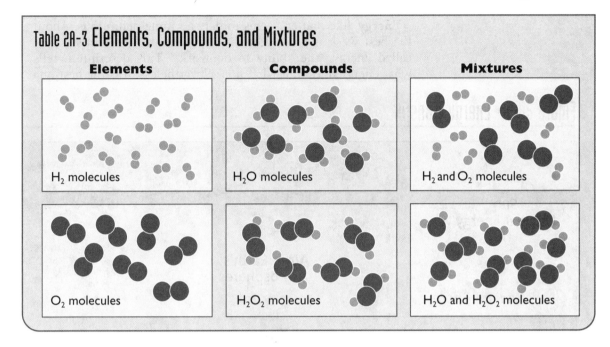

Table 2A-3 Elements, Compounds, and Mixtures

Elements: H_2 molecules; O_2 molecules
Compounds: H_2O molecules; H_2O_2 molecules
Mixtures: H_2 and O_2 molecules; H_2O and H_2O_2 molecules

Section Review Questions 2A

1. How do physical changes differ from chemical changes? Give two examples of each type of change.
2. Describe the difference between heterogeneous and homogeneous matter.
3. Give the number of atoms in each of the following compounds: NaCl, H_2SO_4, $Ca(OH)_2$, 4 Li_2O.

2B Energy in Matter

Although chemistry is the study of matter, the subject demands a basic understanding of energy. Every chemical reaction either releases or absorbs energy. Reactions that liberate energy, such as the burning of natural gas, are called **exothermic***. Reactions that absorb energy are called **endothermic***. Chemical cold packs used to treat sprains feel cold because the reaction that is occurring is absorbing heat energy from the surroundings. Matter that is not undergoing chemical changes can contain energy stored in the form of heat, magnetic fields, electrical energy, chemical energy, nuclear energy, or mechanical energy.

---

*exothermic: exo (Gk. – outside of) + thermic (Gk. therme – heat)
*endothermic: endo (Gk. – within) + thermic (Gk. therme – heat)

Forms of Energy

Energy, like matter, seems to defy accurate definitions. About the best we can do is describe it. Traditionally scientists have called energy "the ability to do work." This description tells what energy does, not what it is. Nevertheless, it is the best one

Figure 2B-1 ▪ Energy Transfer

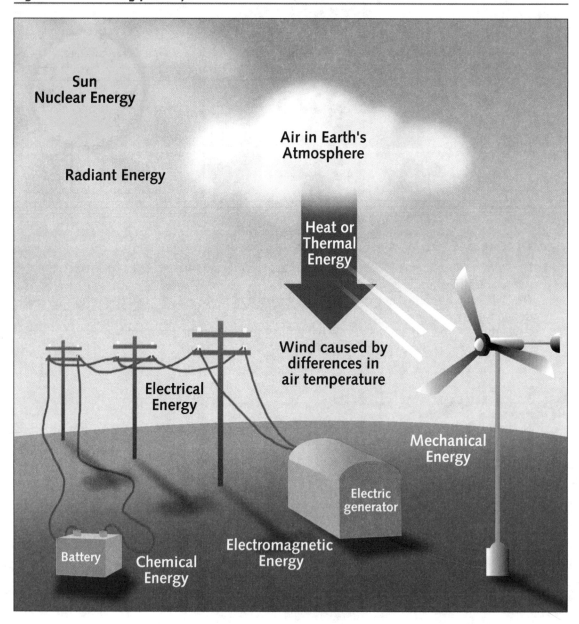

available. All of the commonly recognized forms of energy can be harnessed in some way to do work. The subject of chemistry is concerned mainly with the relationships among chemical, thermal, electrical, and nuclear energy, but it relates in some way to all the different kinds of energy. For example, windmills are used to generate electrical energy, but there is more to the story than just the wind's turning the blades of the windmill. A nuclear reaction occurs in the sun, releasing radiant energy. The radiant energy is transmitted through space and is converted to heat (thermal) energy, which is transmitted by the air molecules. This thermal energy causes different pressure gradients that make the wind blow (wind energy). The wind moves the blades of the windmill (mechanical energy), and the motion of the shaft is transferred to an electrical generator (electromagnetic energy) to produce electricity. The electrical energy is then transmitted over power lines and could be used to perform work or could be stored in a battery (chemical energy).

Energy Conservation Despite Change

Thermodynamics* (ther mo die NAM iks) is the study of the flow of energy, especially heat energy. Several laws that have a great impact on chemistry have been formulated as a result of thermodynamic studies. No exceptions to these laws have been observed; they apply to every field, at all times, in every instance.

These windmills transform wind energy into electrical energy.

You learned in the last section that energy can be converted from one form to another, but in these changes, energy is always conserved. The results of many careful measurements by scientists have shown that energy can be neither created nor destroyed, assuming that *all the forms* of energy are considered. This statement is known as the **law of energy conservation.**

Since Einstein showed in 1900 that matter could be converted into energy, and energy into matter, the law of energy conservation is sometimes called the **law of conservation of mass-energy:** matter and energy can neither be created nor destroyed, only converted from one form into another. This law is known as the **first law of thermodynamics.** Apart from divine intervention, the total amount of mass and energy has remained constant since the sixth day of creation (Gen. 2:2-3). All energy changes that have occurred since that time, either by natural processes or by man, have merely changed one form of energy or matter into another form of energy or matter.

---

thermodynamic: *thermo* (Gk. *therme* – heat) + *dynamic* (Gk. *dunamis* – power, energy)

The first law of thermodynamics presents a problem for the evolutionist: either (1) the universe is eternal in its present form, or (2) there was some period of creation in the past. The first option is not supported by other scientific evidence. The second option necessitates a Creator—something the evolutionist is not willing to accept. The evolutionist's talk of "creation" by the "big bang" ignores the fundamental question of where all the matter-energy involved came from in the first place! But Genesis 1:1 clearly tells us that it was God Who "in the beginning . . . created the heaven and the earth."

The second law of thermodynamics deals with the tendency of the universe to "wax old as doth a garment" (Heb. 1:11). It states that during any energy transformation, some energy goes to an unusable form. Sometimes this law is quoted as saying that energy is "lost." Not so. That would contradict the first law of thermodynamics. When a battery discharges its electrical energy through a coil of wires, the energy is not "lost." It is changed into heat energy, or possibly light energy if the wires glow. These new forms of energy ultimately become dispersed and unusable. Given enough time and a sufficient number of energy transformations, all the energy of the universe will eventually become dissipated and totally unusable.

Only a tiny fraction of the sun's energy bombarding a solar collector can be turned directly into electrical energy. Hence, the great interest of chemists in improving the efficiency of solar panels and photo cells. However, because of the second law, we will never achieve 100 percent conversion efficiency.

The second law of thermodynamics is also expressed in terms of a quantity known as **entropy** (EN tro pee)—the randomness or disorder of a system. There is a tendency for all natural processes to occur with an increase in the entropy (disorder) of the universe. Thus, the universe is headed toward a state of maximum entropy, or "death." (This idea will be further discussed in Chapter 13.) You know this from experience. Disorder in your room happens without any effort; automobiles rust, wear out, and fall apart; marbles dropped on the floor will not land in an organized pile or pattern. Life itself is a constant battle against the natural tendency toward maximum entropy; we age ("wear out"), die, and decay. We might state the second law by saying that *every system left to itself will tend toward maximum disorder (entropy)*. Any decrease in entropy in one part of the universe comes at the expense of increasing entropy somewhere else in the universe. For example, the organization that comes during the growth of our bodies produces complex molecules and results in the production of simpler, more disordered waste products and heat energy.

Heat, Energy, and Temperature: There Is a Difference!

The energy possessed by matter because of its motion is called **kinetic energy.** All matter contains particles that are moving. The kinetic energy (KE) of an object depends on both its mass (m) and its velocity (v), according to the relationship $KE = \frac{1}{2} mv^2$. You can see that a less massive object must have a greater velocity in order to have the same kinetic energy as a more massive object. In addition, two objects of different masses moving at the same velocity (speed) will have different kinetic energies. For example, a tractor-trailer moving at 55 mph has a much greater kinetic energy than a sports car moving at the same speed. In matter, particles move in a variety of ways. They can move in straight lines (translational movement), they can oscillate back and forth in one plane (vibratory motion), or they can spin (rotational movement). The sum total of all of these motions of the particles (kinetic energy) in an object is its **thermal energy.** **Temperature** measures the *average* kinetic energy of all the particles in a sample. Which contains more thermal energy: a teaspoon of boiling water or a bathtub full of lukewarm water? The teaspoon of water obviously has the greater temperature, but we can see that because of its greater mass, the tub full of water possesses the greater thermal energy.

Heat and temperature are related but are definitely not the same thing. **Heat** is thermal energy in transit. As thermal energy naturally flows from a hot object to a colder object, the temperature of both objects will change. Suppose that a pair of gloves were fabricated from a substance that was a perfect insulator. If someone wearing these gloves then grasped a hot coal, there would be no heat (no thermal energy transferred from the coal to the hand), and as a result, the temperature inside the gloves would remain unchanged. The amount of heat transferred between the two objects is related not only to the temperature difference between them, but also to the mass of the hotter object. In Figure 2B-2, the larger branding iron has the greater mass and

will therefore possess the greater thermal energy. When the larger branding iron is pressed against the steer's hide, a larger amount of thermal energy will be transferred (heat), resulting in a greater temperature at the site, causing the steer to launch.

The Measurement of Energy: Important Units

The standard unit of measurement for energy in the SI (after the French *Système Internationale*), or International System, is the **joule** (jule). The joule is named for the English physicist James Prescott Joule who distinguished himself for his scientific integrity and his precise measurements. The English unit for thermal energy is the **BTU,** or **British Thermal Unit,** which is the amount of heat required to raise one pound of water by one degree Fahrenheit. Air conditioning units are often advertised with BTU ratings. Another extensively used energy unit is the **calorie (cal)***—the amount of energy required to raise the temperature of one gram of water one degree Celsius. Joules and calories are related by the definition: 1 cal = 4.184 J. Larger units called kilocalories, which are 1000 times as large, are used for most chemical applications. The kilocalorie is equivalent to the Calorie (note the capital C), which is used in reference to the energy content of foods.

Temperature scales measure the average kinetic energy of the particles in a sample. Even in solids, the atoms and molecules are moving. The two temperature scales used most often in chemistry are the Celsius scale and the Kelvin scale. The **Celsius** (SEL se us) **scale** uses the freezing point of water as its zero point and the boiling point of water as 100°C. The **Kelvin scale** uses **absolute zero** as the zero point. Absolute zero is the temperature at which theoretically all molecular and atomic movement would cease. The "size" of a degree on the Kelvin scale is the same as the size of a degree on the Celsius scale. Therefore, 100 Celsius degrees equal 100 Kelvins (0° C equals 273.15 K). It should be noted that the degree symbol is not used when writing Kelvin temperatures. See Figure 2B-3 to see the comparisons between the Fahrenheit, Celsius, and Kelvin scales.

As you can see from the relationships between the Kelvin and the Celsius scales in Figure 2B-3, Kelvin temperatures differ from Celsius by 273.15; that is the only correction needed for converting between the scales. Note that the Kelvin scale is *never* negative, and that it is *always* higher than the Celsius temperature. This observation should help you when working conversion problems during chemistry calculations. The equations for

---

*calorie: *calor* (L. – heat)

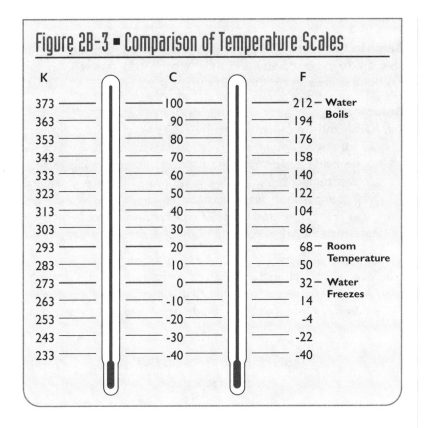

Figure 2B-3 ▪ Comparison of Temperature Scales

conversion between Celsius and Kelvin are as follows:

$$K = °C + 273.15 \quad \text{and} \quad °C = K - 273.15$$

Although the Fahrenheit scale is usually not used for scientific purposes, it is the scale most commonly used by Americans, and you should be familiar with how to make conversions between it and Celsius. Note first from Figure 2B-3 that the freezing point of water is 0°C and the boiling point is 100°C. Thus, there are 100 divisions (degrees) between these defining points on the Celsius scale. On the Fahrenheit scale, water freezes at 32° and boils at 212°, making 180 divisions between these points on the Fahrenheit scale. Since the same "distance" is divided into more parts on the Fahrenheit scale, each part must be smaller. Specifically, there are 1.8 Fahrenheit degrees for every Celsius degree (1.8°F/°C). As you can see, on the Fahrenheit scale water has a 32° head start on the Celsius scale; thus 32 must be subtracted from any Fahrenheit reading and added to any Celsius reading to correct for the starting point differences. The equations for converting between the two scales are given below:

$$°F = (1.8 \times °C) + 32 \quad \text{and} \quad °C = \frac{°F - 32}{1.8}$$

Sample Problem The weatherman announces that the high for the day is expected to be 33°C. What is this temperature on (a) the Kelvin scale and (b) the Fahrenheit scale?

Solution

a. Kelvin and Celsius temperature scales have divisions that are the same size, so only a correction for "starting points" needs to be made. Kelvin temperatures will always be *higher* than Celsius temperatures, so we must *add* 273. Note: Since the given temperature was measured to the nearest whole degree, 273 is used instead of 273.15 because the conversion is determined by the precision of the given measurement.

$$K = °C + 273 = 33°C + 273 = 306 \text{ K}$$

b. To convert Celsius temperature to Fahrenheit temperature, we need to use the equation that will both *add* 32 and *multiply* by 1.8. Note that the addition of 32 *follows* the correction for the difference in degree size.

$$°F = (1.8 \times °C) + 32° = (1.8 \times 33°C) + 32° =$$
$$59.4° + 32° = 91.4°F$$

Section Review Questions 2B

1. What is the first law of thermodynamics? How does it relate to the biblical model of creation?
2. Is energy ever lost? Explain.
3. Describe the difference between heat, thermal energy, and temperature.
4. It is 95°F outside. Convert this into degrees Celsius and Kelvins.

2C   The States of Matter

Scientists theorize that all matter is composed of submicroscopic particles (atoms, molecules, and ions) that are in constant motion. If energy is added to these particles, their motions increase; and the greater the amount of energy added, the faster the resulting motion will be. If this motion were not in some way limited, matter would simply fly apart. Electrical forces attract atoms toward each other and inhibit the movement of the particles. You will learn more about these in Chapter 6.

Scientists call the above set of ideas the **kinetic theory** (kinetic = motion), since it describes the motion of particles in matter. According to the kinetic theory, particles in solids possess relatively little energy compared to the attractive forces that are present. These attractive forces overpower the movements of the particles. The forces keep the particles of a **solid** in fixed positions with set distances between them.

When sufficient thermal energy is applied to a solid, its particles gain enough energy to partially overcome their attractive forces, and the solid becomes a **liquid.** While the attractive forces retain their ability to hold the particles closely together, the energy that the particles possess allows limited motion or "flow."

As the particles of the liquid state acquire additional thermal energy, they are able to completely overcome their attractive forces, as well as the atmosphere pushing them together, and they exist as a **gas.** The particles of a gas possess a large amount of kinetic energy. They move rapidly and randomly across great distances. The average velocity of an oxygen molecule is 480 meters per second, which is equivalent to a rate of speed just over 1000 miles per hour.

If even more thermal energy is applied to the atoms in the gaseous state, the electrons can be stripped away from the nucleus, and a fourth state of matter is formed—plasma. **Plasma** is the most abundant form of matter and consists of a gaseous sea of electrons, ions, and neutral atoms that travel at extremely high speeds. It is this form of matter that

is used in the fusion process of the sun and in controlled thermonuclear fusion research at temperatures over 100 million degrees. Plasmas are also seen in the ionized gas of neon signs, fluorescent lighting, and lightning.

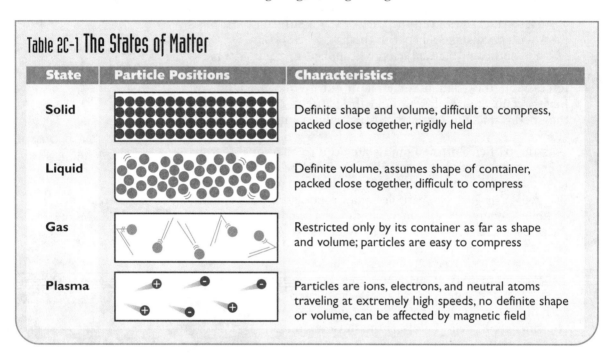

Table 2C-1 **The States of Matter**

State	Particle Positions	Characteristics
Solid		Definite shape and volume, difficult to compress, packed close together, rigidly held
Liquid		Definite volume, assumes shape of container, packed close together, difficult to compress
Gas		Restricted only by its container as far as shape and volume; particles are easy to compress
Plasma		Particles are ions, electrons, and neutral atoms traveling at extremely high speeds, no definite shape or volume, can be affected by magnetic field

In the previous section, the concept of absolute zero was discussed. What would happen to matter if that temperature could be reached? In the late 1920s and early 1930s Albert Einstein, building on the work of Indian physicist Satyendra Bose, predicted that a strange kind of matter would exist at temperatures that approached absolute zero. He theorized that at that temperature, atoms would exist in the same quantum-mechanical state. (Quantum theory will be discussed in Chapter 4.) In other words, they would all act as one single atom. In 1995, researchers lead by Carl Wieman and Eric Cornell were able to super-cool rubidium atoms to a temperature of just 200 billionths of a degree Celsius above absolute zero, and observed the phenomenon that Einstein had predicted. This new hypothetical phase of matter was termed the **Bose-Einstein condensate.** Researchers in this field are continuing to study Bose-Einstein condensates to determine their place in chemistry and physics.

As a gas is cooled, the particles begin to lose thermal energy, and they slow. Lacking the necessary energy to resist the attractive forces, the particles pull together into the liquid state. This phase change from gas to liquid is called liquefaction or, more commonly, **condensation.** Condensation occurs frequently in our homes during the winter months when the warm, moist air inside the house comes into contact with a cold windowpane.

If the particles in a liquid continue to lose thermal energy, they solidify. **Freezing,** or solidification, occurs when the liquid molecules lose enough energy to allow the attractive forces to hold the particles in tightly packed, fixed positions. **Melting** is the phase change from solid to liquid; it is the opposite of freezing. **Sublimation** (sub li MA shun) is a phase change that may be less familiar. Under certain conditions, a solid may change directly into a gas, or a gas may change directly into a solid, without passing through the liquid phase. Examples of substances that sublime directly from solid to gas are moth crystals and solid air fresheners. These gradually convert into the gaseous state and eventually "disappear." Conversely, the sublimation of ice onto aircraft wings has long been a problem of high-altitude flight. When moisture-laden air contacts the cold surface of the metal wing, gaseous water can change directly into ice.

Section Review Questions 2c

1. Matter exists in four states. What are they?
2. How does the kinetic theory relate to the states of matter?

Chapter Review

Coming To Terms

matter
physical property
density
malleability
ductility
conductivity
chemical change
physical change
chemical reaction
pure substance
element
solution
atom
monatomic element

diatomic element
polyatomic element
symbol
compound
formula
subscript
coefficient
molecule
mixture
heterogeneous mixture
homogeneous mixture
exothermic
endothermic
energy

thermodynamics
First law of thermodynamics
Second law of thermodynamics
law of mass/energy conservation
kinetic energy
thermal energy
entropy
heat
temperature
calorie, Calorie
BTU
Celsius scale
Kelvin scale
Fahrenheit scale
absolute zero
kinetic theory
solid
liquid
gas
plasma
Bose-Einstein Condensate
condensation
freezing
melting
sublimation

Review Questions

1. State the best descriptions of *matter* and *energy*.
2. Why is Creation *ex nihilo* a religious belief and not a scientific fact?
3. Tell whether each of the following properties is a physical property or a chemical property.
    a. color
    b. density
    c. ability to conduct electricity
    d. corrosiveness
    e. magnetism
    f. ability to burn rapidly
4. Tell whether each of the following processes involves a physical change, a chemical change, or both.
    a. A block of silicon chips being sliced into wafers to be used in microcomputer chips
    b. A glacier melting
    c. Steam condensing on the bathroom mirror as someone takes a shower
    d. Dynamite exploding and destroying an old building
    e. A burning candlewick melting wax
    f. The corroding of spokes on a bicycle's wheels
    g. The growth of a child
5. Why do some chemical symbols for elements consist of two letters and others of only one?

6. Some chemical symbols are apparently unrelated to the names of the elements to which they refer. Why? Give an example.
7. How many of each kind of atom is present in each of the following? Example: A single H_2O unit has one O atom and two H atoms.
   a. CsBr
   b. $NaNO_3$
   c. $2\ LiH_2PO_4$
   d. $KC_2H_3O_2$
   e. $K_2Cr_2O_7$
   f. $Al(C_2H_3O_2)_3$
   g. $2\ H_3PO_4$
   h. $2\ Ba_3(PO_4)_2$
8. Use the classification scheme to classify the following substances as either elements, compounds, homogeneous mixtures, or heterogeneous mixtures. Example: $MgSO_4$ is a compound.
   a. vegetable soup
   b. air
   c. oxygen
   d. gasoline
   e. sulfuric acid (H_2SO_4)
9. The ideas of men are not always in harmony with the way the universe operates. Identify the law that says that each of the following is impossible.
   a. A perpetual motion machine produces more energy than it consumes.
   b. A frictionless wheel revolves without ever slowing.
   c. The universe spontaneously created itself.
10. Convert the following temperatures from the scale given to the scale indicated in parentheses.
    a. The weatherman says that today's high temperature will be 75°F. (°C)
    b. A batch of cookies bakes at 325°F. (°C)
    c. Pure water freezes at 32°F. (°C)
    d. The temperature on the surface of the sun could be 5820°C. (°F)
    e. At atmospheric pressure, water boils at 100°C. (°F)
    f. Nitrogen gas can be liquefied at a temperature of 77 K. (°C, °F)
    g. Outside temperature is -53°C. (K)

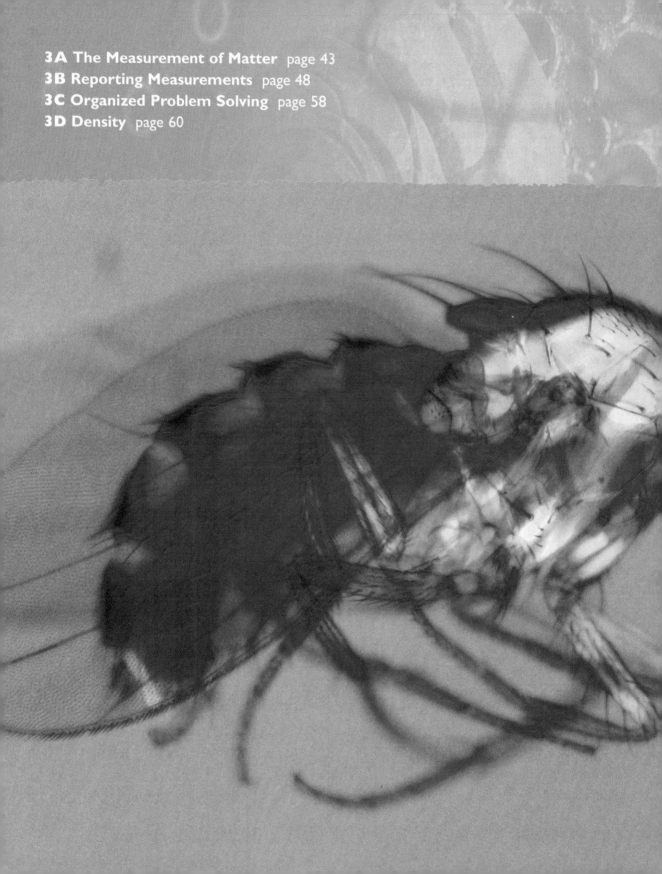

3A The Measurement of Matter page 43
3B Reporting Measurements page 48
3C Organized Problem Solving page 58
3D Density page 60

Math: The Central Language of Science 3

Calculate. Calculate. Calculate. Formulas, equations, and problem-solving techniques—these are just a few examples of the resources scientists use in their calculations. Chemistry is a science dependent on mathematical calculations, but this does not have to be intimidating to a new chemistry student. Mathematical equations and problem-solving techniques are actually helpful tools for calculating the answers to difficult questions. These mathematical equations can be understood and even mastered with time and effort. Galileo, the Italian astronomer and physicist, said, "Mathematics is the alphabet with which God has written the universe." Countless discoveries in science are waiting to be made by those who will take the time to understand the central language of science—math.

3A The Measurement of Matter

Individuals using chemistry must do a large amount of measuring, calculating, and working with numbers. It is impossible to have a working knowledge of God's creation until it can be quantified.

Chapter 2 discussed the characteristics and general classification of matter. Chemistry, however, involves not only describing matter but also measuring it. Measurements consist of numbers that tell "how many" and units that tell "what." A number without a unit is meaningless. Imagine being told that the distance to a shopping mall was 7 without being told if it was 7 blocks, 7 kilometers, or 7 miles. Measurements consist of two essential parts: A pure number and the unit of measurement. A **unit of measurement** is one increment of the system being used. The **pure number** is the number of units determined by the act of measuring. A **measurement** then is the product of the pure number and the unit of measurement. For example, in 6.5 grams, 6.5 is the pure number and grams is the unit of measurement. Understanding this concept of measurements will be important later in unit analysis. Being skillful in chemistry involves knowing how to work with numbers, equations, and equipment in order to perform and communicate accurate measurements.

Precision and Accuracy

No measurement is ever perfect. Tools such as thermometers, balances, and graduated cylinders have limited accuracy. Changes in laboratory conditions such as temperature and humidity cause slight variations in measurements. Measurements are also prone to human error and variability. When two different people read a fluid level in a graduated cylinder, they are likely to report slightly different measurements. Therefore, measurements must be as accurate and precise as possible.

Two important terms, *accuracy* and *precision*, describe the quality of measurements. Sometimes these words are used interchangeably, but they do not have the same meaning. **Accuracy** refers to how close a measurement is to the accepted reference or theoretical value. An accurate measurement has a small percent error; that is, it varies by only a small amount from what it should have been. The following formula shows how to calculate the percent error of a measurement. In the formula, O is the observed value, and A is the actual value.

$$\text{Percent error} = \left| \frac{O-A}{A} \right| \times 100\%$$

Note: The | | symbols mean that you should take the absolute value of the number. Consequently, percent error is always positive.

Precision is the agreement between two or more measurements. The term refers to the reproducibility of data. The term *precise* is used when referring to how finely divided the measurement scale is on an instrument.

Sample Problem You measure the mass of a product in a chemical reaction to be 3.80 g. Theoretical calculations predict that you should have obtained 3.92 g. What is the percent error?

Solution

$$\text{Percent error} = \left|\frac{O-A}{A}\right| \times 100\% = \left|\frac{3.80 \text{ g} - 3.92 \text{ g}}{3.92 \text{ g}}\right| \times 100\%$$

$$= |-3.06\%| = 3.06\%$$

Repeated measurements should yield nearly identical values. A set of measurements can be accurate without being precise, precise without being accurate, neither accurate nor precise, or both accurate and precise. These relationships are illustrated in Figure 3A-1. If you measured the product of a chemical reaction to be 1.70 g, 1.71 g, and 1.69 g on three separate trials, your measurements would be precise. If the mass is actually 3.92 g, your measurements would still be precise, but they would be terribly inaccurate.

Figure 3A-1

Metric Units

Measurements in chemistry demand clearly understandable units. Chemists around the world use metric units because they are common and easy to use. The metric system is an international decimal system of weights and measures that was adopted first in France in 1795.

The SI system, a particular group of units within the metric system, includes units for time, temperature, electrical current, and length. Table 3A-1 lists the units for these quantities as well as each unit's symbol. Note that this table lists only the SI base units that apply to chemistry. Units of measurement that are derived from the basic SI units are called **derivative units.**

Table 3A-1 SI Units Used in Chemistry

Quantity	Unit Name	Unit Symbol
Length (l)	meter	m
Mass (m)	kilogram	kg
Time (t)	second	s
Electric current (I)	ampere	A
Thermodynamic temperature	kelvin	K
Amount of substance (n)	mole	mol

The unit for volume is a derivative unit because it is derived from the SI unit for length, the meter. A simple way to compute volume is to find the product of width, length, and height—all in units of meters. Thus, a base unit of volume could be 1 m³. A cubic meter is too large to be a useful unit of volumetric measurement, so the cube of 0.1 m, or a cubic decimeter, is used instead. This volume is commonly referred to as a liter (L). The unit mL (milliliter) is also used frequently and is equivalent to the cubic centimeter (cm³). A cubic centimeter can also be written cc.

As shown in Table 3A-1, the SI unit of mass is the kilogram (kg). However, in this text we will treat the gram (g) as the basic unit of mass. It is also important to note that mass and weight are not the same. **Mass** is the measure of the amount of matter in a given substance. **Weight** is the gravitational pull exerted on an object. The mass of an object is always constant, whereas weight can change based on the location of the object. The farther an object is from the source of gravity, in this case the earth, the less gravitational pull it experiences and the less it weighs. This is why astronauts are weightless in space, or why you would weigh slightly less on the top of a mountain than at sea level.

The SI system modifies its basic units by putting prefixes on them. See Table 3A-2. You will see that some of the prefixes combined with base units produce a smaller unit, while others produce a larger one. A given prefix will have the same effect on any base unit. For example, since milli- (m) means 0.001, a millisecond is 0.001 second, a millimeter is 0.001 meter, and a milligram is 0.001 gram.

Note that the metric system is the standard system used in science. Therefore, the English system should be converted to the metric system. Difficulties occur when English units are not properly converted to metric units. NASA's $94,000,000 Mars Climate Orbiter was destroyed just before entering an orbit around Mars. The reason—the orbiter was given commands in English units and was programmed to receive commands in metric units.

SI units are related to each other in multiples of 10; thus, it is a decimal system. SI prefixes serve as multipliers for the base unit. A 1 is always placed in front of the prefixed unit, while the decimal equivalent of the prefix is written in front of the base unit. For example, 1 mL = 10^{-3} L and 1 µs = 10^{-6} s (see Table 3A-2). Such a procedure will be particularly helpful when working through unit conversion problems discussed later in the chapter.

Table 3A-2 SI Prefixes

Prefix	Symbol	Meaning	Numerical Meaning	Form
tera-	T	trillion	1,000,000,000,000	10^{12}
giga-	G	billion	1,000,000,000	10^{9}
mega-*	M	million	1,000,000	10^{6}
kilo-*	k	thousand	1,000	10^{3}
hecto-	h	hundred	100	10^{2}
deka-	da	ten	10	10
deci-*	d	tenth	0.1	10^{-1}
centi-*	c	hundredth	0.01	10^{-2}
milli-*	m	thousandth	0.001	10^{-3}
micro-*	µ	millionth	0.000001	10^{-6}
nano-*	n	billionth	0.000000001	10^{-9}
pico-	p	trillionth	0.000000000001	10^{-12}

*Should be memorized

Section Review Questions 3A

1. State why math is the central language of science.
2. Defend the importance of units in chemistry measurements.
3. Evaluate why science is limited in regard to precision and accuracy.
4. Determine which number in each group is the most precise.
   a. 302.1 m, 1.22 m
   b. 1 mm, 1 cm
   c. 34.555 mm, 23.1 nm
5. Explain the differences between weight and mass.
6. Summarize why the SI system is considered a decimal system.
7. Identify the various units of measurement.
   a. The unit equivalent to 1000 m
   b. The unit equivalent to 10^{-3} m
   c. The unit equivalent to 10^{-9} m

3B Reporting Measurements

Measurements can never be perfect, no matter how meticulous the scientists or how good the equipment. In most cases, the tools of the scientists, not careless mistakes, limit precision and accuracy. Consequently, the numbers that scientists use and report must indicate the accuracy of their measurements. Therefore, the way in which you report both the numbers and the units that you get in your work is extremely important. There are several methods of properly reporting numbers and units. These methods include significant digits, scientific notation, and unit analysis.

Significant Digits: Telling the Truth with Numbers

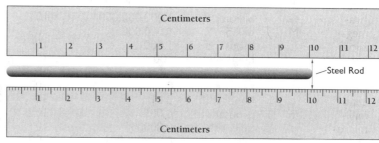

Figure 3B-1

If asked to measure the steel rod in Figure 3B-1, you could see that the accuracy of the measurement depends upon the ruler used. Using the top ruler, a length of 10 cm would seem appropriate; however, the first decimal place would have to be estimated. The end appears to be about one-tenth of the way between 10 cm and 11 cm, and we could record it as 10.1 cm. Nevertheless, someone else might estimate 10.3 centimeters. Since there is no calibration mark for the tenths place, it is uncertain or estimated. However, it is still significant. Using the lower ruler, there is certainty about the 10 and the tenths place; but the second decimal place would have to be estimated. It appears that the end of the rectangle is about three-tenths of the way between 10.1 and 10.2; we would thus record its length as 10.13 cm. Here, the 3 is uncertain but is still included as a significant digit. In this example, the lower ruler is more precise than the top ruler.

The entire concept of significant digits is based on being faithful when presenting information. A Christian especially must be careful to report the proper number of significant digits at all times, since accurate reporting reflects on the honesty of the individual. It is also a matter of scientific integrity. Some laboratories consider the misuse of significant digits grounds for dismissal.

Significant digits indicate the precision of measurements and help scientists report measurements honestly. For instance, if someone measured his own weight with a standard bathroom scale, he could not honestly and accurately report his weight to be 135.67942 pounds. A bathroom scale cannot measure weight that accurately. A reasonable measurement on a bathroom scale might be only as precise as 136 pounds.

In chemistry, the proper number of significant digits indicates the precision of the instrument. The measurements recorded should include all the numbers that are known with certainty plus one digit that is estimated. It is assumed that any measured quantity will have its uncertainty in the rightmost digit. The other digits are known with certainty. The certain digits and the uncertain digit together make up the **significant digits.** The following table shows the normal precision of the laboratory equipment you may use. In this table, the position of the nonzero digit signals the doubtful digit. Use Table 3B-1 as a guideline when making measurements. For instance, do not report a mass of 35.582 g when using a triple-beam balance. The measurement 35.58 g reflects the limited precision of this kind of balance.

A. Analytical balance
B. Electronic balance with digital readout
C. 100 mL glass graduate cylinder

Table 3B-1 The Precision of Instruments

Instrument	Typical Uncertainty
platform balance	±0.1 g
triple-beam balance	±0.01 g
analytical balance	±0.0001 g
100 mL graduated cylinder	±0.1 mL
10 mL graduated cylinder	±0.1 mL
50 mL burette	±0.01 mL
thermometer (10°C–110°C)	±0.1°C

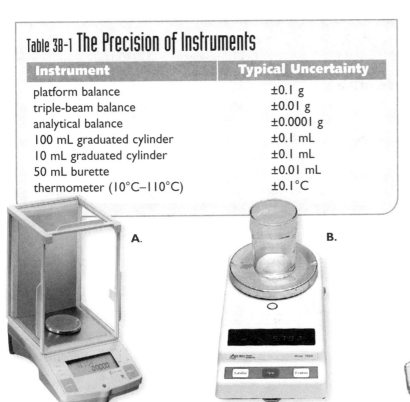
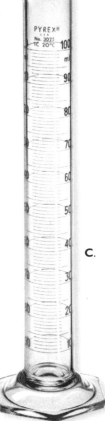

A.   B.   C.

Rules for Significant Digits

1. Significant digits apply only to measurements. Significant digits do not apply to
   a. Counted numbers.
      (Ex: half, twice, dozen, 30 students, 10 books, etc.)
   b. Definitions.
      (60 seconds = 1 minute, 12 inches = 1 foot, etc.)
2. All nonzero digits are significant.
   Example: 375.42 cm has 5 significant digits.
   22.3 m has 3 significant digits.
3. All zeros between two nonzero digits are significant.
   Example: 7008 m has 4 significant digits.
   1,400,002 cm has 7 significant digits.
4. All other zeroes are significant unless they are placeholders.
   Example: 27.30 mm has 4 significant digits.
   0.00025 cm has 2 significant digits.
   0.0002500 cm has 4 significant digits.
5. To indicate the precision of a measurement, use scientific notation. All digits in the decimal portion are significant.
   Example: 380,000 m that was estimated to the nearest 1000 meters should be written 3.80×10^5 m.

Sample Problem Determine the number of significant digits in the following quantities:
   a. 9.370 kg   c. 705.06 mL   e. 0.0001 s
   b. 63,000 g   d. 12 cookies   f. 2300.000

Solution
   a. All nonplaceholding zeros in a measurement having a decimal point are significant, so the number has four significant digits.
   b. No placeholding zeros in a measurement without a decimal point are significant; therefore, the number has two significant digits.
   c. All zeros between two nonzero digits are significant. The number, therefore, contains five significant digits.
   d. This is a counted number; therefore, significant digits do not need to be expressed.
   e. Placeholding zeros are not significant; therefore, the number has one significant digit.
   f. All nonplaceholding zeros in a measurement having a decimal point are significant; therefore, this number has 7 significant digits.

The correct number of significant digits in a result must be maintained during mathematical operations. Three rules guide the process of determining significant digits during addition, subtraction, multiplication, and division.

1. After addition and subtraction, the answer cannot be more precise than the least precise measurement. The addition of 50.23 m, 14.678 m, and 23.7 m yields a preliminary answer of 88.608 m. Since the least precise addend (23.7 m) has its estimated digit in the tenths place, the answer must be rounded to 88.6 m.
2. In multiplication or division the answer cannot contain more significant digits than the measurement with the least number of significant digits. Although 0.238 m × 0.31 m = 0.07378 m^2, this number must be rounded off to 0.074 m^2 because 0.31 m has only two significant digits.
3. The rule for rounding is as follows: If the digit to be dropped is less than 5, simply drop that digit. If it is 5 or greater, increase the preceding digit by one. For example, 12.4489 expressed with 3 significant digits would be 12.4; however, expressed with 4 significant digits it would be 12.45.

Please note that these rules are guideposts. In some calculations more or fewer significant digits are appropriate depending on relative size or precision of the measurement or calculation.

Sample Problem Perform the following operations and express your answers with the proper number of significant digits.

a. 33.153 g
    3.2 g
+ 8.70 g

b. $\dfrac{10.10 \text{ g}}{4.04 \text{ mL}}$

c. 6.00 cm × 3.228 cm × 0.037 cm

Solution
a. 45.1 g—The answer must be expressed to the tenths place since the least precise addend (3.2 g) is expressed to the tenths place.
b. 2.50 g/mL—The answer must have 3 significant digits since the least precise number (4.04 mL) has three significant digits.
c. 0.72 cm^3—The answer must have 2 significant digits since 0.037 cm has only two significant digits. Remember that place-holding zeros are not significant.

Scientific Notation: Long Numbers in Short Forms

Chemists constantly work with very large and very small numbers. For example, a certain wavelength of light is 0.0000006 m. This small number can be written more conveniently in scientific notation as 6×10^{-7} m. One milliliter of liquid water contains 33,400,000,000,000,000,000,000 water molecules. Using scientific notation, scientists express this number as 3.34×10^{22}. **Scientific notation** is a number expressed between 1 and 10 multiplied by an **exponential factor,** which is 10 raised to some power, or an **exponent.** Expressing numbers in terms of powers of ten has several benefits.

- Scientific notation makes it easier to read and write very large and very small numbers.

- Numbers in scientific notation clearly show the number of significant digits because all the digits used are significant.

- Numbers expressed in scientific notation are easier to work with in multiplication and division problems.

A positive exponent or power of 10 indicates how many times the coefficient is multiplied by 10. For example, 3.61×10^2 means the same thing as 361. Also, 5.45×10^4 is equivalent to 54,500. Thus, a positive exponent is always associated with a number that is greater than 1. It tells you how many places you will move the decimal point to the right in order to write the full, nonexponential form of the number.

Calculators can be used to solve problems involving scientific notation.

A negative exponent or power of 10 indicates that 10 is in the denominator and tells how many times the coefficient is divided by 10. For example, 1×10^{-2} is equivalent to 0.01, and 2.66×10^{-4} is equivalent to 0.000266.

A negative exponent is always associated with a number less than 1 and greater than 0, and tells how many places to move the decimal point to the left in order to write the full, nonexponential form of the number. Also, a positive exponent in the denominator can be written as a negative exponent in the numerator; and a negative exponent in the denominator can be written as a positive exponent in the numerator. This can be seen in examples from the previous paragraph. Multiplication by 10^{-2} is the same as division by 100 or 10^2, and multiplication by 10^{-4} is the same as division by 10^4.

Table 3B-2 shows powers of 10 and their numerical equivalents.

Table 3B-2 Powers of 10 and Numerical Equivalents

$1 = 10^0$	
$10 = 10^1$	$\frac{1}{10} = 10^{-1}$
$100 = 10^2$	$\frac{1}{100} = 10^{-2}$
$1000 = 10^3$	$\frac{1}{1000} = 10^{-3}$
$10,000 = 10^4$	$\frac{1}{10,000} = 10^{-4}$
$100,000 = 10^5$	$\frac{1}{100,000} = 10^{-5}$

Sample Problem Write the following exponential numbers in full, non-exponential form.

        a. 5.30×10^{-3} m   **b.** 7.5×10^{5} g

Solution

a. The exponent 10^{-3} tells you to move the decimal point 3 places to the left of where it is in the original number. In many cases, zeros will have to be added to "hold" decimal places. Note that a negative exponent correlates with a number that is less than 1—one that is a fractional part of 1. Both numbers have 3 significant digits.

        5.30×10^{-3} m $= 0.00530$ m

b. The positive exponent tells you to move the decimal point to the right—which in this case is five places. Both numbers have 2 significant digits since the added zeros are only placeholders.

        7.5×10^{5} g $= 750{,}000$ g

Significant Digits in Scientific Notation

How are numbers converted so that they are in proper scientific notation? Consider a number that is greater than 1. The number 36,300 can be written as the product $3.63 \times 10{,}000$ or 3.63×10^{4}. The decimal point has to be moved 4 places to the left from its original position to give a number less than 10 but greater than 1. The exponent is thus a positive 4 because the original number was greater than 1.

A number that is less than 1 (i.e., a fractional part of 1) can be factored. For example, 0.000058 can be factored to 5.8×0.00001 or 5.8×10^{-5}. The decimal point was moved 5 places to the right to obtain a number that was between 1 and 10. Thus, a negative exponent is associated with a number that is smaller than 1.

The issues above apply to positive and negative numbers. For example, -361 can be factored into -3.61×10^{2}. Ignoring the negative sign, we had to move the decimal point two places to the left to get a number that was between 1 and 10. The difference is only the negative sign on the number; the exponential factor remains the same. Thus, positive or negative numbers can be converted to scientific notation if their absolute value is considered.

Sample Problem Write the following numbers in scientific notation.

    a. 0.078 m   **b.** 78,000 cm   **c.** -0.0078

Solution

a. The decimal point must be moved 2 places to the right to obtain a number that is between 1 and 10. Thus, the exponent is -2. Since the number has two significant digits, proper scientific notation is as follows:
$$0.078 \text{ m} = 7.8 \times 10^{-2} \text{ m}$$

b. The number to be converted is greater than 1, so the exponent will be positive. The decimal point is moved to the left 4 places to have a number that is between 1 and 10. Thus, the number is written 7.8×10^4. Since 78,000 has 2 significant digits, proper scientific notation is as follows:
$$78,000 \text{ cm} = 7.8 \times 10^4 \text{ cm}$$

c. A negative number is treated like a positive number when writing scientific notation. Thus, the decimal point must be moved 3 places to the right. Since the absolute value of the number is less than one, there will be a negative exponent. Again, there are two significant digits.
$$-0.0078 = -7.8 \times 10^{-3}$$

Scientific Notation in Mathematical Functions

Adding and subtracting numbers in scientific notation reduces bulky numbers to a workable size. Even though electronic calculators can perform these functions automatically, an understanding of what is happening will help determine proper numbers of significant digits and guard against absurd answers. The exponential factors must be the same before you can add or subtract numbers. If measurements have the same exponential factors, the numbers are simply added or subtracted, while the exponential factors remain the same. If measurements have different exponential factors, convert one of the exponential factors to match the other one. The smaller exponent and corresponding number are changed to match the larger. In this case, change 1×10^2 to 0.1×10^3 so that both of the exponents are the same:

$$\begin{array}{r} 1 \times 10^2 \text{ m} = 0.1 \times 10^3 \text{ m} \\ + 1.0 \times 10^3 \text{ m} = 1.0 \times 10^3 \text{ m} \\ \hline 1.1 \times 10^3 \text{ m} \end{array}$$

In multiplication, the numbers are multiplied, and the exponents are added. In division, the numbers are divided, and the exponents are subtracted instead of added.

Sample Problem Perform the following operations, observing significant digits.

    a. $(3.55 \times 10^4 \text{ g}) - (3.55 \times 10^3 \text{ g})$
    b. $(6.22 \times 10^{-2} \text{ m}) + (8.4 \times 10^{-4} \text{ m})$
    c. $(3.55 \times 10^4 \text{ cm}) \times (3.55 \times 10^3 \text{ cm})$
    d. $(6.22 \times 10^{-2} \text{ kg}) \div (8.4 \times 10^{-4} \text{ kg})$

Solution

a. Since the exponents are not the same, you must convert the one with the smaller exponent to match the one with the larger exponent: $3.55 \times 10^3 \text{ g} = 0.355 \times 10^4 \text{ g}$.

$$\begin{array}{r} 3.55 \times 10^4 \text{ g} \\ - 0.355 \times 10^4 \text{ g} \\ \hline \end{array}$$

$3.195 \times 10^4 = 3.20 \times 10^4$ g (3 significant digits)

b. You must once again convert one of the numbers so both of their exponents match. Note that -4 is a smaller exponent than -2.

$$\begin{array}{rll} 6.22 \times 10^{-2} \text{ m} & = & 6.22 \times 10^{-2} \text{ m} \\ + 8.4 \times 10^{-4} \text{ m} & = & 0.084 \times 10^{-2} \text{ m} \\ \hline 6.304 \times 10^{-2} & = & 6.30 \times 10^{-2} \text{ m (3 significant digits)} \end{array}$$

c. Multiply the coefficients and add the exponents of the exponential factor.

    $(3.55 \times 10^4 \text{ cm}) \times (3.55 \times 10^3 \text{ cm}) =$
    $(3.55 \times 3.55) \times (10^4 \times 10^3) =$
    $12.6025 \times 10^{(4+3)} = 12.6 \times 10^7 = 1.26 \times 10^8$ cm

                                   (3 significant digits)

d. Divide the coefficients and subtract the exponents in the denominator from the exponents in the numerator.

    $(6.22 \times 10^{-2} \text{ kg}) \div (8.4 \times 10^{-4} \text{ kg}) =$
    $(6.22 \div 8.4) \times (10^{-2} \div 10^{-4}) =$

$$\frac{6.22}{8.4} \times \frac{10^{-2}}{10^{-4}} = 0.74 \times 10^{(-2)-(-4)} = 7.4 \times 10^{-2+4}$$

    $= 0.74 \times 10^2 = 7.4 \times 10^1$ kg (2 significant digits)

Since electronic calculators are used to solve most of the problems in this course, you must become familiar with how to enter numbers in scientific notation, as well as the way in which your calculator displays them. Most calculators have an EXP or EE button that is used to enter numbers written in scientific notation. To display numbers in scientific notation, many calculators have a SCI key or mode. To enter negative exponents, the +/- key, (–) key, or − key will have to be used. You must take the responsibility to learn the capabilities and functions of your calculator so that you are comfortable with it. Practicing sufficiently can help you do this.

Unit Analysis: The Chemist's Secret Weapon

Chemistry problems often involve converting a measured value from one unit to another. **Unit analysis,** also called dimensional analysis, is an excellent tool for solving this type of problem. It is a powerful, versatile, and organized method of changing units on numbers.

The addition and subtraction of measurements require that the numbers have identical units. Three meters can be added to 5 meters to get 8 meters, but 3 meters cannot be added to 5 centimeters until the units match. In multiplication and division the units are multiplied or divided along with the numbers. For example, 3 meters times 5 meters equals 15 square meters (m^2). A car that travels 60 kilometers in one hour moves at a rate of 60 kilometers per hour (km/hr). The way in which units multiply and divide is the key to unit analysis.

Changing a number with a unit to an equivalent number with another unit involves multiplying the given number and unit by a conversion factor. A **conversion factor** is the ratio of two measurements and their units that represent the same quantity. Since the numerator and the denominator of the ratio are the same measurements expressed in different units, the value of the ratio is equal to 1. Any quantity multiplied by one is unchanged. Below are some examples of ratios that could be used as conversion factors.

$$\frac{1 \text{ yd}}{3 \text{ ft}} \quad \frac{3 \text{ ft}}{1 \text{ yd}} \quad \frac{60 \text{ min}}{1 \text{ hr}} \quad \frac{1000 \text{ mL}}{1 \text{ L}} \quad \frac{1000 \text{ m}}{1 \text{ km}}$$

How do you know which conversion factor to use in a given product? The key to changing a number's units is constructing a conversion factor that allows the undesired units to cancel out in division. For instance, the conversion factor of 1 yard/3 feet is used to change 15 feet into yards.

$$\frac{15 \cancel{\text{ft}} \,\bigg|\, 1 \text{ yd}}{\bigg|\, 3 \cancel{\text{ft}}} = 5 \text{ yd}$$

Since the feet units appear on both the top and bottom of the fraction bar, they cancel out. After the numerical operations are done, the resulting number has the unit *yards*.

Notice that by using units in the calculation, the correct result is assured. If the conversion factor of 3 feet/1 yard had been used, the units of feet would not have canceled out. Instead, the unit *ft* is multiplied by ft.

$$\frac{15 \text{ ft} \mid 3 \text{ ft}}{\mid 1 \text{ yd}} = \frac{45 \text{ ft}^2}{\text{yd}}$$

This answer has meaningless units and is therefore wrong. Hopefully, you can now see the importance of always using units with a conversion factor. In the above problem, if only the numerical values were present, you would have had no idea the answer was wrong. Always use units!

When performing unit analysis determine the following:

- What is the given quantity and its unit?
- What is the desired unit?
- Is there a relationship between the given unit and the desired unit?

It is often necessary to perform more than 1 unit conversion to reach the desired unit. In general, the unit analysis setup will be as follows:

$$\frac{\cancel{\text{given unit}} \mid \cancel{\text{intermediate unit}} \mid \text{desired unit}}{\mid \cancel{\text{given unit}} \mid \cancel{\text{intermediate unit}}} = \text{desired unit}$$

Sample Problem Perform the following conversions
      a. 132,546 cm to km   b. 350 mg to kg

Solution

a. Starting with the given information—132,546 cm—use conversion factors to obtain the unit of kilometers. Doing the conversions one after the other shortens the operation.

$$\frac{132{,}546 \text{ cm} \mid 1 \text{ m} \mid 1 \text{ km}}{\mid 100 \text{ cm} \mid 1000 \text{ m}} = 1.32546 \text{ km}$$

b. The given unit is mg, and the desired unit is kg. Both units are related to the basic unit, grams. Thus, we can use the relationships 1 mg = 1 × 10⁻³ g and 1 kg = 1 × 10³ g in conversion factors as the "bridge" between mg and kg.

$$\frac{350 \text{ mg} \mid 1 \text{ g} \mid 1 \text{ kg}}{\mid 1000 \text{ mg} \mid 1000 \text{ g}} = 350 \times 10^{-6} \text{ kg} = 3.5 \times 10^{-4} \text{ kg}$$

Section Review Questions 3B

1. List at least two problems that could occur if scientists did not use significant digits when reporting measurements.
2. Describe how significant digits in a measurement can indicate the precision of the instrument used to take the measurement.
3. Determine how many significant digits are in the following numbers:
    a. 20.009 g
    b. 0.00876 m
    c. 45.000 cm
    d. 34500 kg
    e. 234.614 mg
4. Work the following problems and express your answer with the proper number of significant digits.
    a. $(4.62 \times 10^3 \text{ g}) \times (6.42 \times 10^4 \text{ g})$
    b. 6.234 m + 30.1 m + 1.024 m + 0.0020 m
    c. $(5.61 \times 7.891) \div 9.1$
    d. $(8.91 \times 10^6 \text{ L}) - (6.435 \times 10^4 \text{ L})$
    e. 4.2 mL − 58.0 mL − 00.0044 mL
5. Discuss the importance of using scientific notation in mathematical calculations.
6. Compare and contrast positive and negative exponents in scientific notation.
7. Summarize why units are absolutely necessary in unit (dimensional) analysis.
8. Perform the following conversions using unit analysis.
    a. 243 mg to kg
    b. 456 cm to km
    c. 532 nm to cm
    d. 467 ng to g

3C Organized Problem Solving

Problem-solving skills are necessary for daily living. Buying a car, building a house, balancing a checkbook, and even baking cookies all require problem-solving skills. Chemistry provides many opportunities to gain experience in problem solving. Most chemistry problems are word problems. Word problems scare some students, but they can be conquered and even enjoyed if they are attacked systematically.

A Systematic Approach

A systematic approach will not always lead to an immediate solution, but it will give direction in finding the answer. One systematic approach for chemistry has seven steps:

Step 1: Identify what is given. The essential pieces of information are usually easy to spot; organizing them is the main task. To do this, collect all the numerical information and any important non-numerical (qualitative) information, and record it in an organized way. Occasionally you must weed out pieces of irrelevant data from a problem. Don't forget your units.

Step 2: Decide what to do. After jotting down the given information, look for key words that tell you what to do. *Find, how many, name,* and *identify* point you on the path toward the desired solution.

Step 3: Decide how to do it. As you become more experienced in solving problems in chemistry, you will begin to recognize patterns and familiar routes to solutions. Most of the problems will require one of two general approaches: unit analysis or the use of a formula.

Step 4: Use Unit Analysis. If you do not have the right kind of units in the answer, you have the wrong answer! In this course, you should be able to solve most quantitative problems using this method.

Step 5: Estimate the Answer. Do not attempt to solve problems by memorizing formulas and plugging in the proper numbers. This is not productive for learning chemistry and provides no basis for the reasonability of the answer that "pops" out of the calculator. Always do the calculations in your head using round numbers before trying to solve the problem. This method will give you a reasonable way to judge your answers.

Step 6: Apply rules for significant digits. Make sure the answer uses the proper number of significant digits.

Step 7: Use scientific notation if the answer is unusually large or small.

Once we cover the next topic, you will see how to put these steps into practice. Having several "models" to follow should help as you practice using these steps for problem solving. Seeing how it is done, however, is no substitute for doing it yourself. You must practice!

Section Review Questions 3c

1. Why is it important to reason out a problem instead of just plugging numbers into a memorized equation?
2. List 3 ways in which you will have to use problem-solving techniques in your daily life.

Chapter 3D

Relative Density—shown are a cork and three liquids: safflower oil, water, and methylene chloride. Water (light blue) is less dense than methylene chloride (dark purple) and floats on it. Safflower oil (clear) is less dense than water and floats on it. A cork is less dense than safflower oil and floats on it.

3D Density: Applying Mathematical Concepts

As you learned in Section 2A, one of the physical properties of matter is its density (d or the Greek letter rho (ρ)—in this text d will be used to represent density). Density is defined as the ratio of mass (m) to volume (V). This ratio is shown by the following equation:

$$d = \frac{m}{V}$$

The unit of mass is usually grams, and the unit of volume can be mL = cm^3 = cc, or L. Density is a property of matter that does not depend on the amount of matter present. For example, a cup of water has a density of about 1.0 g/mL, but so does a bathtub of water.

People sometimes incorrectly compare the densities of two substances by saying that one is "heavier" than the other. What they actually mean is that one is denser than the other. You may have heard someone say that lead is heavier than water, but 1 kg of water is just as heavy as 1 kg of lead. What is meant is that lead is denser than water, or that it is more compact, having more matter packed into a given volume than does water.

In this section, density will be used to illustrate how to use the problem-solving steps given previously. It is important to understand that density is a relationship between the mass of a substance and its volume. Thus, in a conversion between mass and volume, density serves as the "bridge." Table 3D-1 lists densities of several important industrial chemicals. One practical lesson that you could learn from the table below would be to head for high ground if a cloud of chlorine gas heads your way, but run downhill if a cloud of ammonia gas threatens. Since the density of air averages 1.29 g/L, chlorine gas is denser (not "heavier") than air, and ammonia gas is less dense (not "lighter") than air. Thus, chlorine gas will sink in air and ammonia gas will float. Hundreds of people in Africa were killed by clouds of CO_2 (1.98 g/L) released by volcanoes, that eventually settled into nearby populated valleys.

Sample Problem Use a systematic approach to solve the following problem. A lead fishing sinker that has a density of 11.3 g/mL has a mass of 51 g. What volume will the sinker occupy?

Solution

The problem will be solved by using the systematic approach from Section 3C.

Step 1: Identify what is given. d (density) = 11.3 g/mL; m (for mass) = 51 g

Step 2: Decide what to do. First, determine the volume.

Step 3: Decide how to do it. Since you are given a mass and desire a volume, density (defined as mass/volume) serves as the bridge between them. Density will be used as the conversion factor.

Step 4: Use unit analysis. The desired unit of the conversion factor is in the numerator, and the old unit to be cancelled is in the denominator.

$$\frac{51 \text{ g} \cdot \text{mL}}{11.3 \text{ g}} = 4.5132 \text{ mL}$$

The given unit has been converted to the one desired for volume.

Step 5: Estimate the answer. The setup amounts to dividing the given mass by the density. You can round 51 g to 50 g and 11.3 g/mL to 10 g/mL. Thus, 50 ÷ 10 = 5. Your answer should be about 5 mL.

Step 6: Correct for the proper number of significant digits. The actual calculator answer for the quotient is 4.5132 mL. Since 51 g has 2 significant digits and 11.3 g/mL has 3 significant digits, the answer can have only 2 significant digits. Round the answer to 4.5 mL.

Step 7: If the answer is unusually large or small, report it in scientific notation. In this case, you need not report the answer in scientific notation, but if you did, it would be 4.5×10^0 mL. The answer is also reasonable, since it is very close to the estimated value of 5 mL and the unit of volume is the desired unit.

Table 3D-1 Densities of Important Industrial Chemicals

Name	Description at Room Temperature	Density
Sulfuric Acid	Clear, colorless, oily liquid	1.84 g/mL
Nitrogen	Colorless, odorless, tasteless gas	1.250 g/L (STP)*
Oxygen	Colorless, odorless, tasteless gas	1.429 g/L (STP)
Ammonia	Colorless gas, pungent odor	0.7714 g/L (STP)
Chlorine	Greenish-yellow gas, suffocating odor	3.214 g/L (STP)

* STP represents a set of agreed-upon conditions used in gas laws where standard temperature and pressure are 0°C and 1 atm. STP is discussed in Chapter 9.

Sample Problem A bottle of isopropyl alcohol has a volume of 473.0 mL and a density of 0.785 g/mL. What is the mass?

Solution

Step 1: Identify what is given. d (for density) = 0.785 g/mL and
V (for volume) = 473.0 mL

Step 2: Decide what to do. First, determine the mass.

Step 3: Decide how to do it. Since you are given the volume and desire the mass, density (defined as mass/volume) serves as the bridge between them. Density will be used as the conversion factor.

Step 4: Use unit analysis. The desired unit of the conversion factor is grams, and the old unit is mL.

$$\frac{0.785 \text{ g}}{\text{mL}} \times \frac{473.0 \text{ mL}}{} = 371.305 \text{ g}$$

Step 5: Estimate the answer: The setup amounts to multiplying the density by the volume to get the mass. You can round 0.785 g to 0.80 g and 473.0 mL to 470 mL. Thus, 470 × 0.80 = 376. Your answer should be close to 376 g.

Step 6: Correct for the proper number of significant digits. The actual calculator answer for the quotient is 371.305 g. Since 0.785 g has 3 significant digits and 473.0 mL has 4 significant digits, the answer can have only 3 significant digits. Round the answer to 371.

Step 7: If the answer is unusually large or small, report it in scientific notation. It is not necessary to report the answer in scientific notation, but if you did, it would be 3.71×10^2. A good rule you may wish to use is to report scientific notation when a number is greater than 1×10^3 or less than 1×10^{-3}.

While these sample problems involved density calculations only, they serve as models for setting up and solving problems of all types. The steps involved will be illustrated throughout the text as other types of problems occur. Learn these steps well!

Section Review Questions 3D

1. 1 kg of lead and 1 kg of packaging foam have the same weight but different densities. Explain.
2. Use Table 3D-1 to describe what happens to the nitrogen and oxygen components in the air.
3. Toluene is a colorless, aromatic liquid derived from coal tar. It is used in explosives and high octane gasoline. A sample of

toluene has a volume of 35.1 mL and a mass of 30.5 g. Determine the density.
4. An experiment requires 45.6 g of isopropyl alcohol. Instead of being measured on a balance, the liquid is poured into a graduated cylinder. The density of the isopropyl alcohol is 0.785 g/mL. What volume of isopropyl alcohol should be used?
5. A liquid with a volume of 10.7 mL has a mass of 7.51 g. The liquid is either octane, ethanol, or benzene, the densities of which are 0.702 g/mL, 0.789 g/mL, and 0.879 g/mL respectively. What is the identity of the liquid?

Chapter Review

Review Questions

1. Evaluate the precision and accuracy of the measurement in the following situations.
   a. In three separate identical experiments, you measured the same product of a chemical reaction to be 6.20 g, 4.96 g, and 5.12 g. The predicted mass of the product was 3.81 g.
   b. While measuring temperature during three consecutive experiments you obtained 45.2°C, 44.9°C, and 45.1°C. The actual temperature was 45.0°C.
   c. In three separate trials, you measured the density of a gas to be 2.865 g/L, 2.852 g/L, and 2.860 g/L at STP. The actual density of the gas was 3.214 g/L at STP.
2. Determine the percent error for the following equations.
   a. The density of mercury is 13.6 g/mL. What would be your percent error if you performed an experiment and determined the density to be 13.0 g/mL?
   b. The density of oxygen gas is 1.429 g/L (STP). What would be your average percent error if in two experiments you obtained values of 1.112 and 1.069 g/L (STP)?
3. Rename the following prefixes with their decimal equivalent. (Example: deka- = ten)
   a. deci-
   b. milli-
   c. kilo-
   d. nano-

Coming to Terms

unit of measurement
pure number
measurement
accuracy
precision
derivative unit
mass
weight
significant digit
counted number
scientific notation
exponential factor
exponent
unit (dimensional) analysis
conversion factor

4. Give the number of significant digits in each of the following quantities.
   a. 2.40 cm
   b. 0.0101 g
   c. 20.0 mi
   d. 102 students
   e. 130.01200 m
   f. 300.0 mm

5. Add, subtract, multiply, and divide the following measurements, and give your answer with correct units and the correct number of significant digits. Round off answers when necessary.
   a. 5.85 g + 0.032 g + 6.19 g
   b. 4.718 cm − 3.94 cm
   c. (6.98 cm − 2.83 cm) × 1.7 cm²
   d. (5.45×10^6 km) × (3.22×10^3 km)
   e. (2.1 × 3.14) × 6.83 m²
   f. 3.0 m × 2.54 m
   g. 1.75 cm ÷ 2
   h. 4.020 m + 00.23 m + 40.25 m + 10.00 m

6. Write the following numbers in full, non-exponential form, observing significant digits.
   a. 6.08×10^4 g
   b. 2.070×10^{-3} m

7. Write the following numbers in proper scientific notation. Make sure to use significant digits properly.
   a. -6001 kg
   b. 0.03000 g
   c. 450×10^4 cm³

8. Perform the following operations, observing significant digits.
   a. (4.08×10^{-3} g) − (7.2×10^{-4} g)
   b. (7.04×10^{-4} m) × (4.1×10^4 m)
   c. (3.40×10^{-2} cm³) ÷ (5.08×10^{-3} cm³)
   d. (8.95×10^3 kg) + (9.8×10^1 kg)

9. Write an equation that states the relationship between the two given units, and then write the conversion factor (or conversion factors) that could be used to convert the first unit to the second. Example: Millimeters and meters are related by the equation 1000 mm = 1 m. The conversion factor 1 m/1000 mm should be used.
   a. grams and milligrams
   b. nanometers and meters
   c. kilograms and grams
   d. liters and milliliters
   e. megahertz and hertz

10. Convert the following measurements from the units given to the units requested with the use of the conversion factors in the previous problem. Example: How many meters wide is 35 mm camera film?

    Solution: $\dfrac{35 \text{ mm}}{} \Big| \dfrac{1 \text{ m}}{1000 \text{ mm}} = 0.035$ m

a. How many grams are in 5,280 mg?
b. The wavelength of blue light is about 475 nanometers (nm). How many meters is this?
c. What is the mass in grams of a 72.6 kg man?
d. How many milliliters of ginger ale are contained in a 2 L bottle?

11. Use unit analysis to convert the following measurements from the given units to the requested units.
    a. The earth's mass is 5.9763×10^{27} g. How much is its mass in mg?
    b. The distance from the earth to the sun is an astronomical unit (AU) and is defined as 149,599,000 km. How many nm is this?
    c. The volume of the earth is $1,083.1579 \times 10^9$ km^3. How many liters is this?
    d. A box is 0.2 m high, 0.050 m wide, and 0.100 m long. How many liters will this box hold?

12. Use unit analysis and a systematic approach to solve the following density problems. Give all answers with the correct units and the correct number of significant digits.
    a. Mercury, which is often used in thermometers, has a density of 13.6 g/mL at room temperature. What volume of mercury contains 10.0 g?
    b. Copper has a density of 8.92 g/mL. What mass of copper will occupy 45 mL? What volume of copper will have a mass of 1.0 kg?
    c. Ethyl alcohol has a density of 0.789 g/cm^3. What volume of ethyl alcohol must be poured into a graduated cylinder to give 17.6 g of alcohol?
    d. The density of quartz was determined by adding a weighed sample to a graduated cylinder containing 56.3 mL water. After submerging the quartz, the water level was 68.2 mL. The mass of the quartz was 36.5 g. What was the density of the quartz?
    e. A solid will float on any liquid that is more dense than it is. The volume of a piece of calcite (used in the production of chalk) is 12.5 mL, and the mass is 30.2 g. On which of the following liquids will the calcite float: acetone (density = 0.792 g/mL), water (density = 0.998 g/mL), propane (density = 1.26 g/mL), methylene bromide (density = 2.50 g/mL), bromine (density = 3.10 g/mL), mercury (density = 13.55 g/mL)?

4A The Development of Atomic Models page 67
4B The Quantum Model page 80
4C Numbers of Atomic Particles page 88
FACETS: Spectroscopy: Fingerprinting Atoms page 76

Atomic Structure 4

Small-Scale Architecture

It has been only recently that scientists have been able to "see" atoms on surfaces using a technique known as scanning tunneling microscopy (STM). Therefore, most of the evidence about atoms has been collected indirectly. Scientists ingeniously used bits of fragmented evidence to form the theories that are now called atomic models. These theories about atomic structure serve as the foundational ideas for studies of bonding and reactions, and the rest of chemistry.

4A The Development of Atomic Models: A Historical Perspective

The quantum model, the Bohr model, the plum pudding model—you'll note that each of these theories about the structure of atoms is called a model. **Models** are working representations of experimental facts. They are not exactly correct on all points, but then, they do not claim to be. They are mental pictures or simplifications of what is being studied. To further understand models, consider the concept of eternity, which is sometimes likened to traveling in a circle: "There is no start, and there is no stop. It just goes on forever." This analogy is like a scientific model. It is used to visualize an idea and to explain a complex concept. Although this model is crude and does not hold true in many aspects, it is still useful because it helps the human mind to grasp some of the concept.

Our Lord Jesus often used models during His teaching ministry on earth. We call these models *parables*. For example, Jesus taught the doctrine of salvation in Matthew 13:3-8 with the parable of the sower, a simple model to illustrate a complex truth.

The Origin of the Atomic Concept: Greek Ideas

A Greek philosopher named Democritus (b. 465 B.C.) is credited with being the first to say that matter is discontinuous: it is made of separate, discrete particles. He said matter contained definite particles and that it could not be divided infinitely without losing its properties. Democritus used the term *atomos** to name these smallest particles. Today we use the term **atom** to describe the smallest part of an element that retains the properties of that element.

When it became possible to precisely measure the mass of each element that had combined to form a compound, the way was paved for full acceptance of Democritus's ideas. Chemists observed that a compound always contained a set mass ratio of elements. For instance, when a 9.00 gram sample of water is analyzed, 8.00 grams of oxygen and 1.00 gram of hydrogen are found. The mass ratio here is 8:1. An analysis of an 18 gram sample reveals 16.0 grams of oxygen and 2.00 grams of hydrogen. Again the ratio holds true ($\frac{8}{1} = \frac{16}{2}$). Any sample of water from any source always contains 8 grams of oxygen for every 1 gram of hydrogen.

Further study showed that every compound has its own unique and definite mass composition. This finding is called the **law of definite composition.** This law states that every compound is formed of definite particles combined in definite numbers. God has caused compounds to form in an orderly way; they do not combine in a haphazard or random fashion. This fact allows the scientist to expect reproducibility in his work. What works today will work tomorrow.

The First Experimental Model: Dalton's Atomic Theory

An English schoolteacher named John Dalton (1766-1844) was the first to frame an atomic model based on sound experimental evidence instead of "mental gymnastics" or philosophy. With incisive logic and new knowledge such as the law of definite composition,

*atomos: atom: a (Gk. *a–* not or unable) + tom (Gk. *–tomos*, cutting) therefore, unable to be cut or divided

Dalton formed his remarkable model. The following statements summarize his model:

1. Elements are made of minute particles called atoms, which are tiny, indestructible spheres.
2. Atoms of different elements have unique sizes and properties.
3. An atom of one element cannot be changed into an atom of another element.
4. Atoms form compounds by combining with each other.
5. A certain compound always contains the same relative number and kinds of atoms.

The major points of Dalton's theory were his ideas that different combinations of atoms form compounds and that atoms of different elements have different masses. Dalton proceeded to assign relative masses to atoms of various elements. He assigned relative masses because he could determine the masses of the atoms only in relation to each other. Through painstaking analysis, Dalton and other chemists of his day determined that "oxygen is much heavier than hydrogen, and slightly more massive than carbon" and other important facts about atoms.

The exact values in Dalton's table were not accurate, and in some cases, even the order of elements from smallest to largest was not correct. Nevertheless, the fact that Dalton was able to start such an ambitious project is remarkable, given that he was handicapped by a lack of information. His theory started a trickle of experimentation that soon brought a flood of new information. Dalton's model still serves as the foundation for present theories.

Discovery of the Electron: Thomson's Model

Like Democritus, Dalton and others had assumed that atoms were tiny spheres. They "just knew" (or thought they did) that atoms were indivisible, as the Greek word *atomos* implied. Until the 1880s, these ideas about atoms were common. Then new clues led men to believe that atoms were made of even smaller particles and that they could be divided.

The invention of batteries sparked the study of electricity. Scientists found that gases at atmospheric pressure did not conduct an electrical current. However, if a gas was sealed in a glass tube under very low pressure, it could carry a current between two electrical contacts. The glass tube was called a gas discharge tube.

During their efforts to reduce the amount of gas in the discharge tubes, scientists noticed a strange new phenomenon. When most of the gas molecules were removed by a vacuum pump, the current decreased. This was understandable; fewer molecules could carry less current. When even more gas was removed,

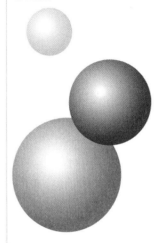

Figure 4A-1

Dalton's model of the atom was based on differences in mass.

scientists saw surprising things happen. As the gas molecules were further removed by a vacuum pump, the current gradually began to *increase* and the tube glowed with an eerie green light. (Later it was determined that the color of the light depended on what gas was in the tube.) Removal of even more gas molecules resulted in a disappearance of the glow, but a continued current flow. Obviously, something other than the gas allowed the current to flow. Because the "something" came from the electrical contact called the cathode, it was named **cathode rays.**

The use of cathode-ray tubes opened the way for the study of subatomic particles.

Although many men worked with cathode rays, an English physicist named J. J. Thomson (1856-1940) did the most extensive experiments. Thomson carried out a series of experiments to examine cathode rays in detail. The first observation he made was that cathode rays travel in straight lines, unaffected by gravity. This could be explained only if the cathode rays were made of waves (similar to light) or composed of particles moving at incredible speeds. Second, he found out that a magnet could deflect cathode rays. This fact told him that the rays were actually tiny charged particles. When he passed a cathode ray between two electrically charged plates, the ray bent toward the positively charged plate. Since opposite charges attract, Thomson knew that the particles carried a negative charge. Finally, he deflected the particles by a combination of magnetic and electrical fields.

English physicist J. J. Thomson (1856-1940)

By manipulating the strength of the fields, he was able to determine the charge-to-mass ratio (e/m) of the particle. He found that this ratio was surprisingly large. Compared to the charge, the mass was almost nothing.

Thomson further proved that the same type of particle was emitted from atoms of every element. Regardless of the gas in the tube or the type of cathode material, he obtained the same results. Gold, silver, iron, and copper all gave off the same particle. After his experiments were complete, Thomson had proved that every atom contained smaller negatively charged particles. These particles were named **electrons.**

Dalton's atomic model—the indestructible sphere concept—could not explain these new facts. It was time for a new model. This new model had to explain (1) how electrically charged particles could exist in an atom, (2) why the atom as a whole was neutral, and (3) how negative charges, but not positive charges, could leave an atom. Thomson explained these observations by postulating a new atomic model. He said that

1. Negatively charged electrons exist embedded in a positively charged substance that completely surrounds them.
2. The positive material balances out the negative charges on the electrons so that the atom is neutral.
3. Under certain conditions, electrons could be removed from the atom.

A sketch of these ideas looked like the then-common English plum pudding. The electrons were negatively charged "plums" in a positively charged "pudding." Thus, Thomson's model is often known as the "plum pudding model" of the atom (Figure 4A-2). A more contempory example would be a chocolate chip cookie—the chocolate chips would be the electrons, just as the plums represented the electrons in Thomson's plum pudding. Since Thomson's first experiments, other scientists have defined the charge on an electron as -1, and determined the mass of an electron to be a scant 9.11×10^{-31} kilograms.

Figure 4A-2

Thomson's plum pudding (chocolate chip cookie) model

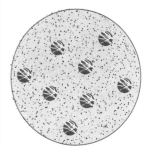

Discovery of the Proton: Rutherford's Model

While working with cathode ray tubes in 1896, Wilhelm Roentgen accidentally discovered X rays. This discovery triggered further experiments that led to the discovery of radiation and detailed studies of the inner parts of the atom. In the twenty years before World War I, the best minds in the scientific world pioneered the fascinating field of radioactive particles. One of these brilliant scientists was Ernest Rutherford (1871-1931), a professor at Cambridge University in England. His experiments with radioactive particles known as alpha particles led the world to new insights about the atom.

Alpha particles are massive, positively charged ions (one alpha particle has the mass of 7350 electrons) that are emitted by some radioactive elements at very high speeds. Their large mass and high speed combine to give alpha particles a great amount of energy. Rutherford found that he could mark the passage of one of these particles as it struck a zinc sulfide-coated screen, producing a brief flash of light.

During his work, Rutherford examined a beam of alpha particles as they struck a thin gold foil; most of the particles went straight through the foil. However, a few particles were slightly deflected by the gold atoms. Some even ricocheted back toward the source. According to Thomson's model, this was impossible. No part of the atom should have enough density or electrical charge to withstand, let alone repel, a speeding alpha particle. It was time for a new atomic model.

Cambridge professor Ernest Rutherford (1871-1931)

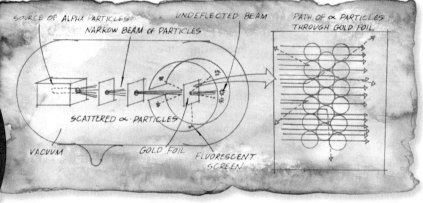

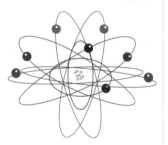

Figure 4A-3

Rutherford's model of the atom

Rutherford reasoned that atoms must have an extremely dense, positively charged central portion. What else could deflect the massive alpha particle? Because only a few alpha particles were greatly affected (about 1 in every 20,000), there must be little likelihood of an alpha particle's colliding with a dense region of an atom. From this fact, Rutherford reasoned that the **nucleus,** as he called it, must be very small. Subsequent calculations indicated that the diameter of a nucleus was only 1/100,000 the size of an entire atom—atoms and matter are mostly empty space. Rutherford's model is sometimes known as the nuclear model of the atom (Figure 4A-3).

Rutherford named the small, positively charged particles that make up the nucleus **protons.** A proton has an electrical charge of +1, exactly opposite the electron's charge. The mass of a proton, 1.67×10^{-27} kilograms, is approximately 1836 times greater than the mass of an electron.

Discovery of the Neutron: Chadwick's Work

Scientists soon realized that atoms were much more massive than expected. The protons and electrons alone could not supply all the mass that was being observed. There had to be something else. Rutherford had suggested that a neutral particle could supply the extra mass without bringing an extra positive charge. Unfortunately, since the particle was so small and had no electrical charge, it eluded detection for many years. It was not until 1932 that an English physicist named Sir James Chadwick (1891-1974) identified neutral particles that existed in the nuclei of atoms. Appropriately enough, these particles were named **neutrons.** A neutron has a mass of 1.68×10^{-27} kilograms, just slightly greater than the mass of a proton. Neutrons and protons together make up the nucleus and contribute almost all of the mass of an atom.

Energy Levels for Electrons: Bohr's Model

Rutherford's experiments supplied much information about the nucleus. But what about the electrons? Rutherford had simply guessed that electrons must exist at great distances from the nucleus. Many questions were left unanswered. How were the various electrons arranged? Were they moving? If so, in what manner?

Niels Bohr (1885-1962), a young Danish physicist working in Rutherford's laboratory, took up the task of answering some of the nagging questions. He knew that the nucleus had a positive charge. He also knew that the electrons had negative charges. What he did not know was why the electrons did not fall into the nucleus. The attraction between positive and negative charges should pull the electrons in and collapse the atom. It was assumed that the electrons moved around the nucleus and that their momentum kept them from falling into the nucleus just as the earth's momentum balances the gravitational attraction of the sun.

Danish physicist Niels Bohr (1885-1962)

The science of spectroscopy (the study of how atoms absorb and emit light) helped Bohr devise a model that describes the movement of the electrons around the nucleus. Light from the sun and from incandescent bulbs is in the form of a **continuous spectrum;** that is, the light separates into all the colors of a rainbow as it passes through a prism. In contrast, light that is emitted from excited electrons forms spectra that contain definite bands of colored light. Furthermore, each element has its own unique set of colored lines, or a **line spectrum.**

The explanation of these bright lines eluded scientists for several decades, but in 1913 Niels Bohr, while working with

Figure 4A-4

Bohr's "solar system" atomic model

Figure 4A-5

A continuous spectrum and the bright-line spectrum of hydrogen

Table 4A-1 Capacities of Principal Energy Levels

Principal Energy Level (n)	Maximum Number of Electrons ($2n^2$)
1	2
2	8
3	18
4	32
5	50* (32)
6	72* (18)
7	98* (8)

*Values are theoretical. The elements in the universe do not have enough electrons to completely fill these levels. The smaller numbers are the observed capacities for known elements in their ground state.

hydrogen, devised an atomic model that helped explain them. He said that the electrons could exist only in definite energy levels outside the nucleus. Energy levels close to the nucleus correspond to lower levels of potential energy. Levels at greater distances correspond to higher potential energy levels.

The movement of electrons between the various energy levels could explain the bright-line spectra of gaseous elements. Normally electrons exist in the lowest energy level (the ground state). When the right energy and intensity of light, heat, or electricity "excites" the atom, the electrons jump up to a higher energy level. This change in energy levels is definite and quantized. There is no stopping between levels. Once the electrons are at their higher energy level, they immediately seek to return to their lower energy state. As soon as possible, the electrons fall back into a lower level. The potential energy the electrons possessed at the higher level is converted into light (electromagnetic) energy when they return to a lower level. The observed colors in bright-line spectra give valuable clues about the energy levels in atoms. Bohr was actually able to predict the exact wavelengths (which correspond to colors) of the lines in the hydrogen spectrum.

Bohr's energy levels are called **principal energy levels.** They were temporarily envisioned as a set of concentric tracks on which the electrons orbited. However, a more accurate picture soon replaced this concept. Theoretically, many principal energy levels exist. In practice, however, six or seven can be measured. Figure 4A-6 shows that upper energy levels are spaced so closely together that measuring differences between them soon becomes nearly impossible. Transitions between such close energy levels can account for virtually any possible energy difference and can result in any color of light, giving a continuous spectrum. Each principal energy level has a maximum number of electrons that it can hold at one time. Table 4A-1 gives capacities of the first seven energy levels.

Figure 4A-6

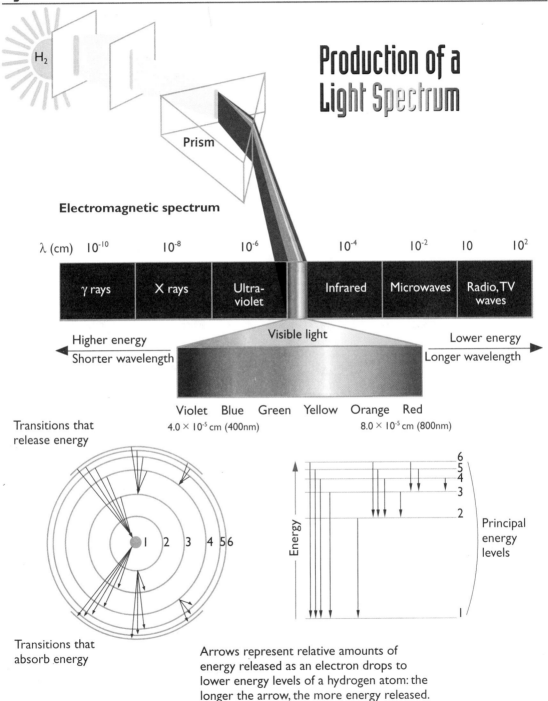

Production of a Light Spectrum

Arrows represent relative amounts of energy released as an electron drops to lower energy levels of a hydrogen atom: the longer the arrow, the more energy released.

Perhaps an analogy would help you to understand the concept of quantized energy levels. Picture a building with two stories (or floors) and a set of six stairs between them. The first story could represent the lowest potential energy state—the ground state—and each succeeding step would give you additional potential energy. The highest energy state is at the top of the steps, i.e., on the second floor. Since you cannot stand between steps (with both feet together), there are only specific quantities of potential energy associated with your body as you climb the stairs; that is, your energy is **quantized.** Even if you could climb the stairs two

FACETS of CHEMISTRY: Spectroscopy: Fingerprinting Atoms

Where was the element helium first discovered? If you were to guess that it was isolated from air, found in some mineral, or detected anywhere else on this earth, you would have missed the mark by some 93 million miles. Surprisingly, scientists discovered helium in the sun, not the earth. Even more surprisingly, the discovery occurred in 1868—a full century before interplanetary space probes ever ventured from NASA's launching pads. Scientists made this remarkable discovery by using one of the most powerful tools known to science: spectroscopy.

Spectroscopy is the study of how matter interacts with electromagnetic radiation. When atoms are highly energized, they can release light and other forms of radiation. In other instances atoms can absorb electromagnetic radiation. Every element has its own characteristic set of interactions. In a way, the types of light that elements emit and absorb serve as the "fingerprints" of atoms.

Atoms can absorb and emit all kinds of electromagnetic radiation: infrared, visible, ultraviolet, and X ray. Visible light is the easiest to observe, so this branch of spectroscopy is the most common. When an atom receives a small amount of heat, electricity, or light of the proper wavelength, one or more of its outer electrons becomes "excited." It jumps from its original position to a higher orbital. Electrons in higher orbitals are extremely unstable, so within a fraction of a second, they fall back to some lower orbital. When they fall, they give off a burst of light whose wavelength (or color) depends on the energy difference between the higher and the lower orbitals.

Early spectroscope

If electrons in an atom always jumped to the same higher orbital and always fell directly back to their "home" orbital, atoms would emit only one type of light. Electrons, however, are not limited to only one transition. On their initial jump upwards, they can go to any one of several orbitals. On their return to a low-energy, more stable state, they can fall all the way back to the original orbital in a single step. They can also cascade downward, landing temporarily in some or all of the in-between orbitals. As a result, atoms can release an entire set of wavelengths.

Normally these individual colors of light cannot be observed because they are all mixed together. A prism, however, can separate the colors of light so that they can be seen easily. When light from energized atoms is analyzed, distinct bands of colors appear at specific locations. The set of bright lines is called a bright-line spectrum.

No two elements have exactly the same set of orbitals or the same arrangement of outer-level electrons. As a result, every element has its own or three (or more) at a time, the changes in your potential energy are still determined by the spacing of the stairs and their height above the first floor; your energy is still quantized. Now suppose a young child or an elderly person needs to be able to pass between floors but finds the distance between the stairs too great. An alternate way is to use a ramp, which does not restrict the user to specific places for his feet. A person can take steps as small or as large as he desires; his energy can have any possible value. We could say the energy of a person on the ramp is continuous—having all possible values—rather than quantized.

Robert Bunsen

Gustav Kirchhoff

unique set of colored lines in its spectrum. Just as police detectives use the fact that no two people have the same fingerprint, scientific detectives rely on the fact that no two elements have the same spectrum.

The simplest device for studying spectra, called the prism spectroscope, was invented in the 1850s by Gustav Kirchhoff and Robert Bunsen at the University of Heidelberg. It consisted of four parts—a flame; an arrangement of lenses; a prism; and a small, movable telescope. The flame energized the atoms in the sample to be tested. The lenses directed the light rays from the flame into a parallel beam. The prism took the unified beam of light and broke it up into its component wavelengths. The telescope swiveled around and detected the colors of light that came from the prism at various angles. Using this early device, Kirchhoff observed a strong yellow line when table salt was placed in the flame. He correctly identified the line as an emission line of the sodium atom. Later Bunsen and Kirchhoff jointly discovered cesium and rubidium by observing emission lines from vaporized mineral waters.

Time has brought numerous improvements in the basic spectroscope. Instead of prisms, scientists now use diffraction gratings to separate the beams of light into their separate components. More sensitive optical detection systems have been developed, and it is now possible to study the spectral lines in great detail. In many cases what was thought to be one line has turned out to be a compilation of many lines. Cameras have replaced the telescope of the old spectroscope, and it is now possible to photograph an entire

Robert Bunsen's experiments

spectrum instantaneously. No longer must scientists spend hours searching for tiny lines of color by inching a telescope through all the various angles. New instruments have allowed scientists to study infrared and ultraviolet emissions. Armed with these new tools and the basic theory of spectroscopy, scientists can quickly determine what kinds of elements are in a sample of water, in clay from an archeological relic, or in the plasma of a distant star.

Evidence for the Quantum Model: New Physics

Bohr's model works well for atoms with one electron: for hydrogen. However, it does not work well for larger atoms that have more electrons. While Bohr's model was a great accomplishment, he soon saw that modifications were necessary. The scientists of the early twentieth century (Bohr included) set about to do this task.

The familiar laws that govern the motion of large objects such as balls, cars, and trains do not describe well the movement of electrons. Electrons are so small and move so rapidly that new, more fundamental rules were formulated to describe their movement. Werner Heisenberg (1901-1976) developed one of these rules in 1927. **The Heisenberg (HI zen berg) uncertainty principle** states that it is impossible to know both the energy (momentum) and the exact position of an electron at the same time. This is a fundamental property of all submicroscopic systems and cannot be avoided. All communication between atomic matter and an observer (you) has to be mediated by **photons** (light particles) or by other electrons. Therefore, the mere act of trying to look at an electron in order to see its location actually changes its position by means of a subatomic collision.

Since according to Heisenberg's uncertainty principle we cannot know exactly where the electron is positioned unless it stops moving, the idea of electrons positioned in set tracks like planets orbiting the sun had to be modified. Bohr's precise orbits were replaced by orbitals, or three-dimensional regions of probable position. An **orbital** is actually a four-dimensional map of where an electron is most likely to be found at any given time. It is four-dimensional because it contains location information on three geometrical planes (x, y, and z), as well as the probability of finding the electron at every point, as shown in Figure 4A-7.

An orbital is a general area where an electron is likely to exist.

In diagrams, orbitals look like fuzzy clouds without definite boundaries. The areas where electrons are most likely to be found are shaded the darkest. Away from the main part of the cloud, the shading becomes lighter as the chances of finding an electron decrease. As in the Bohr model, principal energy levels are arranged around the nucleus. Orbitals in which the electron's average distance from the nucleus is small have low energies. Orbitals in which the electron's average distance from the nucleus is great have high energies.

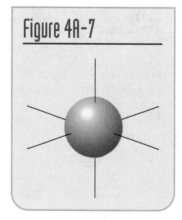

Figure 4A-7

The developments in physics that brought about the idea of orbitals also changed what scientists thought about electrons. Are electrons particles or waves? Although Thomson proved long ago that electrons act like particles, a quarter of a century after his work, scientists began to think that electrons could often act like waves. In 1924, Louis de Broglie stated that the matter of an electron was not concentrated at one point but was spread out over the entire orbital. Although the concept of matter-waves sounds like science fiction, it is firmly rooted in theory and observations.

So, is an electron a particle, or is it a wave? It is both! The electron is said to have a dual nature. This concept of a dual nature is often difficult to grasp because the human mind is limited. Many things are difficult to understand, but that difficulty does not keep us from studying about them, using them, or appreciating them. For example, your inability to explain or understand exactly how a computer handles information does not keep you from using one. The same could be said about electricity, meteorology, or even your brain. What difficult concepts should do is bring us face-to-face with the reality that an infinitely wise Creator made and designed the universe according to His plan. If we understood it all, we would be as wise as God! Isaiah 55:8-9 tells us, "For my thoughts are not your thoughts, neither are your ways my ways, saith the Lord. For as the heavens are higher than the earth, so are my ways higher than your ways, and my thoughts than your thoughts." The limitations of the mind are even more apparent when spiritual matters are considered. Concepts like the Trinity, predestination, free will, eternity, and the new nature of Christians after salvation are simply beyond understanding. In Romans 7:19 the apostle Paul pointed out that Christians have a dual nature: "For the good that I would I do not: but the evil which I would not, that I do." Christians have a new nature that strives to obey Christ and an old nature that longs to serve sin. This fact is true but not easy to understand fully. Likewise, the fact that an electron can act like a particle and a wave at the same time is not easy to understand but must be accepted. The idea has become a cornerstone of the very successful quantum model of the atom.

Section Review Questions 4A

1. The law of definite composition states that every compound has a definite composition by mass. What does that mean?
2. What did Thomson's work with cathode rays demonstrate?
3. Discuss how the theories of Heisenberg and de Broglie differed from previous models of electron activity.
4. Name the three particles of which atoms are composed, list their differences, and name their discoverers.

4B The Quantum Model: Where Are the Electrons?

The latest, but not necessarily the last, idea on atomic structure is the quantum model. Not only does this model focus on the locations of electrons, but it also incorporates parts of all the previous atomic models. The wave concept of electrons, along with a branch of higher mathematics called wave mechanics, allowed scientists to develop this new model of the atom.

Sublevels and Orbitals: Home of the Electron

Like the Bohr model, the **quantum model** states that electrons exist in principal energy levels. Unlike the Bohr model, the quantum model subdivides all but the first principal energy level into sublevels. These **sublevels** contain other divisions called orbitals, and the orbitals contain electrons. There are four sublevels: *s, p, d,* and *f*.

The s *sublevel*. This sublevel is the simplest of all. Its probability map has a spherical shape. Unlike the other sublevels, it contains only one orbital, which can hold two electrons. An orbital never contains more than two electrons.

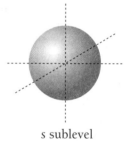

s sublevel

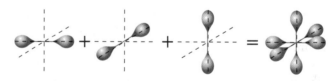

p sublevel

The p *sublevel*. The probability shape of a *p* sublevel looks like three barbells that intersect in the middle. It contains three orbitals: one in the *x*-direction on a graph, one in the *y*-direction, and one in the *z*-direction. Since each orbital can hold a maximum of two electrons, a complete *p* sublevel can hold no more than six (3 × 2) electrons.

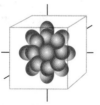

d sublevel

The d *sublevel*. A *d* sublevel probability map has a complicated shape. It is a combination of five orbitals and has a capacity of ten (5 × 2) electrons.

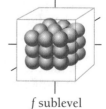

f sublevel

Figure 4B-1

Orbitals come in different sizes. Large orbitals surround small orbitals.

The f sublevel. An *f* sublevel probability map is even more complicated in shape than a *d* sublevel. It has seven orbitals and a capacity of fourteen (7 × 2) electrons. You must never lose sight of the fact that the shapes of these electron clouds are mathematical descriptions—models—of electron behavior that better enable us to visualize something invisible and account for some of the properties of electrons.

Not every principal energy level has room for an *s*, *p*, *d*, and *f* sublevel. Each principal energy level has specific types of sublevels (Table 4B-1). The first principal energy level has only enough room for the one *s* orbital. The second level contains an *s* and a *p* sublevel. The third level contains *s*, *p*, and *d* sublevels. The fourth and fifth principal levels contain all four types of sublevels. The higher principal energy levels do not normally contain all the sublevels they could because the known elements have only a certain number of electrons. You can determine the electron capacity for each energy level as follows. The first energy level contains one *s* sublevel containing one orbital, which can hold two electrons. The total capacity of the first principal energy level is thus two electrons. The second principal energy level has an *s* and a *p* sublevel. Since the *p* sublevel can hold six electrons, the total capacity of the second principal energy level is the sum of the two electrons from the *s* sublevel and the six electrons from the *p* sublevel—

Table 4B-1 Sublevels in Energy Levels

Principal Energy Level	Possible Types of Sublevels*
1	s
2	s p
3	s p d
4	s p d f
5	s p d f
6	s p d
7	s p

*For the first 105 elements

Table 4B-2 Capacities of Energy Levels

Principal Energy Level	Sublevels	Orbitals in Each Sublevel	Electron Capacity of Each Sublevel	Total Electron Capacity
1	s	1	2	2
2	s	1	2	
	p	3	6	8
3	s	1	2	
	p	3	6	
	d	5	10	18
4	s	1	2	
	p	3	6	
	d	5	10	
	f	7	14	32
5	s	1	2	
	p	3	6	
	d	5	10	
	f	7	14	32
6	s	1	2	
	p	3	6	
	d	5	10	18
7	s	1	2	
	p	3	6	8

Figure 4B-2 ▪ Energies of Sublevels

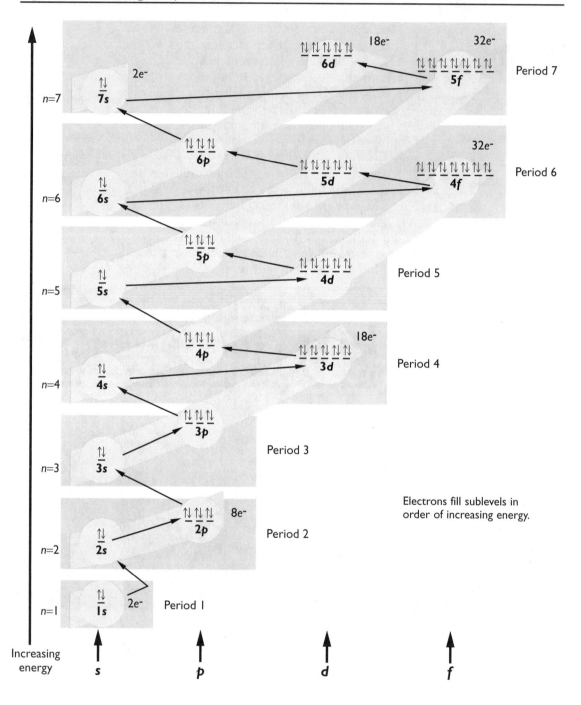

Electrons fill sublevels in order of increasing energy.

a total of eight electrons. The total capacities of the other principal energy levels can be determined by a similar procedure.

Energies of Sublevels

All sublevels can be ranked according to their energies. As expected, the 1s sublevel has the least energy. The 2s and 2p sublevels have more energy since they are in the second principal energy level—one that is farther from the nucleus. Note that all three 2p orbitals have the same energy, and the electrons in these orbitals have more energy than those in the 2s sublevels. Above them are the 3s and the 3p sublevels. A careful look at Figure 4-B2 reveals that the 3d orbitals do not come next as might be expected. The 4s has less energy than the 3d does. Figure 4B-2 ranks the sublevels (from bottom to top) in order of increasing energy.

1s, 2s, 2p, 3s, 3p, 4s, 3d, 4p, 5s, 4d, 5p, 6s, 4f, 5d, 6p, 7s, 5f, 6d, 7p

Figure 4B-3

7s	7p		
6s	6p	6d	
5s	5p	5d	5f
4s	4p	4d	4f
3s	3p	3d	
2s	2p		
1s			

Notice that several of the higher principal energy levels "overlap." Memorizing the exact order of the sublevels is tedious, so a mnemonic device (memory aid) called the **diagonal rule** has been devised. To use this device, make a diagram like the one to the right (top), which shows which types of sublevels exist in each principal energy level.

Starting at the lower left corner, draw a diagonal arrow upward through the 1s sublevel. The arrow immediately hits the left border of the structure, so return to the lower portion of the diagram but more to the right. Draw the next diagonal arrow through the 2s sublevel and the third through the 2p and then the 3s sublevels. Continue this process of drawing parallel diagonal arrows from the lower right to the upper left of the chart. The order in which the diagonal arrows hit the sublevels reveals the energy order. (Read the arrows from bottom to top, tail to head.) When completed, your diagonal-rule mnemonic device will look like Figure 4B-3.

Mnemonic device for the filling order

The Aufbau Principle: How to Build an Atom

The **Aufbau principle** (OUF-bough) states that the arrangement of electrons in an atom may be determined by the addition of electrons to a smaller atom. The word *aufbau* is a German word meaning "a building up." In the progression from hydrogen up to the larger elements, each successive element has one additional proton and one additional electron. As a rule, electrons add to the least energetic orbital possible. They must fill low-energy orbitals before they can occupy high-energy orbitals.

Chapter 4B

Hydrogen's one electron normally resides in the 1s sublevel. The **electron configuration,** or arrangement of electrons, in hydrogen is shown as $1s^1$. The coefficient 1 is the principal energy level and the letter s is the type of sublevel in which the electron resides. The superscript above the 1s tells how many electrons occupy that sublevel. In this case, the superscript 1 means that one electron is in the 1s sublevel. Chemists often use a picture called **orbital notation** to convey the same meaning. A horizontal line represents the one orbital in the 1s sublevel. An arrow pointing upward represents the first electron in the orbital. The orbital notation for hydrogen is written as follows:

$$\text{H} \quad \underline{\uparrow}^{1s}$$

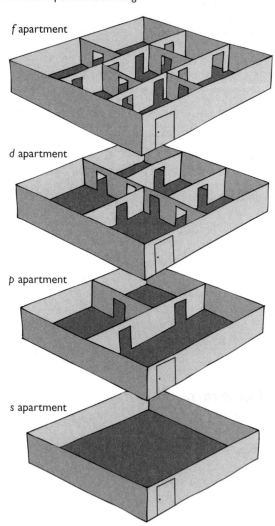

Figure 4B-4
Electron apartment building

The second electron in any orbital will be written as an arrow pointing downward. This is based on a property of electrons known as "spin" which results from their behaving as if they were tiny tops spinning about on their axes. Thus, paired electrons in any orbital must *always* have opposite spins represented by arrows pointing in opposite directions. The rule that mandates that only two electrons with opposing spins can reside in one orbital is called the **Pauli exclusion principle.**

You can think of an electron configuration as a means of determining "where" electrons of an atom reside. Imagine principal energy levels as floors in a building: i.e., the farther from the ground floor (the first level) the higher the energy. The sublevels correspond to apartments on each floor, and the orbitals correspond to rooms of the apartments. All apartments have an odd number of rooms; s apartments have one room, p apartments have three rooms, d apartments have five rooms, and f apartments are grand suites with seven rooms. Only two people are allowed in any room, corresponding to electron spins: up and down. Therefore, when we write $1s^1$, it is like writing an address for one person in the only room of the only apartment on the first floor.

Helium has two electrons. The second electron joins the first to fill the 1s sublevel. Helium's electron configuration is thus $1s^2$.

$$\text{He} \quad \underline{\uparrow\downarrow}_{1s}$$

According to the same rules, lithium's electron configuration would be $1s^2\,2s^1$ and beryllium's electron configuration is $1s^2\,2s^2$. The orbital notations would be as follows:

$$\text{Li} \quad \underline{\uparrow\downarrow}_{1s} \; \underline{\uparrow}_{2s}$$

$$\text{Be} \quad \underline{\uparrow\downarrow}_{1s} \; \underline{\uparrow\downarrow}_{2s}$$

Now that electrons fill the 1s and 2s sublevels, the next electrons must go into the 2p orbitals since they are the next higher in energy. The electron configuration of boron (with five electrons) is $1s^2\,2s^2\,2p^1$ and the orbital notation is

$$\text{B} \quad \underline{\uparrow\downarrow}_{1s} \; \underline{\uparrow\downarrow}_{2s} \; \underline{\uparrow}\,\underline{}\,\underline{}_{2p}$$

Note that three lines are used for the p sublevel because it has 3 orbitals, even though there are no electrons occupying the last two.

When determining the electron configuration of carbon, an option arises. The electron configuration is $1s^2\,2s^2\,2p^2$. But what about the orbital notation? Are the two electrons in the 2p sublevel in the same orbital, or are they in different orbitals? In orbital notation, is carbon

$$\text{C} \quad \underline{\uparrow\downarrow}\;\underline{\uparrow\downarrow}\;\underline{\uparrow\downarrow}\,\underline{}\,\underline{} \quad \text{or} \quad \underline{\uparrow\downarrow}\;\underline{\uparrow\downarrow}\;\underline{\uparrow}\,\underline{\uparrow}\,\underline{}\;?$$

Hund's rule states that as electrons fill a sublevel, all orbitals receive one electron before any receive two. Using the apartment building analogy—if you have three people in a p apartment, each gets his own room. Electrons are like unsociable people; they prefer to be as far away from each other as possible because they have the same charge—they will room together if they have different spins. By Hund's rule, carbon's orbital notation is the second option. The orbital notation of nitrogen ($1s^2\,2s^2\,2p^3$) is

$$\text{N} \quad \underline{\uparrow\downarrow}_{1s} \; \underline{\uparrow\downarrow}_{2s} \; \underline{\uparrow}\,\underline{\uparrow}\,\underline{\uparrow}_{2p}$$

However, the sixth electron in the second energy level in oxygen is forced to pair up.

$$\text{O} \quad \underline{\uparrow\downarrow}_{1s} \; \underline{\uparrow\downarrow}_{2s} \; \underline{\uparrow\downarrow}\,\underline{\uparrow}\,\underline{\uparrow}_{2p}$$

The Aufbau principle and Hund's rule can help reveal the ground state electron configuration of any element. The Aufbau principle suggests the order in which electrons fill sublevels. Hund's rule tells whether electrons in the same sublevel will be paired or unpaired.

Sample Problem Give the ground state electron configuration of manganese and draw its orbital notation.

Solution

Using the mnemonic for filling the orbitals, we find that the twenty-five electrons in a manganese atom fill the 1s, 2s, 2p, 3s, 3p, 4s, and 3d sublevels. We can then obtain the following configuration:

$$1s^2\ 2s^2\ 2p^6\ 3s^2\ 3p^6\ 4s^2\ 3d^5$$

Always check to make sure that the sum of the superscripts equals the total number of electrons in the atom.

All sublevels prior to the final one are filled, but the 3d sublevel is only partly filled. Since there are 5 electrons in the 3d sublevel and 5 orbitals in any d sublevel, each orbital is occupied by one electron; no pairing is needed (Hund's rule). The complete orbital notation is as shown.

1s	2s	2p	3s	3p	4s	3d
↑↓	↑↓	↑↓ ↑↓ ↑↓	↑↓	↑↓ ↑↓ ↑↓	↑↓	↑ ↑ ↑ ↑ ↑

Hund's rule and the Aufbau principle are not always followed precisely. Chromium and copper are notable exceptions. In these cases the electrons tend to remain unpaired, thus overcoming the Aufbau principle. One electron that normally appears in the 4s sublevel shifts up to a 3d orbital so that it can remain unpaired, resulting in two half-filled sublevels and a number of unpaired electrons. Ground state electron configurations for all the elements may be found in Appendix C.

Quantum Numbers: Addresses for Electrons

Chemists use numbers called **quantum numbers** to describe the locations and energies of electrons. Each electron in an atom has its own set of four numbers that serves as a "ZIP code" for the electrons. Just as the numbers in a ZIP code have a specific meaning, so do the quantum numbers for an electron.

Figure 4B-5 ▪ Quantum Numbers

An "address" for electrons like ZIP codes

Example: 29614 = BJU
2 / 96 / 14

2 = second region of USA
96 = major city
14 = section of that city

The first quantum number (n) identifies the principal energy level. The number can have an integral value of 1, 2, 3, 4, or higher. As you saw in Table 4A-1, you can find the maximum number of electrons that can occupy a principal energy level by the formula $2n^2$, where n is the first quantum number. It can also tell you how many types of sublevels are possible (theoretically) in a given main energy level. For example, if $n = 2$, there are 2 types of sublevels (s and p).

The second quantum number (l) identifies the type of sublevel in which an electron exists (s, p, d, or f). The sublevels receive the following numbers: $s=0$; $p=1$; $d=2$; $f=3$.

The third quantum number (m) specifies the electron's orbital. If an electron exists in an s sublevel, there is only one possible value: 0. If an electron resides in a p sublevel (second quantum number = 1), there are three possibilities: -1, 0, or 1. All d sublevels have five orbitals, so there are five possible values for the third quantum number. They are -2, -1, 0, 1, 2. You can see that there will be $2l + 1$ orbitals in a sublevel whose value is l. In addition, each orbital within a given sublevel has a different numerical value or "address." Actually, what is described by the m quantum number is the orbital's orientation in space with respect to the x-, y-, and z-axes.

Table 4B-3 Values for the Third Quantum Number

Second Quantum number (l)	Number of Orbitals ($2l+1$)	Possibilities for Third Quantum Number (m)
0 (s sublevel)	1	0
1 (p sublevel)	3	-1, 0, 1
2 (d sublevel)	5	-2, -1, 0, 1, 2
3 (f sublevel)	7	-3, -2, -1, 0, 1, 2, 3

The fourth quantum number (m_s) differentiates between the two electrons in a given orbital. One electron is assigned a value of $+\frac{1}{2}$, and the other is assigned a value of $-\frac{1}{2}$. Scientists have described this difference in electrons as the "spin" of the electrons, although there is no actual spinning going on. One important result of the Pauli exclusion principle is that no two electrons can have the same four quantum numbers in the same atom. For example, 2 electrons may be in the same principal energy level, sublevel, and orbital, and given the same first three quantum numbers, but they must have opposite signs for m_s.

Sample Problem (1) Tell the first two quantum numbers (n and l) for electrons in the following sublevels and (2) give the possible values for the third quantum number (m). **a.** 2p **b.** 3s **c.** 4d

Solution
First, remember what each of the quantum numbers represents. The first quantum number (n) is the principal energy level. The second quantum number (l) designates the type of sublevel. From Table 4B-3, you can find the values assigned to the various sublevels. The third quantum number (m) specifies the electron's orbital. You may again refer to Table 4B-3. It would be advantageous to commit to memory the values in that table.

Answer: a. (1) $n = 2, l = 1$ (2) $m = -1, 0, 1$
b. (1) $n = 3, l = 0$ (2) $m = 0$
c. (1) $n = 4, l = 2$ (2) $m = -2, -1, 0, 1, 2$

Section Review Questions 4B

1. List the four sublevels and their electron capacities.
2. What do the Aufbau principle and Hund's rule tell about the ground state electron configuration?
3. For each of the following atoms, (1) give the ground state electron configuration, (2) draw the orbital notation, and (3) find the set of four quantum numbers for the last electron.
    a. silicon   b. vanadium
4. Tell whether each of the following sets of quantum numbers, given in the order n, l, m, m_s, is possible or impossible. For any that are impossible, tell why.
    a. $(0, 1, 0, \frac{1}{2})$ b. $(4, 1, 0, -\frac{1}{2})$ c. $(2, 0, 1, \frac{1}{2})$

4C Numbers of Atomic Particles: Things We Can Count On

Even though the quantum model could be changed in the future, scientists are generally in agreement on the "basic" particles in atoms—protons, neutrons, and electrons. The nucleus carries a positive charge and contains most of the mass of an atom. It consists of positively charged protons and uncharged neutrons. The number of protons in an atom determines the identity of the atom; the number of neutrons affects its mass; and the number of electrons affects its electrical charge. Counting the particles in atoms gives valuable information.

Atomic Mass

Once the chemists of Dalton's day started to analyze chemical compounds, they tried to determine the masses of individual atoms. They knew that carbon atoms were twelve times as heavy as hydrogen atoms and that oxygen atoms were sixteen times as heavy as hydrogen atoms. A few elements with their relative masses are listed in Table 4C-1. The values given as the mass for each element are not the mass of any particular atom, but are a weighted average of all the naturally occurring forms of the element. (You will learn more about this later.)

Chemists express these masses in **atomic mass units** (abbreviated *amu*). An amu is defined as $\frac{1}{12}$ the mass of a carbon-12 atom, or as 1.66×10^{-27} kilograms. It is approximately the same size as a proton or a neutron (one proton = 1.0073 amu; one neutron = 1.0087 amu). The electron is a scant 0.00055 amu. The sum of the protons and neutrons in an atom is that atom's **mass number.** Note that it is not really a mass; it is a number of nuclear particles and must always be a whole number.

mass number = number of protons + number of neutrons

Since both protons and neutrons have masses that are very close to 1 amu, the mass number can be used as an approximation of the mass of an atom. Electrons are ignored because they add a negligible amount of mass. The number of protons in an atom is vitally important. This number, called the **atomic number,** determines the identity of the atom. If an atom has twenty-nine protons, it must be a copper atom. Conversely, all copper atoms have twenty-nine protons. You might think of the atomic number as the atom's ID number, unique to a particular atom.

atomic number = number of protons

Table 4C-1

Element	Average Atomic Mass* (in amu)	Mass Number	Atomic Number
H	1.008	1	1
C	12.01	12	6
O	16.00	16	8
Na	22.99	23	11
S	32.06	32	16
Ca	40.08	40	20
U	238.0	238	92

*Given to 4 significant figures

Isotopes: Count Those Neutrons!

Even though all atoms of an element must have the same number of protons (atomic number), they do not all have the same mass. Some atoms of the same element have a different number of neutrons, resulting in a different mass number. Atoms with the same atomic number but different numbers of neutrons are called **isotopes**.

Isotopic notation is often used to specify the exact composition of an atom. This notation includes the atom's symbol, atomic number, and mass number. A boron atom that has five protons and six neutrons has a mass number of 11 and would be written as

$$^{11}_{5}B$$

Given an atom's isotopic notation, it is easy to calculate the number of protons, neutrons, and electrons. The atomic number at the lower left of the symbol signifies that this atom has five protons. To be electrically neutral, the atom must also have five electrons. The number of neutrons in the atom is the difference between the mass number and the atomic number.

mass number − atomic number = number of neutrons

(protons + neutrons) − protons = neutrons

11 − 5 = 6 neutrons

If this boron atom had five neutrons, its mass number would be 10 (5 protons and 5 neutrons), and its isotopic notation would be

$$^{10}_{5}B$$

Sample Problem Determine the number of protons, neutrons, and electrons in $^{51}_{24}Cr$.

Solution
The atomic number tells the number of protons: 24. The number of electrons equals the number of protons in a neutral atom: 24. The number of neutrons is the difference between the mass number and the atomic number.

51 − 24 = 27 neutrons

Sample Problem An atom has nineteen protons and twenty-two neutrons. Write its isotopic notation.

Solution
Looking at the periodic table located on the back cover of the book, the element with an atomic number of 19 is potassium (K). Remember, the identity of an element is determined by the number of protons in its nucleus. Add the number of protons and neutrons to find the mass number.

$$19 \text{ protons} + 22 \text{ neutrons} = 41$$

The isotopic notation is

$$^{41}_{19}K$$

When writing the isotopes of an element, first write the name of the element followed by a hyphen (-), then the number of neutrons in the element. In the examples above, boron with five neutrons would be written boron-5. The isotope of boron with six neutrons would be boron-6. In the second sample problem, the isotope of potassium with twenty-two neutrons would be written potassium-22. Alternatively, the isotopes can be written using the element's symbol, followed by the number of neutrons, as in B-5 or K-22.

Typically, samples of elements are mixtures of isotopes. **Atomic masses** are the weighted averages of isotopes. They show the average mass of an atom in a sample. For example, a naturally occurring sample of lithium is composed of two lithium isotopes: 7.42 percent lithium-6 and 92.58 percent lithium-7. The weighted average will be closer to 7 than to 6, since a greater percentage of the sample is lithium-7. This explains why the listed atomic mass of lithium is 6.941.

The situation with the isotopes can be illustrated by calculating the average grade on a test in a class. Let us say that 10 students took a 5-question test (20 points/question) and scored as follows: 3 got all the questions right (100 points each), 6 missed one question (80 points each), and one missed 2 questions (60 points). How would you find the "weighted" average test grade? You would easily be able to predict that the average would be closest to the 80 points because the largest number of students made that score. The actual answer is:

$$3 \times 100 \text{ points} = 300 \text{ pts}$$
$$6 \times 80 \text{ points} = 480 \text{ pts}$$
$$+ 1 \times 60 \text{ points} = 60 \text{ pts}$$

840 pts divided by 10 students = 84 pts

Note that this average does not represent any particular student's test score; it is simply the "weighted" average for the class.

Now suppose that you want to calculate the average atomic mass of lithium atoms. The 7.42 percent of lithium atoms have a mass of 6.015 amu per atom. The 92.58 percent of lithium atoms have a mass of 7.016 amu per atom. To find the weighted average, first convert the percents to decimals; then multiply the mass of each type of atom by its "contribution" to the total. Finally, add the results to get the weighted average.

Mass of Li-6 atoms: 0.0742×6.015 amu $= 0.446$ amu
+ Mass of Li-7 atoms: 0.9258×7.016 amu $= 6.495$ amu
***
weighted mass of 100% of atoms $= 6.941$ amu

The average mass, or the atomic mass, of an atom in any sample consisting of 7.42% Li-6 and 92.58% Li-7 is 6.941 amu.

Sample Problem The element chlorine, Cl, has two naturally occurring isotopes: $^{35}_{17}$Cl (atomic mass 34.969 amu) with a natural abundance of 75.77%, and $^{37}_{17}$Cl (atomic mass 36.966 amu) with a natural abundance of 24.23%. Determine the average atomic mass of chlorine.

Solution
First, express each percentage as a decimal; then multiply each by the atomic mass of the corresponding atom. Finally, add the results.

$$(34.969 \text{ amu} \times 0.7577) + (36.966 \text{ amu} \times 0.2423)$$
$$26.50 \text{ amu} \quad + \quad 8.957 \text{ amu}$$
$$35.46 \text{ amu}$$

Remember also that your answer can have no more significant digits than the *least* precise measurement given in the problem. There are five significant digits in 34.969, but only four in 75.77%; therefore, the answer can have only four significant digits.

The identity of a particular isotope *does* make a difference. For example, the element hydrogen has three known isotopes: $^1_1$H; $^2_1$H (or D) called deuterium; and $^3_1$H (or T) called tritium.

The most important compound of deuterium is D_2O, or heavy water, which in large quantities is poisonous to human beings. Its only large-scale use is in nuclear reactors. Tritium is unstable and therefore radioactive. It is used as a reactant in nuclear fusion devices, such as hydrogen bombs.

Table 4C-2 Naturally Occurring Isotopic Abundances of Common Elements

Isotope	Percent	Mass (in amu)
H-1	99.9885	1.007825
H-2	0.0115	2.0141
H-3	trace	3.0160
Li-6	7.42	6.01512
Li-7	92.58	7.01600
B-10	19.9	10.0129
B-11	80.1	11.00931
C-12	98.93	12.0000
C-13	1.07	13.00335
Si-28	92.23	27.97693
Si-29	4.7	28.97649
Si-30	3.09	29.97376
P-31	100	30.97376
Br-79	50.69	78.9183
Br-81	49.31	80.9163
Au-197	100	196.9665
U-234	0.0055	234.0409
U-235	0.72	235.0439
U-238	99.27	238.0508

Figure 4C-1

Isotopes of hydrogen: hydrogen, deuterium, and tritium

Valence Electrons: Last, but Not Least

The electrons in the outermost energy level are the most important electrons of an atom. Although all electrons in an atom can be excited into a higher energy, the outer energy level electrons are the ones most likely to be involved in chemical bonding. They are also the ones that give elements their physical properties. The electrons in the outermost energy level of an atom are given a special name: **valence electrons.**

Sample Problem   How many valence electrons do these atoms have?

  a. argon   **b.** nickel

Solution

a. The electron configuration for argon is $1s^2\ 2s^2\ 2p^6\ 3s^2\ 3p^6$. The outermost energy level—the one with the largest value for n—contains both s and p electrons. The total in this energy level is eight.

b. Even though the electron configuration obtained by using the mnemonic shows 4s filling before 3d, valence electrons are defined as those which occupy the *outermost* principal energy level—the fourth:

$$1s^2\ 2s^2\ 2p^6\ 3s^2\ 3p^6\ 4s^2\ 3d^8$$

Thus, nickel has two valence electrons that most often participate in chemical reactions.

Electron-Dot Symbols

Since valence electrons determine how atoms bond, the inner electrons can usually be ignored. An **electron-dot symbol** (sometimes called a Lewis Dot Symbol) is a shorthand way of representing only the valence electrons in an atom. Dots that represent the valence electrons are placed around the element's symbol. The first two dots are paired since they represent the two electrons in the s sublevel. The other electrons remain unpaired if possible (Hund's rule). See Table 4C-3 for the conventional arrangement of dots for each number of valence electrons.

Table 4C-3 Electron-Dot Symbols

Valence Electrons	Example
1	Na·
2	Mg:
3	·Äl·
4	·S̈i·
5	·P̈·
6	·S̈·
7	·C̈l·
8	:Är:

Ions: Charged Atoms

Electrons move between energy levels when they gain or lose energy. Given enough energy, an electron can jump away from an atom. This "defection" results in an atom that has an unbalanced electrical charge. Such atoms are called **ions.** Remember, changing the number of electrons does not change the identity of the atom. It is the number of *protons* that determines the identity of the atom.

Ions may have positive or negative charges. If an atom *loses* an electron, it will have a positive charge because the number of protons (+ charge) is unaffected. There are now fewer electrons (- charge) to balance them. Positive ions are called **cations** (KAT i ons). If an atom *gains* an electron, it will have an extra negative charge—again the number of protons is unaffected. Negative ions are called **anions** (AN i ons). Note that ions cannot be formed by the removal of a proton, only by the addition or loss of electrons. Electron-dot symbols can be used to represent ions. If a fluorine atom gains an electron, it acquires a negative charge. The 1- superscript reminds us of the charge.

$$·\ddot{F}: +\ 1\ \text{electron} \longrightarrow :\ddot{F}:^{1-}$$

A magnesium atom that lost two electrons would have a +2 charge.

$$Mg: \longrightarrow Mg^{2+} + 2\ \text{electrons}$$

Sample Problem Tell how many protons, neutrons, and electrons are present in $^{23}_{11}Na^+$

Solution
The left subscript gives the element's atomic number—the number of protons (11). The left superscript tells you the total number of protons and neutrons, from which the number of neutrons may be obtained. Thus, since 23 − 11 = 12, there are 12 neutrons. The positive charge indicates one less negative charge (electron) than positive charges; therefore, there are 10 electrons.

The position of the dots around the symbol is not important. For example, the following symbols for magnesium (Mg), having two valence electrons, are all equivalent:

$$\ddot{Mg} \quad Mg: \quad \underset{..}{Mg} \quad :Mg$$

The important thing is that paired valence electrons are shown.

Sample Problem Write the electron-dot symbols for the following.
  a. Oxygen   b. Boron

Solution
  a. Oxygen has the electron configuration $1s^2\ 2s^2\ 2p^4$, and therefore 6 valence electrons. Its electron-dot symbol is $:\ddot{O}\cdot$ (Any arrangement of two electron pairs and two unpaired electrons is correct; e.g., $:\dot{O}:\ \cdot\dot{O}\cdot$)
  b. Boron's electron configuration is $1s^2\ 2s^2\ 2p^1$, showing that there are three valence electrons. This results in an electron-dot symbol with three electrons, such as $\ddot{B}\cdot$.

Section Review Questions 4C

1. Explain the difference between a positive ion and a negative ion.
2. Write the orbital notation for the following: F, Ca, P.
3. Define valence electrons. What is their significance?
4. Determine the number of protons, neutrons, and electrons for the following: $^{28}_{14}Si$, $^{81}_{35}Br$.

Chapter Review

Coming to Terms

atom
model
law of definite composition
quantum
cathode rays
electron
nucleus
proton
neutron
continuous spectrum
line spectrum
quantized
principal energy level
Heisenberg uncertainty
  principle
Pauli exclusion principle
orbital
sublevel

diagonal rule
Aufbau principle
electron configuration
orbital notation
photon
Hund's rule
quantum number
atomic mass unit
mass number
atomic number
isotope
isotopic notation
atomic mass
valence electron
electron-dot symbol
ion
cation
anion

Review Questions

1. What relatively new technique do scientists use in order to image atoms?

2. Point out one flaw in each of the following statements:
   a. Scientific models do not deal with facts.
   b. Scientific models make predictions, but whether these predictions are reasonable makes no difference.
   c. Scientific models merely organize facts.
   d. Models are used only in science.

3. Briefly describe each of the following atomic models:
   a. Dalton's
   b. Thomson's
   c. Rutherford's
   d. Bohr's
   e. Quantum

4. Describe the discoveries or advancements that made each of these atomic models obsolete:
   a. Dalton's
   b. Thomson's
   c. Rutherford's
   d. Bohr's
5. How did observations of spectra lead to conclusions that the energy given off by excited atoms is quantized?
6. Describe what happens to the outermost electron in a sodium atom when that atom is heated and made to give off a line spectrum.
7. Explain how atoms can be electrically neutral even though they contain charged particles.
8. What is the electron capacity of
   a. an orbital?
   b. an *s* sublevel?
   c. a *p* sublevel?
   d. the second principal energy level?
   e. the fourth principal energy level?
9. How many orbitals are in
   a. an *s* sublevel?
   b. a *d* sublevel?
   c. the second principal energy level?
   d. the fourth principal energy level?
10. For each of the following atoms,
    i. write out the ground state electron configuration;
    ii. write out the orbital notation;
    iii. show the number of electrons in each principal energy level.

    Example: Carbon

    i. $1s^2\ 2s^2\ 2p^2$

    ii. C $\underset{1s}{\uparrow\downarrow}$ $\underset{2s}{\uparrow\downarrow}$ $\underset{2p}{\uparrow\ \uparrow\ \_}$

    iii. 2, 4

    a. oxygen, $_8$O
    b. sulfur, $_{16}$S
    c. potassium, $_{19}$K
    d. titanium, $_{22}$Ti
    e. bromine, $_{35}$Br
    f. barium, $_{56}$Ba

11. Draw a blank orbital notation chart that contains all seven energy levels (through the 7p subshell). Instead of drawing in arrows, number the blanks according to the filling order. Follow Hund's rule. Example: The first two levels should be labeled as follows:

$$\underset{1s}{\underline{1,\ 2}} \quad \underset{2s}{\underline{3,\ 4}} \quad \underset{2p}{\underline{5,\ 8}\ \underline{6,\ 9}\ \underline{7,\ 10}}$$

12. What general information can quantum numbers give?
13. What are the possible values for the third quantum number (m) for an electron in the following sublevels?
    a. 1s
    b. 3p
    c. 4f
    d. 3d
14. If the first quantum number (n) for an electron is 2, what possible values may the other three quantum numbers have?
15. How is the number of electrons in a neutral atom related to the atomic number of the element?
16. What information does the mass number of an isotope provide?
17. For each of the following atoms, tell how many protons, neutrons, and electrons are present.
    a. $^{9}_{4}Be$
    b. $^{45}_{21}Sc$
    c. $^{127}_{53}I$
    d. $(^{132}_{55}Cs)^{+}$
18. Fill in the blanks in the following chart:

Element	Symbol	Atomic Number	Mass Number	Electrons	Protons	Neutrons
hydrogen	___	___	___	___	___	0
___	___	___	___	___	10	___
___	___	29	65	___	___	___
___	___	___	104	44	___	___

19. Write isotopic notation for the atoms or ions that have these compositions:
    a. 14 protons, 14 neutrons
    b. 1 proton, 1 neutron
    c. 80 protons, 120 neutrons
    d. 26 protons, 30 neutrons, 23 electrons
    e. 52 protons, 78 neutrons, 54 electrons
20. Natural boron contains both B-10 and B-11 isotopes. Considering that boron's listed atomic mass is 10.81 amu, which isotope is more common? Explain.
21. A naturally occurring sample of the element gallium is a mixture of two isotopes. Calculate the atomic mass of a natural gallium sample given that 60.4 percent of the atoms have a mass of 68.9257 amu and that 39.6 percent of the atoms have a mass of 70.9244 amu. (Hint: see problem example in text.)
22. State the number of valence electrons in the following elements:
    a. oxygen
    b. sulfur
    c. potassium
    d. bromine
    e. barium
    f. zinc
23. Draw electron-dot structures for each of the elements in question 22.
24. State the charge that would be on the ion if a neutral atom were to
    a. gain one electron.
    b. lose one electron.
    c. gain two electrons.
    d. lose two electrons.

5A The Periodic Table page 101
5B Periodic Trends page 110
5C Descriptive Chemistry page 117
FACETS: The Case of the Unknown Chemical page 116

Elements 5

Organization and Properties

The key to learning chemistry is understanding the framework of the course. You have learned about the scientific method, the general forms of matter, and the specific structure of the atom. However, before you accumulate any more facts about chemistry, you need to understand another part of the framework: how atoms of one element differ from atoms of another. This information is found in an unexpectedly simple form—the **periodic table.**

5A The Periodic Table

Early Organizational Attempts: The Table Shapes Up

Imagine a list of all the elements, their atomic structures, and their properties. The great number of facts would be overwhelming even to someone blessed with a photographic memory. Clearly this essential information needs to be organized and presented so that all the needed facts can be at anyone's fingertips.

Chemists began this process long ago by searching for properties that were common to the known elements. One of the first scientists to discover such common properties was Johann Döbereiner. In 1829, this German chemist announced that he had observed several triads, or groups of three similar elements, among the known elements. One of Döbereiner's triads contained Cl, Br, and I. Each of these elements forms a gas with a distinct color and has similar properties.

As more elements were discovered, Döbereiner's concept of chemical element triads did not hold up. Elements with similar properties joined the triads to form quartets and quintets. Apparently the number three was not significant. Grouping chemical elements into families according to similar chemical properties, however, was an important step.

In 1864, John Newlands presented a classification scheme that added another idea to Döbereiner's groupings. Newlands arranged the known elements by their increasing atomic masses.

John Newlands's table aligned similar elements in vertical columns.

						H
Li	Be	B	C	N	O	F
Na	Mg	Al	Si	P	S	Cl
K	Ca	Cr	Ti	Mn	Fe	Co, Ni
Cu	Zn	Y	In	As	Se	Br
Rb	Sr	La, Ce	Zr	Nb, Mo	Ru, Rh	Pd
Ag	Cd	U	Sn	Sb	Te	I
Cs	Ba, V					

It appeared that every eighth element had similar properties. The known elements fell into similar groups when arranged into seven columns. Having both musical and scientific training, Newlands saw a correlation between the two worlds of music and science. The arrangement he had uncovered paralleled the octaves in music. He labeled this observed relationship the Law of Octaves. His ideas, however, were not well received, and his parallel to music was subjected to much ridicule. Newlands's idea that atomic mass and chemical properties might be related was correct, but he did not undertake a detailed study of each element's mass and characteristics.

The Modern Periodic Table: Getting It All Together

Credit for the development of the modern periodic table goes mostly to the Russian chemist Dmitri Ivanovich Mendeleev (men duh LAE uhf). Like Newlands, Mendeleev (1834–1907) arranged elements by their atomic masses. When an element did not seem to fit into a column, he noted that the atomic mass jumped significantly from that of the last element. He reasoned that undiscovered elements belonged in the gaps, so he left blanks in the chart and placed the known elements in columns in which they fit. Using information about the physical and chemical characteristics of elements, he predicted the properties of the elements that would fit into the blanks. Mendeleev's table was ridiculed for five years until one of these missing elements was discovered.

Tabelle II

Reihen	Gruppe I — R^2O	Gruppe II — RO	Gruppe III — R^2O^3	Gruppe IV RH^4 RO^2	Gruppe V RH^3 R^2O^5	Gruppe VI RH^2 RO^3	Gruppe VII RH R^2O^7	Gruppe VIII — RO^4
1	H = 1							
2	Li = 7	Be = 9,4	B = 11	C = 12	N = 14	O = 16	F = 19	
3	Na = 23	Mg = 24	Al = 27,3	Si = 28	P = 31	S = 32	Cl = 35,5	
4	K = 39	Ca = 40	— = 44	Ti = 48	V = 51	Cr = 52	Mn = 55	Fe = 56, Co = 59, Ni = 59, Cu = 63
5	(Cu = 63)	Zn = 65	— = 68	— = 72	As = 75	Se = 75	Br = 80	
6	Rb = 85	Sr = 87	?Yt = 88	Zr = 90	Nb = 94	Mo = 98	— = 100	Ru = 104, Rh = 104 Pd = 106, Ag = 108
7	(Ag = 108)	Cd = 112	In = 113	Sn = 118	Sb = 122	Te = 125	J = 127	
8	Cs = 133	Ba = 137	?Di = 138	?Ce = 140	—	—	—	— — — —
9	(—)							
10	—	—	?Er = 178	?La = 160	Ta = 182	W = 184	—	Os = 195, Ir = 197 Pt = 198, Au = 199
11	(Au = 199)	Hg = 200	Ti = 204	Pb = 207	Bi = 208	—	—	
12	—	—	—	Th = 231	—	U = 240	—	— — — —

Mendeleev's periodic table contained gaps for undiscovered elements.

Another innovation of Mendeleev's chart involved elements called transition metals. These elements did not fit into the major families of the chart, but they all had similar characteristics. Mendeleev put them in the chart but did not let them interfere with the groupings of the other elements. He then summarized his discoveries in a periodic law: the properties of the elements vary with their atomic masses in a periodic way.

Even after most of Mendeleev's "missing" elements were found, the table still had some problems. Arranging the elements in order of increasing atomic masses did not always produce a table with similar elements below each other. For example, nickel resembles palladium and platinum, and is therefore grouped with those elements. However, nickel's atomic mass should place it in the group with rhodium and iridium.

A young Englishman, Henry Moseley, discovered a new technique that eventually cleared up discrepancies in the periodic table. In 1912 he developed a way to count the protons in a nucleus. He found that if the elements were arranged in order of increasing atomic numbers, the problems in the table disappeared. Moseley's work led to a revision of the **periodic law.** After this revision it read as follows: the properties of an element vary with their atomic numbers in a systematic way.

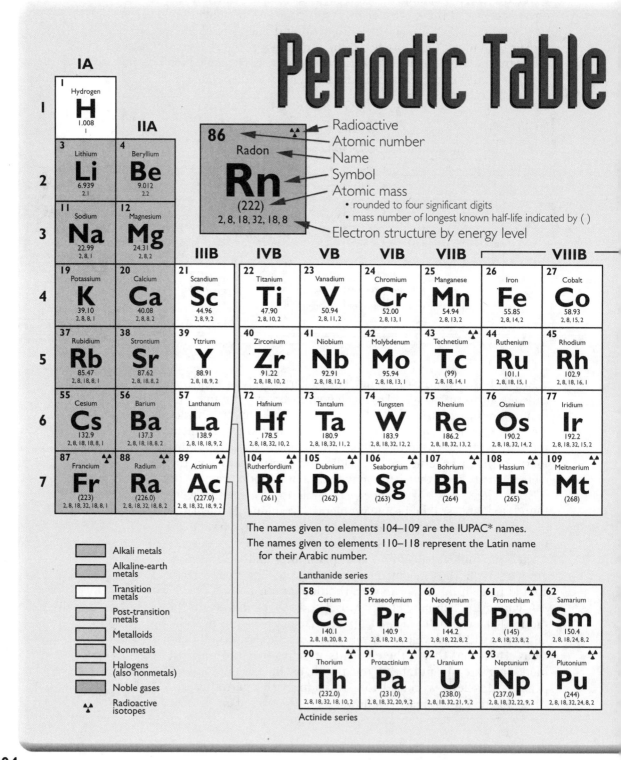

of the Elements

			IIIA	IVA	VA	VIA	VIIA	VIIIA
								2 Helium **He** 4.003 2
			5 Boron **B** 10.81 2,3	6 Carbon **C** 12.01 2,4	7 Nitrogen **N** 14.01 2,5	8 Oxygen **O** 16.00 2,6	9 Fluorine **F** 19.00 2,7	10 Neon **Ne** 20.18 2,8
IB	IIB		13 Aluminum **Al** 26.98 2,8,3	14 Silicon **Si** 28.09 2,8,4	15 Phosphorus **P** 30.97 2,8,5	16 Sulfur **S** 32.06 2,8,6	17 Chlorine **Cl** 35.45 2,8,7	18 Argon **Ar** 39.95 2,8,8
28 Nickel **Ni** 58.71 2,8,16,2	29 Copper **Cu** 63.55 2,8,18,1	30 Zinc **Zn** 65.38 2,8,18,2	31 Gallium **Ga** 69.72 2,8,18,3	32 Germanium **Ge** 72.59 2,8,18,4	33 Arsenic **As** 74.92 2,8,18,5	34 Selenium **Se** 78.96 2,8,18,6	35 Bromine **Br** 79.90 2,8,18,7	36 Krypton **Kr** 83.80 2,8,18,8
46 Palladium **Pd** 106.4 2,8,18,18	47 Silver **Ag** 107.9 2,8,18,18,1	48 Cadmium **Cd** 112.4 2,8,18,18,2	49 Indium **In** 114.8 2,8,18,18,3	50 Tin **Sn** 118.7 2,8,18,18,4	51 Antimony **Sb** 121.8 2,8,18,18,5	52 Tellurium **Te** 127.6 2,8,18,18,6	53 Iodine **I** 126.9 2,8,18,18,7	54 Xenon **Xe** 131.3 2,8,18,18,8
78 Platinum **Pt** 195.1 2,8,18,32,17,1	79 Gold **Au** 197.0 2,8,18,25,18,1	80 Mercury **Hg** 200.6 2,8,18,32,18,2	81 Thallium **Tl** 204.4 2,8,18,32,18,3	82 Lead **Pb** 207.2 2,8,18,32,18,4	83 Bismuth **Bi** 209.0 2,8,18,32,18,5	84 Polonium **Po** (209) 2,8,18,32,18,6	85 Astatine **At** (210) 2,8,18,32,18,7	86 Radon **Rn** (222) 2,8,18,32,18,8
110 Ununnilium **Uun** (269)	111 Unununium **Uuu** (272)	112 Ununbium **Uub** (269)		114 Ununquadium **Uuq** (285)		116 Ununhexium **Uuh** (289)		118 Ununoctium **Uuo** (293)

63 Europium **Eu** 152.0 2,8,18,25,8,2	64 Gadolinium **Gd** 157.3 2,8,18,25,9,2	65 Terbium **Tb** 158.9 2,8,18,27,8,2	66 Dysprosium **Dy** 162.5 2,8,18,28,8,2	67 Holmium **Ho** 164.9 2,8,18,29,8,2	68 Erbium **Er** 167.3 2,8,18,30,8,2	69 Thulium **Tm** 168.9 2,8,18,31,8,2	70 Ytterbium **Yb** 173.0 2,8,18,32,8,2	71 Lutetium **Lu** 175.0 2,8,18,32,9,2
95 Americium **Am** (243) 2,8,18,32,25,8,2	96 Curium **Cm** (247) 2,8,18,32,25,9,2	97 Berkelium **Bk** (247) 2,8,18,32,26,9,2	98 Californium **Cf** (251) 2,8,18,32,28,8,2	99 Einsteinium **Es** (254) 2,8,18,32,29,8,2	100 Fermium **Fm** (257) 2,8,18,32,30,8,2	101 Mendelevium **Md** (258) 2,8,18,32,31,8,2	102 Nobelium **No** (259) 2,8,18,32,32,8,2	103 Lawrencium **Lr** (260) 2,8,18,32,32,9,2

*IUPAC – International Union of Pure and Applied Chemistry

Chapter 5A

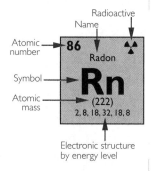

Parts of the North American Convention Periodic Table

The periodic table contains a small block for each element, and each block contains basic information about the element. An element's atomic number appears at the top of the block. This number specifies the number of protons in the nucleus. Some tables give the names of the elements along with the symbol. Located under the symbol is the atomic mass. This number gives the average mass of the atom in amu. Some tables also show the electron structure of the neutral atom.

Together the various blocks form columns and rows. A column of elements is called a **group** or a **family** because the elements usually have similar physical and chemical properties. The similar properties of these elements are due to their similar electron configurations. Rows of elements are called **periods** or **series.**

In the North American convention, a Roman numeral I above the first column indicates that all elements in this family have one valence electron. Likewise, a Roman numeral II, which shows that each member possesses two valence electrons, heads the second family of elements. The Roman numerals above the families give a reasonably accurate indication of valence electrons. Transition metals, however, are noted for being the exceptions to almost every rule. Each Roman numeral has an accompanying A or B that tells whether the families are **main groups** (A) or transition metals (B).

The middle and left side of the table contains metals. **Metals** are usually hard, lustrous (shiny), malleable, and ductile, and are good conductors of heat and electricity. However, there are exceptions to these characteristics, such as mercury, which is a liquid. The heavy, stair-step line marks the home of the metalloids. **Metalloids*** are elements having characteristics of both metals and nonmetals. The far right side of the table contains the nonmetals. **Nonmetals** are generally gases or soft solids, although there are exceptions, such as bromine, which is a liquid.

Sample Problem
  a. What element belongs to family VA and the second period?
  b. To what family and period does the element with the atomic number 10 belong?

Solution
  a. Nitrogen
  b. Neon, the tenth element, appears in the far right column and on the second row. It belongs to family VIIIA and the second period.

---

*metalloid: *metall* (Gk. – metal) + *oid* (Gk. – like)

Two rows of elements, called the lanthanide series and the actinide series, have been placed at the bottom of the table. The **lanthanide series** fits into the table immediately after lanthanum, and the **actinide series** fits into the table after actinium. If those elements were in their proper places within the periodic table, the table would be expanded into an outsized, unmanageable shape. Figure 5A-1 shows this layout.

Figure 5A-1

An awkward shape results when the inner transition metals are inserted into the main body of the Periodic Table.

Other Forms of Periodic Tables

There are three forms of periodic tables that are used in the scientific community today. Each one of these periodic tables presents the same information and general location of the elements, but they differ in their approach to numbering and labeling. The three systems are as follows:

- The North American Convention
- The European Convention
- IUPAC Periodic Table

The **North American Convention** is the main periodic table that you will see in this book. It is the periodic table that you are probably most familiar with. Note the following exceptions:

- It is common to see the Roman numerals replaced by Arabic numerals. For example, IIA = 2A.
- Group VIIIA, the noble gases, has also been replaced in some periodic tables by the number 0. This format is not as popular as the Arabic numeral format.

The **European Convention** is another form of periodic table that uses A and B designations; however, these designations do not place elements in the same A and B categories as the North American Convention. The system starts at the left group or family and labels IA through VIIA, then labels the next three columns 8, and

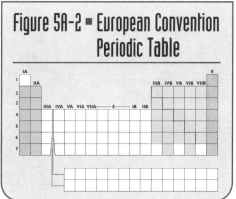

Figure 5A-2 ▪ European Convention Periodic Table

then continues on with IB through VIIB. This system remains popular in the European scientific community. It has the same exceptions as the North American Convention in regard to Arabic numerals; however, 0 is often used to represent the noble gases.

The **IUPAC Periodic Table** is the periodic table recommended by IUPAC, the International Union of Pure and Applied Chemistry. This system was placed into effect in 1984, but has met great opposition because of its lack of A and B designations. This format uses the Arabic numerals 1-18 to number the groups.

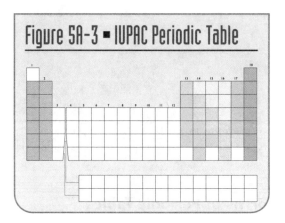

Predicting Electron Configurations

In its own way, the periodic table gives the electron configurations of the elements. Keeping the filling order of the sublevels in mind, follow the order of the elements in the periodic table. The electrons of hydrogen and helium occupy the 1s sublevel. After that, the electrons of both lithium and beryllium begin filling the 2s sublevel, and the electrons of boron, carbon, nitrogen, oxygen, fluorine, and neon fill the 2p sublevel. The outermost electrons of the next ten elements occupy the 3s, the 3p, and the 4s sublevels. Then a group of ten transition metals have highest energy electrons filling the 3d sublevel. Figure 5A-4 labels all the regions of the periodic table according to the sublevels that are being filled. Note that the widths (number of elements per row) of the regions match the capacities of the sublevels.

Each s sublevel can hold two electrons, and the s regions on the periodic table are two elements wide. The p sublevels can hold six electrons, and the series of six elements in the p regions fill these positions. The center of the periodic table contains a series of ten elements whose last electrons fill the ten positions in d sublevels. The lanthanide and actinide series have fourteen members whose electrons fill the f sublevels. Thus, you can see

Figure 5A-4

The periodic table reveals the order in which sublevels are filled.

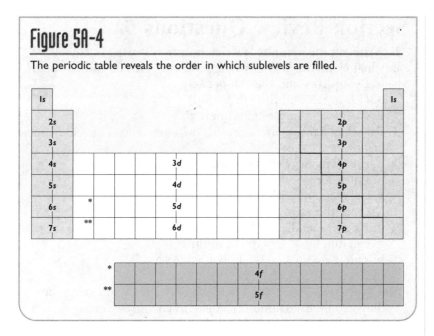

how the periodic table can be thought to consist of an "s block," a "p block," a "d block," and an "f block" based on the type of sublevel that is being filled. Such an observation will enable you to use the periodic table easily to predict both the electron configuration and orbital notation, simply by noting an element's position on the table.

Sample Problem Using the periodic table, predict the electron configuration and orbital notation of calcium.

Solution
To get to calcium from the beginning of the periodic table, you must pass through the 1s, 2s, 2p, 3s, 3p, and 4s regions. Since all these sublevels are filled, the electron configuration of Ca is

$$1s^2\ 2s^2\ 2p^6\ 3s^2\ 3p^6\ 4s^2$$

The orbital notation of Ca is

```
  1s   2s     2p     3s      3p     4s
  ↑↓   ↑↓   ↑↓ ↑↓ ↑↓   ↑↓   ↑↓ ↑↓ ↑↓   ↑↓
```

Section Review Questions 5A

1. Write out the periodic law that is currently used in chemistry.
2. What elements belong to the following groups and periods?
   a. Group IIIA and the 5th period
   b. Group IIA and the 4th period
   c. Group VIA and the 2nd period
3. To what family and period do the following elements belong?
   a. the element with the atomic number of 17
   b. potassium
   c. the element with the atomic mass of 32.06
4. Classify the following elements as a metal, nonmetal, post-transition metal, metalloid, or noble gas.
   a. calcium   d. aluminum
   b. radon   e. silicon
   c. tungsten   f. oxygen
5. Compare the three different periodic tables, and explain how they can be used simultaneously in chemistry today.
6. Using the periodic table, predict the electron configuration and the orbital notation of the following elements.
   a. zinc   b. iodine   c. tungsten

5B Periodic Trends

Each period of elements shows the same general progression in electron configurations. As more electrons join the outermost sublevel, elements become less metallic in character. In addition, the sizes of the atoms and their ions result directly from electron configurations. The forces between nuclei and electrons also change regularly as the atomic numbers increase. Periodic trends help to reveal the characteristics of the elements.

Atomic and Ionic Radii

Though the model of the atom we use today is not the hard sphere model of Dalton's day, it is often helpful to visualize the atom as if it looked like a marble. You must always keep in mind, however, that the boundaries of an atom are not as clearly defined as they are for an object such as a marble. This uncertainty can lead to variations in data depending on the assumptions made and the methods used to obtain measurements. In any case, an atom's size is determined by the dimensions of the electron cloud that surrounds it. One of the properties of elements that vary in a periodic way is the size of their atoms, described by the term **atomic radius**.

In general, the radii of atoms decrease in size as you move from left to right across a period of the periodic table. At first, this may seem strange since each successive atom in a period has an additional electron that should add to the size of the cloud. However, the decrease in size occurs because of the increasing nuclear charge and its greater attraction for the outer-level electrons. For example, as we proceed from lithium to neon, the nuclear charge (the number of protons) increases from +3 to +10, which is balanced overall by additional electrons. But since the electrons are entering the same energy level at about the same distance from the nucleus, they do not "shield" one another from the increased nuclear charge and thereby prevent each other from being affected by it. Hence, each successive negative electron that is added is attracted more strongly to the positive nucleus by **electrostatic attractions** between opposite charges. The result is that the electron is pulled toward the nucleus more tightly, and the size decreases. You will notice that the atomic radii for the transition metals remain quite constant. This is because they are filling an inner *d* sublevel, and the outermost energy level—an *s* sublevel—remains fairly constant in the number of electrons it contains. Further, the electrons that enter an inner level essentially balance the effect of the increasing nuclear charge by shielding the outer electrons from it.

The atomic radius is observed to increase when moving down a group or family in the periodic table. This is due to the addition of another principal energy level with each subsequent period. Because there is a greater number of principal energy levels, there is also a larger cloud of electrons and a larger atomic radius. Figure 5B-1 illustrates the general trends we have just discussed for atomic radii.

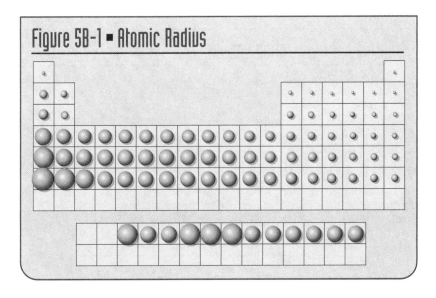

Figure 5B-1 ▪ Atomic Radius

We can extend our discussion of radii to ions. Remember, ions are atoms that have lost or gained electrons. Cations—positive ions—are smaller than their parent atoms because they have lost electrons, making the electron cloud smaller. Usually, atoms that have only a few electrons in their outermost energy levels (e.g., metals) will lose them all when they react; this will result in one less energy level in the cation than in the neutral atom—a significant reduction in size! On the other hand, atoms that gain electrons will be larger than their parent atoms. These additional electrons are generally added to the same main energy level to make a larger cloud. Also, without an additional nuclear charge to balance the added electrons, there is more repulsion between the like-charged electrons and they tend to spread out more.

Ionization Energy: The Electron "Rip-off"

Some elements lose their electrons easily, whereas others stubbornly hold on to theirs. As you will further learn in Chapter 6, the formation of compounds from elements involves valence electrons. The atoms that make up chemical compounds either give up or take electrons to some extent. Therefore, the ease with which atoms either acquire or donate electrons is a very important property for predicting chemical reactivity. The minimum energy required to remove a neutral atom's outermost electron and make that atom an ion is that atom's **first ionization energy.** Figure 5B-2 shows the trend for ionization energy.

Note what happens to the values of each period. Ionization energies increase from left to right across the periods because the electrostatic attractions increase. That is, it becomes more difficult to remove an electron that is more strongly attracted to a more highly charged nucleus. The electrons in the outermost en-

Figure 5B-2 ▪ Ionization Energies

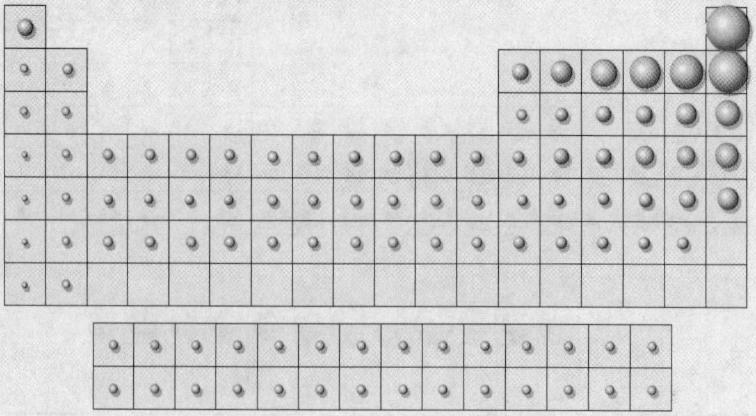

ergy level are also closer to the nucleus. (See Figure 5B-1.) Ionization energies decrease from the top of the table to the bottom. The reason for this decrease is twofold. First, the outer electrons are in higher energy levels and are thus farther away from the nucleus. Second, the outer electrons are somewhat shielded from the positive charges in the nucleus by the electrons in the lower levels.

Notice that the largest ionization energies are for removal of an electron from a full energy level. The smallest ionization energies are for metals that can achieve a full outermost energy level by the loss of a single electron. These trends can be explained on the basis of atomic radius, for it is easier to remove an outer electron from a larger atom than from a smaller one. Thus, if you know the atomic radius trend, you can predict the ionization energy trend. A large atomic radius means a small ionization energy, and a small atomic radius means a large ionization energy. For example, helium has the largest ionization energy, but the smallest atomic radius.

Electron Affinity: Electrons Anyone?

Whereas ionization energy is the amount of energy required to remove an electron and form a positive ion, **electron affinity** is the amount of energy released when an electron joins an atom to form a negative ion. Electron affinity measures the degree of attraction that an atom has for additional electrons. The factors that affect ionization energy also affect electron affinity. Electron affinities increase from left to right on the periodic table and decrease from top to bottom. Figure 5B-3 shows this trend by using color intensities in the periodic table. The stronger the color intensity the stronger the electron affinity.

Figure 5B-3 • Electron Affinity

H																	He
Li	Be											B	C	N	O	F	Ne
Na	Mg											Al	Si	P	S	Cl	Ar
K	Ca	Sc	Ti	V	Cr	Mn	Fe	Co	Ni	Cu	Zn	Ga	Ge	As	Se	Br	Kr
Rb	Sr	Y	Zr	Nb	Mo	Tc	Ru	Rh	Pd	Ag	Cd	In	Sn	Sb	Te	I	Xe
Cs	Ba	Lu	Hf	Ta	W	Re	Os	Ir	Pt	Au	Hg	Tl	Pb	Bi	Po	At	Rn
Fr	Ra	Lr	Rf	Db	Sg	Bh	Hs	Mt	Uun	Uuu	Uub		Uuq		Uuh		Uuo

La	Ce	Pr	Nd	Pm	Sm	Eu	Gd	Tb	Dy	Ho	Er	Tm	Yb
Ac	Th	Pa	U	Np	Pu	Am	Cm	Bk	Cf	Es	Fm	Md	No

Electronegativity: A Tug of War with Electrons

Electronegativity values reveal the tendency of atoms to attract electrons when the atoms are bonded to other atoms. Electronegativities are related to both ionization energies and electron affinities; thus large electronegativities invariably accompany large ionization energies and high electron affinities. The correlation is logical. Atoms that have large electron affinities tend to readily acquire electrons, and atoms that have large ionization energies tend to hang on to the electrons they already have. Thus, these characteristics of the uncombined atom can be expected to be present in the combined atom. This results in a large electronegativity—a strong attraction for electrons in a compound. Notice that we do not report values for He, Ne, or Ar since these elements are still considered to be chemically inert. Figure 5B-4 represents the trend of the electronegativities of the elements.

The values on the electronegativity chart were not determined experimentally. They were selected arbitrarily on the basis of comparisons with other elements. A Nobel Prize-winning chemist named Linus Pauling devised a scale with a maximum value of 4. The element fluorine received the highest number, since it is the most electronegative element. Pauling then assigned the values of the other elements.

Of the three measures of electrostatic attraction between electrons and the nucleus, electronegativity has the widest use. It plays a central role in predicting how atoms combine chemically with each other.

Figure 5B-4 ▪ Electronegativities of the Elements

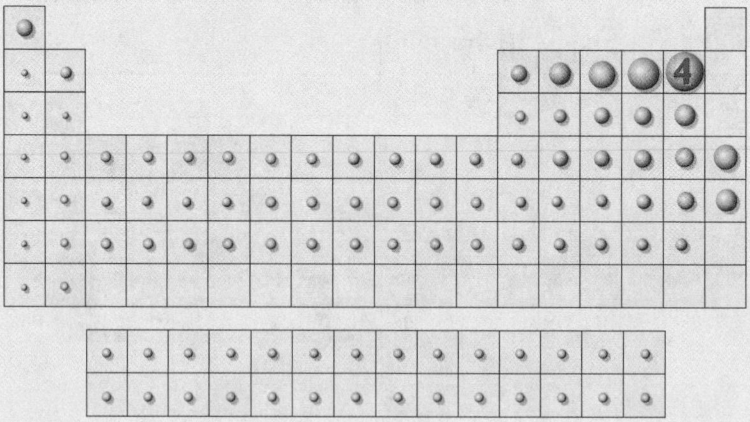

Sample Problem For each of the following pairs of elements, use a periodic table and your knowledge of atomic radius to predict which element has the largest electronegativity.

        a. N or P   **b.** Rb or I

Solution

a. Nitrogen and phosphorus are in the same group; atomic radii increase from top to bottom down a group, because of additional principal energy levels. As a result,
- P has a larger atomic radius than N.
- It will be more difficult for the smaller N to lose one of its outermost electrons since they are closer to the nucleus, resulting in a larger first ionization energy.
- The larger atom, P, will have a smaller attraction for additional electrons, resulting in a smaller electron affinity.
- The smaller atom, N, will have a greater attraction for electrons in a compound—i.e., a larger electronegativity—since they will be closer to the nucleus.

b. Rubidium and iodine are in the same period; atomic radii decrease across a period since there are the same number of energy levels but more protons in the nucleus. As a result,
- Rb has a larger atomic radius than I.
- It will be more difficult to move an outer electron from the smaller atom, I; thus, I has a larger first ionization energy.
- The larger atom, Rb, will have a smaller attraction for additional electrons, resulting in a smaller electron affinity.
- The smaller atom, I, will have a greater attraction for electrons in a compound and therefore be more electronegative.

Section Review Questions 5B

1. Match the following terms with their correct definition: first ionization energy, electronegativity, electron affinity, and atomic radius.
   a. energy released when a negative ion is formed
   b. the relative size of an atom
   c. energy required to make a neutral atom a positive ion
   d. the ability of an atom to draw electrons to itself
2. Which direction would you have to travel across the periodic table to go from larger (or higher) to smaller (or lower) for the following:
   a. atomic radius   c. electron affinity
   b. ionization energy   d. electronegativity
3. Of the elements in group IIIA, which has (1) the largest atomic radius, (2) the largest electronegativity, and (3) the smallest ionization energy? Explain.
4. If an element has a large ionization energy, what kind of atomic radius, electron affinity, and electronegativity would the same element have?

Facets of Chemistry: The Case of the Unknown Chemical

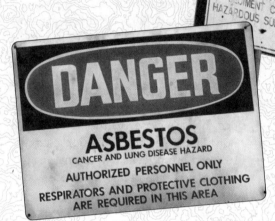

A farmer asks, "Does my soil contain enough minerals and nutrients for my crops?"

A home renovator asks, "Does the home I am renovating have lead-containing paint or asbestos-containing insulation?"

A rural homeowner wonders, "Are there any pollutants in my well water?"

An archaeologist thinks, "I wish I knew whether the metal alloy used in this ancient plowshare contains tin."

A government worker wonders, "How many phosphate ions are polluting Reedy River?"

In a way, each of these people needs a detective. They need someone who knows how to search for the right clues, examine the evidence, and identify the "suspect" chemicals. Although police detectives are good at tracking down missing people, they do not track down elusive chemicals. A different kind of detective, called an *analytical chemist*, solves these cases.

The need for analytical chemistry has never been greater. The demand in modern society for safe food and water, abundant energy resources, the health of families in the environment, and advanced technology all depend in some part on analytical chemistry. Undesirable substances in our food, air, water, and soil must first be identified in order to be removed or neutralized.

Identifying the contents of a substance is the main goal of analytical chemistry. One part of analytical chemistry, called *qualitative analysis*, identifies but does not measure the elements or grouping of elements in an unknown substance. The techniques used in qualitative analysis vary in complexity depending on the sample that the chemist is working with. Specific tests are available for each individual element.

Qualitative analysis is an important science in the protection of families. Suppose a family is thinking about buying and renovating a home that was built prior to the 1970s. During the home inspection, lead-base paint and asbestos ceiling tiles and insulation are found. In order to prevent future illness to the family, specially trained contractors must be employed to remove the lead-base paint and asbestos. Why go to all this trouble and expense to remove paint and asbestos? Lead paint causes lead poisoning, which is especially dangerous to children since it affects the development of the brain and the child's intelligence. Asbestos causes asbestosis, a serious lung disease, which leads to diminished lung capacity.

Sometimes knowing which chemicals are present in a substance is not enough—the amounts must also be known. Once an element has been separated, it is further examined to find the amount that is present. This process is called *quantitative analysis*. There are two common methods of quantitative analysis. The gravimetric (or weight) method of analysis measures chemicals by their mass. The second method, colorimetry, examines the color intensity of solutions containing a specific chemical.

These two techniques make up only a small percentage of the tools used by chemical detectives. Newly invented instruments and recently discovered reactions constantly provide analytical chemists with more tools. Along with the many traditional techniques, the new methods equip analytical chemists to help the many people who require their services.

5C Descriptive Chemistry

"Science is built up with facts, as a house is built up with stones, but a collection of facts is no more science than a heap of stones is a house." This statement by Henri Poincaré accurately describes the interaction between facts and theories in chemistry. While facts can serve as the foundation or even the polishing touches of chemistry, theories provide the framework by which the facts fit together. Most of this course deals with general theories involving the fundamental questions of chemistry. But the facts needed to answer these fundamental questions are provided by **descriptive chemistry**—the study of elements and the compounds they form.

Hydrogen: A Family by Itself

Hydrogen, the simplest and most abundant of all the elements, has an electron configuration similar to that of the IA metals. Yet because it displays unique properties, it is often considered to be a family by itself. Although other scientists probably prepared hydrogen gas earlier, an Englishman named Henry Cavendish was the first to collect and study it. In 1766, he prepared the gas by reacting a metal with an acid. The gas burned rapidly, even explosively, so Cavendish called it "inflammable air." Lavoisier later renamed it hydrogen, which means "water-former," because burning it in the presence of oxygen produces water.

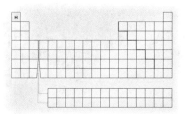

Physical Properties. As a gas, hydrogen is colorless, odorless, and tasteless. Because hydrogen is the least dense of all gases, its molecules move at high speeds and diffuse more quickly than other gases. Hydrogen molecules have little attraction for each other, so they stay in the gaseous state even when temperatures go down to -253°C. There are predictions that under extreme pressure hydrogen assumes a metallic structure.

Chemical Properties. Hydrogen has unique chemical properties. On the one hand, it has a single electron in its lone occupied energy level, so it can act like an alkali metal. On the other hand, it is only one electron shy of filling the first energy level, so it can act like halogens (halogens are discussed later in this chapter). It is sometimes shown as a member of both families on periodic tables.

Hydrogen atoms do not float freely in the atmosphere. If they are given the opportunity, they immediately bond to other hydrogen atoms to form diatomic molecules (H_2). At room temperature hydrogen molecules usually refuse to react with other elements. But at high temperatures and pressures molecular hydrogen can split apart and become highly reactive. Sparked by a flame or electrical discharge, hydrogen molecules can combine

with oxygen to form water. Controlled combustion reactions between hydrogen and oxygen are also used to power space vehicles. The explosion of the space shuttle *Challenger* in 1986 demonstrated the reactive nature of hydrogen when control is lost.

Bonded to nitrogen atoms, hydrogen forms ammonia (NH_3). Acids, including the well-known hydrochloric acid (HCl), form when hydrogen reacts with the halogens. Occasionally hydrogen reacts with active metals to form compounds called **metallic hydrides.** Lithium hydride (LiH), sodium hydride (NaH), magnesium hydride (MgH_2), and calcium hydride (CaH_2) are examples of these compounds.

Commercial Sources. Although hydrogen is found abundantly in the earth's atmosphere, it must be isolated and made usable. Large quantities of hydrogen are produced from various fuel gases. In electrolysis reactions (the use of electricity to force a non-spontaneous reaction to occur) involving these fuel gases, hydrogen is a by-product. Hydrogen is also produced by the electrolysis of water or by passing steam over coke (impure carbon obtained from coal).

Uses. Ammonia manufacturers use a majority of the hydrogen gas produced; however, the space program consumes large amounts of hydrogen gas for rocket fuel. Hydrogen atoms bond with liquid vegetable oils to form semi-solid fats in a process called hydrogenation. Manufacturers of cooking shortening carefully monitor the number of hydrogen atoms in their products. In the future we may see more engines powered by hydrogen since it is the ultimate clean fuel—producing only water when it is burned.

The explosion of the space shuttle *Challenger* resulted from an uncontrolled hydrogen and oxygen reaction.

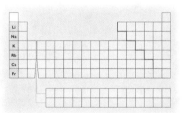

Group IA: Alkali Metals

Group IA metals are so reactive that they never exist by themselves in nature. Not until 1807 did Sir Humphrey Davy isolate one of the **alkali metals** from a compound. The alkali metals consist of lithium, sodium, potassium, rubidium, cesium, and francium.

Physical Properties. Like typical metals, alkali metals conduct electricity well and have a bright luster when they are freshly cut. Unlike the traditional stereotypes of metals, they have low densities and can be easily cut with a pocketknife because they are soft.

Chemical Properties. A solitary, easily lost outermost electron makes this group of elements very reactive. Alkali metals lose their electrons readily to "electron-hungry" elements. Because the largest atoms have the smallest electronegativities, the alkali metals are the most reactive metallic family. These metals react violently with water to produce strong, caustic bases. The metals are so reactive that they are usually immersed in an oil so that they will not react with oxygen or moisture in the air. None are found in their metallic form in nature.

Lithium differs from the other alkali metals because of its small size. Its small atomic radius allows the nucleus to hold on tightly to its one valence electron. Some of lithium's chemical and physical properties resemble those of magnesium, a member of the adjoining family, more than those of sodium. This similarity between small elements and elements in neighboring families is called a **bridge relationship.**

Uses. The most visible application of elemental sodium is in sodium vapor streetlights, characterized by their yellow glow. Human bodies depend on a balance of sodium and potassium ions to carry electrical signals through the nerves and to trigger muscle contractions. Sodium is also used in the manufacturing of baking soda, soap, rayon, and paper. Today's controversy about the amount of salt in diets is really concerned with the level of sodium ions. Commercial "salt substitutes" are actually a potassium salt instead of the more common sodium salt. Potassium compounds serve as main ingredients for many industrial processes such as the production of fertilizers, soaps, glass, explosives, and fireworks. Rubidium is found in small quantities in tea and coffee. Cesium is used in atomic clocks and in ion propulsion systems. One kilogram of cesium in outer space could propel a vehicle 140 times as far as the burning of the same amount of any other known liquid or solid. Lithium compounds make modern lubricants water-resistant and able to withstand extreme temperatures. Lithium battery cells have been extensively developed for a wide variety of uses. Francium has no known uses.

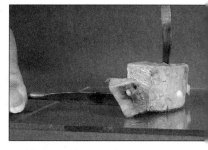

Alkali metals are soft enough to be cut with a knife.

Group IIA: Alkaline Earth Metals

The chemical term *earth* originally applied to metal-and-oxygen compounds that dissolved slightly in water. Some of these compounds were similar to compounds of alkali metals, so they were given the more specialized name **alkaline earth metals.** Today the term applies to the metals in group IIA. These metals are beryllium, magnesium, calcium, strontium, barium, and radium. Most people come into contact with alkaline earth metals each time they turn on a faucet. Hard water contains dissolved ions that inhibit soap from producing suds. Water that percolates through underground deposits of limestone picks up calcium and magnesium ions (Ca^{2+} and Mg^{2+}) that combine with soap to produce an insoluble scum. Water softeners prevent the suds-inhibiting action of these ions.

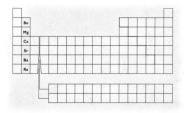

A. Stalactites
B. Stalagmites
C. *Blue Angels* in formation

Physical Properties. Fresh cuts into these metals reveal a common shiny luster under the dull-gray coatings. Densities range slightly higher than those of the alkali metals, but these metals are much harder. All the alkaline earth metals are malleable. These elements are found widely distributed in rock structures of the earth's crust, but not in elemental form, since they are reactive.

Chemical Properties. Each alkaline earth metal has two electrons in its outermost s sublevel. The alkaline earth metals release two electrons when they react, and the elements that hold electrons loosely react the best. Reactivities increase as the alkaline earth metals get larger. Beryllium does not react with water, magnesium can react with hot water, and calcium will react vigorously with even cold water. Like lithium, beryllium resembles a member of an adjacent family. Being a small element, it has properties similar to many of the chemical properties of aluminum, the second member of the family IIIA.

Uses. Elemental beryllium is used in X ray tubes because of its transparency to low energy X rays. Magnesium is often used to make lightweight alloys for airplanes, automobiles, and boats. Military flares and some fireworks burn magnesium because it produces a brilliant white light. Magnesium compounds in the mineral water of Epsom, England, make up Epsom salts, and magnesium hydroxide gives milk of magnesia its chalky taste and stomach-soothing ability. Although calcium is a necessary part of a balanced diet, its major use is for building materials. Limestone, or calcium carbonate, serves as the foundational material for everything from concrete to finishing plaster. This material also has an important aesthetic value in nature. The beautiful formations of stone known as stalactites and stalagmites consist of limestone. When water containing dissolved carbon dioxide trickles through limestone, it dissolves some of the stone. If that water happens to fall into a cave, its evaporation results in the deposition of limestone. Calcium carbonate also forms the skeletons of the brilliantly colored animals that form coral reefs. Calcium oxide (lime) is used to manufacture iron from its ores. Strontium is used for fireworks and flares since it emits a brilliant red flame. Barium is used in fireworks, rat poison, rubber, linoleum, and in X ray examination of the gastrointestinal tract. Radium is used to treat specific forms of cancer.

The B Groups: Transition Metals

All the B groups on the periodic table belong to the group of elements called **transition metals.** The name for this group of elements stems from a former belief that the properties of these elements progressed in regular graduations to the nonmetals. Today it is known that these elements have characteristics all their own. The term *transition* still has meaning, though, as a label for the physical location in the center of the periodic table.

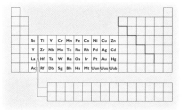

Physical Properties. The most distinct property of the transition elements is that they are the "typical" metals. Unlike the first two families, many of these metals have high densities and considerable strength. They also have a shiny luster, conduct well, and conform to desired shapes when pounded or stretched within reasonable limits. Mercury exists as a liquid at room temperature, but all the others are solids.

Chemical Properties. Because their highest energy electrons occupy *d* sublevels, transition metals have many different chemical properties. Unreactive metals, such as gold, silver, and platinum, can resist corrosion for centuries. Others, like iron and copper, corrode quickly when exposed to moist air. Iron, copper, and nickel are the only elements known to produce a magnetic field. Because the last few energy sublevels of these elements are close to each other, electrons have little difficulty jumping from one sublevel to another, and the same atoms can bond in a variety of ways. The chemistry of the transition metals involves just as many exceptions as common trends.

Uses. Society uses transition metals in countless ways. Trusses that span gymnasiums, I-beams that support skyscrapers, artificial hip joints, coins, jewelry, and electrical wires all testify to the diversity of uses. In addition, human bodies require certain trace amounts of iron, chromium, cobalt, copper, manganese, and vanadium to function normally. Hemoglobin, an iron-containing complex in red blood cells, is responsible for oxygen transport in the blood. Transition elements are also important catalysts in a variety of industrial processes.

A. Artificial hip joints
B. Copper wire rope
C. Red blood cells
D. United States coins

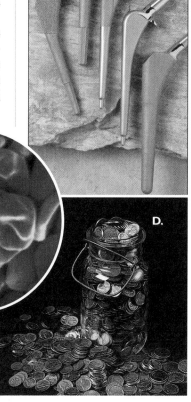

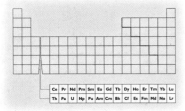

Inner Transition Metals

The lanthanide and actinide series are called the **inner transition metals.** The lanthanide series was once called the "rare-earth elements" because these elements were seldom found and difficult to remove from their ores. They are now found in high concentrations in a considerable number of minerals and in low concentrations throughout the earth's crust. The actinide series has no nickname from the early days of chemistry because most of these elements are manmade. Only the first four elements in the actinide series have been found in nature. The **transuranium elements** are the chemical elements with atomic numbers greater than that of uranium. These elements consist of many radioactive isotopes that are produced artificially and are characterized by instability.

Physical Properties. The lanthanide series extends from lanthanum, atomic number 57, through lutetium, atomic number 71. Scandium and Yttrium are also sometimes included in this "rare-earth" grouping because of similar properties. The majority of the lanthanide series is strongly paramagnetic. **Paramagnetism** describes a substance that is weakly attracted by a magnetic field because of unpaired electrons. The lanthanide series elements all occur naturally with the exception of promethium. The pure forms of these rare-earth metals are bright and silvery. The actinide series extends from actinium, atomic number 89, through lawrencium, atomic number 103. Only minute amounts of some actinide series elements are obtainable because of their instability, and whatever quantities may be produced decay too fast to accumulate.

Chemical Properties. The inner transition metals occupy the 4f and 5f energy levels. Although lanthanum and actinium contain no 4f or 5f electrons, they resemble the atoms of the lanthanide and actinide elements closely. The names lanthanide and actinide literally mean "lanthanum-like" and "actinium-like." Unlike the lanthanide series, the actinide series displays a variety of oxidation states. The danger of the actinide series lies in their radioactivity. They are emitters of tissue-destroying and cancer-producing rays. Once ingested they tend to remain in the body indefinitely, and less than one-millionth of a gram can be fatal.

Uses. The lanthanides are used as catalysts in chemical reactions and in the glass and television industries. Actinide series elements are valuable because of their use in fission reactions. Fission is the process by which an element's nucleus is split, producing large amounts of energy. Both uranium and plutonium have been used in atomic bombs because of their explosive power and also in nuclear plants for the production of electricity.

Groups IIIA, IVA, and VA: The Post-Transition Metals and the Metalloids

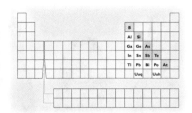

Post-transition metals, as their name implies, follow the series of transition metals on the periodic table. They are found in groups IIIA, IVA, and VA. This group of metals includes well-known elements such as tin, lead, and aluminum as well as obscure elements like thallium, indium, and gallium.

Metalloids have properties of both metals and nonmetals. Aluminum is distinctly metallic, but it also has several metalloid characteristics. Boron, silicon, arsenic, and antimony are the most typical of the metalloids, but germanium, tellurium, and probably astatine fit in as well. The classification of astatine is somewhat arbitrary. Like metals, the metalloids have a metallic luster. They can conduct electricity, but not very well. Consequently, they are called **semiconductors.**

The post-transition metals and metalloids will be addressed further within their specific periodic groups.

Group IIIA

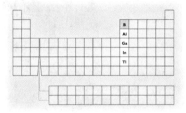

This group consists solely of post-transition metals and metalloids. There is a significant difference between the chemical properties of boron, a metalloid, at the top of the column and the four metals below it. These four metals are aluminum, gallium, indium, and thallium. Aluminum has some metalloid properties, but the characteristics we associate with metalloids are more recognizable in other groups.

Physical Properties. The two most important elements of Group IIIA are the elements boron and aluminum. Boron, found combined with oxygen and sodium in the mineral borax, has interesting optical characteristics and is an excellent semiconductor material. Aluminum, on the other hand, is an important metal because it combines high strength and low density. Unlike the metalloid boron, aluminum is nontoxic, easily machined, and highly conductive. Aluminum is the most common metal in the earth's crust, found mostly in clays.

Chemical Properties. Aluminum is too reactive to be found free in nature; it is usually bonded to oxygen atoms in aluminum ore, which is called bauxite. Like iron, aluminum reacts with oxygen if it is exposed to the atmosphere. How then did aluminum get its reputation for being a durable, corrosion-resistant metal? The difference lies in the nature of the corrosion. Whereas iron oxide is porous and allows new oxygen molecules to penetrate deeper into the metal, aluminum oxide forms an impenetrable shield against further oxidation.

Boron is a "bridge" element. It has the electron configuration of the elements in group IIIA but has several properties of silicon, which is in group IVA. The small size of the atom allows the nucleus to hold electrons as if the nucleus had a greater positive charge than it actually does.

Uses. Metallurgists use electrical currents to pull pure aluminum from aluminum oxide molecules. Before this process was invented, aluminum was so expensive that a large chunk was included with the crown jewels of England. Napoleon III even ordered a rattle of this priceless metal for Prince Louis. Today, this metal vies with steel for the title of "most-used metal." Aluminum can be valuable when its oxides form rubies and sapphires. Fuller's earth, a compound of silicon and aluminum, can remove spots and stains from textiles by absorbing them. A boron compound called borax softens water and helps clean clothes in the laundry. Pyrex glass contains fused sand, aluminum oxide, and borax. The borax gives the glass a low thermal expansion so that sudden changes in temperature do not tear the glass apart. Enamel, a compound of soft glass with colored metallic oxides added, has borax added for the same reason. Antiseptics and bleaching agents can be formed if borax is treated with sodium hydroxide (NaOH) and hydrogen peroxide (H_2O_2). Boric acid, an ingredient in eye-soothing medicines, also contains boron.

Top: sapphires
Bottom: Borax is found in many common household items.

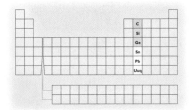

Group IV

As in group IIIA, a wide range of properties is present in this group. The only nonmetal—carbon—is at the top, followed by two metalloids, silicon and germanium, and then three metals at the bottom—tin, lead, and ununquadrium. Little is known about ununquadrium. It was recognized by Russian physicists in 1999.

Carbon could be a group by itself since its compounds form the basis of life. Even though carbon constitutes only 0.027 percent of the earth's crust, it is one of the most important elements. There are so many compounds of carbon that the study of these compounds has become a specialized branch of chemistry—organic chemistry. There are other more complex carbon compounds that are studied in biochemistry, and important carbon compounds such as the carbon oxides (CO, CO_2), bicarbonates ($NaHCO_3$), and cyanides (NaCN) that form the major inorganic carbon compounds.

Carbon was well known in the ancient world. Charcoal from fire was probably the earliest recognized form of carbon. The word carbon probably comes from the Latin *carbo*, meaning "coal," "charcoal," or "ember." The two naturally occurring forms of elemental carbon, diamond and graphite, were also aptly named. Diamond is named from the Greek *adamas*, meaning "invincible," and graphite comes from the Greek *graphein*, meaning, "to write." Graphite, most familiar as a component of pencil lead, is also used in tennis rackets, golf clubs, and fishing poles, and as a powdered lubricant.

Just as carbon is the building block for the animal and vegetable worlds, so silicon serves as the foundational element in the mineral world. Silicon forms about 28 percent of the earth's crust, but it does not occur alone. Nearly 40 percent of all common minerals contain silicon. The silicate minerals are the basis of rocks, soils, clays, and sands.

Physical Properties. The carbon atoms in diamonds form a tight, interlocking pattern. Diamond is the hardest substance known, has a high melting point, and does not conduct electricity. In graphite, the atoms are arranged in layers that slide across one another very easily. These sheets of molecules make graphite soft and slippery. Graphite also conducts electricity—a rare property for a nonmetal. The other elements in this group exhibit the same physical properties as the nonmetals and metalloids.

Chemical Properties. Carbon, when found in diamonds and graphite, is relatively unreactive. Graphite, however, does oxidize slowly in the presence of nitric acid and sodium chlorate. Silicon does not react with air, water, or acids at low temperatures.

Uses. Diamonds can be used in a wide variety of ways. Valuably cut diamonds are used to create beautiful jewelry. Many diamonds with imperfections are used as polishing and grinding abrasives, and in rotary dental instruments, such as a drill. Graphite is also used in the manufacturing of paints. Charcoal, another form of carbon, is used in water and air filters to strain out organic impurities that cause objectionable smells and tastes. It is also used in deodorant shoe inserts. The semiconducting property of silicon and germanium has made them vital materials in microchips. Microchips of semiconductors control the flow of electrons in digital watches, calculators, and computers. Tin, when combined with copper, forms the alloy known as bronze. Bronze is often used in the production of statues. Lead is used mainly to make ammunition, but can still be found in the soldering of stained glass.

Graphite is often used to make sporting equipment. Silicon and germanium are key elements in microchips.

Group VA

This group exhibits the most dramatic changes in properties. There is a gas at the top (nitrogen) and a solid metal at the bottom (bismuth). Nitrogen and phosphorus are nonmetals, and arsenic and antimony are metalloids.

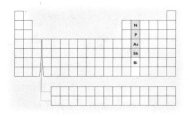

In 1772, a Scottish physician named Daniel Rutherford first recognized nitrogen as an element. Rutherford also showed that nitrogen gas could not support life the way oxygen does for animals or the way carbon dioxide does for plants. The gas was appropriately named *azote,* meaning "lifeless."

About one hundred years before the discovery of nitrogen, a German alchemist had discovered phosphorus. During his experiments, he distilled a substance that glowed in the dark. In later years, Lavoisier recognized this substance as an element and named it *phosphorus,* from a Greek word meaning "light bearer."

Physical Properties. Nitrogen normally exists as diatomic N_2 molecules in the gaseous state. The gas has no taste, no color, and no odor, and accounts for approximately 78 percent of the earth's atmosphere. Phosphorus exists in one of four or more forms that result from different arrangements of the atoms. All the forms are solids, but they have different colors. Some common forms of this element are white (or yellow), red, and black (or violet) phosphorus.

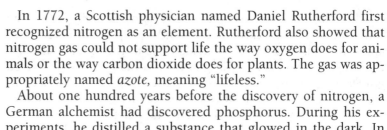

Chemical Properties. Nitrogen molecules rarely enter into chemical reactions. These molecules are tightly bonded together and are difficult to split up. In contrast, one of the most important properties of phosphorus is its high chemical reactivity, especially with oxygen. Phosphorus is so reactive that it cannot be found pure in nature.

Top: White and red phosphorus
Bottom: Nitrogen is a key ingredient in many fertilizers.

Uses. Plants and animals require a constant supply of nitrogen atoms. It might seem that the atmosphere would provide an unlimited supply, but this is not the case. The nitrogen molecules in the air do not usually react to provide the necessary nitrogen atoms. Some nitrogen compounds exist in the soil, and others are provided through artificial fertilizers or decaying organic matter. Some plants coexist with bacteria that can transform atmospheric nitrogen molecules into compounds with ready-to-use nitrogen atoms. Both nitrogen and phosphorus form a variety of gaseous compounds. Phosphorus is used in match heads and on safety match boxes. Ammonia (NH_3) can be used in its liquefied form as a fertilizer. Nitrogen and oxygen can bond together to produce a variety of compounds such as laughing gas (N_2O) and smog-producing air pollutants (NO and NO_2). Although arsenic is usually thought of as a potent poison, it serves our industries in several important ways. Its compounds find applications in the preserving of animal skins and in glass manufacturing.

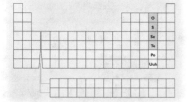

Group VIA

Oxygen, a gaseous nonmetal, is the most easily recognized element in this group. Sulfur and selenium are also nonmetals.

Tellurium is a highly reactive metalloid with few industrial uses. Polonium is believed to be a highly reactive metal. It was discovered by Pierre and Marie Curie and has been studied very little. Ununhexium is thought to be metallic and was recognized at the Lawrence Berkley National Laboratory in 1999.

In 1774, an English clergyman named Joseph Priestley discovered oxygen during a series of experiments in which he decomposed substances with the aid of the sun and a strong lens. When he focused the rays of the sun on mercury oxide (HgO), it decomposed into mercury and a gas that made flames burn brighter. Lavoisier eventually gave the new element the name *oxygine*, which means "acid producer." When either sulfur or phosphorus was burned in this gas, it produced a gaseous compound that formed acids in water.

Men of ancient times knew of sulfur because it occurs uncombined in nature. The Bible refers to sulfur fifteen times when it describes brimstone (meaning "burning stone"). These references associate sulfur with hell, the lake of fire, and God's judgment. The acrid fumes and searing blue flame of burning sulfur give a sobering picture of what separation from God for eternity may be like.

Physical Properties. Oxygen, a colorless, odorless, tasteless gas that is slightly soluble in water, is the most abundant element in the earth's crust. Through divine design, enough oxygen dissolves in lakes, rivers, and oceans to sustain fish and aquatic plants. Oxygen liquefies to a pale blue liquid between -183°C and -218°C and solidifies to a pale blue solid below -218°C. Atmospheric oxygen exists in two forms: O_2 gas and ozone (O_3). Lightning can convert odorless O_2 to the pungent O_3, which can often be detected after an electrical storm. God has graciously created a protective layer of ozone in the upper atmosphere to shield the earth from harmful amounts of ultraviolet radiation from the sun.

Sulfur exists in a variety of forms. Native sulfur is a yellow solid, but when it is heated to 113°C, it melts into a straw-colored liquid that can crystallize into another form. If the molten sulfur is quickly cooled by being poured into water, it forms amorphous globs with a plastic-like consistency.

Chemical Properties. While sulfur and oxygen do not resemble each other physically, their relationship becomes apparent through their chemical properties. Oxygen is one of the most reactive elements, forming compounds called **oxides** with all other elements except those in group VIIIA. Oxygen's electronegativity ranks second only to fluorine, and its strong pull on electrons can be expected to result in chemically active properties. Sulfur is reactive at room temperatures, but it does not match the reactivity of oxygen. Metals such as zinc, calcium, and

Top: Mounds of sulfur
Bottom: The effects of acid rain

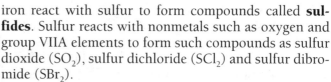

iron react with sulfur to form compounds called **sulfides**. Sulfur reacts with nonmetals such as oxygen and group VIIA elements to form such compounds as sulfur dioxide (SO_2), sulfur dichloride (SCl_2) and sulfur dibromide (SBr_2).

Uses. As Lavoisier pointed out, oxygen allows combustion and supports animal life. Ozone, which has some undesirable effects, can also be beneficial: it can kill bacteria by oxidizing them.

Sulfur atoms bonded into rubber help make today's rubber supple, strong, and pliable in a wide range of temperatures. Sulfur dioxide has a variety of uses ranging from bleaching agent to disinfectant. The majority of the sulfur used in industrial countries goes into the production of sulfuric acid. This thick, syrupy, colorless liquid serves as a workhorse of chemical-related businesses. Compounds of sulfur and oxygen released from coal-burning factories have been discovered to be the culprits in the formation of acid rain. The sulfur compounds from the smokestacks mix with water in the clouds and form dilute sulfuric acid that rains down on the earth. Selenium is used in photocopying and in the treatment of dandruff, acne, eczema, and other skin diseases. Tellurium is used for glass tinting and as an antiknock compound for gasoline. In printing and photography equipment, polonium is used in devices that ionize the air to eliminate accumulation of electrostatic charges. Polonium is also a thermoelectric power source in space satellites.

Group VIIA: Halogens

Group VIIA elements carry the name **halogens**,* because they form salts when they react with active metals. Because these elements are so reactive, they are difficult to obtain in their elemental forms. The halogens are fluorine, chlorine, bromine, iodine, and astatine.

Fluorine's electronegativity is greater than that of the other halogens; thus this element gains electrons in chemical reactions more readily than the other group VIIA elements. Because of its higher reactivity, fluorine was one of the last halogens to be discovered. Chlorine, the second halogen on the periodic table, was first recognized in 1771 when someone combined hydrochloric acid (HCl) and manganese oxide (MnO_2). The greenish yellow gas that escaped was named *chlorine*, from the Greek word for "green." The next element in the family, bromine, stands alone as

*halogen: *halo* (salt) + *gen* (producer, former)

the only nonmetallic liquid at room temperature. Its irritating, and even poisonous, vapor has a very pungent odor. For this reason it was named after the Greek word *bromus*, meaning "stench." Iodine was discovered when seaweed treated with concentrated sulfuric acid gave off a violet-colored vapor that crystallized when it was cooled. Joseph Gay-Lussac, a French chemist and physicist, soon identified the substance as an element and named it after the Greek word for "violet." Astatine, whose name means "unstable," is a highly radioactive element: it has no stable isotopes. It was first isolated in 1940 when it was synthesized from bismuth through a nuclear reaction. Astatine is the only metalloid in the halogen family. All other halogens are nonmetals.

Physical Properties. The halogens show a definite trend in their physical properties. As their atomic numbers increase, their densities, melting points, and boiling points increase and their colors exhibit increasingly darker hues. For example, fluorine is a pale yellow gas with a low density, chlorine is a greenish yellow gas, bromine is a deep, reddish brown liquid, and iodine is a grayish black crystalline solid.

Chemical Properties. Halogens have relatively high reactivities because of their strong electronegativities. Each element exists as a diatomic molecule when it is pure and forms an acid when it reacts with hydrogen. They all form salts when they react with metals.

Since nonmetals share or gain electrons when they react, the ones with strong attractions for electrons exhibit high reactivities. Fluorine is extremely reactive and ignites many substances on contact. The resulting reactions often take the form of spectacular releases of heat and light energy. When fluorine reacts with metals, it often forms a protective layer of metallic fluoride that prevents all the metal from reacting. Violent reactions result when fluorine reacts with hydrogen-containing compounds such as water or organic compounds. As might be expected, the chemical properties of the other halogens are similar to those of fluorine, though less reactive. Note this significant difference between nonmetals and metals: the most reactive nonmetals are those whose atoms are small; the most reactive metals are those whose atoms are large. This difference is due to the fact that metals usually react by losing electrons and nonmetals usually react by gaining them. Large metal atoms can lose electrons easily, but larger nonmetal atoms do not gain electrons as easily as the smaller family members do.

Uses. The human body benefits from small amounts of halogens, but larger doses are harmful. Fluorine compounds help develop decay-resistant teeth, and chlorine kills algae and bacteria in drinking water and swimming pools. Even iodine has a use in

the body: small amounts keep the thyroid gland working properly.

Most industrial uses of halogens are obscure, but some items have become familiar household products. Teflon-coated cookware utilizes the nonstick properties of a fluorine compound. Aerosol cans once contained fluorocarbons that served as propellants. Because these compounds were suspected of decomposing the ozone layer in the upper atmosphere, most products now use other propellants. Laundry bleach consists of a chlorine compound. This is why a load of wash with bleach may smell somewhat like a swimming pool. Chlorine dioxide is used to bleach wood pulp in paper production. Chlorine is also a component of table salt (NaCl). Bromine is used in photographic compounds and in natural gas and oil production. Iodine is important from a medical standpoint: lack of iodine causes stunted growth and goiter. Iodine is also used with alcohol as a disinfectant and as an oxidizing agent. Astatine has no primary uses since it is highly carcinogenic.

Group VIIIA: The Noble Gases

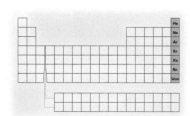

The group VIIIA elements were formerly called inert gases but now carry the name **noble gases**. They normally do not react with other elements. Lord Raleigh discovered one of these elements by carefully measuring the constituents of air. He noticed that the nitrogen he isolated from the atmosphere had more mass than the nitrogen he separated from pure ammonia. Suspecting that the atmospheric nitrogen contained some unknown substance, he separated the residual gas and found that it would not react with other elements. Because of its chemical sluggishness, this gas was called argon, meaning "the lazy one." This was the first noble gas to be discovered. More noble gases were discovered, some in significant amounts in the atmosphere. Helium was first discovered in a spectrogram of the sun. Neon was discovered during a purification of oxygen. Scientists also isolated and identified krypton, meaning "hidden element," xenon, meaning "stranger," and radon. Radon is the heaviest of the noble gases and is a gaseous product of radioactive decay.

Physical Properties. All noble gases are colorless, odorless, and tasteless. Extremely low boiling points and freezing points indicate that the individual atoms of these gases have little attraction for each other.

Chemical Properties. As their name implies, noble gases tend to be separate from the other elements. After many attempts to get noble gases to react with other elements, scientists have been able to force only xenon, radon, and krypton to bond. These elements formed several short-lived compounds such as xenon difluoride (XeF_2), xenon trioxide (XeO_3), krypton tetrafluoride (KrF_4), and radon fluoride (RnF).

Uses. Balloonists prefer helium as their source of lifting power because it has an extremely low density and does not react with oxygen. This may seem unimportant, but helium's inertness rules out the possibility of a disastrous fire. Deep-sea divers also use helium to dilute the amount of nitrogen in the air they breathe. The high pressures at great depths can force nitrogen to be absorbed by the lipids in the body, leading to nitrogen narcosis. Nitrogen narcosis affects primarily the brain and nervous system because of their high lipid content. Symptoms of nitrogen narcosis vary from light-headedness and numbness to convulsions and unconsciousness. Many colored lights that advertise along busy streets consist of glass tubes filled with neon. When an electric current passes through the gas, it gives off a bright orange light. Fluorescent lights contain a mixture of argon and mercury vapor. Incandescent light bulbs contain inert gases, but not because these gases give off light. The ability of noble gases to resist a reaction makes them perfect candidates for surrounding the fragile filament to keep it from reacting with oxygen. An isotope of radon is used to treat malignant tumors. Liquid neon is used as a refrigerant. Krypton is used alone or in combination with argon and neon in incandescent bulbs. Xenon is used primarily in lighting devices such as high-speed photographic tubes.

The Goodyear blimp and deep-sea divers both depend on helium.

Section Review Questions 5C

1. Explain why hydrogen is often considered a family by itself.
2. Which element has the greatest predominance in chemistry, as well as an influence in other branches of science? Explain.
3. Which element
   a. is predominant in the earth's crust?
   b. is predominant in the earth's atmosphere?
4. What is the one major benefit of the noble gases? Why?
5. Given the elements Ag, Al, Ar, Au, C, Ca, Cl, Cu, Fe, Hg, O, Pb, U, and W, which
   a. is used as a source of electricity in nuclear power plants?
   b. fills the tubes of fluorescent lights?
   c. helps your body construct strong bones?
   d. can form diamonds, graphite, or charcoal?
   e. is used to disinfect the water in swimming pools?
   f. was once included among the crown jewels?

Chapter Review

Coming to Terms

periodic table
periodic law
group or family
period or series
main group
metal
metalloid
nonmetal
lanthanide series
actinide series
North American Convention Periodic Table
European Convention Periodic Table
IUPAC Periodic Table
atomic radius
electrostatic attraction
ionization energy
electron affinity
electronegativity
descriptive chemistry
metallic hydride
alkali metal
bridge relationship
alkaline earth metal
transition metal
inner transition metal
paramagneticism
post-transition metal
semiconductor
oxide
sulfide
halogen
noble gas

Review Questions

1. Identify the chemist who
   a. is given credit for developing the modern periodic table.
   b. formulated the concept of triads.
   c. proposed that elemental properties varied in octaves.
   d. predicted the existence of several "missing" elements.
   e. determined atomic numbers of many elements.
   f. devised the commonly used electronegativity scale.
2. What was wrong with Döbereiner's classification of the elements? What good was it?
3. What is the purpose of the periodic table?
4. Why were several elements in the wrong places in Mendeleev's table? How was the problem corrected?
5. Identify each of the following elements as an actinide, an alkali metal, an alkaline earth metal, a halogen, a lanthanide, a metalloid, a noble gas, or a transition metal.
   a. lithium
   b. antimony
   c. bromine
   d. tungsten
   e. iron
   f. cesium
   g. cerium
   h. calcium
   i. argon
   j. uranium
6. Give two names for horizontal rows in the periodic table.
7. Give two names for vertical columns in the periodic table.
8. What is the general location of metallic elements in the periodic table? of nonmetals? of metalloids?
9. Use the periodic table to complete the following chart.

Name	Symbol	Atomic Number	Period	Family
cadmium				
		56		
	Sn			
			4	IA

10. Contrast the three different forms of periodic tables.
11. Use the periodic table to give the electron configurations of silicon, germanium, and calcium.
12. Use the periodic table to determine which elements have the following electron configurations:
    a. $1s^2\ 2s^2\ 2p^6\ 3s^1$ b. $1s^2\ 2s^2\ 2p^6\ 3s^2\ 3p^6$
13. Explain why sodium ions are smaller than sodium atoms and why chlorine ions are larger than chlorine atoms.
14. Of the eighteen elements in the fourth series, which one
    a. has the largest atomic radius?
    b. has the smallest ionization energy?
    c. has the largest electron affinity?
    d. has the lowest electronegativity?
    e. has the highest electronegativity?
    f. is the most reactive metal?
    g. is a semiconductor?
    h. is one of the least reactive elements?
15. Tell which of the stable alkaline earth metals (Be, Mg, Ca, Sr, and Ba) fit the following descriptions. (Use Figures 5B-1 through 5B-4; one answer will be an exception to a rule, and you may need to extrapolate.)
    a. has the largest atomic radius.
    b. has the smallest ionization energy.
    c. has the largest electron affinity.
    d. has the smallest electronegativity.
16. Why were metals like gold, silver, and copper known in Old Testament times, but metals like sodium, aluminum, and potassium not discovered until recently?
17. Why was chlorine isolated before fluorine? Can the reason for this be applied to other elements? If so, to which ones?
18. Given the elements Al, Au, Br, Ca, F, H, He, Hg, I, K, Mg, Na, O, P, S, and Si, which …
    a. are gaseous elements at room temperature?
    b. are liquids at room temperature?
    c. are soft metals?
    d. is a constituent of table salt?
    e. is found in salt replacements?
    f. are responsible for making water hard?
    g. is a relatively unreactive solid metal?
    h. is found in glass and many minerals?
    i. is found in a compound called bauxite?
    j. glows in the dark?
    k. is called brimstone in the Bible?
    l. is the most electronegative element?
    m. helps keep thyroid glands working properly?
    n. is used for filling balloons?
    o. is a gas that can behave as an alkali metal?
    p. is a major constituent of stalagmites and stalactites?

6A How and Why Atoms Bond page 135
6B The Quantum Model and Bonding page 148
FACETS: Molecular Orbital Theory page 157

Chemical Bonds 6

When Atoms Stick Together

Few atoms exist by themselves. Powerful electrostatic attractions called **chemical bonds** link them with other atoms. This chapter addresses some important questions about chemical bonds. Why do they form? How do they hold atoms together? What are the properties of the resulting atomic alliances?

6A How and Why Atoms Bond

Why Bond?

The world is decaying. Dead limbs fall to the ground from the treetops; air escapes from tires if nails puncture them; and lava cools as it oozes down a volcano's slope. Objects do not gain energy by rising against gravity, compressing themselves, or getting hot spontaneously. We can deduce, therefore, that matter goes to low-energy, less reactive states. Atoms and molecules follow the same trend. The tendency for atoms to seek stable, low-energy states is the key to explaining why they bond. Most unbonded atoms have high-energy, low-stability electron configurations. Bonded atoms have achieved greater stability by losing some of their energy. Favorable bonding processes release energy—often in the form of heat or light.

Figure 6A-1 ▪ Bonding Processes

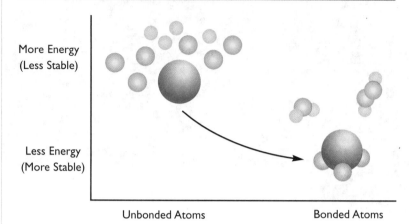

Favorable bonding processes produce low-energy, stable molecules from high-energy, unstable atoms.

Atoms seek a certain arrangement of electrons, with eight valence electrons being the most favorable. Any arrangement that provides less than eight valence electrons is generally unsatisfactory. The tendency for atoms to react with other atoms in such a way that they have eight outer-level electrons is called the **octet rule.** This rule serves as a general guide to chemical bonding processes; however, it does *not* apply to the d and f block elements. How do atoms gain a full octet? They can gain, lose, or even share electrons.

Sample Problem State several ways in which sodium atoms and oxygen atoms could get eight outer-level electrons.

Solution
Sodium, with one outer-level electron (Group IA), could get eight outer-level electrons by gaining seven electrons. Another way would be to lose the one valence electron in the third energy level, leaving the second energy level with its eight electrons as the outermost energy level. Oxygen atoms (Group VIA), with six valence electrons, would likely gain two electrons, since the loss of oxygen's six valence electrons would leave only the first energy level, which contains only two electrons.

Bonding processes rearrange the smallest number of electrons possible. Atoms with fewer than four valence electrons tend to lose electrons. Atoms with more than four tend to gain electrons.

Bond Character: What Kind of Bond Am I?

The way atoms get an octet depends on the types of atoms that are involved—particularly on their electronegativities. Three combinations of atoms can be bonded together: metals and nonmetals, nonmetals and nonmetals, and metals and metals. In the first case, highly electronegative nonmetals greatly attract electrons toward their nuclei. The valence electrons of the metal transfer to the nonmetal. The metal then becomes a cation, the non-metal becomes an anion, and an ionic bond forms between the two. In the second case, the large electronegativities of both atoms work against each other so that neither atom gives up its electrons. The electrons are shared between the atoms in a covalent bond. In the third case, neither metal atom strongly attracts the electrons that are involved in the bond. A communal sharing of electrons called a metallic bond occurs.

Actually, atomic bonding is not this straightforward. Elemental sodium represents a case of purely metallic bonding. Chlorine gas, Cl_2, is an example of a purely covalent bond and CsF, cesium fluoride, is a classic case of ionic bonding. Yet, a molecule such as BeO, beryllium oxide, possesses all three bond types, though the ionic type is dominant.

It is easy to think of bonds as falling into one of the categories described previously—ionic, covalent, or metallic—as if they each have very distinct boundaries. Although it is possible to have bonds that are 100 percent covalent if they are between two of the same atoms, virtually no bond can be 100 percent ionic. That would imply a *total* transfer of electrons from one atom to another, with no attraction for them whatsoever by the donor atom. A more accurate way of referring to bonds in compounds is to designate them as *predominantly* covalent or *predominantly* ionic. Such a distinction is often made based on differences in electronegativity values between atoms. For simplicity's sake, we will categorize a bond as metallic, covalent, or ionic when that is its predominate characteristic. However, you should be well aware that many covalent bonds have some ionic character, and many ionic bonds have some covalent character.

Table 6A-1 Types of Bonds

Combination	Electronegativities	Electron Action	Predominant Type of Bond
metal/nonmetal	small/large	transfer	ionic
nonmetal/nonmetal	large/large	tight sharing	covalent
metal/metal	small/small	loose sharing	metallic

Sample Problem Predict the predominant type of bond that will form when the following atoms bond.
  a. N and O   **b.** Ag and Cu   **c.** Cs and F

Solution
a. Both of these elements are nonmetals with large electronegativities. The electrons will be tightly shared in a covalent bond.
b. Both of these atoms are metals with small electronegativities, so they bond metallically.
c. Cesium's electronegativity is extremely small, whereas fluorine's is extremely large. Together they form an ionic bond.

Ionic Bonds: Transferring Electrons

Picture the way in which metals and nonmetals react with each other. Before reacting, nonmetals are just short of having full octets of electrons in their outermost energy levels. Metals have one to three electrons beyond their last full energy levels. Both kinds of atoms can obtain an octet through a simple transfer of electrons. Nonmetal atoms can gain the electrons they need from metals, and metal atoms can lose their valence electrons and use the full energy levels beneath as their octets. When electrons transfer, ions form. The electrostatic attractions between oppositely charged ions in a solid are called **ionic bonds.**

Consider how sodium and chlorine atoms form sodium chloride (table salt). Neither atom has an octet before it bonds: sodium has one valence electron, and chlorine has seven. In the reaction that occurs, chlorine acquires a full octet by gaining sodium's one valence electron. In doing so, it becomes a negative ion. Sodium's loss of its valence electron exposes the full second energy level and produces a positive ion. Both atoms benefit because their electron configurations become more stable—both achieve an octet. Since opposite charges attract, the two ions stick together. Because compounds are electrically neutral, the net electric charge of all the ions must be zero. Therefore one Na^+ ion combines with one Cl^- ion.

Atoms before bonding:	Na˟ ·C̈l:
Ions after transfer:	$Na^+ \longrightarrow [:\ddot{C}l:]^-$
Resulting compound:	$Na^+[:\ddot{C}l:]^-$ or NaCl

In the formation of calcium chloride, calcium atoms transfer two electrons to chlorine atoms. But it takes two chlorine atoms to receive the electrons from every one calcium atom.

Atoms before bonding: $Ca{:}$ $\cdot\ddot{Cl}{:}$
$\cdot\ddot{Cl}{:}$

Ions after transfer: Ca^{2+} $\begin{matrix}[{:}\ddot{Cl}{:}]^{-}\\ [{:}\ddot{Cl}{:}]^{-}\end{matrix}$

Resulting compound: $Ca^{2+}\,[{:}\ddot{Cl}{:}]_2^{-}$ or $CaCl_2$

Ionic Compounds: Opposites Attract

Do not think that only one or two ions are involved in ionic bonding. Millions of ions are formed, and each of these ions interacts with its neighbors. Each positive ion attracts the negative ions in its immediate vicinity. Each negative ion in turn attracts all the neighboring positive ions. As rows and columns of these alternating ions are added, a three-dimensional pattern develops. Solids whose particles are arranged in orderly patterns are called **crystals.** Once these crystals are formed, energy is required to separate them. This energy is called **lattice energy.** Every time you dissolve table salt (NaCl) in water, you are separating the crystals and expending lattice energy, resulting in an endothermic reaction. Add a large amount of salt to a glass of water. As the salt dissolves, you will feel the glass get cooler as the surroundings provide the lattice energy needed to separate the salt crystals.

Sodium and chloride ions pack together in a simple repeating pattern. Each sodium ion has chloride ions on its six sides. The formula for another ionic compound, calcium fluoride, is CaF_2. There must be two fluoride ions for every calcium ion that is present in the crystal. Scientists have found that in this case the ions alternate, but not in the same simple pattern that sodium chloride uses.

Figure 6A-2 ▪ Crystalline Structure NaCl and CaF$_2$

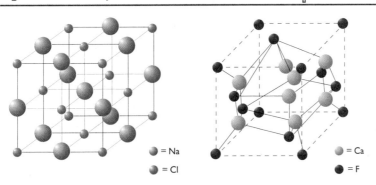

NaCl and CaF_2 both have ionic bonds, but their crystalline structures are different.

The exact way that the ions pack together depends on their charges and their sizes. The ions position themselves to be as near as possible to oppositely charged particles and as far away as possible from similarly charged particles. Ions take the most stable, lowest energy positions possible.

Each cation and anion in an ionic compound interacts with all its neighbors. The attractions between the positive and negative ions hold the compound together. Since the individual ions interact with all of the ions in the substance equally, the term *molecule* is not appropriate for ionic compounds because there are no separate, distinct units. There can be no justification for picking one Na^+ and one Cl^- ion out of a large salt crystal and calling it a sodium chloride (NaCl) molecule. It would be just as valid to form a Na_2Cl_2 or a Na_7Cl_7 molecule. Instead of using the term *molecule*, scientists use **formula units,** which tell the relative numbers of atoms in a compound by giving the simplest whole-number ratio between the atoms in the compound. The formula unit for sodium chloride contains one sodium and one chlorine ion because there is a one-to-one relationship between the ions. Calcium fluoride's formula unit contains one calcium and two fluorines because there are two F^- ions for every Ca^{2+} ion.

Ionic bonds give the compounds they form some very distinct properties. Since ions are held in place by strong electrostatic forces on every side, the resulting compounds are hard solids. The melting points of many of these compounds are 800°C, or

Large salt crystals are composed of many smaller salt crystals.

higher, illustrating just how strong these bonds are. Because of their orderly structures, crystals of ionic compounds can usually be split, or cleaved, along a straight line.

Ionic compounds are often soluble in substances such as water. They are also good conductors of electricity as long as they are not in the solid state; melting or dissolving the crystals helps free the ions so that they can move. Normally, ions bound in solid crystals cannot carry electrical current. However, not all crystalline materials are poor conductors. Superconductors are crystalline structures that allow the flow of electrical current without resistance below a certain critical temperature.

Covalent Bonding: Sharing Electrons

In general, nonmetals have large electronegativities. When they bond with each other, neither one will release the electrons needed for there to be ionic bonding. Instead of transferring electrons to gain a full outer energy level, they share them. Bonds consisting of one or more shared pairs of electrons are called **covalent bonds.**

Chlorine gas molecules use covalent bonds to hold two chlorine atoms together. Before bonding, each atom has seven valence electrons. Each needs one more for a full octet. Since both atoms have the same electronegativity, neither one can pull electrons from the other to itself. Instead, both atoms share one of their outer-level electrons. By sharing a pair of electrons, each chlorine atom can effectively have eight outer-level electrons. Because both nuclei strongly attract the shared pair of electrons, the molecule stays together. This covalent bond is often represented by a single line and is called a single covalent bond. We can represent this process by using electron dot symbols:

$$:\ddot{Cl}\cdot + {}_\times^\times\!\ddot{Cl}{}_\times^\times \longrightarrow :\ddot{Cl}{}_\times^\times\!\ddot{Cl}{}_\times^\times \text{ or } Cl-Cl$$

Hydrogen atoms also bond covalently. They do not seek an octet, since the first energy level holds only two electrons. They fill the first energy level when they bond.

$$H\cdot + {}^\times H \longrightarrow H{}_\times^. H \text{ or } H-H$$

Whenever nonmetal atoms bond together, they can be expected to have covalent bonds. Hydrogen and oxygen bond to form water. The oxygen lacks two electrons, so it shares the electrons of two hydrogen atoms. Each hydrogen gains its second valence electron in the process.

$$H^\times + H^\times + \cdot\ddot{O}: \longrightarrow \underset{H}{H{}_\times^.\ddot{O}:} \text{ or } \underset{H}{H-O}$$

Double covalent bonds form when atoms need two additional electrons. Sulfur monoxide (SO) molecules contain a double bond between an oxygen and a sulfur atom. The atoms share two pairs of electrons in a double covalent bond. **Triple covalent bonds** form when atoms share three pairs of electrons. The nitrogen molecules in the air consist of two triply bonded nitrogen atoms.

$$:\!\ddot{S}\cdot \;+\; {}^{\times\!\times}_{\times}\!\ddot{O}{}^{\times} \longrightarrow \;:\!\ddot{S}\!:\!\!:\!\ddot{O}{}^{\times}_{\times} \text{ or } :\!\ddot{S}=\ddot{O} \text{ or } S=O$$

$$:\!\dot{N}\cdot \;+\; {}^{\times}\!\dot{N}{}^{\times}_{\times} \longrightarrow \;:\!N\!:\!\!:\!\!:\!N{}^{\times}_{\times} \text{ or } N\equiv N$$

The following set of rules can be a helpful guide for writing electron-dot structures, especially when the structures are complex.

1. Sharing occurs when one nonmetal bonds with another (e.g., C–O), or when two identical nonmetal atoms bond (e.g., Cl–Cl).
2. Each atom generally reacts so as to form enough covalent bonds to have 8 electrons in the valence shell, including shared and unshared electrons. Remember the octet rule from Chapter 4. Hydrogen is the one exception.
3. Electrons are normally shared in pairs.
4. Hydrogen will share only one pair of electrons; thus H can never be the central atom in a covalent bond in a polyatomic molecule. The closer the atom is to the center of the periodic table (Group IVA), the more likely it is to be the central atom.
5. The number of shared pairs of electrons is frequently equal to the difference: 8 – family number. For example, 8 – VI (or 8 – 6) = 2 electron pairs shared.
6. There are a few exceptions to these rules that will be discussed as we go along.

Steps for drawing electron-dot (Lewis) structures:

1. Write the symbol for each element in the compound, showing the valence (or outer energy level) electrons with dots, circles, or Xs.

    Phosphorous is $\cdot\ddot{P}\cdot$
    Sulfur is $\cdot\ddot{S}:$
    Hydrogen is   H·
    Oxygen is $:\!\ddot{O}\cdot$

2. Decide how many pairs need to be shared to give 8 electrons total in the valence shell. (Two for hydrogen)

Phosphorous needs (8 − 5) = 3.
Sulfur and oxygen need (8 − 6) = 2.
Hydrogen needs (2 − 1) = 1.

3. Put the elements together with the less electronegative nonmetals surrounded by the other atoms.

For PH$_3$
```
        H
      H P H
```

For SO$_2$
```
      O S
         O
```

4. Add all the electrons: PH$_3$ = 5 + 3 × (1) = 8 total
(P = 5, 3Hs = 3, or 8 total)

(The symbols • and x are used simply to show where the electrons originated.)

$$H \overset{\overset{H}{x}}{\underset{..}{P}} H \quad \text{or} \quad H-\underset{|}{\overset{H}{P}}-H$$

SO$_2$ has 6 electrons for each atom, or 18 total; so possible structures could be:

:Ö:S:Ö: or   :Ö:S:Ö: or   :Ö:S:Ö:

However, none of these is satisfactory for SO$_2$, since in each example one of the atoms has only 6 electrons total. Step 5 is needed to remedy this problem.

5. If necessary, move *pairs* of electrons until every atom has 8 shared or unshared electrons. (Remember—only 2 for hydrogen.)

Redraw SO$_2$ from the first picture:

:Ö::S:Ö: or   :Ö=S−Ö: Either is correct.

6. Sharing 1, 2, or 3 pairs between two atoms is acceptable. See SO$_2$ in the above example.

7. You may move electrons liberally; however, always check to see that
   a. every atom has 8 electrons total (2 for hydrogen).
   b. the *total* number of electrons is correct for the molecule.

For example, :Ö::C:Ö: looks fine for CO$_2$, but it has too many electrons. There are 9 electron pairs. The correct electron dot symbol would be

:Ö=C=Ö: or   :Ö::C::Ö:

Now do NO_3^-. First determine the total number of electrons for the ion. N = 5; 3 × O = 3 × 6 = 18; -1 charge = 1 additional electron. Total electrons = 24.

(Step 3) (Step 4) Total e⁻ = 24, but N has only 6.

```
      O              :Ö:
    O N O         :Ö:N:Ö:
```

(Step 5) reshuffle any pair from O to N :Ö:N:Ö: with :Ö: above (This is no better because one O now has only 6 e⁻.)

Do Step 5 again, but this time *double*-bond the N with one of the Os, since both atoms need this pair of electrons in order to have 8 e⁻ in the outer energy level. Since the double bond can be with any of the oxygen atoms, all of these structures are correct.

$$\left[\begin{array}{c} :\ddot{O}: \\ | \\ :\ddot{O}-N=\ddot{O}: \end{array} \right]^- \text{ or } \left[\begin{array}{c} :\ddot{O}: \\ | \\ :\ddot{O}=N-\ddot{O}: \end{array} \right]^- \text{ or } \left[\begin{array}{c} \ddot{O} \\ \| \\ :\ddot{O}-N-\ddot{O}: \end{array} \right]^-$$

Sample Problem Draw the electron-dot structures for
  a. $CHCl_3$ **b.** SiO_2

Solution

a. The electron-dot symbols for carbon, hydrogen, and chlorine are as follows: ×C̤×   H·   ·C̈l:
   Carbon, with its smallest electronegativity and greatest need for bonds (8 − 4 = 4), will likely be at the center. Since H and Cl each need one bond (2 − 1 = 1 and 8 − 7 = 1, respectively), they would not bond to each other, or they could not then bond to carbon. The electron-dot structure is therefore

```
         H                    H
         |                    |
   :Cl:C:Cl:      or    :Cl - C - Cl:
      :Cl:                    |
                            :Cl:
```

   The H atom can be placed on any side of the carbon atom. Each atom has an octet and hydrogen has a duet, as required.

b. The electron-dot symbols for silicon and oxygen are ×S̤i×   ·Ö:
   Silicon needs 8 − 4 = 4 shared pairs of electrons, and each oxygen needs 8 − 6 = 2 shared pairs. Putting the less electronegative element in the center of the two atoms gives

   :Ö:Si:Ö:

However, none of the atoms has an octet, or the correct number of shared pairs. If each oxygen double-bonds its single electron with silicon's single electrons, the result is

$$:\!\ddot{O}\!:\!\!\times\!\!Si\!\!\times\!\!:\!\ddot{O}\!: \quad \text{or} \quad :\!\ddot{O}\!=\!Si\!=\!\ddot{O}\!:$$

We see now that all requirements are met, so this must be the correct structure.

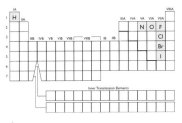

Figure 6A-3

Diatomic Molecules: What a Pair!

The atoms of halogens (F, Cl, Br, and I), hydrogen, oxygen, and nitrogen are not stable by themselves. In their uncombined forms, these elements are not made of individual atoms but of two-atom, covalent molecules. Molecules that contain two atoms are called **diatomic molecules.** The diatomic molecules that are formed from the elements listed above are F_2, Cl_2, Br_2, I_2, H_2, O_2, and N_2. The halogens and hydrogen have single bonds, oxygen has a double bond, and nitrogen is triply bonded. Be familiar with this list; in future chapters it will be necessary to know that atoms of these elements automatically form molecules when they exist in their uncombined elemental form.

Polyatomic Ions: Covalent, Yet Ionic

Some atoms combine to form multi-atom ions. These groups of atoms have electrical charges because electrons must be gained or lost in order for every atom to have a complete octet. Just as a Cl⁻ ion has one more electron than a chlorine atom, an OH⁻ ion has one more electron than the oxygen and hydrogen atoms that participate in the covalent bonding.

$$:\!\ddot{Cl}\!\cdot + \overset{\text{one}}{\underset{(\times)}{\text{electron}}} \rightarrow [:\!\ddot{Cl}\!:\!\times]^- \text{ or } Cl^-$$

$$:\!\ddot{O}\!:\!H + \overset{\text{one}}{\underset{(\times)}{\text{electron}}} \rightarrow [:\!\ddot{O}\!:\!H]^- \text{ or } OH^-$$

The electron-dot structures of polyatomic ions can be written according to the same rules that were used for covalent compounds, as you saw from the example of NO_3^- ion discussed earlier. Just remember that ions have different numbers of electrons than the uncharged atoms have. If an ion has a +1 charge, there is one *less* electron to work with. If an ion has a -2 charge, two *extra* electrons are present. Square brackets should be placed around electron-dot structures of the ions for clarity, and the charge written as a superscript outside of the brackets.

Sample Problem Write the electron-dot structure of the ammonium ion $[NH_4]^+$.

Solution
Logical structure:

```
          H
      H   N   H
          H
```

When counting electrons, remember that the +1 charge indicates that there is one less electron than would be calculated based on valence electrons. The nitrogen atom has five valence electrons, and each hydrogen has one. The total number of electrons (nine) is decreased by one to account for the +1 charge. Drawing one pair of electrons for each bond results in the following equivalent structures:

$$\left[\begin{array}{c} H \\ H \overset{\times}{\underset{\times}{N}} H \\ H \end{array} \right]^+ \text{ or } \left[\begin{array}{c} H \\ H : \overset{..}{N} : H \\ H \end{array} \right]^+ \text{ or } \left[\begin{array}{c} H \\ | \\ H - N - H \\ | \\ H \end{array} \right]^+$$

Polyatomic ions act as units in many chemical reactions. Their stable octets of electrons allow them to go through reactions and form compounds without splitting up. They can even bond with other oppositely charged polyatomic ions to form ionic compounds. When polyatomic ions bond to form ionic compounds, the ratio of positive ions to negative ions must still be such that the overall charge on the compound is zero. When sodium ions (Na^+) and nitrate ions (NO_3^-) combine, they do so in a one-to-one ratio. If sodium and sulfate ions (SO_4^{2-}) combine, there must be two Na^+ ions for every SO_4^{2-} ion.

Metallic Bonding: Share and Share Alike

What holds the atoms in a piece of metal together? Ionic bonds certainly do not form; all metal atoms have small electronegativities. None of them can pull electrons from other atoms to form ions. What about covalent bonds? The answer is no again. Metal atoms do not have enough valence electrons to share them with other needy atoms. A different type of bond must form. The theory that explains how metal atoms stick together also explains why metals have the properties they have.

The theory used to explain the characteristics of metals is called the **electron-sea theory** (or free electron theory). In this theory, metals are pictured as an extended array of positive ions surrounded by unattached electrons—a "sea" of electrons. Remember that metals easily lose their valence electrons to form cations. The "lost" valence electrons are mobile and are now shared by all of the cations; they are said to be **delocalized.** The metallic bond is thus a modified form of the covalent bond. Electrons are shared, but not between a mere one, two, or three atoms. All the atoms share them. The model for metallic bonding can be likened to a "sea" of electrons surrounding "islands" of positive charges.

If we compress H_2 gas that is below 33 K, it will eventually liquefy and then solidify into a molecular solid in which the H_2 molecules are very weakly coupled. Applying further pressure causes the H_2 molecules to move closer together. The coupling becomes stronger, and the hydrogen becomes metallic and will conduct electricity. Apparently, it is the close-packed nature of metal ions that allows them to share electrons freely.

The electron-sea theory explains many characteristics of metals. The electrons in the sea are free to carry electrical current. The luster of metals can be explained by the combination of the ideas of spectroscopy and free electrons. Since electrons are delocalized, or relatively free, they are able to absorb and emit many wavelengths of light rather than being restricted to a few specific energy levels. All the wavelengths of light, when reflected together, make up the shiny luster. Metallic bonds allow pieces of metal to be formed into new shapes without shattering. The widespread sharing of electrons allows the cations to shift when under stress. Their positions can be "rearranged" by a swift blow from a hammer.

Section Review Questions 6A

1. Contrast the three major bond types.
2. Predict the predominant type of bond that will form when the following atoms combine:
    a. Al and F   b. Hg and Au   c. H and I
3. Draw the electron-dot structures for the following:
    a. $[PO_3]^{3-}$ b. $COBr_2$ c. $[SeO_4]^{2-}$ d. $SiHCl_3$

6B The Quantum Model and Bonding

So far, electron-dot structures have been used to study covalent bonds. These simple sketches show the bonding process clearly, but they are so simple that they leave out valuable information about the compounds that form. For instance, they can tell nothing about the three-dimensional shape of a molecule. Is water

$$H\!:\!\ddot{O}\!:\!H \quad \text{or is it} \quad H\!:\!\ddot{O}\!: \atop H \quad ?$$

From the electron-dot structures alone, there is no way to tell. For a more complete description of chemical bonds—especially covalent bonds—the quantum model must be used.

Orbital and Bonding: The Quantum Model Strikes Again

As we saw in Chapter 4, in the quantum model, electrons are said to exist in various types of sublevels. These sublevels, named *s*, *p*, *d*, and *f*, contain varying numbers of orbitals. An **orbital** is a region in which the probability of finding an electron is great. Each orbital can hold a maximum of two electrons.

In this model, a covalent bond is pictured as an overlap of partially filled orbitals. When two orbitals mesh, the overlapping region is available to both nuclei. Effectively, both atoms have another electron. The term **valence bond theory** is given to the idea that covalent bonds are formed when orbitals of different atoms overlap.

Bonds: When Orbitals Overlap

The hydrogen molecule (H:H) is a familiar diatomic molecule. According to valence bond theory, the *s* orbital of each hydrogen atom overlaps with the other. A region of high electron probability forms between the two atoms. This area is shaded because the probability of finding an electron there is great.

Figure 6B-1a

A single sigma bond forms when *s* orbitals overlap end to end.

Not only can valence s orbitals overlap to form a single bond, but valence p orbitals can also overlap similarly. Fluorine (F₂) forms a single bond by the overlap of each atom's unfilled 2p orbital.

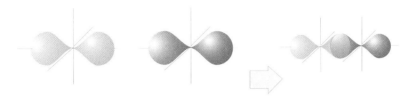

Figure 6B-1b

Sigma bonds can be formed when p orbitals bond end to end.

In both of these cases, there is an increased probability of finding the shared electrons between the two nuclei. Both of these bonds involve an end-to-end or "head-on" type of overlap of the orbitals and are known as *sigma bonds*. In addition, a sigma bond forms if an s and a p orbital overlap in an end-to-end fashion.

Double and triple bonds are formed when more than one set of orbitals overlap. In either case, one of the bonds is always due to the end-to-end overlap that is common in single bonds. The second bond is different. The *sides* of orbitals pointing in the same direction overlap. Note that a side-by-side overlap results in two regions of high electron probability. Do not let the two regions confuse you. They still represent one bond. This type of side-to-side bond is called a *pi bond*.

Figure 6B-2

The second bond in a double bond forms when two orbitals overlap side by side.

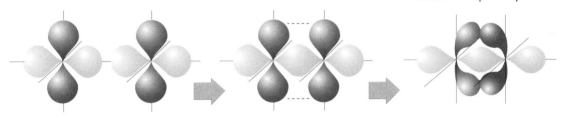

A triple bond is made up of one end-to-end overlap (a sigma bond) and two side-by-side overlaps (pi bonds). A second set of p orbitals is involved in the second side-by-side overlap.

Figure 6B-3

The third bond in a triple bond forms when another set of orbitals overlaps side by side.

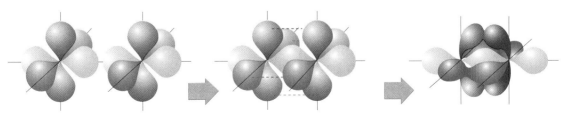

Hybridization:
What Atoms Go Through to Bond

Many observations forced chemists to make bonding theories more complicated than they originally were. One of these observations concerns the bonding of carbon. The ground state electron configuration shows that carbon has two partially filled p orbitals.

It would seem probable that carbon would fill these orbitals by forming two bonds, but it does not. In the CH_4 molecule, carbon forms four equivalent bonds. A modification of bonding theory was necessary to account for this observation.

In order for carbon to bond this way, the electron configuration of the atom must change. It probably shifts from something like

$$\begin{array}{ccc} 1s & 2s & 2p \\ \uparrow\downarrow & \uparrow\downarrow & \uparrow\;\uparrow\;\_ \end{array} \quad \text{to} \quad \begin{array}{ccc} 1s & 2s & 2p \\ \uparrow\downarrow & \uparrow & \uparrow\;\uparrow\;\uparrow \end{array}.$$

The four singly occupied orbitals—one s and three p—combine to form four orbitals of equal energy. The process by which new kinds of orbitals with equal energies are formed from a combination of orbitals of different energies is called **hybridization.** Since one s and three p orbitals are involved in forming the new orbital type, they are known as sp^3 hybrid orbitals. Scientists believe that many other atoms besides carbon hybridize when they bond.

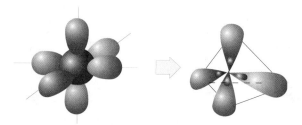

Figure 6B-4
The hybrid orbitals that form before atoms bond point in new directions and have different shapes from the original orbitals.

If a carbon atom forms bonds to three other atoms, its s and p orbitals will hybridize in such a way that one s and two p orbitals are involved, thus forming three sp^2 hybrid orbitals. The third p orbital that is "left over" (unhybridized) will overlap with another atom's orbital to form a pi bond. Similarly, when carbon bonds to two other atoms, it forms two sp hybrid orbitals: the two remaining p orbitals form two pi bonds to other atoms.

Bonding and Molecular Shape:
Keep Your Distance

The **Valence Shell Electron Pair Repulsion Theory** (abbreviated VSEPR, which is pronounced "vesper") assumes that regions of electron concentrations in molecules are arranged so that they are separated by the maximum distance possible. This makes sense, since electrons repel each other. The application of this rule is the key to determining the shapes of molecules.

The electron-dot structure can be used to determine the number of electron concentrations around an atom. An electron concentration can be an unbonded pair, a single bond, a double bond, or a triple bond. CH_4 has four electron concentrations around the carbon atom. CF_2O has three electron concentrations: one double bond and two single bonds.

sp^3 hybrid: tetrahedral 109.5°

```
      H                    
   H:C:H       :F:C::O:    
      H          :F:       
```

If an atom has four electron concentrations (sp^3 hybrid) around it, the orbitals point away from each other as much as possible. The orbitals point toward the four corners of a tetrahedron (a triangular pyramid), with each angle measuring 109.5°. This three-dimensional setup allows more space between the orbitals than any other arrangement does.

sp^2 hybrid: trigonal 120°

The hybridized orbitals around an atom with three electron concentrations (sp^2 hybrid) point toward the corners of a triangle. The angles between the orbitals all measure 120°, and all the orbitals lie in the same plane. This arrangement is called trigonal planar.

sp hybrid: linear 180°

The hybridized orbitals around an atom with two electron concentrations (sp hybrid) point in opposite directions, 180° from each other, and are called linear. Determining the shapes of molecules requires knowing the directions in which hybridized orbitals point and knowing which orbitals bond to other atoms. Consider the following molecules. All of them are sp^3 hybrids and have four regions of electrons around the central atom.

CH_4. The electron-dot structure shows that the molecule has four regions of electrons and must therefore have four hybridized orbitals that point to the corners of a tetrahedron. Hydrogen nuclei are located at the end of each orbital. The resulting shape is called **tetrahedral.**

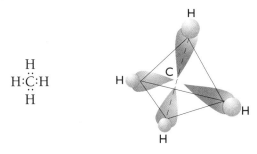

NH₃. Like CH₄, this molecule has four regions of electrons around the central atom. Unlike CH₄, it does not have an atom at the end of one of its orbitals. That orbital is occupied by an unbonded pair of electrons. This molecule's shape is not a full tetrahedron; it is **pyramidal.** The hybridized orbitals still point to the four corners of a tetrahedron, but only three of the corners are occupied, giving it a different shape from CH₄. Remember, the shape of a molecule is determined by the positions of the atoms that are bonded to the *central atom*, not by the position of the unbonded electron pairs.

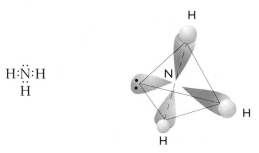

H₂O. Water molecules also have four regions of electrons. The shape of the molecule is determined by the locations of the two hydrogen atoms and the oxygen atom. The positions of the nuclei form a **bent** line. The angles between the bonds are about 105° since the unshared electron pairs exert more of a repulsive force than the bonded pairs do. However, you can still think of the water molecule as forming two legs of a tetrahedral.

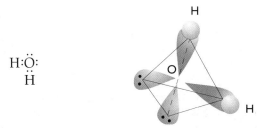

HF. The fluorine atom can be called the central atom. It has four electron concentrations around it. The single bond to the hydrogen atom results in a linear shape.

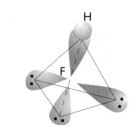

Table 6B-1 Molecular Shapes

Example		Number of Electron Regions	Number of Bonds	Diagram	Geometry	Structure
NH_4^+	Ammonium Ion	4 (sp^3 hybrid)	4		Tetrahedral	$\left[\begin{array}{c} H \\ H-N-H \\ H \end{array}\right]^+$
H_3O^+	Hydronium Ion		3		Pyramidal	$\left[H-\overset{..}{O}-H \atop H \right]^+$
H_2S	Hydrogen Sulfide		2		Bent 109.5° (Approx.)	$H-\overset{..}{\underset{..}{S}}-H$
HCl	Hydrochloric Acid		1		Linear	$H-\overset{..}{\underset{..}{Cl}}:$
NO_3^-	Nitrate Ion	3 (sp^2 hybrid)	3		Trigonal Planar	$\left[\begin{array}{c} :\overset{..}{O}: \\ \| \\ :\overset{..}{\underset{..}{O}}-N-\overset{..}{\underset{..}{O}}: \end{array}\right]^-$
SO_2	Sulfur Dioxide		2		Bent 120°	$:\overset{..}{\underset{..}{O}}-\overset{..}{S}=\overset{..}{\underset{..}{O}}:$
SO	Sulfur Monoxide		1		Linear	$:\overset{..}{S}=\overset{..}{\underset{..}{O}}:$
BeH_2	Berylium Hydride	2 (sp hybrid)	2		Linear	$H-Be-H$
CO	Carbon Monoxide		1		Linear	$:C\equiv O:$

The same basic principles hold true when determining the shapes of the sp^2 and sp hybrid compounds. Table 6B-1 provides examples for you to study. Remember the shape is determined by the position of the atoms that are bound to the central atom.

Sample Problem Predict the shapes of the following molecules.

a.  :Ö:Ö::Ö: b. 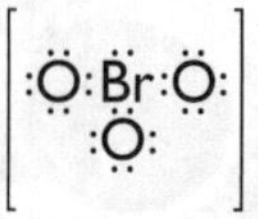 [:Ö:Br:Ö: / :Ö:]⁻   c. H:C:::N:

Solution
a. From the number of electron concentration regions, you know that the three concentrations of electrons (sp^2 hybrid) around the oxygen point 120° from each other. Since only two of the electron concentrations are involved in bonding to atoms, the bent 120° shape results.
b. Four regions of charge mean that hybridized orbitals point to the four corners of a tetrahedron (sp^3 hybrid). Since only three bonds have been formed, the ion has a pyramidal shape.
c. Two regions of electrons (sp hybrid) mean that the hybridized orbitals point in a straight line. This shape is called linear.

Polar Covalent Bonds: Unequal Partnerships

Nonmetal atoms with similar electronegativities form covalent bonds. But what about atoms with slightly different electronegativities? Do they transfer, or do they share electrons? The answer is that they share, but not equally.

The shared pair of electrons in a hydrogen chloride (HCl) molecule has a unique position. Compared to hydrogen, the chlorine atom is very electronegative. It pulls the shared pair closer to itself. The shifting of electrons results in a "semi-ionic" condition. The proximity of the electrons to the chlorine atom gives it a partial negative charge. Since the electrons are a greater distance from the hydrogen nucleus, it is left more exposed. Therefore, it has a partial positive charge. These are not full charges because there has not been a complete transfer of electrons from one atom to another. They are shared, although unequally.

Partial charges on molecules are shown by the lower-case Greek letter delta (δ). The hydrogen chloride molecule should be thought of as

$$\overset{\delta^+ \ \ \delta^-}{\text{H :}\ddot{\text{C}}\text{l:}}$$ instead of as completely covalent H:$\ddot{\text{C}}$l:

or completely ionic H$^+$ $[:\ddot{\text{C}}\text{l}:]^-$

Another way of showing the partial charges is with a crossed arrow. The head of the arrow points toward the negative area of the molecule—the more electronegative atom—and the cross is located near the positive area—the less electronegative atom.

$$\overset{\longmapsto}{\text{H :}\ddot{\text{C}}\text{l:}}$$

Objects with two different electrical ends, or poles, are said to be **polar.** Bonds with shared but shifted electrons (unequal sharing) are called **polar covalent bonds.** These bonds form whenever atoms with different electronegativities share electrons. All bonds formed between atoms of different elements are polar at least to some extent. The magnitude of the bond polarity depends on the difference in electronegativity of the two elements. The greater the difference in electronegativity, the more polar the molecule.

Dipole Moment: A Measure of Polarity

Does the presence of polar covalent bonds always result in a polar molecule? Though it may be tempting to think that this is true, another factor is involved. In order for a molecule as a whole to be nonpolar, there must be a symmetrical arrangement of polarity. It is helpful to visualize the concept of polar molecules as individuals or teams in a game of tug-of-war. (Figure 6B-5). If they are pulling on the ropes with equal strength in opposite directions (*i.e.*, symmetrical), there is no "winner." However, if the forces are unequal, or unsymmetrical, then there will be a movement of the object being pulled.

Figure 6B-5

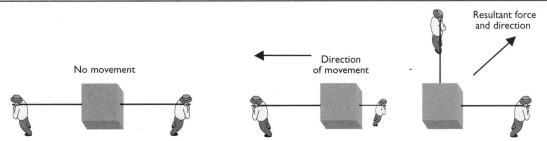

Polar bonds often make entire molecules polar. Linear hydrochloric acid molecules have a positively charged end and a negatively charged end. There is an unsymmetrical arrangement of polarity, resulting in a polar molecule. The tetrahedral carbon tetrachloride (CCl_4) has bonds that are quite polar, but the bonds are arranged symmetrically. As a result the molecule has no discernible regions of charge. The polar bonds cancel each other out. Each chlorine atom draws electrons away from the carbon atom. But since the chlorine atoms are arranged symmetrically around the carbon, no end of the molecule is more negative than another. The geometric shapes that are considered symmetrical are as follows: tetrahedral, trigonal planar, and linear. Both the bent and the pyramidal shapes are always unsymmetrical and result in polar molecules. Symmetrical shapes are nonpolar if all the outer atoms (*i.e.*, not the central one) have the same electronegativities: they are polar if the outer atoms are different. Table 6B-2 summarizes the guidelines given.

Table 6B-2 When Molecules Are Polar

Shapes for Four Regions of Electrons	Shapes for Three Regions of Electrons	Shapes for Two Regions of Electrons
Tetrahedral—if outer atoms have different electronegativities		
Pyramidal—always polar	Trigonal planar—if outer atoms have different electronegativities	
Bent (109.5°)—always polar	Bent (120°)—always polar	
Linear—if outer atoms have different electronegativities	Linear—if outer atoms have different electronegativities	Linear—if outer atoms have different electronegativities

Molecular Orbital Theory

The head researcher gathered his team of assistants together and informed them of an upcoming experiment. "According to the accepted theories of bonding, oxygen should form a colorless liquid that is not affected by a magnetic field. As you know, a molecule must have at least one unpaired electron before it can have a color or can respond to a magnetic field. The Lewis structure of O_2 predicts that all the electrons in the molecule are paired with other electrons. In our experiments, we shall liquefy oxygen and test the validity of these predictions. Our aim is to prove that the accepted Lewis structure for the O_2 molecule is indeed the correct one."

Soon the researcher and his team had pressurized and cooled a sample of O_2 enough to form a liquid. To their surprise, the liquid O_2 had a pale blue color and could be held in place by a magnetic field. What started as a simple verification of the accepted valence bond theory turned out to provide contradictory, disturbing data.

The lack of a completely satisfactory model of molecules has always frustrated chemists. Many different models have been proposed, but no single model fully explains all that is known about molecules. The Bohr model of the atom explains how atoms can lose and gain electrons to form ions. It also helps to explain how covalent bonds form and why molecules contain the atoms they do. Unfortunately, this solar-system model of the atoms cannot explain why molecules have the shapes they do.

The more advanced valence bond (VB) theory explains why molecules have the shapes they do. The orientation of the hybridized orbitals determines the general shape of the resulting molecules. Many chemists, however, are not satisfied. According to this theory, oxygen molecules should be colorless and unresponsive to magnetic fields. Since they are colored and they do respond, the VB theory must not be the final word in bonding theories.

A more recent and more complicated theory of bonding fills the gaps left by the VB theory. The molecular orbital (MO) theory holds that the orbitals of the individual atoms disappear when a molecule forms. Totally new orbitals form in their place. Each molecule has a unique set of orbitals. Some orbitals encircle two, three, four, or even more atoms and often the entire molecule. The MO theory ranks the resulting orbitals in order of increasing energy. Electrons fill low-energy molecular orbitals before they fill high-energy orbitals. The arrangement of electrons in these orbitals tells chemists whether bonds will form; whether they will be single, double, or triple bonds; and whether the bonds will contain unpaired electrons.

Liquid oxygen will be held by a magnetic field.

The most interesting idea of the MO theory is the concept of antibonding orbitals. When a molecule forms, two things can happen to the electrons in the atomic orbitals. Because the electrons behave like waves, combining orbitals can either reinforce or interfere with each other. When they reinforce each other, bonding orbitals form. Bonding orbitals are located between the nuclei. The electrons in these orbitals stabilize the molecule because they can be shared by both atoms. Antibonding orbitals form when atomic orbitals combine in an unfavorable manner. If the electron waves interfere with each other, an orbital forms on the outside of the molecule, far from the two nuclei. The electrons in an antibonding orbital spend little time between the two nuclei, so these antibonding orbitals destabilize the molecule.

The ability to predict color and magnetic properties of molecules is a major triumph of the MO theory.

All three models of molecules have a place in chemistry. Since none of them is completely right, scientists use the one that works best in a particular application. It would, of course, be better to have a single theory that would explain everything. But because human minds are limited and God's creation is complex, men do not have that luxury.

A **dipole moment** is a measure of polarity used to rate or compare the polarity of molecules and bonds. The Greek letter μ (mu) is used as the abbreviation for this quantity. The larger the value for μ, the more polar the bond or molecule.

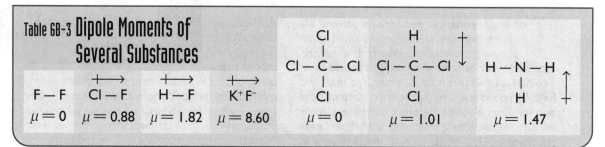

Table 6B-3 Dipole Moments of Several Substances

Sample Problem Determine whether the following molecules are polar. If they are, indicate the direction of their dipole moments with a crossed arrow.

a. CO_2 b. H_2S c. ClBr

Solution
The first thing you must do in each case is to draw the molecule's electron-dot structure. This can be used to predict its shape using the VSEPR theory. Then, knowing its shape, you can determine the polarity of the molecule.

a. CO_2 has the structure $\ddot{\text{O}}::\text{C}::\ddot{\text{O}}$. Because there are two regions of electrons, CO_2 is linear—a symmetrical shape. Since both outer atoms have the same electronegativity, they are "pulling" on the carbon equally, resulting in a nonpolar molecule.

b. H_2S has the electron-dot structure $\begin{smallmatrix}H:\ddot{S}:\\H\end{smallmatrix}$, from which we predict its shape to be bent (4 electron regions but only two are bonded to atoms). Since all bent shapes are unsymmetrical, H_2S is polar with the more electronegative sulfur pulling toward itself. The dipole moment is as follows:

$\begin{smallmatrix}H:\ddot{S}:\\H\end{smallmatrix}$ ↗

c. ClBr must be linear since it is a diatomic molecule. Since chlorine has a greater electronegativity than bromine (it is above Br in Group 17), it pulls electrons toward itself, resulting in the following dipole moment:

:C̈l:B̈r:
⟵┼

Section Review Questions 6B

1. Predict the shapes of the following molecules or ions.

   a.   b.   c.

2. Determine whether the following molecules are polar or non-polar. If they are polar, indicate the direction of the dipole moment with a crossed arrow.

   a. $COBr_2$ b. NI_3 c. SI_2 d. CF_4

3. What is VSEPR? Explain its effect on molecular shape.
4. Why was the hybridization bonding model introduced?

Chapter Review

Coming to Terms

- chemical bond
- octet rule
- ionic bond
- lattice energy
- crystal
- formula unit
- covalent bond
- double covalent bond
- triple covalent bond
- valence bond theory
- hybridization
- electron-sea theory
- delocalization
- Valence Shell Electron Pair Repulsion theory
- tetrahedral
- pyramidal
- bent
- sp^3 sp^2 and sp hybrids
- linear
- trigonal planar
- sigma and pi bonds
- polar
- polar covalent bond
- dipole moment

Review Questions

1. What is the general reason that atoms form bonds?
2. State the most probable way in which the following atoms could obtain eight outer-shell electrons.

   a. K
   b. Ca
   c. Ga
   d. Ge
   e. As
   f. Se
   g. Br
   h. Kr
   i. C

3. Identify the types of atoms (metal or nonmetal) in the following substances, and then tell whether the bonds are predominantly ionic, covalent, or metallic. Example: P_2O_5 combines nonmetal P and nonmetal O atoms; the bonds are covalent.
   a. NaCl
   b. bronze (an alloy of tin, copper, and zinc)
   c. CO_2
   d. $MgBr_2$
   e. brass (an alloy of zinc and copper)

4. Draw electron-dot structures of the atoms in the following compounds, the ions that result from the electron transfers, and the resulting ionic compounds. Example: NaCl

   Atoms before bonding: Na·   ·Cl:

   Ions from the transfer: Na^+ ⟶ [:Cl:]$^-$

   Resulting compound: Na^+ [:Cl:]$^-$

   a. LiI
   b. MgO
   c. $CaBr_2$
   d. $AlCl_3$
   e. SrF_2

5. Why is it said that ionic compounds do not consist of molecules?

6. Give the correct formula unit for each of the following ionic crystals.
   a. $K_{20}Br_{20}$
   b. $Al_{15}Cl_{45}$
   c. $Na_{12}(PO_4)_4$
   d. $Al_{24}O_{36}$

7. The following covalent compounds contain only single covalent bonds. Draw their electron-dot structures.
   a. H_2
   b. HCl
   c. CH_4
   d. CF_2Cl_2
   e. H_2S

8. The following covalent compounds or ions each contain at least one double or triple bond. Draw their electron-dot structures.
   a. H_2CO
   b. CS_2
   c. CN^-
   d. CO_2
   e. C_2H_4
   f. C_2H_2

9. Draw electron-dot structures of the polyatomic ions below. Example: SO_4^{2-}

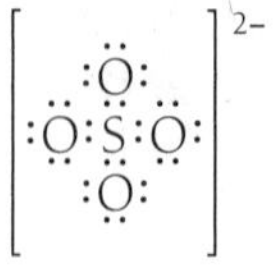

a. OH⁻
b. PO₄³⁻
c. CN⁻
d. ClO₄⁻
e. SO₃²⁻

10. Draw electron-dot structures for each of the following compounds containing polyatomic ions. Example: K₂SO₄

$$K_2^+ \left[:\!\overset{..}{\underset{..}{O}}\!:\!\overset{..}{\underset{..}{S}}\!:\!\overset{..}{\underset{..}{O}}\!: \right]^{2-}$$

a. NaOH
b. Na₃PO₄
c. KCN
d. Mg(ClO₄)₂
e. Li₂SO₃

11. Explain how the electron-sea theory accounts for the following properties of metals.
    a. electrical conductivity
    b. luster
    c. ductility and malleability

12. Describe how the orbitals in the s and the p sublevels hybridize when carbon atoms form bonds.

13. How many regions of electron concentrations exist around the central atoms of the molecules in problem 8?

14. Make simple drawings that show how the hybridized orbitals around an atom point when two, three, and four orbitals are present.

15. Predict the molecular shapes for the compounds in problem 7.

16. Predict the molecular shapes for the compounds in problem 8, sections a–d.

17. How many polar bonds are found in each of the molecules below? (Consider a double or triple bond as one bond.)
    a. H₂
    b. HCl
    c. CO₂
    d. CH₄
    e. CF₂Cl₂
    f. H₂S
    g. H₂CO
    h. C₂H₄
    i. C₂H₂
    j. CS₂
    k. CO

18. Which of the molecules in problem 17 are polar (i.e., have a dipole moment)?

161

7A Oxidation Numbers page 163
7B Nomenclature page 170
7C The Mole page 181
FACETS: Quantitative Analysis Solves a Mystery page 180

Describing Chemical Composition 7

Numbers, Names, and Something Called a Mole

Compounds can be effectively described by formulas, by names, and by measurements of their masses. Formulas serve as a convenient yet powerful shorthand system of describing compounds. Systematic names provide additional information about the composition of chemical compounds. Chemists use a concept called the mole to describe masses of compounds.

7A Oxidation Numbers: Help in Predicting Formulas

Chemists use **oxidation numbers** to keep track of electrons during bonding. These numbers tell whether electrons are gained, lost, or unequally shared. Proficiency with oxidation numbers enables chemists to predict the formulas of chemical compounds.

Rules for Assigning Oxidation Numbers

Oxidation numbers are essentially a way to track the relative changes in "possession" of electrons by elements that are reacting. These numbers are, in general, the result of assigning balance electrons to the more electronegative element in a bond, thereby giving it a negative oxidation number. The less electronegative element then has a positive oxidation number. The following set of rules governs the assigning of oxidation numbers to the elements in most compounds:

Rule 1 The oxidation number of free atoms (atoms in their uncombined states) and of atoms in pure elements is zero. Individual atoms such as Fe, Na, Ar, and He have oxidation numbers of zero. This rule also applies to Cl_2, H_2, S_8, and other polyatomic elements. The electrons in the covalent bonds of these elements are neither transferred nor shifted, because both atoms have equal electronegativities.

Rule 2 The oxidation number of a monatomic ion is equal to the charge of the ion. When a Br atom gains an electron to become a Br^{1-} ion, it has an oxidation number of -1. The -1 shows that one electron has been gained. When a Mg atom loses two electrons to become a Mg^{2+} ion, its oxidation number becomes +2.

Rule 3 The sum of the oxidation numbers of all the atoms in a compound must be zero. Compounds are not electrically charged.

When ionic bonds form, negative and positive charges are evenly balanced. Ions with a +1 charge and ions with a -1 charge combine in a 1:1 ratio. A +2 ion combines with two -1 ions to have its charge equalized.

$$\overset{+1}{Na^+} + \overset{-1}{Cl^-} \longrightarrow \overset{+1\ -1}{NaCl}$$

$$\overset{+2}{Mg^{2+}} + \overset{-1}{2\ Cl^-} \longrightarrow \overset{+2\ -1}{MgCl_2}$$

This rule applies to covalent bonds as well. Remember that shared electrons can be shifted toward or away from atoms, depending on the relative electronegativities. If an atom attracts electrons toward itself, that atom is assigned a negative oxidation number. Atoms that lose influence over electrons during bonding become partially positive. They receive positive oxidation numbers.

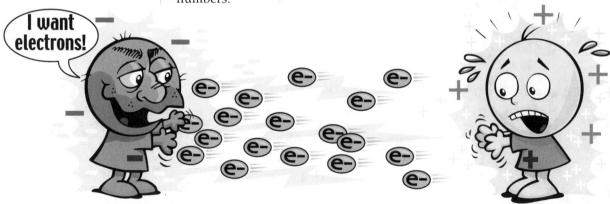

Rule 4 Alkali metals (Group IA) always have a +1 oxidation number when they are not free elements. These metals have small electronegativities, so they release their one valence electron when they react.

Rule 5 Alkaline earth metals (Group IIA) always have a +2 oxidation number. These atoms form +2 ions when they bond.

Rule 6 Certain elements have the same oxidation number in almost all their compounds.
   a. Halogens (Group VIIA) have an oxidation number of -1 when they are bonded to metals. Since they have seven valence electrons and large electronegativities, they gain an extra electron and form -1 ions. When halogens are bonded to other nonmetals, the element with the higher electronegativity is assigned the negative number.
   b. Hydrogen is assigned a +1 oxidation number in most of its compounds. Only when hydrogen is bonded to metals in compounds called metallic hydrides is hydrogen's number -1. This is because hydrogen is more electronegative than any other metal.
   c. Oxygen is assigned a -2 oxidation number in most of its compounds. Only when oxygen is bonded to highly electronegative elements such as fluorine or another oxygen does it not have its usual -2 number. Oxygen is so electronegative that it pulls electrons from most other elements. A major exception is the peroxide ion, O_2^{2-}, where oxygen has a -1 oxidation state.

Applications of Oxidation Number Rules

The few rules just given can be used to determine oxidation numbers in multitudes of compounds. In a situation where one rule contradicts another, the rule that was listed first should be followed. If the oxidation number of an atom is unknown, an algebraic equation can be used to solve for it. This is based on the fact that the sum of all the oxidation numbers in a compound must equal zero.

Sample Problem Determine the oxidation number of each element in the following compounds.
        a. Na_2O **b.** $KMnO_4$

Solution

a. The sum of the oxidation numbers of all the atoms in the compound must add up to zero (Rule 3). The sodium atom has a +1 oxidation number always (Rule 4), and the oxygen atom is -2 as expected (Rule 6c). The sum of two +1s and a single -2 is zero.

b. In $KMnO_4$ the only unknown oxidation number is that of manganese. The oxidation numbers of the oxygen and potassium atoms are already known to be -2 and +1 respectively. An algebraic equation can be used to solve for the oxidation number of manganese.

K's number + Mn's number + 4(O's number) = 0
(+1) + Mn's number + 4(-2) = 0
Mn's number + (-7) = 0
Mn's number = +7

The oxidation numbers of all the atoms in the compound are $K^{+1}Mn^{+7}O_4^{-2}$.

Scientists use oxidation numbers as an aid in writing chemical formulas. Given the elements that are in an ionic compound, they can determine how many atoms of each element must be present in order to have a neutral compound. The key idea to keep in mind is that there must be as many negative charges as positive ones if there is to be a compound without any net charge. For ionic compounds, a simple way to make this happen is to use the numeric value of the charge (or oxidation number) of one ion as the subscript for the other one. (This is sometimes called the "criss-cross method"). Since the formulas for ionic compounds must be the simplest ratios, a formula obtained by this method may need to be simplified by dividing each subscript by a common factor. One important exception to the simplification step is the peroxide ion, O_2^{-2}. For example, sodium peroxide has the formula Na_2O_2 not NaO because the peroxide ion exists as a diatomic ion that must not be simplified.

Sample Problem Write the formula for the ionic compound that results when

a. barium and iodine combine.
b. calcium and oxygen combine.

Solution

a. Barium, being an alkaline earth metal, has an oxidation number of +2. Iodine has a -1 oxidation number because it is a halogen bonding with a metal. In order for the sum of the oxidation numbers to equal zero, two iodine atoms must be included in the formula.

$$\text{BaI} = \overset{+2\ -1}{\text{Ba}}\overset{(+2)(-1)}{\text{I}} = \text{Ba}_1\text{I}_2 = \text{BaI}_2$$

b. Calcium is also an alkaline earth metal, so it has an oxidation number of +2. Oxygen has a -2 oxidation number when combined with metals (Rule 6C). Thus,

$$\text{CaO} = \overset{+2\ -2}{\text{Ca}}\overset{(+2)(-2)}{\text{O}} = \text{Ca}_2\text{O}_2 = \text{CaO}$$

Though the first formula, Ca_2O_2, has equal negative and positive charges, it is incorrect because it is not in the simplest possible ratio.

Atoms With Multiple Oxidation States

Some atoms can have more than one oxidation number. Electrons behave differently, depending on the particular atoms in the bond. Transition metals are especially notorious for having more than one oxidation number. Because their outer energy levels are very close to each other, the bonding circumstances determine how many electrons participate in chemical bonds. For example, iron can form $FeCl_2$ as well as $FeCl_3$. Nonmetals often have more than one oxidation state. Table 7A–1 shows nitrogen-plus-oxygen compounds. Each of these compounds is very different. N_2O and NO_2 can both make you go to sleep; however, NO_2 would do so quickly and permanently.

Table 7A-1 Compounds of Nitrogen Plus Oxygen

Compound	Name	Oxidation State of the Nitrogen	Properties
NO	nitric oxide	+2	Colorless gas, used to make nitric acid
N_2O	nitrous oxide	+1	Laughing gas (anesthetic)
N_2O_5	dinitrogen pentoxide	+5	White crystalline solid
NO_3	nitrogen trioxide	+6	Bluish gas, decomposes at room temperature
NO_2	nitrogen dioxide	+4	Yellow liquid, used in explosives, poison
N_2O_3	dinitrogen trioxide	+3	Reddish brown gas that decomposes at 4°C

If nitrogen, sulfur, phosphorus, or carbon bond to highly electronegative elements, they lose a measure of control over their electrons and are assigned a positive oxidation number. If they bond to a less electronegative element, they can attract electrons and are assigned a negative oxidation number.

Clear-cut rules rarely work for metals that do not belong to the IA or IIA families. The oxidation numbers of the most common elements must either be memorized or looked up in tables. Figure 7A–1 shows the oxidation states of commonly encountered elements. The shaded elements on the chart are the ones used most frequently in this course. If you are not familiar with which elements have multiple oxidation states, you will have difficulty correctly naming compounds that contain them.

Figure 7A-1 ▪ Oxidation Numbers of Common Elements

H +1, -1																	He
Li +1	Be +2											B +3	C +4, +2, -4	N +5, +4, +3, +2, +1, -3	O -1, -2	F -1	Ne
Na +1	Mg +2											Al +3	Si +4, +3, -4	P +5, +3, -3	S +6, +2, +4, -2	Cl +7, +1, +5, -1, +3	Ar
K +1	Ca +2	Sc +3	Ti +4, +3, +2	V +5, +3, +4, +2	Cr +6, +3, +2	Mn +7, +4, +6, +3, +2, +5	Fe +3, +2	Co +3, +2	Ni +2	Cu +2, +1	Zn +2	Ga +3	Ge +4, -4	As +5, +3, -3	Se +6, +4, -2	Br +5, +1, -1	Kr +4, +2
Rb +1	Sr +2	Y +3	Zr +4	Nb +5, +4	Mo +6, +4, +3	Tc +7, +4, +3	Ru +8, +4, +6, +3	Rh +4, +3, +2	Pd +4, +2	Ag +1	Cd +2	In +3	Sn +4, +2	Sb +5, +3, -3	Te +6, +4, -2	I +7, +1, +5, -1	Xe +6, +4, +2
Cs +1	Ba +2	La +3	Hf +4	Ta +5	W +6, +4	Re +7, +6, +4	Os +8	Ir +4, +3	Pt +4, +2	Au +3, +1	Hg +2, +1	Tl +3, +1	Pb +4, +2	Bi +5, +3	Po +2	At -1	Rn
Fr +1	Ra +2	Ac +3															

Oxidation Numbers and Polyatomic Ions: The Charge Makes the Difference

Polyatomic ions are covalently bonded groups of atoms that carry charges. Since polyatomic ions are found in many compounds and reactions, they are included in the oxidation number rules.

Rule 7 The oxidation numbers of all the atoms in a polyatomic ion add up to the charge on the ion.

An examination of the hydroxide ion (OH$^-$) shows how this rule works. The oxygen atom has an oxidation number of -2, and the hydrogen atom has a +1 number. The sum of these numbers is -1, the charge on the ion. Polyatomic ions remain intact throughout most chemical reactions. Using the charge of the ion as you would use an oxidation number simplifies many problems.

Sample Problem What is the oxidation number of the lead atom in Pb(OH)$_2$?

Solution

$$\text{Pb's number} + 2(\text{OH}^-\text{'s number}) = 0$$
$$\text{Pb's number} + 2(-1) = 0$$
$$\text{Pb's number} = +2$$

The same technique used to find formulas of two-element ionic compounds can be used for polyatomic ionic compounds. The polyatomic ion is treated as a single unit and will require parentheses around it if more than one is present in the formula.

Sample Problem What is the formula of the compound that contains ammonium (NH$_4^+$) ions and phosphate (PO$_4^{3-}$) ions?

Solution
The oxidation numbers of all the atoms in a compound must add up to zero. Three ammonium ions are required to balance the -3 of the phosphate ion. Using the "criss-cross" method we get the following.

$$\overset{+1}{\text{NH}_4}\overset{-3}{\text{PO}_4} = \overset{(+1)}{\text{NH}_4}\overset{(-3)}{\text{PO}_4} = \overset{+1}{(\text{NH}_4)_3}\overset{-3}{(\text{PO}_4)_1} = (\text{NH}_4)_3\text{PO}_4$$

Algebraic equations can be used to find the oxidation numbers of individual atoms in polyatomic ions. The following equation yields the oxidation state of the S atom in a sulfate ion (SO$_4^{2-}$). Both rule 6c and 7 are needed here.

$$\text{S's number} + 4 (\text{O's number}) = \text{charge of ion}$$
$$\text{S's number} + 4(-2) = -2$$
$$\text{S's number} + (-8) = -2$$
$$\text{S's number} = +6$$

Two Important Exceptions

Like the rules of grammar, oxidation number rules have exceptions. It is helpful to become familiar with the following significant exceptions:

- Compounds such as LiH and NaH, in which hydrogen is bonded to a metallic atom, are called metallic hydrides (Chapter 5). In these compounds the hydrogen atom is the more electronegative element present. Consequently, it attracts the electrons and receives an oxidation number of -1.
- **Peroxides** are a class of compounds in which two oxygens are bonded together. The simplest and most common peroxide is hydrogen peroxide, H-O-O-H. In this compound each oxygen has an oxidation number of -1 instead of the usual -2.

Section Review Questions 7A

1. Write the formula for the ionic compound that results when
   a. aluminum and chlorine combine.
   b. lithium and fluorine combine.
2. What is the formula of the compound that contains iron (II) (Fe^{2+}) and sulfate (SO_4^{2-}) ions?
3. Determine the oxidation number of each element in the following compound: $KClO_3$, potassium chlorate.
4. Contrast elements that have been assigned negative oxidation numbers with those that have been assigned positive oxidation numbers.
5. What are the two exceptions to the oxidation rules?

7B Nomenclature: Naming Compounds

What is in a name? In our society today we do not generally associate a personality trait with someone's name. Of course, there are nicknames that are sometimes given to a child to describe something about him. However, this was different in Old Testament times. In Genesis 25, we see how the twins Esau and Jacob received their names. Esau was born first and appeared red and hairy, so he was named Esau, which means "hairy" in Hebrew. He was also called Edom, which means "red." Both names described his appearance. During birth, Jacob's hand grabbed Esau's heel, so his parents named him Jacob, which means "supplanter" or "heel-grabber."

Though you may not put a lot of significance on the meaning of a person's name today, that name still remains significant. If you don't think so, just see how you react when someone calls you by the wrong name or mispronounces your name. Why is that? Because your name represents you. It is important that every person have a unique name to avoid confusion. When two people in the same class have the same name, a teacher must devise some way to differentiate between them. So it is with chemical compounds—each one must be given a unique name so there is not confusion. Names are extremely important!

Do the names *soda ash* and *epsomite* mean anything to you? If you can recall the properties of these compounds or the elements they contain, you are relying on a good memory. The names themselves certainly do not give you any clues. Chemical compounds have been given many names throughout history. Some names, like soda ash, came about because they described how a compound looked or how it acted. Other names told where a compound came from. One particular compound was called epsomite because it was often found near the English town of Epsom. In America this compound is called Epsom salts. Although the backgrounds of these names are interesting, the names do not give us very much information. You can see the difficulty in determining the formulas for many compounds based on their common names by looking at the examples in Table 7B-1.

Table 7B-1 Common Names of Some Industrial Chemicals

Common Name	Systematic Name	Formula
Oil of vitriol	Sulfuric acid	H_2SO_4
Lime	Calcium hydroxide	$Ca(OH)_2$
Ammonia	Nitrogen hydride	NH_3
Soda ash	Sodium carbonate	Na_2CO_3
Potash	Potassium sulfate*	K_2SO_4

*Note: Can also be potassium chloride–KCl

Epsom salt has many practical uses. Soda ash (sodium carbonate) is often used to raise the pH in swimming pools and spas.

As more and more compounds were discovered and synthesized, chemists realized that they could not continue to rely on memorized names. In the twentieth century the International Union of Pure and Applied Chemistry (IUPAC) developed a systematic way to name compounds. This system of names, called a **nomenclature,** allows all compounds to be named according to a common set of rules. The IUPAC names of compounds are packed with information. They tell which elements are present in the compound, and indirectly, they tell about the types of bonds, the intermolecular attractions, and the general properties of the compound.

Today the term *soda ash* is used very little. Instead, people refer to this compound as sodium carbonate. The name *epsomite* has given way to a more informative name, magnesium sulfate. Some common names are still used, but only when the compound is familiar to many people. One such example is ammonia, which is nitrogen hydride.

Using the Greek Prefix System

In this discussion of nomenclature, a graphic device called a **flow chart** will help you determine which rules apply to which compounds. The entire flow chart, presented as Figure 7B-1, provides an organized method for identifying an unnamed formula. This discussion will first consider covalent compounds that are not acids. Remember that covalent compounds are those

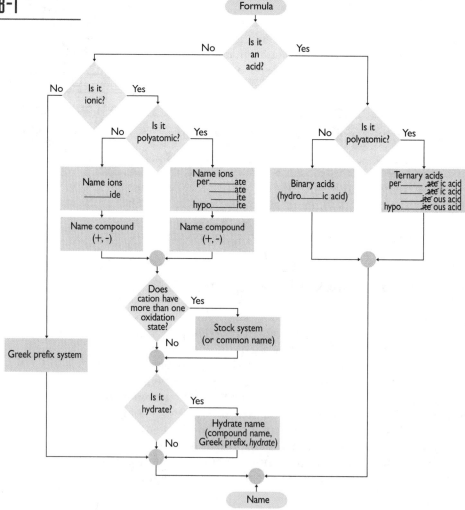

Figure 7B-1

composed of only nonmetals. For now, any covalent compound whose formula begins with hydrogen can be considered to be an acid. Covalent compounds that are not acids are named according to a system of Greek prefixes.

The **Greek prefix system** indicates how many atoms of each element are included in a covalent compound. The commonly used prefixes are listed in Table 7B-2. This system omits the prefix *mono* unless it is needed for emphasis or clarity. When it is used, any extra vowels are omitted. (For instance, CO is named carbon monoxide, not carbon mono-oxide.) The least electronegative element is stated first and then the more electronegative element. The ending of the last element is changed to *-ide*.

Table 7B-2 Greek Prefixes

Prefix	Number
mono-	1
di-	2
tri-	3
tetra-	4
penta-	5
hexa-	6
hepta-	7
octa-	8
nona-	9
deca-	10

Sample Problem Name PCl_3.

Solution

PCl_3 is not an acid, since its formula does not begin with hydrogen. It is not ionic, because all its elements are nonmetals. It should thus be named according to the Greek prefixes. Phosphorus trichloride is the accepted name. The prefix *mono* is generally not used for the first element in the compound.

Binary Ionic Compounds

Ionic compounds are not normally named with Greek prefixes. **Binary compounds** (two-element compounds) that are ionic are named according to each of the two ions that are involved. Positive ions use the same name as their parent atoms (sodium atoms form sodium ions), but negative ions have an *-ide* ending (chlorine atoms form chloride ions). The accepted *-ide* endings for some common compounds are listed in Table 7B-3. Table 7B-4 lists accepted *-ide* endings for nonmetals. For a binary ionic compound, the name of the positive ion appears first, followed by that of the negative ion. The names of metallic hydrides start with the active metal and end with hydride because hydrogen is the more electronegative element in these compounds.

Table 7B-3 -Ide Endings for Some Common Compounds

Compound	Name
NaCl	sodium chloride
MgO	magnesium oxide
Al_2O_3	aluminum oxide
K_2S	potassium sulfide
MgH_2	magnesium hydride

Table 7B-4 Names of Nonmetallic Ions

B boride	C carbide	N nitride	O oxide	F fluoride
	Si silicide	P phosphide	S sulfide	Cl chloride
		As arsenide	Se selenide	Br bromide
		Sb antimonide	Te telluride	I iodide
				At astatide

Table 7B-5 Common Oxyanions

Charge	Greatest Number of Oxygens per___ate	Base Number of Oxygens ___ate	Fewer Number of Oxygens ___ite	Fewest Number of Oxygens hypo___ite
-1		acetate $C_2H_3O_2^-$		
	perbromate BrO_4^-	bromate BrO_3^-	bromite BrO_2^-	hypobromite BrO^-
	perchlorate ClO_4^-	chlorate ClO_3^-	chlorite ClO_2^-	hypochlorite ClO^-
		cyanate OCN^-		
		hydrogen carbonate (bicarbonate) HCO_3^-		
		hydrogen sulfate (bisulfate) HSO_4^-	hydrogen sulfite (bisulfite) HSO_3^-	
	periodate IO_4^-	iodate IO_3^-		hypoiodite IO^-
		nitrate NO_3^-	nitrite NO_2^-	
	permanganate MnO_4^-			
-2		carbonate CO_3^{2-}		
		oxalate $C_2O_4^{2-}$		
		chromate CrO_4^{2-}		
		dichromate $Cr_2O_7^{2-}$		
		silicate SiO_3^{2-}		
		sulfate SO_4^{2-}	sulfite SO_3^{2-}	
		thiosulfate $S_2O_3^{2-}$		
-3		arsenate AsO_4^{3-}	arsenite AsO_3^{3-}	
		borate BO_3^{3-}		
		phosphate PO_4^{3-}	phosphite PO_3^{3-}	

Sample Problem Name CaI_2.
Solution
Since this is a compound of a metal and a nonmetal, it is ionic. With only two elements it is binary and is named by listing the ions: calcium iodide. Note that the name must end in *-ide* since it is a binary compound. Iodine is changed to iodide. Note further that CaI_2 would not be named "calcium diiodide" because prefixes are used only in binary covalent compounds.

Why is the prefix *di-* not needed and incorrect? Since calcium has only a +2 oxidation number and iodine is -1, there will be two iodide ions for each calcium ion in order to have a neutral compound. No other formula is possible, so there is no need to specify that there are two iodide ions by using *di-*. Remember, the Greek prefix system is for covalent nonacids.

Polyatomic Ions and Their Compounds

Tables 7B-5 and 7B-6 list the formulas and names of common polyatomic ions. Some generalizations can help make many of these ions easier to learn. The only positive polyatomic ions are NH_4^+, the ammonium ion, and Hg_2^{2+}, the mercurous ion. Anions composed of oxygen and one other element are known as **oxyanions** and often have two or more forms for the same element. If there are only two forms, the one with the fewer oxygen atoms ends in *-ite* and the one with more oxygen atoms ends in *-ate*. For example, SO_3^{2-} is sulfite and SO_4^{2-} is sulfate. For some elements, such as the halogens, there are more than two oxyanions. In such cases, the one that has fewer oxygen atoms than the *-ite* ion has a prefix of *hypo-*, and the one with more oxygen atoms than the *-ate* ion has the prefix *per-*. You can see how this works for the oxyanions of chlorine and bromine in Table 7B-5.

Once you have determined the names of the polyatomic ions, you can name **polyatomic compounds** that contain them. Simply name the cation and then the anion. Do not use prefixes to express numbers (*di-*, *tri-*, etc.) Table 7B-7 gives some examples of polyatomic compounds and their names.

You will notice that some anions in Tables 7B-5 and 7B-6 contain hydrogen. You should treat the entire combination of two or three elements as one ion. That is, when they are in combination with metal ions, simply follow the same procedure as you would for naming other polyatomic compounds—name the cation and the anion. For example, $NaHCO_3$ is called sodium hydrogen carbonate (or sodium bicarbonate) and $Mg(HS)_2$ is named magnesium hydrogen sulfide.

Table 7B-6 Other Polyatomic Ions

Name	Formula
Ammonium	NH_4^+
Mercurous	Hg_2^{2+}
Amide	NH_2^-
Azide	N_3^-
Cyanide	CN^-
Hydrogen sulfide	HS^-
Hydroxide	OH^-
Peroxide	O_2^{2-}
Thiocyanate	SCN^-

Table 7B-7

Compound	Name
Na_2SO_4	sodium sulfate
KOH	potassium hydroxide
$Ba(ClO_3)_2$	barium chlorate
$Ca(ClO)_2$	calcium hypochlorite
$Ba(NO_2)_2$	barium nitrite
NH_4Cl	ammonium chloride

Sample Problem Name NH_4BrO_3.

Solution

The compound NH_4BrO_3 is not an acid, and it is ionic, so we simply name the ions. The name *ammonium bromate* is formed from the names of the two polyatomic ions included in the formula.

Sample Problem What is the formula for potassium dichromate?

Solution

The dichromate ion ($Cr_2O_7^{2-}$) carries a -2 charge. Two potassium atoms (oxidation number of +1) for every dichromate ion can form a neutral compound. The formula must be $K_2Cr_2O_7$.

Handling Atoms with Multiple Oxidation States

If the metallic element in an ionic compound is one that can have more than one oxidation number (often a transition metal), a Roman numeral is placed after the element's name to show the oxidation number. This convention is called the **Stock system** or sometimes the **Roman numeral system.** The older way of identifying the oxidation state of a transition metal is with the suffixes *-ous* and *-ic*. The *-ous* ending refers to the smaller oxidation number, and the *-ic* ending refers to the larger oxidation number. For example, Cu^+ is called the cuprous ion and Cu^{2+} is called the cupric ion. We will primarily use the Stock system in this book. Though the older system is still prevalent, it is not the preferred system. Table 7B-8 gives examples of ionic compounds containing metals having more than one oxidation number, and their names in both systems. You will need to learn which metals occur in more than one oxidation state so that you will know when to use Roman numerals in the name and when they are not needed.

Sample Problem Name $Hg(BrO_3)_2$ according to the Stock system and the common system.

Solution

Following the flow chart, you see that $Hg(BrO_3)_2$ is not an acid, is ionic, contains a polyatomic ion, and involves a metal that can have more than one oxidation number. The BrO_3 portion of the formula corresponds to a BrO_3^- ion, so the last part of the name will be *bromate*. Since two -1 ions are present, the mercury atom must have a +2 oxidation number. The Stock system name, therefore, is *mercury (II) bromate*. Since the +2 is the larger of mercury's two oxidation numbers, the common name of the compound is *mercuric bromate*.

Table 7B-8

Compound	Stock System Name	Common Name
Hg_2I_2	mercury (I) iodide	mercurous iodide
HgI_2	mercury (II) iodide	mercuric iodide
CuBr	copper (I) bromide	cuprous bromide
$CuBr_2$	copper (II) bromide	cupric bromide
$FeCl_2$	iron (II) chloride	ferrous chloride
$FeCl_3$	iron (III) chloride	ferric chloride
CoC_2O_4	cobalt (II) oxalate	cobaltous oxalate
$Co_2(C_2O_4)_3$	cobalt (III) oxalate	cobaltic oxalate
SnO	tin (II) oxide	stannous oxide
SnO_2	tin (IV) oxide	stannic oxide
$PbSO_4$	lead (II) sulfate	plumbous sulfate
$Pb(SO_4)_2$	lead (IV) sulfate	plumbic sulfate

Sample Problem What is the formula of lead (II) phosphate?

Solution

The Roman numeral II identifies the +2 oxidation state of lead. Phosphate ions have a -3 charge. (You need to memorize this.) A combination of three lead atoms for every two phosphate ions results in an electrically neutral compound. The formula is $Pb_3(PO_4)_2$.

Hydrates

Hydrates are compounds that have water molecules in their crystalline structures. These compounds hold a characteristic amount of water called the "water of hydration." Formulas of these compounds indicate the presence and number of water molecules by a centered dot followed by the number of water molecules ($Na_2CO_3 \cdot 7\ H_2O$). The word *hydrate* with a Greek prefix is added to the usual name of an ionic compound that normally incorporates water molecules into its structure. In order to distinguish between the hydrates of a compound and the form that does not contain water, the term **anhydrous** is used. *Anhydrous* designates a compound that has no water of hydration in its crystalline structure. Table 7B-9 lists several compounds and their status as anhydrous or a hydrate.

Table 7B-9

Formula	Name (compound's name, Greek prefix, hydrate)
Na_2CO_3	sodium carbonate (anhydrous)
$Na_2CO_3 \cdot H_2O$	sodium carbonate monohydrate
$Na_2CO_3 \cdot 7 H_2O$	sodium carbonate heptahydrate
$Na_2CO_3 \cdot 10 H_2O$	sodium carbonate decahydrate

Note: Plaster of Paris is a well-known hydrate. Its formula, $CaSO_4 \cdot \frac{1}{2} H_2O$, is systematically named calcium hemihydrate (hemi for $\frac{1}{2}$).

Binary Acids

Binary "two-element" compounds that show hydrogen as the first element in their formulas are called **binary acids.** These acids can have names based on the general rules used for other binary compounds. For example, as a gas, HCl is called hydrogen chloride. When these binary compounds are dissolved in water, however, they form acids and are given different names. These names include the prefix *hydro-* (referring to the hydrogen), the root name for the nonmetal with an *-ic* ending, and the word acid. In an aqueous solution HCl is called hydrochloric acid. Table 7B-10 gives some examples of these name changes.

Table 7B-10

Formula	Common Name (gases)	Acid Name (hydro___ic acid)
HCl	hydrogen chloride	hydro*chloric* acid
HBr	hydrogen bromide	hydro*bromic* acid
H_2S	hydrogen sulfide	hydro*sulfuric* acid

Note: For sulfur the entire name is used as the root, rather than the shortened *sulf* that was used in the anion name.

Ternary Acids

Ternary acids contain three elements: hydrogen, oxygen, and another nonmetal. The oxygen and the nonmetal are often bound together in a polyatomic ion. The names of ternary acids are derived from the anions in the acids. If the anion's name ends in *-ate*, the ending changes to *-ic* and the word *acid* is added. If the anion's name ends in *-ite*, the ending changes to *-ous* and the word *acid* is added. Table 7B-11 lists several examples.

Table 7B-11

Anion in Ternary Acid		Ternary Acid	
Formula	Name	Formula	Name
ClO_4^-	perchlorate	$HClO_4$	perchloric acid
ClO_3^-	chlorate	$HClO_3$	chloric acid
ClO_2^-	chlorite	$HClO_2$	chlorous acid
ClO^-	hypochlorite	$HClO$	hypochlorous acid
NO_3^-	nitrate	HNO_3	nitric acid
NO_2^-	nitrite	HNO_2	nitrous acid
SO_4^{2-}	sulfate	H_2SO_4	sulfuric acid
SO_3^{2-}	sulfite	H_2SO_3	sulfurous acid

Once again you will note that the entire name for sulfur is used as the root; also, the root for phosphorus is *phosphor* instead of the shortened *phosph*, which is used in the oxyanion name.

Sample Problem Name $HBrO_4$.

Solution

The formula $HBrO_4$ starts with hydrogen and is therefore an acid. Since the compound contains a polyatomic ion, its name is derived from the name of the ion. The BrO_4^{1-} ion is the perbromate ion. The *-ate* ending changes to *-ic*, and the correct name, *perbromic acid*, results. Note that this does not contain the prefix *hydro-* because that is used only for binary acids.

Section Review Questions 7B

1. Explain the necessity of using an established chemical name in place of a common name.
2. Name the following compounds:
   a. P_2S
   b. Al_2O_3
   c. $CaSO_4$
   d. FeC_2O_4
3. Give the names for the following polyatomic ions:
   a. NH_4^+
   b. CN^-
   c. $C_2H_3O_2^-$
   d. NO_3^-
   e. MnO_4^-
4. Give a formula for the compound copper (II) sulfate pentahydrate.
5. Write the correct ternary acid (formula and name) that would result from the ion BO_3^-.

Quantitative Analysis Solves a Mystery

Just a few strands of hair—most people would consider them insignificant. But to three European scientists, they were the key to unlocking a century-old secret. The mystery involved the untimely death of Napoleon Bonaparte. Although many historians say Napoleon died of cancer or an ulcer, some scientists believe he was murdered. What do they have as proof? A few strands of hair.

For many years Sten Forshufvud had been convinced that the circumstances behind Napoleon's death were not all known. Forshufvud spent many years studying the memoirs of those people associated with Napoleon during his years of exile. Of particular interest to Forshufvud were the memoirs of Louis Marchand. Marchand, a chief valet, took over the complete care of Napoleon during his last months. In his account Marchand gave many details of Napoleon's deteriorating physical condition. From these observations Forshufvud recognized twenty-two out of over thirty generally accepted symptoms of chronic arsenic poisoning. He also concluded that Napoleon's continually changing health from good to bad indicated that the former emperor was poisoned with small doses of arsenic over a long period of time.

In spite of Forshufvud's careful research, he still needed more scientific proof. He soon learned of a new method devised by Hamilton Smith for detecting arsenic and other elements in hair. By determining the level of arsenic in hair, it was possible to calculate the amount of arsenic in the entire body. Fortunately for Forshufvud, Napoleon's valet had shaved off several locks of Napoleon's hair before the emperor was buried in 1821. These locks had been saved and passed down through families, and the researchers were able to obtain permission to use several strands of hair. The process involved a powerful tool called neutron activation analysis (NAA). With this tool, scientists can detect traces of an element in doses as small as one-billionth of a gram. To determine the arsenic content of Napoleon's hair, Smith placed a single strand of hair about 5 inches long and a standard arsenic solution (included for comparison) in the NAA reactor. As neutrons bombarded the samples, many of the nuclei became radioactive and began emitting gamma rays. After a twenty-four-hour period, Smith removed the sample from the reactor and analyzed the gamma rays. Since gamma rays from different elements display characteristic wavelengths and strengths, Smith could identify which elements were present as well as their respective amounts.

The results of these tests showed an extremely high level of arsenic. In the first hair sample, for instance, the tests revealed an average of 10.38 parts per million (ppm). The normal amount of arsenic in hair is about 0.5 ppm. Napoleon's hair contained over twenty times the normal amount of arsenic. If Napoleon had been poisoned, the arsenic would have permeated all his body tissues, including his hair.

Later Smith refined his tests to analyze arsenic levels in specific portions of hair. Since he knew when the locks had been shorn, he could match the arsenic level with specific dates in Napoleon's life. Interestingly, Forshufvud found that the arsenic was not distributed evenly along the hair. It was concentrated in points every few millimeters. Doses must have been administered periodically during the four months before Napoleon died. The results showed that the arsenic content of the emperor's hair ranged from a low of 1.06 ppm to a high of 76.6 ppm. These peaks and valleys coincided perfectly with Napoleon's ever-changing health.

Neutron activation analysis is only one of several modern methods of chemical analysis. Unlike traditional methods of analytical chemistry, modern techniques offer greater speed, a wider range of possible tests, and the ability to use small samples. These improvements allow scientists to gain a better understanding of the present as well as the past.

7C The Mole: A Unit Tailor-Made for Tiny Things

Describing the mass of atoms is similar to what a manufacturer might face when buying small ball bearings. When a shipment arrives, how can he know whether it includes all the ball bearings he needs? It would take an exceedingly long time to count them one by one. Instead, he could work in terms of mass. For instance, if 1 kilogram contained 500 ball bearings, a 60-kilogram shipment would contain 30,000 bearings. Large units can be used to measure many small items quickly. If we are going to work with measurable quantities of matter in the laboratory, we will need a very large number of atoms or formula units. Thus, the unit that is needed to conveniently represent such large numbers must also be exceedingly large. Common units such as dozen, gross, or ream are just too small. Even referring to billions or trillions of atoms is not adequate. The large unit that chemists use is called the mole. It is so large that it is useful only for very small items such as atoms, molecules, and ions.

What Is in a Mole? Avogadro's Number

A **mole** is the amount of substance contained in 6.022×10^{23} units. The number 6.022×10^{23} is called **Avogadro's number** in honor of the Italian physicist Amedeo Avogadro (1776-1856). This number is used so often that it is abbreviated N_A. The mole's value was changed from 6.023 to 6.022 because the latter worked in mathematical equations, whereas the former did not.

1 mole of He atoms = 6.022×10^{23} He atoms
1 mole of H_2O molecules = 6.022×10^{23} H_2O molecules
1 mole of NaCl formula units = 6.022×10^{23} NaCl formula units

Because a mole contains such a huge number of items, it is used to measure only very small objects. The reason that no one has ever heard of a mole of bricks or a mole of golf balls is that no one has ever manufactured 602 sextillion of them. Smaller units, such as a dozen, are used to measure large objects. Moles are used for measuring objects on the atomic and molecular scale. If an atom were the size of a marble and one mole of marbles were spread out over the surface of the earth, a 50-mile-deep layer of marbles would cover the earth. Table 7C-1 gives several comparisons of objects and their groupings.

Table 7C-1

Objects	Common Group	Number of Items in Group
Socks, shoes	pair	2
Eggs, doughnuts	dozen	12
Pencils	gross	144
Sheets of paper	ream	500
Atoms, molecules, and ions	mole	6.022×10^{23}

Although there are 6.022×10^{23} units in one mole, it is not a counted or defined number as the units dozen and gross. This is important when considering significant figures in calculations. In this book we will use 4 significant figures for mole calculations.

Sample Problem How many atoms are in a 4.5-mole sample of helium?

Solution

The fact that one mole of anything contains 6.022×10^{23} items justifies the formation of a conversion factor that can be used to change 4.5 moles into the number of atoms.

$$\frac{4.5 \text{ mol} \mid 6.022 \times 10^{23} \text{ atoms}}{\mid 1 \text{ mol}} = 2.7 \times 10^{24} \text{ atoms}$$

Atoms and the Mole

Since the mole is such a large number of particles, it is out of the question to measure amounts of atoms or molecules by counting them individually. Just as the manufacturer can "count" ball bearings by weighing them if he knows their relative mass, so then can the chemist "count" atoms if he knows their relative masses.

Consider this analogy: suppose an average plum has a mass five times the mass of an average cherry. As long as you had a 5:1 mass ratio of plums to cherries, you would know that you had an

One mole each of various substances. Each sample contains 6.022×10^{23} atoms but has a different mass.

equal number of each fruit. Regardless of the unit—whether lb., kg, or ton—the same number of fruit would be present in both groups as long as the mass ratio for the individual fruit applied to the group of fruit. You still, however, would not know how many of each fruit you had unless you knew at least one of the individual fruit's masses. Instead, you would know you had equal numbers of each fruit.

Scientists use Avogadro's number to relate atomic mass units to the larger, more practical unit of grams. They can find the mass of a mole of objects by expressing an object's mass in atomic mass units and changing the unit to grams. A hydrogen atom has a mass of about 1 amu. Scientists have experimentally proved that 6.022×10^{23} hydrogen atoms have a mass of 1 gram. ✶ Avogadro's number was specifically chosen as the number of particles in a mole so that the atomic mass of an element and the mass of a mole of the element have the same numeric value, just different units. A carbon atom has a mass of 12 amu's; 6.022×10^{23} carbon atoms have a mass of 12 grams. If a molecule has a mass of 130 amu's, a mole of these molecules will have a mass of 130 grams. Thus, the periodic table can be used to find not only the masses of individual atoms, but also the masses of moles of atoms. These principles are shown in Table 7C-2.

Table 7C-2

Chemical Unit	Mass of One Unit (amu's)	Number of Units in One Mole	Mass of One Mole (grams)
He atom	4.003	6.022×10^{23}	4.003
H_2O molecule	18.016	6.022×10^{23}	18.016
NaCl formula unit	58.44	6.022×10^{23}	58.44

How can one mole of hydrogen atoms have a mass different from one mole of carbon atoms? For the answer, compare atoms to varieties of fruits and vegetables. Suppose that all fruits come in cartons of twelve (one dozen). A dozen apples have a larger mass than a dozen cherries. A dozen cantaloupes have an even larger mass, and the mass of a dozen watermelons is greater still. In the same manner, a mole of carbon atoms has a larger mass than a mole of hydrogen atoms. A mole of iron atoms has an even larger mass, and a mole of uranium atoms has the greatest mass of them all. Remember that a mole, like a dozen, always contains the same number of objects. The mass of the individual objects involved determines the mass of one mole.

Sample Problem Calculate the mass of 0.500 mole of helium atoms.

Solution

On average, a helium atom has a mass of 4.003 amu's. This number, expressed in grams, is the mass of 1 mole of helium atoms.

$$\frac{0.5000 \text{ mol He}}{1} \times \frac{4.003 \text{ g He}}{1 \text{ mol He}} = 2.002 \text{ g He}$$

Note: Any time you scale up the number of atoms from one atom to one mole of atoms, you will scale up the unit from amu to g.

How many copper atoms are in one penny? What is the mass of a single atom? These questions can be answered with the help of the mole concept. The mole also makes it possible to convert between the mass of a chemical sample, the number of moles present, and even the number of chemical units in the sample.

The arrows between mass, moles, and number of units in Figure 7C-1 show possible routes to take when solving a problem. The quantities near the arrows are the ones that should be used for the particular operation.

Figure 7C-1

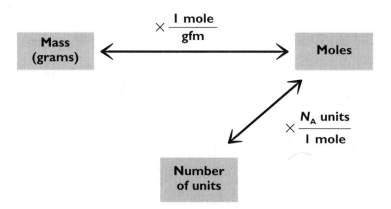

What is the mass of 5.00 moles of potassium atoms? The given value is expressed in moles (5.00 moles K), and the value to be calculated is the mass (how many grams). The flow chart shows that the conversion using gram-formula mass (gfm) is the only step that needs to be taken. Solving this problem on the basis of units alone, you know from the periodic table that 1 mole K = 39.10 g K. This relationship contains both the given unit (mole) and the desired unit (g); thus it forms the only conversion factor needed for this problem.

$$\frac{5.00 \text{ mol K}}{} \Bigg| \frac{39.10 \text{ g K}}{1 \text{ mol K}} = 196 \text{ g K}$$

How many copper atoms are in a 4.00-gram copper coin? A look at Figure 7C-1 shows that there is no step that will directly convert a mass to the number of units in that mass. The number of moles must be calculated as an intermediate step. Knowing from the periodic table that 1 mole of copper atoms is contained in 63.55 grams allows the following conversion factor.

$$\frac{4.00 \text{ g Cu}}{} \Bigg| \frac{1 \text{ mol Cu}}{63.55 \text{ g Cu}} = 0.0629 \text{ mol Cu}$$

Once the number of moles has been found, the last step on the flow chart can be taken.

$$\frac{0.0629 \text{ mol Cu}}{} \Bigg| \frac{6.022 \times 10^{23} \text{ atoms Cu}}{1 \text{ mol Cu}} = 3.79 \times 10^{22} \text{ atoms Cu}$$

Note: Mole is always the "bridge" between mass of a substance and the number of particles (atoms, molecules, formula units) of that substance.

Sample Problem How many atoms are in 33 mg (3.3×10^{-3} g) of gold? (This is an amount about the size of the period at the end of this sentence.) Figure 7C-1 shows that grams should first be changed to moles; then the moles should be changed to atoms.

Solution

$$\frac{3.3 \times 10^{-3} \text{ g Au}}{} \Bigg| \frac{1 \text{ mol Au}}{197.0 \text{ g Au}} \Bigg| \frac{6.022 \times 10^{23} \text{ atoms Au}}{1 \text{ mol Au}}$$

$$= 1.0 \times 10^{20} \text{ atoms Au}$$

Compounds and the Mole

Chemical compounds contain two or more atoms chemically combined so as to behave as one unit. The masses of the units composing compounds can be found by simply adding the masses of the atoms contained in them. Water (H_2O) has a mass of $2(1.008 \text{ amu}) + 1(16.00 \text{ amu}) = 18.02$ amu. A sodium chloride formula unit (NaCl) has a mass equal to the sum of one atom of sodium and one atom of chlorine:

$$1(22.99 \text{ amu}) + 1(35.45 \text{ amu}) = 58.44 \text{ amu}.$$

Several terms are used to refer to the mass of a mole of a substance. Each term that is used specifies the object in the mole. For example, the mass of a mole of atoms is called the **gram-atomic mass**. The mass of a mole of molecules is called the **gram-molecular mass**. The mass of a mole of formula units in an ionic compound is called the **gram-formula mass.** (Remember that ionic compounds are made of formula units, not molecules.) Gram-atomic masses, gram-molecular masses, and gram-formula masses all have units of *grams per mole*. As shown in Table 7C-3 the value of relative mass and the value mass of a mole are the same.

Table 7C-3

Object	Relative Mass (amu's)	Mass of One Mole (grams/mole)
He atom	4.003	gram-atomic mass = 4.003
H_2O molecule	18.02	gram-molecular mass = 18.02
NaCl formula unit	58.44	gram-formula mass = 58.44

Sample Problem Find the gram-molecular mass of ammonia (NH_3).

Solution

A mole of NH_3 molecules contains 1 mole of nitrogen atoms and 3 moles of hydrogen atoms. The gram-molecular mass is the sum of all the gram-atomic masses.

$$\frac{1 \text{ mol N} \mid 14.01 \text{ g N}}{1 \text{ mol N}} = 14.01 \text{ g N}$$

$$\frac{3 \text{ mol H} \mid 1.008 \text{ g H}}{1 \text{ mol H}} = 3.024 \text{ g H}$$

One mole of NH_3 contains 14.01 grams of nitrogen atoms and 3.024 grams of hydrogen atoms. The gram-molecular mass is the sum of the two masses.

$$1 \text{ mol } NH_3 = 14.01 \text{ g} + 3.024 \text{ g} = 17.03 \text{ g}$$

Sample Problem Find the gram-formula mass of $Al_2(SO_4)_3$.

Solution
Each formula unit contains two aluminum, three sulfur, and twelve oxygen atoms. A mole of $Al_2(SO_4)_3$ consists of 2 moles of aluminum atoms, 3 moles of sulfur atoms, and 12 moles of oxygen atoms.

$$\frac{2 \text{ mol Al}}{} \left| \frac{26.98 \text{ g Al}}{1 \text{ mol Al}} \right. = 53.96 \text{ g Al}$$

$$\frac{3 \text{ mol S}}{} \left| \frac{32.06 \text{ g S}}{1 \text{ mol S}} \right. = 96.18 \text{ g S}$$

$$\frac{12 \text{ mol O}}{} \left| \frac{16.00 \text{ g O}}{1 \text{ mol O}} \right. = 192.0 \text{ g O}$$

$$1 \text{ mol } Al_2(SO_4)_3 = 342.1 \text{ g}$$

Types of Formulas and Percent Composition

There are several ways to describe the composition of a chemical substance. **Structural formulas** show the types of atoms involved, the exact composition of each molecule, and the arrangement of chemical bonds. These formulas are informative, but they can also be difficult to draw, and often take up large amounts of space. One simple structural formula is for water and is drawn H-O-H. However, the following structure for acetic acid (CH_3COOH) shows how complex they can become.

$$\begin{array}{c} H \quad O \cdots H - O \quad H \\ | \quad \diagup \qquad \qquad \diagdown | \\ H-C-C \qquad\qquad\quad C-C-H \\ | \quad \diagdown \qquad\qquad \diagup | \\ H \quad O - H \cdots O \quad H \end{array}$$

Molecular formulas show the types of atoms involved and the exact composition of each molecule. These formulas are more convenient than structural formulas, but they do not show the shapes of the molecules, the locations of the bonds, or the types of bonds present.

Molecular Formulas

H_2O	C_2H_4	Cl_2
Water	Ethene	Chlorine

Empirical formulas tell what elements are present and give the simplest whole-number ratio of atoms in the compound. Empirical formulas have already been used to describe the composition of ionic compounds. When used for a molecular compound, the empirical formula might represent the actual

FINE, BOYS—TOMORROW WE'LL WORK ON ACETIC ACID.

molecular composition if the molecular formula contains a simple ratio—as in H_2O. On the other hand, the empirical formula of a molecular compound may not represent the make-up of one molecule. Ethene's empirical formula is CH_2 because there are twice as many hydrogen atoms as there are carbon atoms. In this case the molecular formula is not the simplest ratio, so the empirical and molecular formulas are different.

Empirical Formulas

H_2O	CH_2	Cl
Water	Ethene	Chlorine

All of the compounds listed in Table 7C-4 have the same empirical formula—CH_2O—but have very different molecular formulas. Notice that for formaldehyde, the empirical formula and the molecular formula are the same.

Table 7C-4 Compounds with the Empirical Formula CH_2O

Molecular Formula	Name
CH_2O	Formaldehyde
$C_2H_4O_2$	Acetic Acid
$C_3H_6O_3$	L-Lactic Acid
$C_4H_8O_4$	D-Threose
$C_5H_{10}O_5$	D-Ribose
$C_6H_{12}O_6$	D-Fructose
$C_7H_{14}O_7$	D-Mannoheptulose (from avocados)

Percent composition describes the mass composition of a compound. All the other formulas describe the numbers of atoms in substances. The percent composition deals with the masses of the atoms. This is an important difference.

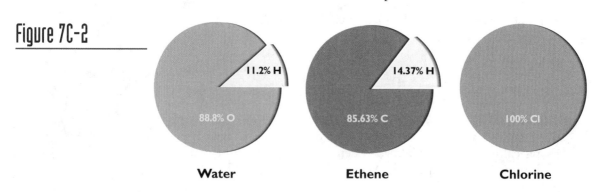

Figure 7C-2

Percent compositions often look totally different from empirical formulas. For example, although water has only one oxygen atom, this one atom contains the majority of the compound's mass. The two hydrogen atoms do not contribute much mass because they are so small. Likewise, ethene has more hydrogen atoms than carbon atoms, but more mass comes from carbon because the carbon atoms are much more massive.

A laboratory analysis of a substance is usually expressed as a percent composition. It is essential to understand that "percent" literally means "per hundred." A general setup to calculate any percent is part/whole × 100%. Suppose that a 60.00-gram sample of water were decomposed into its elements and that 53.28 grams of oxygen and 6.72 grams of hydrogen gas resulted. The percent composition of water could be calculated as follows:

Hydrogen: There are 6.72 grams of hydrogen in 60.00 grams of water.

$$\frac{\text{part}}{\text{whole}} \times 100\% = \frac{6.72 \text{ g H}}{60.00 \text{ g H}_2\text{O}} \times 100\%$$

$$= 0.112 \times 100\% = 11.2\%$$

Oxygen: There are 53.28 grams of oxygen in 60.00 grams of water.

$$\frac{\text{part}}{\text{whole}} \times 100\% = \frac{53.28 \text{ g O}}{60.00 \text{ g H}_2\text{O}} \times 100\%$$

$$= 0.8880 \times 100\% = 88.80\%$$

Sample Problem A laboratory analysis of a 30.00-gram sample of $Al_2(SO_4)_3$ showed that it contained 4.731 grams of aluminum, 8.433 grams of sulfur, and 16.836 grams of oxygen. What is the percent composition of this compound?

Solution

Al: $\dfrac{4.731 \text{ g Al}}{30.00 \text{ g Al}_2(SO_4)_3} \times 100\% = 0.1577 \times 100\% = 15.77\%$

S: $\dfrac{8.433 \text{ g S}}{30.00 \text{ g Al}_2(SO_4)_3} \times 100\% = 0.2811 \times 100\% = 28.11\%$

O: $\dfrac{16.836 \text{ g O}}{30.00 \text{ g Al}_2(SO_4)_3} \times 100\% = 0.5612 \times 100\% = 56.12\%$

> **Sample Problem** How many grams of oxygen would a 65.00-gram sample of $Al_2(SO_4)_3$ contain?
>
> **Solution**
> Because of the law of definite composition, any other sample of $Al_2(SO_4)_3$ will have the same percent composition that the 30.00-gram sample did in the previous problem. The problem could be worked by unit analysis as in the preceding sample problem.

Calculations with Empirical Formulas

Empirical formulas contain the information necessary to calculate the percent composition of compounds. To find percent composition, the mole ratio in the empirical formula must be converted to a mass ratio through a series of calculations.

The formula H_2O means that there are exactly 2 moles of hydrogen atoms for every 1 mole of oxygen atoms. One mole of water contains 2 moles of hydrogen atoms and 1 mole of oxygen atoms. Expressed in masses, 1 mole of water contains 2.016 grams of hydrogen atoms and 16.00 grams of oxygen atoms.

$$\frac{2 \text{ mol H}}{} \bigg| \frac{1.008 \text{ g H}}{1 \text{ mol H}} = 2.016 \text{ g H}$$

$$\frac{1 \text{ mol O}}{} \bigg| \frac{16.00 \text{ g O}}{1 \text{ mol O}} = 16.00 \text{ g O}$$

The total amount of mass being considered is 2.016 grams of hydrogen plus 16.00 grams of oxygen, or 18.02 grams of water. The percent composition of water can now be found.

$$\frac{2.016 \text{ g H}}{18.02 \text{ g H}_2\text{O}} \times 100\% = 0.1119 \times 100\% = 11.19\%$$

$$\frac{16.00 \text{ g O}}{18.02 \text{ g H}_2\text{O}} \times 100\% = 0.8879 \times 100\% = 88.79\%$$

Laboratories often do the reverse process. They first determine the mass composition of an unknown compound and then calculate an empirical formula. Suppose that a chemist is given 100.0 grams of an unknown compound and is told to determine its empirical formula. After a careful analysis in an analytical laboratory, the chemist concludes that 75.00 grams (75.00%) of the sample's mass is carbon. The other 25.00 grams (25.00%) is hydrogen. To determine the relative number of atoms that are present, he must use the mole concept. Remember that empirical formulas are based on mole ratios, not mass ratios.

Finding the masses of substances before and after reactions can reveal the substances' empirical formulas.

Because carbon atoms have much more mass than hydrogen atoms, the empirical formula is nothing like $C_{75}H_{25}$ or C_3H_1. The key questions are "How many moles of carbon atoms are in the sample?" and "How many moles of hydrogen atoms are present?" To determine the mole composition, the mass composition of the sample must be used.

$$\frac{75.00 \text{ g C} \mid 1 \text{ mol C}}{\mid 12.01 \text{ g C}} = 6.245 \text{ mol C}$$

$$\frac{25.00 \text{ g H} \mid 1 \text{ mol H}}{\mid 1.008 \text{ g H}} = 24.80 \text{ mol H}$$

There are 6.245 moles of carbon atoms for every 24.80 moles of hydrogen atoms. The empirical formula could be written as $C_{6.245}H_{24.80}$. Although this formula is numerically accurate, it is not in its final form. Empirical formulas are written as ratios of simple whole numbers. Dividing both numbers by the smaller number gives the simplest form of the ratio and guarantees that one of the subscripts will be a 1.

$$\text{mol C: mol H} = \frac{6.245}{6.245} : \frac{24.80}{6.245} = 1:3.97 \text{ or } 1:4$$

The empirical formula of the compound is C_1H_4 or CH_4

The steps of the process can be written in a flow chart.

Table 7C-5

Percent composition	Mass composition	Mole composition	Mole ratio	Empirical formula
75.00% C	75.00 g C	6.245 moles C	1 mole C	CH_4
25.00% H	25.00 g H	24.80 moles H	4 moles H	

In some problems the mole ratios might not be small whole numbers. If a ratio of 1:1.99 resulted from a calculation, it would be logical to round it off to 1:2. Errors in measuring the mass composition are the most common cause of these minute discrepancies. Sometimes ratios such as 1:1.5 or 1:1.33 result. These should not be rounded off. In general, if the decimal portion of the noninteger part of the ratio is greater than 0.1 and less than 0.9, you should not round off. The ratio 1:1.5 is equivalent to the whole-number ratio of 2:3. The ratio 1:1.33 is equivalent to the whole-number ratio 3:4. Insight and practice are necessary for knowing when to round off and when to seek another form of the ratio.

Sample Problem A laboratory analysis of an unknown gas has determined that the gas is 72.55% oxygen and 27.45% carbon by mass. What is the empirical formula of the compound?

Solution
Percent composition: 72.55% O
  27.45% C

Mass composition: Any quantity of sample could be selected; the percentages would be the same. But if we choose the sample size to be exactly 100 grams, then the "%" unit can simply be changed to "g," and the calculation is thus simplified. Therefore, in a 100-gram sample of the gas, there will be 72.55 grams of oxygen and 27.45 grams of carbon.

Mole composition of sample:

$$\frac{72.55 \text{ g O}}{} \cdot \frac{1 \text{ mol O}}{16.00 \text{ g O}} = 4.534 \text{ mol O}$$

$$\frac{27.45 \text{ g C}}{} \cdot \frac{1 \text{ mol C}}{12.01 \text{ g C}} = 2.286 \text{ mol C}$$

Mole ratio:
  mol O:mol C = 4.534 : 2.286 reduced to lowest terms is

$$\text{mol O:mol C} = \frac{4.534}{2.286} : \frac{2.286}{2.286} = 1.983:1.000$$

The slight difference between 1.983 and 2.000 can be attributed to experimental error.

Notice that in these problems we use the gram-atomic masses of elements, even for those that naturally exist as diatomic molecules in their uncombined form. Such elements are combined with one or more other elements; we therefore do not use their gram-molecular masses.

Empirical formula: For every 2 moles of oxygen, there is 1 mole of carbon. The empirical formula must be CO_2.

Sample Problem A 5.000-gram sample of an unknown compound contains 1.844 grams of nitrogen and 3.156 grams of oxygen. Find the empirical formula.

Solution

The mass composition of the sample is already known.

Mass composition: 1.844 g N
3.156 g O

Mole composition of sample:

$$\frac{1.844 \text{ g N} \mid 1 \text{ mol N}}{\mid 14.01 \text{ g N}} = 0.1316 \text{ mol N}$$

$$\frac{3.156 \text{ g O} \mid 1 \text{ mol O}}{\mid 16.00 \text{ g O}} = 0.1973 \text{ mol O}$$

Mole ratio:
mol N:mol O = 0.1316 : 0.1973 reduced to lowest terms is

$$\text{mol N:mol O} = \frac{0.1316}{0.1316} : \frac{0.1973}{0.1316} = 1.000 : 1.499$$

The ratio should be 1:1.5 (N:O).

Empirical formula: The ratio 1:1.5 is in its simplest form, but the numbers are not whole numbers. To put the ratio in whole numbers, express the numbers as fractions.

$$1:1.5 = 2/2 : 3/2$$

Eliminate the fractions by multiplying through by the common denominator.

$$(2) \; 2/2 : (2) \; 3/2 = 2:3$$

The empirical formula of the compound is N_2O_3.

Section Review Questions 7C

1. What is Avogadro's number, and what physical significance does it have?
2. What is the difference between an empirical formula and a molecular formula?
3. Give the structural, molecular, and empirical formulas of hydrogen peroxide.
4. How many moles of atoms are in 1.00 lb. (454 g) of lead?
5. Find the gram-molecular mass of vitamin A ($C_{20}H_{30}O$).
6. DDT, an insecticide banned in the United States, has the formula $C_{14}H_9Cl_5$. Find its percent composition.
7. Find the empirical formula for the explosive TNT, which is composed of 37.0% carbon, 2.22% hydrogen, 18.5% nitrogen, and 42.3% oxygen.

Chapter Review

Coming to Terms

oxidation number
peroxide
nomenclature
flow chart
Greek prefix system
binary compound
oxyanion
polyatomic compound
Stock system
Roman numeral system
hydrate
anhydrous

binary acid
ternary acid
mole
Avogadro's number
gram-atomic mass
gram-molecular mass
gram-formula mass
structural formula
molecular formula
empirical formula
percent composition

Review Questions

1. Explain why
   a. the oxidation number of F is always negative.
   b. the oxidation numbers of alkali metals are always positive.
   c. elements such as P, N, and S have positive oxidation numbers in some compounds but negative oxidation numbers in others.

2. When the following pairs of atoms bond, which atom gets the positive oxidation number?
   a. H, O   b. Na, S   c. N, S   d. Na, H

3. Greek Prefixes
   a. Give the oxidation number of each atom in the following compounds:
      1. N_2O_3 3. P_4O_6 5. PCl_3
      2. I_2O_5 4. S_2Cl_2 6. Cl_2O_7
   b. Name each compound in question 3a.
   c. Give formulas for the following compounds:
      1. carbon disulfide
      2. sulfur trioxide
      3. boron trichloride
      4. phosphorus pentabromide
      5. dinitrogen pentasulfide
      6. dibromine monoxide

4. Binary Ionic Compounds
   a. Give the oxidation number of each atom in the following compounds:
      1. LiCl
      2. Mg_3N_2
      3. CaO
      4. NaI
      5. Al_2S_3
      6. CuCl
   b. Name each compound in question 4a.
   c. Give formulas for the following compounds:
      1. zinc chloride
      2. calcium phosphide
      3. potassium chloride
      4. barium chloride
      5. strontium oxide
      6. calcium chloride

5. Polyatomic Compounds
   a. Give the oxidation number of each atom in the following compounds:
      1. $AgNO_3$
      2. NH_4NO_3
      3. $NaNO_2$
      4. $Zn_3(PO_4)_2$
      5. $(NH_4)_2S$
      6. $Zn(C_2H_3O_2)_2$
   b. Name each compound in question 5a.
   c. Give formulas for the following compounds:
      1. ammonium bromate
      2. potassium permanganate
      3. barium phosphate
      4. aluminum acetate
      5. calcium carbonate
      6. barium chromate

6. Atoms with Multiple Oxidation States
   a. Give the oxidation number of each atom in the following compounds:
      1. $PbCl_2$
      2. HgS
      3. CoS
      4. $Fe(OH)_3$
      5. $Pb(CrO_4)$
      6. $Sn(C_2H_3O_2)_2$
   b. Name each compound in question 6a. Give both the Stock system and common name.
   c. Give formulas for the following compounds:
      1. iron (III) oxide
      2. copper (I) hydroxide
      3. lead (IV) chromate
      4. mercury (I) chloride
      5. lead (II) arsenate
      6. tin (IV) chloride

7. Hydrates
   a. Name the following compounds:
      1. $CaSO_4 \cdot 2 H_2O$
      2. $MgSO_4 \cdot 7 H_2O$
      3. $Na_2S_2O_3 \cdot 5 H_2O$
      4. $Na_2SO_4 \cdot 10 H_2O$
      5. $NiSO_4 \cdot 6 H_2O$
      6. $FeSO_4 \cdot 7 H_2O$
   b. Give formulas for the following compounds:
      1. iron (III) bromide hexahydrate
      2. barium chloride dihydrate
      3. lead (II) acetate decahydrate
      4. cobalt (II) chloride hexahydrate
      5. sodium tetraborate decahydrate
      6. magnesium carbonate pentahydrate

8. Binary Acids
   a. Name the following compounds, and give both the common name and the acid name.
      1. HF
      2. H_2Te
   b. Give formulas for the following compounds:
      1. Hydroselenic acid
      2. Hydroiodic acid

9. Ternary Acids
   a. Identify the anion and give the ternary acid name for the following compounds:
      1. H_3AsO_4
      2. $HC_2H_3O_2$
      3. H_2CO_3
      4. $HMnO_4$
      5. $H_2C_2O_4$
   b. Give formulas for the following compounds:
      1. periodic acid
      2. chromic acid
      3. phosphoric acid
      4. cyanic acid
      5. bromic acid

10. Give the gram-atomic mass for H, Sc, As, I, and U. Give your answers with four significant figures and the correct units.

11. How many atoms are in the following?
    a. 12.01 g of C
    b. 16.00 g of O
    c. 1.008 g of H

12. How many moles
    a. of Fe are in 37.0 g of Fe?
    b. of Kr are in 4.58×10^{20} Kr atoms?
    c. of $NaIO_3$ are in 3.25×10^{26} formula units of $NaIO_3$?
    d. of N_2O_4 are in 26.75 g of N_2O_4?
    e. of $Na_2SO_4 \cdot 10\,H_2O$ are in 8.99×10^{24} formula units of Na_2SO_4?
    f. of aspirin are in one aspirin tablet (0.324 g)? The formula for aspirin is $C_9H_8O_4$.

13. Calculate the mass of 1 mole for the following compounds. For each compound, tell whether the mass you calculate is a gram-formula mass or gram-molecular mass.
    a. N_2O_3
    b. Mg_3N_2
    c. $AgNO_3$
    d. $PbCl_2$
    e. H_2Te

14. How many
    a. Fe atoms are in 0.256 mole of Fe?
    b. Kr atoms are in 3.87 g of Kr?
    c. $AlCl_3$ formula units are in 6.17 moles of $AlCl_3$?
    d. $NaIO_3$ formula units are in 8.58 g of $NaIO_3$?
    e. N_2O_4 molecules are in 8.16 g of N_2O_4?
    f. $Na_2SO_4 \cdot 10\,H_2O$ formula units are in 3.87 moles of $Na_2SO_4 \cdot 10\,H_2O$?
    g. H_2O_2 molecules are in 1.00 L of 3% H_2O_2 purchased at the drug store? (Assume a density of 1.00 g/mL.)

15. What is the mass (in grams) of
    a. 6.58 moles Fe?
    b. 8.58×10^{28} Kr atoms?
    c. 1.05 moles $AlCl_3$?
    d. 3.17×10^{18} formula units $NaIO_3$?
    e. 0.0387 mole N_2O_4?
    f. 5.41×10^{26} formula units $Na_2SO_4 \cdot 10\ H_2O$?
    g. 2.51×10^{23} formula units of $(NH_4)_2C_2O_4 \cdot H_2O$?

16. Ferrous sulfate, $FeSO_4$, is a therapeutic agent for iron deficiency anemia. It is administered orally as ferrous sulfate heptahydrate. If 0.300 g of $FeSO_4 \cdot 7\ H_2O$ contains 0.0603 g of Fe, 0.0346 g of S, 0.190 g of O, and 0.0151 g of H, what is the percent composition of $FeSO_4 \cdot 7\ H_2O$?

17. Vitamin E occurs naturally in vegetable oil where it acts as an antioxidant. Some people take 0.500 g of vitamin E daily as a nutritional supplement. Each 0.500 g contains 0.404 g of C, 0.0585 g of H, and 0.0372 g of oxygen. What is the percent composition of Vitamin E?

18. What mass of vitamin E will contain 0.500 g of C? (Use your answers from problem 17 to help you.)

19. Dinitrogen oxide, commonly called nitrous oxide or laughing gas, was once commonly used as an anesthetic. N_2O is 63.65% N. What mass of N_2O contains 4.850 g N?

20. Epsom salts that are used as a laxative consist of $MgSO_4 \cdot 7\ H_2O$. How many grams of Mg are present in 0.0250 mole of $MgSO_4 \cdot 7\ H_2O$, which is 9.86% Mg?

21. Manganese (II) acetate is used to make dyes permanent. $Mn(C_2H_3O_2)_2$ is 27.8% C, 3.50% H, and 37.0% O. What mass of Mn is found in 125 g of $Mn(C_2H_3O_2)_2$?

22. Limestone, which is foundational to cement, consists of calcium carbonate. If 35.80 kg of $CaCO_3$ contain 4.296 kg C, 14.33 kg Ca, and 17.17 kg O, what is the percent composition of $CaCO_3$?

23. Aspirin is the common name for acetyl salicylic acid. If 100 g of aspirin contains 60.00 g C, 4.480 g H, and 35.53 g O, what is its empirical formula?

24. Many aspirin substitutes in the United States contain acetaminophen as the active ingredient. Acetaminophen is 63.56% C, 6.00% H, 9.27% N, and 21.17% O. What is the empirical formula for acetaminophen?

25. Lidocaine is a widely used local anesthetic. A laboratory analysis of lidocaine reveals that a 5.000 g sample of lidocaine contains 3.588 g C, 0.473 g H, 0.598 g N, and 0.342 g O.
    a. What is the percent composition of lidocaine?
    b. What is the empirical formula of lidocaine?

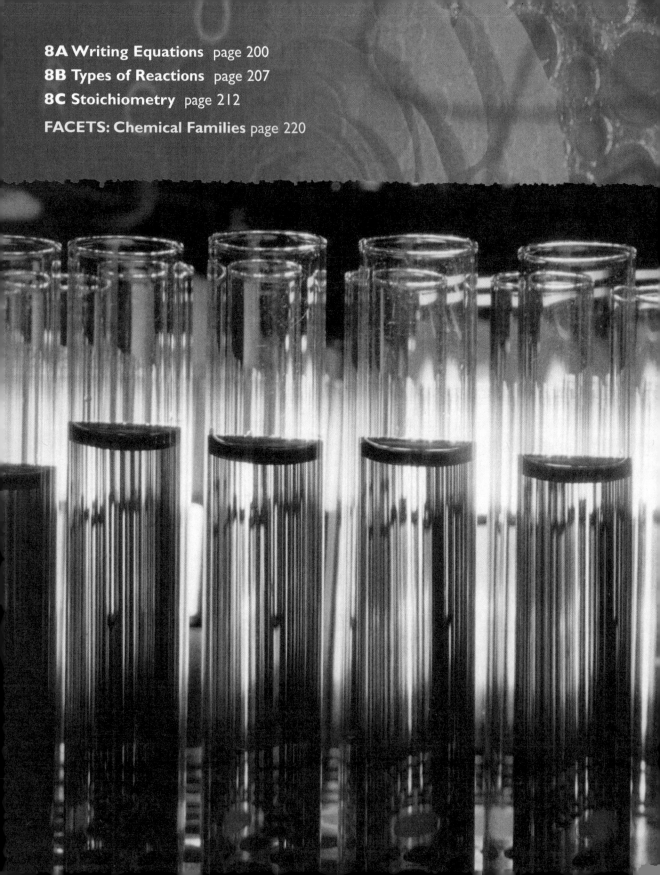

8A Writing Equations page 200
8B Types of Reactions page 207
8C Stoichiometry page 212
FACETS: Chemical Families page 220

Describing Chemical Reactions 8

The dissolving of antacid tablets leads to the escape of carbon dioxide bubbles. Some water softeners cause solid, white calcium carbonate to settle out from hard water. Invisible nitric oxide gas from the exhaust of cars turns into reddish brown nitrogen dioxide when it reacts with atmospheric oxygen. The combination of natural gas and oxygen releases energy to heat a home.

In each of these cases, one set of chemical substances forms from some other set of substances. In the process the characteristics of the chemicals will change, and there will be visible signals telling of the unseen molecular changes.

In a way a chemical reaction is like the change in a person's heart at salvation. When the Holy Spirit acts in a person's heart and he accepts Jesus Christ as his Savior, there are visible changes in that person's life. Second Corinthians 5:17 tells us that "if any man be in Christ, he is a new creature: old things are passed away; behold, all things are become new." Salvation brings a love for God, a desire to follow His commands, and many other evidences that God has changed the heart of a person.

8A Writing Equations: What Goes in Must Come Out

Knowing that a reaction occurred is just the starting point. What substances went into the reaction? What substances came out? How much of each substance is involved? Chemists pack the answers to these questions into shorthand expressions called **chemical equations.**

What Equations Do

Unlike mathematical equations, chemical equations do not show equalities. Instead, they represent processes called *chemical reactions*. To tell about reactions, they must do several things.

1. Equations must identify all the substances involved in the reaction. Some water softeners remove calcium—salts such as calcium hydrogen carbonate—from hard water by adding calcium hydroxide. The two compounds react to form water and calcium carbonate, which settles out of the solution. A word equation shows all the substances that are involved.

    calcium hydrogen carbonate + calcium hydroxide $\longrightarrow$ water + calcium carbonate

2. Equations must show the composition of the substances. Since molecular and unit formulas are more informative than names, they are used in equations.

    $Ca(HCO_3)_2 + Ca(OH)_2 \longrightarrow H_2O + CaCO_3$

3. Equations must account for all the atoms involved in the reaction. The law of mass conservation states that matter cannot be created or destroyed in chemical reactions. Applied to equations, this law says, "What goes in must come out." It must be noted that this is on the atomic level. Different compounds will be formed, but the total number of atoms and the types of atoms will remain constant.

As it now stands, the equation above does not show the conservation of atoms. A total of two calcium atoms, four hydrogen atoms, two carbon atoms, and eight oxygen atoms enter the reaction. After the reaction, the equation shows one calcium atom, two hydrogen atoms, one carbon atom, and four oxygen atoms.

$$2 \text{ Ca atoms} \longrightarrow 1 \text{ Ca atom}$$
$$4 \text{ H atoms} \longrightarrow 2 \text{ H atoms}$$
$$2 \text{ C atoms} \longrightarrow 1 \text{ C atom}$$
$$8 \text{ O atoms} \longrightarrow 4 \text{ O atoms}$$

Can sixteen atoms really turn into eight atoms? Absolutely not! To be correct, the equation must show equal numbers of atoms before *and* after the reaction.

Equations that account for all atoms and indicate that the mass of matter involved does not change are called **balanced chemical equations.** Balancing an unbalanced equation involves adjusting the number of molecules, ions, atoms, or formula units. By placing the number 2 in front of both of the compounds on the right side of the equation, the number of atoms on each side of the equation is the same—it is balanced.

$$Ca(HCO_3)_2 + Ca(OH)_2 \longrightarrow 2\ H_2O + 2\ CaCO_3$$

Now each side of the equation has two calcium atoms, four hydrogen atoms, two carbon atoms, and eight oxygen atoms. The equation does what it is supposed to do; it accurately describes the reaction.

Parts of an Equation

Substances that are present before the reaction are called **reactants.** Substances that emerge from the reaction are called **products.** *Coefficients* tell how many atoms, molecules, ions, or formula units are present. An *arrow* separates the reactants from the products and shows the direction of the reaction.

$$\underset{\text{Reactants}}{Ca(HCO_3)_2 + Ca(OH)_2} \longrightarrow \underset{\text{Products}}{2\ H_2O + 2\ CaCO_3}$$

(Coefficients indicated on the 2's)

Special Symbols in Equations

Additional information can be packed into equations with the use of special symbols. Double half-arrows between reactants and products show that the reaction goes forward as well as backward.

$$3\ Fe + 4\ H_2O \rightleftharpoons Fe_3O_4 + 4\ H_2$$

The physical states of the substances can also be indicated. The gaseous state is indicated by (g) immediately after the formula. If the gas is a product, an upward arrow ($\uparrow$) is sometimes used. A liquid is represented by (l) after the formula, and a solid is shown by (s). If the solid falls out of a solution, it is called a **precipitate.** The process of **precipitation** is sometimes noted with a downward arrow ($\downarrow$). Since steam is gaseous water, we write the above equation as follows.

$$3\ Fe\ (s) + 4\ H_2O\ (g) \rightleftharpoons Fe_3O_4\ (s) + 4\ H_2\ (g)$$
$$\text{or, } 3\ Fe\ (s) + 4\ H_2O\ (g) \rightleftharpoons Fe_3O_4 \downarrow + 4\ H_2 \uparrow$$

If a substance is dissolved in water, (aq), meaning "aqueous," is placed after the formula.

$AgNO_3$ (aq) + NaCl (aq) ⟶ AgCl (s) + $NaNO_3$ (aq)
or, $AgNO_3$ (aq) + NaCl (aq) ⟶ AgCl ↓ + $NaNO_3$ (aq)

Technically speaking, we should put (aq) after every compound we give an acid name. For example, H_2SO_4 is hydrogen sulfate, but H_2SO_4 (aq) is sulfuric acid. Often this rule is ignored, but it should not be forgotten. Symbols above and below the reaction arrow are often used to tell about special reaction conditions. A Δ above the arrow means that the reactants are heated. Other descriptions of pressure, light, or specific temperatures can also be placed above the arrow. **Catalysts** are substances that change the rate of the reaction but do not undergo permanent changes themselves. Their symbols can also be placed above the arrow.

$$2\ KClO_3\ (s) \xrightarrow{\Delta,\ Fe_2O_3} 2\ KCl\ (s) + 3\ O_2\ (g)$$

Table 8A-1 Symbols Used in Chemical Equations

Symbol	Use
+	Between the formulas of individual reactants and products
⟶	Means "yields" or "produces"; separates reactants from products
=	Same as arrow
⇌	Used in place of a single arrow for reversible reactions
(g)	Indicates a gaseous reactant or product
↑	Sometimes used to indicate a gaseous product
(s)	Indicates a solid reactant or product
↓	Sometimes used to indicate a solid product
(l)	Indicates a liquid reactant or product
(aq)	Indicates that the reactant or product is in aqueous solution (dissolved in water)
$\xrightarrow{\Delta}$	Indicates that heat must be supplied to reactants before a reaction occurs
$\xrightarrow{MnO_2}$	An element or compound written above the arrow is a *catalyst*; a catalyst speeds up a reaction but is not consumed in the reaction

Balancing Equations by Inspection

Balancing an equation by inspection involves adjusting coefficients to show that atoms and mass are conserved. The object is to write an equation that has equal numbers of each kind of atom in the reactants and products. Skill in balancing equations will improve with practice, but some general guidelines will be helpful in the meantime.

Consider the reaction that occurs when nitrogen monoxide mixes with oxygen in the atmosphere.

nitrogen monoxide + oxygen ⟶ nitrogen dioxide

1. Write the correct formulas for all reactants and products. The formulas of covalent compounds can often be obtained from their names. (Nitrogen monoxide is NO, and nitrogen dioxide is NO_2.) Remember that some elements exist as diatomic molecules. The formulas of ionic compounds can be found with the use of oxidation numbers. Make sure that all the formulas are correct. Once they are, **do not change them!**

$$NO + O_2 \longrightarrow NO_2$$

2. Count atoms of each kind to see whether the equation is already balanced. If there are polyatomic ions that do not change in the reaction, treat them as single units.

$$1 \text{ N atom} \longrightarrow 1 \text{ N atom}$$
$$3 \text{ O atoms} \longrightarrow 2 \text{ O atoms}$$

3. Adjust coefficients until there are equal numbers of atoms on both sides of the arrow. Remember—change *coefficients*, not *subscripts*. The formulas of the reactants and products have already been determined. Never change a compound's formula in order to balance an equation! It is best to start with the most complicated molecules and save simple molecules like O_2 and individual elements until last. In the example above, more oxygen atoms are needed in the product. The only way to get more oxygen atoms in the product is to add another NO_2 molecule.

$$NO + O_2 \longrightarrow \mathbf{2}\ NO_2$$

Now there are two nitrogen atoms in the product but only one in the reactants. An additional NO molecule will supply the needed nitrogen and will balance the oxygen atoms.

$$\mathbf{2}\ NO + O_2 \longrightarrow 2\ NO_2$$

4. Always check to be sure that the coefficients are all whole numbers and that they are in the simplest ratio possible. If the equation looked like $NO + \frac{1}{2} O_2 \longrightarrow NO_2$, it would be numerically correct, but not all coefficients would be whole numbers. You should multiply all the coefficients by a common denominator so that they become whole numbers.

$$\mathbf{2}\ NO + \mathbf{2}\ (\tfrac{1}{2}\ O_2) \longrightarrow \mathbf{2}\ NO_2$$
$$2\ NO + O_2 \longrightarrow 2\ NO_2$$

If the equation looked like $4\ NO + 2\ O_2 \longrightarrow 4\ NO_2$, it would again be numerically correct, but the coefficients would not be in the simplest possible ratio. Dividing each coefficient by the lowest common factor yields the lowest possible ratio.

$$\tfrac{4}{2}\ NO + \tfrac{2}{2}\ O_2 \longrightarrow \tfrac{4}{2}\ NO_2$$
$$2\ NO + O_2 \longrightarrow 2\ NO_2$$

Sample Problem Iron and gaseous chlorine can react to form iron (III) chloride. Write and balance the equation for this reaction.

Solution
Write the formulas. Iron atoms can react as individual atoms, but gaseous chlorine atoms exist as diatomic molecules. The formula of iron (III) chloride can be found with the aid of oxidation numbers.

$$Fe + Cl_2 \longrightarrow FeCl_3$$

Count the atoms.

$$1 \text{ Fe atom} \longrightarrow 1 \text{ Fe atom}$$
$$2 \text{ Cl atoms} \longrightarrow 3 \text{ Cl atoms}$$

Adjust the coefficients. Start with the element having the largest subscript in the most complex formula—the Cl in the $FeCl_3$. In order to balance an element with subscripts of 2 and 3, you will need to use the lowest common multiple (6). This will give coefficients of 3 and 2, as follows:

$$Fe + 3\ Cl_2 \longrightarrow 2\ FeCl_3$$

The chlorine atoms are balanced, but now the iron atoms are not. Another iron atom is needed.

$$2\ Fe + 3\ Cl_2 \longrightarrow 2\ FeCl_3$$

Check. Atoms are balanced, and all coefficients are whole numbers in the simplest ratio possible.

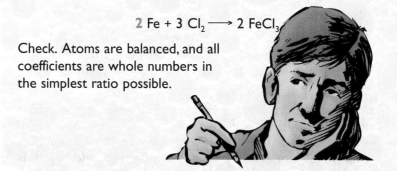

Sample Problem Ethane (C_2H_6) burns in oxygen gas, producing carbon dioxide and water. Write and balance the equation for this reaction.

Solution
When writing the formulas, remember that oxygen exists as diatomic molecules.

$$C_2H_6 + O_2 \longrightarrow CO_2 + H_2O$$

Start with the most complex formula (C_2H_6), and leave any free elements (O_2) till last. Balancing carbon and hydrogen atoms results in an equation that is still unbalanced.

$$C_2H_6 + O_2 \longrightarrow 2\ CO_2 + 3\ H_2O$$

Now seven oxygen atoms are in the products. Since oxygen molecules exist as two-atom units, we divide the number of atoms needed (7) by the number found in each unit (2) to give $\frac{7}{2}$ as a coefficient.

$$C_2H_6 + \frac{7}{2}\ O_2 \longrightarrow 2\ CO_2 + 3\ H_2O$$

The coefficient $\frac{7}{2}$ is not a whole number. All coefficients should be doubled to complete the balancing act.

$$2\ C_2H_6 + 7\ O_2 \longrightarrow 4\ CO_2 + 6\ H_2O$$

Sample Problem Aqueous aluminum nitrate and aqueous sodium carbonate react to form a precipitate of aluminum carbonate and aqueous sodium nitrite. Write a balanced equation for this reaction.

Solution
First, write the correct formulas for all of the reactants and products, using charges and oxidation numbers as aids. Be sure to use parentheses if there is more than one polyatomic ion in the formula.

$$Al(NO_3)_3\ (aq) + Na_2CO_3\ (aq) \longrightarrow Al_2(CO_3)_3 \downarrow + NaNO_3\ (aq)$$

Count the atoms or polyatomic ions. Since the nitrate and carbonate ions are unchanged in the reaction, treat them as units.

$$1\ Al^{3+}\ \text{ion} \longrightarrow 2\ Al^{3+}\ \text{ions}$$
$$3\ NO_3^-\ \text{ions} \longrightarrow 1\ NO_3^-\ \text{ion}$$
$$2\ Na^+\ \text{ions} \longrightarrow 1\ Na^+\ \text{ion}$$
$$1\ CO_3^{2-}\ \text{ion} \longrightarrow 3\ CO_3^{2-}\ \text{ions}$$

Adjust the coefficients. You should begin with the CO_3^{2-} ion in the product, $Al_2(CO_3)_3$. Placing a coefficient of 3 in front of the reactant containing the carbonate ion will result in 6 Na^+ ions on the left; thus, a coefficient 6 is needed for $NaNO_3$ on the right:

$$Al(NO_3)_3\ (aq) + 3\ Na_2CO_3\ (aq) \longrightarrow Al_2(CO_3)_3 \downarrow + 6\ NaNO_3\ (aq)$$
$$\text{(unbalanced)}$$

Now the sodium and carbonate ions are balanced. The coefficient of 6 which balanced the sodium also affected the number of nitrate ions, giving six on the right side. A coefficient of 2 for Al(NO$_3$)$_3$ will balance the nitrate ions as well as the aluminum ions.

$$2\text{ Al(NO}_3)_3 \text{ (aq)} + 3\text{ Na}_2\text{CO}_3 \text{ (aq)} \longrightarrow \text{Al}_2(\text{CO}_3)_3 \downarrow + 6\text{ NaNO}_3 \text{ (aq)}$$

Check. The number of monatomic and polyatomic ions is equal on both sides of the equation. Note that the balancing process often involves a right-left-right movement, since changing a coefficient for one compound alters the number of all the atoms in the compound. Often this requires an adjustment of a coefficient on the other side of the equation. Do not attempt to "fix" everything on one side of the equation before moving to the other side!

Although balanced equations give much information about reactions, they do have several limitations:

1. The fact that an equation can be written does not mean that the reaction can occur. The equation below is balanced, but the reaction it represents will not occur.

$$\text{Ag} + \text{NaCl} \longrightarrow \text{Na} + \text{AgCl}$$

2. Equations do not tell whether a reaction goes to completion. Some reactions leave a mixture of reactants and products.

3. Equations do not show how a reaction occurs. Some reactions involve more than one step. The equation does not show the steps of a reaction or the order in which those steps take place. For example, the reaction

$$4\text{ C (graphite)} + 6\text{ H}_2 \text{ (g)} + \text{O}_2 \text{ (g)} \longrightarrow 2\text{ C}_2\text{H}_5\text{OH (l)}$$

is the sum of three more basic reactions:

$$4\text{ C (graphite)} + 4\text{ O}_2 \text{ (g)} \longrightarrow 4\text{ CO}_2 \text{ (g)}$$
$$6\text{ H}_2 \text{ (g)} + 3\text{ O}_2 \longrightarrow 6\text{ H}_2\text{O (l)}$$
$$4\text{ CO}_2 \text{ (g)} + 6\text{ H}_2\text{O (l)} \longrightarrow 2\text{ C}_2\text{H}_5\text{OH (l)} + 6\text{ O}_2$$

Section Review Questions 8A

1. Balance the following equations, applying the guidelines discussed in this section.
   a. $\text{P}_4 \text{ (s)} + \text{S}_8 \text{ (s)} \longrightarrow \text{P}_4\text{S}_3 \text{ (s)}$
   b. $\text{KClO}_3 \text{ (s)} \xrightarrow{\Delta} \text{KCl (s)} + \text{O}_2 \text{ (g)}$

c. $AgNO_3$ (aq) + Cu (s) $\longrightarrow$ $Cu(NO_3)_2$ (aq) + Ag (s)
  d. H_3PO_4 (aq) + $Ba(OH)_2$ (aq) $\longrightarrow$ $Ba_3(PO_4)_2$ (s) + H_2O (l)
  e. $NaHCO_3$ (s) $\xrightarrow{\Delta}$ Na_2CO_3 (s) + CO_2 (g) + H_2O (g)
2. Write balanced chemical equations for the following word equations.
   a. Aqueous calcium hydroxide reacts with gaseous sulfur trioxide, producing a precipitate of calcium sulfate and liquid water.
   b. Solid aluminum metal reacts with aqueous sulfuric acid, producing hydrogen gas and aqueous aluminum sulfate.
   c. Acetylene gas (C_2H_2) burns in oxygen, producing gaseous carbon dioxide and gaseous water.
   d. Table sugar (sucrose, $C_{12}H_{22}O_{11}$) burns in gaseous oxygen, producing gaseous carbon dioxide and gaseous water.
   e. Aqueous aluminum sulfate reacts with aqueous barium hydroxide, producing precipitates of aluminum hydroxide and barium sulfate.

8B Types of Reactions: A Classification Scheme

Most reactions can be classified on the basis of the chemical change that takes place. Classifying reactions leads to generalizations that in turn lead to a better understanding of the reactions. The major classes used in the scheme are combination, decomposition, single replacement, and double replacement reactions.

Combination Reactions

Combination reactions combine two or more substances into a single product. Since the major effect of these reactions is the production of a new substance, they are often called **synthesis reactions**. The general equation $A + B \longrightarrow AB$ represents combination reactions.

1. Metals and nonmetals other than oxygen can form compounds that are called salts.

$$Mg + F_2 \longrightarrow MgF_2$$

2. Metals can combine with oxygen to form metallic oxides.

$$2\ Mg + O_2 \longrightarrow 2\ MgO$$

A combination reaction occurs when magnesium reacts with atmospheric oxygen.

3. Nonmetals can react with oxygen to form oxides.
$$P_4 + 5\ O_2 \longrightarrow 2\ P_2O_5$$
4. Water and metal oxides can form metal hydroxides.
$$H_2O + CaO \longrightarrow Ca(OH)_2$$
5. Water and nonmetal oxides can combine to form oxyacids.
$$H_2O + SO_3 \longrightarrow H_2SO_4\ (aq)$$

Decomposition Reactions

The opposite of a combination reaction is a decomposition reaction. Whereas a combination reaction *combines* substances, a **decomposition reaction** breaks a substance down into two or more substances. Breaking compounds apart usually requires an input of energy. Decomposition reactions have the general form $AB \longrightarrow A + B$.

1. Oxygen can be driven out of some compounds.
    a. Metal oxides are usually very stable; however, some can be decomposed by heating to high temperatures.
    $$2\ HgO \xrightarrow{\Delta} 2\ Hg + O_2$$
    b. Metal chlorates can be heated to produce oxygen.
    $$2\ KClO_3 \xrightarrow{\Delta} 2\ KCl + 3\ O_2$$
    c. Water can be decomposed by an electric current.
    $$2\ H_2O \xrightarrow{elec} 2\ H_2 + O_2$$

2. Metal hydroxides can release gaseous water when heated.
$$Mg(OH)_2 \xrightarrow{\Delta} MgO + H_2O\ (g)$$

3. Metal carbonates can release carbon dioxide when heated.
$$CaCO_3 \xrightarrow{\Delta} CaO + CO_2$$

4. Some acids can be decomposed into nonmetal oxides and water.
$$H_2CO_3\ (aq) \longrightarrow CO_2 + H_2O$$

A carbonated drink goes flat as this reaction occurs. The carbonic acid that gives the tangy taste decomposes at room temperature.

5. Hydrates can release their water molecules when heated sufficiently.
$$BaCl_2 \cdot 2\ H_2O \xrightarrow{\Delta} BaCl_2 + 2\ H_2O$$

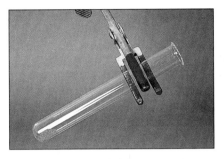
Orange ammonium dichromate can be decomposed by gentle heating.

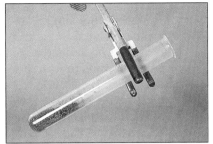
Green chromium (III) oxide, water vapor, and nitrogen gas result when ammonium dichromate decomposes.

Single Replacement Reactions

In **single replacement reactions,** an active element takes the place of a less active element in a compound. These kinds of reactions are also called displacement or substitution reactions. The general equation $A + BZ \longrightarrow B + AZ$ represents several types of single replacement reactions.

1. Atoms of active metals can replace less active ions from solutions and compounds. When a piece of solid zinc is placed in a copper (II) chloride solution, the zinc replaces the copper ions in solution and forces them to precipitate out as solid copper.

$$Zn\ (s) + CuCl_2\ (aq) \longrightarrow Cu\ (s) + ZnCl_2\ (aq)$$

Active metals can replace the hydrogen ions of acids. This is why some metals react to form bubbles of hydrogen gas when placed in acids. In these reactions, metal atoms replace hydrogen ions in acid molecules. The hydrogen bubbles off as H_2 gas, and a salt forms from the acid molecule.

$$Mg\ (s) + 2\ HCl\ (aq) \longrightarrow MgCl_2\ (aq) + H_2\ (g)$$

Very active metals react with water to produce hydrogen gas. In the equation below, HOH is an alternate formula for water that is used to show replacement reactions more clearly.

$$2\ Na\ (s) + 2\ HOH\ (l) \longrightarrow 2\ NaOH\ (aq) + H_2\ (g)$$

2. Active halogens can replace less active halogens that are in solution. When chlorine (an active element) gas is bubbled through a sodium bromide solution, chlorine replaces bromine in the solution. The bromine atoms that are replaced escape as Br_2 gas.

$$Cl_2\ (g) + 2\ NaBr\ (aq) \longrightarrow 2\ NaCl\ (aq) + Br_2\ (g)$$

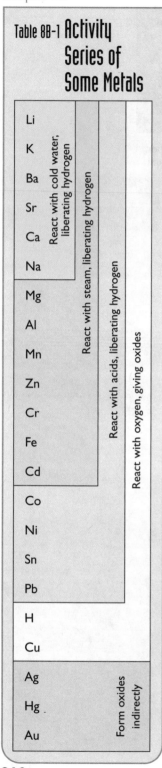

Table 8B-1 Activity Series of Some Metals

When will a single replacement reaction occur? This depends on the activity of the elements that are involved. An active element can be defined as one that has a strong tendency to lose or gain electrons to form bonds. If unbonded atoms are more active than the bonded atoms, a reaction will probably take place. The more active elements can force the less active elements from the bond. The most active metal is lithium; the least active is gold. Other metals can be arranged according to their activities in an **activity series,** as shown in Table 8B-1.

Elements at the top of the list are the most active. (Note that hydrogen is included even though it is not really a metal.) Predictions about whether a reaction is probable can be based on this series. For instance, will a reaction occur if barium is placed into a tin (II) chloride solution? Ba (s) + $SnCl_2$ (aq) ⟶ ? Barium is *higher* in the activity series, so it can replace tin and force it to precipitate.

$$Ba\ (s) + SnCl_2\ (aq) \longrightarrow BaCl_2\ (aq) + Sn\ (s)$$

Will gold metal react when it is placed in a sodium chloride solution? Gold is at the bottom of the activity series. Its tendency to lose electrons and form ionic bonds is nowhere near that of sodium. Since no reaction will occur, a person can safely dip a gold ring into salt water without losing a fortune.

$$Au\ (s) + NaCl\ (aq) \longrightarrow No\ reaction$$

With the aid of an activity series, it is also possible to predict which metals will react with acids. If a metal is more active than hydrogen, it can take the place of hydrogen. Because zinc is *above* hydrogen in the activity series, it will react with an acid such as dilute sulfuric acid.

$$Zn\ (s) + H_2SO_4\ (aq) \longrightarrow ZnSO_4\ (aq) + H_2\ (g)$$

Metals above magnesium readily replace hydrogen in water molecules.

$$Ba\ (s) + 2\ HOH\ (l) \longrightarrow Ba(OH)_2\ (aq) + H_2\ (g)$$

The halogens have their own activity series, which follows the same order in which they are listed in the periodic table. A halogen can replace another halogen that is below it in the series. Iodine, at the bottom of the series, cannot replace any of the other halogens because of its low activity.

$$Cl_2\ (g) + MgBr_2\ (aq) \longrightarrow MgCl_2\ (aq) + Br_2\ (g)$$
but
$$I_2\ (aq) + MgCl_2\ (aq) \longrightarrow No\ Reaction$$

Figure 8B-1 Halogen Activity Series

F_2
Cl_2
Br_2
I_2

Double Replacement Reactions

In **double replacement reactions** two compounds switch partners with each other. Equations of double replacement reactions have the general form $AX + BZ \longrightarrow AZ + BX$. Most double replacement reactions occur in an aqueous mixture of two ionic compounds. A precipitate often indicates that a double replacement reaction has occurred.

When lead (II) nitrate and potassium chromate solutions are mixed, a double replacement reaction occurs. Solid lead (II) chromate, a brilliant yellow compound that has been used as a pigment in paint, falls out of solution.

$$Pb(NO_3)_2 \ (aq) + K_2CrO_4 \ (aq) \longrightarrow PbCrO_4 \ (s) + 2 \ KNO_3 \ (aq)$$

An **ionic equation** represents all the particles present before and after the reaction. Ionic equations can be written only for reactions taking place in solution. It is the aqueous medium that allows the ionization to occur. Soluble ionic compounds are shown as separate ions, whereas insoluble ionic compounds are not.

Ionic equation:
$$Pb^{2+} \ (aq) + 2 \ NO_3^- \ (aq) + 2 \ K^+ \ (aq) + CrO_4^{2-} \ (aq) \longrightarrow$$
$$PbCrO_4 \ (s) + 2 \ K^+ + 2 \ NO_3^- \ (aq)$$

When lead (II) nitrate (colorless) is mixed with potassium chromate (clear orange), lead (II)-chromate precipitates, leaving potassium nitrate in the solution.

Ionic equations can include some non-ionic particles and some ions that remain unchanged. **Spectator ions** appear in the reactants and in the products. In the example above, NO_3^- and K^+ are the spectator ions. They do not precipitate or join other ions. They can be cancelled from both sides of the ionic equation, yielding the net ionic equation. **Net ionic equations** show only the ions that actually react; the spectator ions are left out.

$$Pb^{2+} \ (aq) + \cancel{2 \ NO_3^-} \ (aq) + \cancel{2 \ K^+} \ (aq) + CrO_4^{2-} \ (aq) \longrightarrow$$
$$PbCrO_4 \ (s) + \cancel{2 \ K^+} + \cancel{2 \ NO_3^-} \ (aq)$$

Net ionic equation: $Pb^{2+} \ (aq) + CrO_4^{2-} \ (aq) \longrightarrow PbCrO_4 \ (s)$

Many neutralization reactions between acids and bases can be classified as double replacement reactions.

$$HCl \ (aq) + KOH \ (aq) \longrightarrow HOH \ (l) + KCl \ (aq)$$

Written as an ionic equation, it would be:

$$H^+ \ (aq) + Cl^- \ (aq) + K^+ \ (aq) + OH^- \ (aq) \longrightarrow$$
$$HOH \ (l) + K^+ \ (aq) + Cl^- \ (aq)$$

Net ionic equation: $H^+ \ (aq) + OH^- \ (aq) \longrightarrow HOH \ (l)$. The Cl^- and the K^+ are the spectator ions.

Double replacement reactions usually reduce the number of ions in solution. Solid precipitates, such as lead (II) chromate, and the formation of the largely non-ionizable water molecules in acid-base reactions (to be discussed in Chapter 15) reduce the number of ions.

Section Review Questions 8B

1. Classify each of the reactions listed in question 1 in the 8A Section Review Questions as a combination, decomposition, single replacement, or double replacement reaction.
2. For each of the following equations, identify the spectator ions and write the net ionic equation.
   a. Balanced equation:
   $$3 \text{ HCl } (aq) + \text{Al(OH)}_3 \text{ } (s) \longrightarrow 3 \text{ H}_2\text{O } (l) + \text{AlCl}_3 \text{ } (aq)$$
   Complete ionic equation:
   $$3 \text{ H}^+ + 3 \text{ Cl}^- + \text{Al(OH)}_3 \text{ } (s) \longrightarrow 3 \text{ H}_2\text{O } (l) + \text{Al}^{3+} + 3 \text{ Cl}^-$$
   b. Balanced equation:
   $$\text{Pb(NO}_3)_2 \text{ } (aq) + \text{Na}_2\text{CrO}_4 \text{ } (aq) \longrightarrow \text{PbCrO}_4 \text{ } (s) + 2 \text{ NaNO}_3 \text{ } (s)$$
   Complete ionic equation:
   $$\text{Pb}^{2+} + 2 \text{ NO}_3^- + 2 \text{ Na}^+ + \text{CrO}_4^{2-} \longrightarrow \text{PbCrO}_4 \text{ } (s) + 2 \text{ Na}^+ + \text{NO}_3^-$$

8C Stoichiometry: Predicting How Much

Ninety percent of the phosphorus produced in America is used to manufacture phosphoric acid. To form phosphoric acid, chemical engineers burn elemental phosphorus (P_4) in dry air to make diphosphorus pentoxide (P_2O_5). They then mix the diphosphorus pentoxide with water to form the acid H_3PO_4. The chemical engineers who run this process must keep careful track of all the compounds involved. For instance, if they burn 60 kilograms of phosphorus, they must know what mass of diphosphorus pentoxide will result. They must also know how much phosphoric acid can be produced from the diphosphorus pentoxide. The engineers can calculate quantities of reactants and products by knowing the mole and mass relationships between them—the "stoichiometry" of the reactions. Thus, **stoichiometry*** (stoi kee OM e tree) is the measurement and calculation of the amounts of matter in chemical reactions.

---

*stoichiometry: stoichio- (Gk. *stiocheon* – elements) + -metry (Gk. suffix – to measure)

To determine the quantities of phosphorus, diphosphorus pentoxide, and phosphoric acid, the chemical engineer needs a plan. In stoichiometric calculations, the analogy with planning a trip is helpful. When planning the trip, you know the departure point and the destination, but a map is needed in order to plan the best route from departure to destination. In the following sections, a map for stoichiometic calculations will be developed. The mole-to-mole and mass-to-mole sections and a balanced chemical equation are tools to help navigate this map. The final section, mass-to-mass, incorporates these tools to finalize the stoichiometric map.

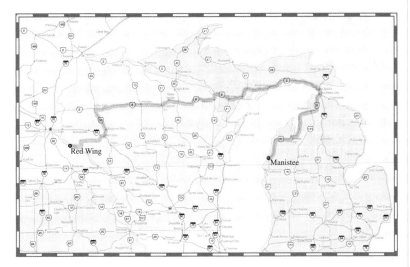

Solving stoichiometric problems is similar to planning a trip.

Mole-to-Mole Conversions

It might be helpful at this point to compare mole-to-mole conversions to the ingredients in a recipe such as the following one:

CRISPED RICE TREATS (Makes 24 2-by-2-inch squares)
3 T margarine
One 10-ounce package of marshmallows (about 40)
6 cups of crisped rice cereal

This recipe gives you the relationship between the ingredients necessary to make 24 treats. If you were really hungry, you could scale the recipe up to make more treats. By the same token, the recipe could be reduced if needed. As long as you keep the relationship between the ingredients equivalent, you can adjust the recipe to suit your needs. In the recipe there is a relationship or equivalency between 3 T of margarine and 10 ounces of marshmallows, between 10 ounces of marshmallows and 6 cups of cereal, between 24 treats and 10 ounces of marshmallows, or between *any* two of the ingredients or the number of the treats (product) formed. Suppose you had only $4\frac{1}{2}$ cups of cereal and sufficient amounts of the other ingredients. How many marshmallows would you need? You could use the recipe relationships and dimensional analysis as follows:

$$\frac{4\frac{1}{2} \text{ cups cereal}}{1} \cdot \frac{40 \text{ marshmallows}}{6 \text{ cups cereal}} = 30 \text{ marshmallows}$$

Further, you could determine how many treats you could make (assuming they were the same size) or how much margarine is needed. All of these can be determined from the relationships given in the recipe.

$$\frac{4\frac{1}{2} \text{ cups cereal}}{1} \cdot \frac{24 \text{ treats}}{6 \text{ cups cereal}} = 18 \text{ treats}$$

$$\frac{4\frac{1}{2} \text{ cups cereal}}{1} \cdot \frac{3 \text{ T margarine}}{6 \text{ cups cereal}} = 2\frac{1}{4} \text{ T margarine}$$

In a similar fashion, chemical equations show what substances are involved in a reaction. Balanced chemical equations contain the numerical information necessary for detailed calculations. The coefficients in front of each substance show how many atoms, molecules, and moles are involved. The equation $P_4 + 5\,O_2 \longrightarrow 2\,P_2O_5$ means that 1 P_4 molecule + 5 O_2 molecules form 2 P_2O_5 molecules. The coefficients can also represent the number of moles of a compound or element.

$$1 \text{ mole } P_4 + 5 \text{ moles } O_2 \longrightarrow 2 \text{ moles } P_2O_5$$

It is the coefficients in balanced equations that are the key to stoichiometric calculations. They give the numerical relationships between the substances in a reaction, i.e., the ingredients in the recipe. The balanced equation shows that

1 mole P_4 reacts to form 2 moles P_2O_5 (a 1:2 ratio)
1 mole P_4 reacts with 5 moles O_2 (a 1:5 ratio)
5 moles O_2 react to form 2 moles P_2O_5 (a 5:2 ratio)

Even if the original quantities were cut in half, the mole ratios would still hold true. In each case, the coefficients from the balanced equation give the ratios between the moles of one substance and the moles of another substance.

0.5 mole P_4 reacts to form 1 mole P_2O_5 (a 1:2 ratio)
0.5 mole P_4 reacts with 2.5 moles O_2 (a 1:5 ratio)
2.5 moles O_2 react to form 1 mole P_2O_5 (a 5:2 ratio)

The ratios can be used as conversion factors for calculating the moles of one substance from a known number of moles of another substance. Figure 8C-1 is a flow chart that illustrates how

Figure 8C-1

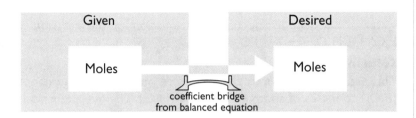

a mole-to-mole stoichiometric conversion can be done. The chart shows that the coefficients from the balanced equation can be used to calculate the number of moles of one substance when given the number of moles of another substance.

Suppose a chemical engineer wanted to produce 25.0 moles of diphosphorus pentoxide. How many moles of phosphorus would he burn? From the previous example, we can see that 1 mole of phosphorus produces 2 moles of diphosphorus pentoxide. Using that information, he can set up the following equation to solve the problem.

$$\frac{25.0 \text{ mol } P_2O_5}{} \left| \frac{1 \text{ mol } P_4}{2 \text{ mol } P_2O_5} = 12.5 \text{ mol } P_4$$

Notice that we have written out the formula for the substances along with the "mol" unit (mol = moles) so that the proper ratio will be used. Be sure to get into the habit of using the complete formula and unit labels, not just "grams" or "moles." Specify the substance in each case so you can see how the units cancel to get the desired result. Without the "P_4" or the "P_2O_5," you could not be sure you had the correct form of the ratio; cancellation of the units would still be possible, but the answer would be wrong, as in the relationship below.

$$\frac{25.0 \text{ mol}}{} \left| \frac{2 \text{ mol}}{1 \text{ mol}} = 50.0 \text{ mol} \quad \text{INCORRECT}$$

The amount of oxygen required could be calculated from either the 12.5 moles of phosphorus or from the 25.0 moles of diphosphorus pentoxide. Both methods use ratios formed from the coefficients in the balanced equation.

$$\frac{12.5 \text{ mol } P_4}{} \left| \frac{5 \text{ mol } O_2}{1 \text{ mol } P_4} = 62.5 \text{ mol } O_2$$

$$\frac{25.0 \text{ mol } P_2O_5}{} \left| \frac{5 \text{ mol } O_2}{2 \text{ mol } P_2O_5} = 62.5 \text{ mol } O_2$$

Sample Problem If 25.0 moles of diphosphorus pentoxide reacts with water to form phosphoric acid, how many moles of water are required? The balanced equation of the reaction is
$$P_2O_5 + 3\ H_2O \longrightarrow 2\ H_3PO_4$$
Solution
This problem amounts to changing from moles of one substance (P_2O_5) to moles of another substance (H_2O). Their relationship is found in the balanced equation, where the coefficients show that 3 moles of water are needed for every 1 mole of diphosphorus pentoxide—the "coefficient bridge."

$$\frac{25.0\ \text{mol}\ P_2O_5}{} \left| \frac{3\ \text{mol}\ H_2O}{1\ \text{mol}\ P_2O_5} \right. = 75.0\ \text{mol}\ H_2O$$

Mass-to-Mole Conversions

Any stoichiometric conversions between substances in a reaction must be done in terms of moles. Why? Because the balanced equation—the recipe—is based on moles, not mass. The coefficients show molar ratios, not mass ratios. The mass of a substance must first be converted to moles before any other stoichiometric conversions can be done. The flow chart for mass-to-mole conversions shows that the moles of the given substance must be calculated as an intermediate step.

Figure 8C-2

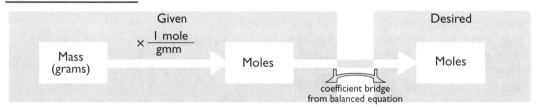

Calculate the moles of diphosphorus pentoxide that would result from the burning of 1.55 kilograms (1550 g) of phosphorus. Before the conversion factor of 2 moles P_2O_5/1 mol P_4 can be used, the given mass of 1550 grams of phosphorus must be converted to moles. Remember from the previous chapter that the mass of one mole of a substance is numerically equal to the atomic mass (gram-atomic mass). The gram-atomic mass of phosphorus is 123.9 (30.97 × 4) grams/mole.

$$\frac{1550\ g\ P_4}{} \left| \frac{1\ \text{mol}\ P_4}{123.9\ g\ P_4} \right. = 12.5\ \text{mol}\ P_4$$

The 12.5 moles of phosphorus can now be converted to moles of diphosphorus pentoxide using the mole ratio obtained from the balanced equation.

$$\frac{12.5 \text{ mol P}_4}{} \left| \frac{2 \text{ mol P}_2\text{O}_5}{1 \text{ mol P}_4} \right. = 25.0 \text{ mol P}_2\text{O}_5$$

Rather than calculating the intermediate quantity, moles of P_4, as a separate step, you can use dimensional analysis to make one equation using both the gram-to-mole and the mole-to-mole conversion factors:

$$\frac{1550 \text{ g P}_4}{} \left| \frac{1 \text{ mol P}_4}{123.9 \text{ g P}_4} \right| \frac{2 \text{ mol P}_2\text{O}_5}{1 \text{ mol P}_4} = 25.0 \text{ mol P}_2\text{O}_5$$

Sample Problem How many moles of phosphoric acid can be formed from 3550 grams of diphosphorus pentoxide?

$$P_2O_5 + 3 H_2O \longrightarrow 2 H_3PO_4$$

Solution

Note that the given value is in grams and the desired value is in moles. Since the conversion factors from the balanced equation are based on a mole ratio, the 3550 grams of diphosphorus pentoxide must first be converted to moles. The gram-molecular mass of P_2O_5 is 141.9 grams/mole.

$$\frac{3550 \text{ g P}_2\text{O}_5}{} \left| \frac{1 \text{ mol P}_2\text{O}_5}{141.9 \text{ g P}_2\text{O}_5} \right. = 25.0 \text{ mol P}_2\text{O}_5$$

Now the moles of diphosphorus pentoxide can be converted to moles of phosphoric acid according to the ratio of the coefficients.

$$\frac{25.0 \text{ mol P}_2\text{O}_5}{} \left| \frac{2 \text{ mol H}_3\text{PO}_4}{1 \text{ mol P}_2\text{O}_5} \right. = 50.0 \text{ mol H}_3\text{PO}_4$$

Or you may set it up with all steps combined using dimensional analysis.

$$\frac{3550 \text{ g P}_2\text{O}_5}{} \left| \frac{1 \text{ mol P}_2\text{O}_5}{141.9 \text{ g P}_2\text{O}_5} \right| \frac{2 \text{ mol H}_3\text{PO}_4}{1 \text{ mol P}_2\text{O}_5} = 50.0 \text{ mol H}_3\text{PO}_4$$

Note that in either method the complete units should be used and that three significant figures are used.

The technicians who control the production of diphosphorus pentoxide and phosphoric acid find mass quantities more convenient to use than molar quantities. Stoichiometric calculations can also be used to solve mass-to-mass conversions.

Mass-to-Mass Conversions: Putting It All Together

When the mass of one substance in a reaction is known, the mass of a second substance can be calculated. In the previous section, the number of moles of diphosphorus pentoxide that formed from 1550 grams of phosphorus was calculated. Only one additional step is needed to calculate the mass of diphosphorus pentoxide that would form. The gram-molecular mass of 141.9 grams/mole can be used to convert moles of diphosphorus pentoxide to grams.

Figure 8C-3

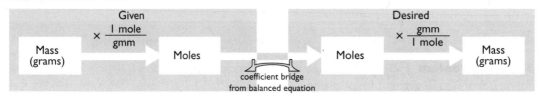

The moles of the known and unknown substances must be determined before the unknown mass can be calculated. Note that the flow chart shows all the previously given stoichiometric conversions (mole-mole, mass-mole, and mass-mass conversions).

$$\frac{1550 \text{ g } P_4}{} \left| \frac{1 \text{ mol } P_4}{123.9 \text{ g } P_4} \right| \frac{2 \text{ mol } P_2O_5}{1 \text{ mol } P_4} \left| \frac{141.9 \text{ g } P_2O_5}{1 \text{ mol } P_2O_5} \right. = 3550 \text{ g } P_2O_5$$

Sample Problem What mass of water will react with 3550 grams of diphosphorus pentoxide?

$$P_2O_5 + 3 H_2O \longrightarrow 2 H_3PO_4$$

Solution
The mass of diphosphorus pentoxide must be expressed as moles before the molar conversion can be done. Once the number of moles of water is known, the mass of water can be calculated.

$$\frac{3550 \text{ g } P_2O_5}{} \left| \frac{1 \text{ mol } P_2O_5}{141.9 \text{ g } P_2O_5} \right| \frac{3 \text{ mol } H_2O}{1 \text{ mol } P_2O_5} \left| \frac{18.016 \text{ g } H_2O}{1 \text{ mol } H_2O} \right.$$

$$= 1350 \text{ g } H_2O$$

The number of atoms or molecules involved in a reaction can be calculated from known molar quantities. Suppose that the sample problem above asked, "How many water molecules will react with 3550 grams of diphosphorus pentoxide?" The solution to the problem is the same until the step in which the number of moles of water is converted to the mass of water. The conversion factor 6.022×10^{23} molecules/1 mol can be used to convert the number of *moles* of water to the number of *molecules* of water.

$$\frac{3550 \text{ g } P_2O_5 \mid 1 \text{ mol } P_2O_5 \mid 3 \text{ mol } H_2O \mid 6.022 \times 10^{23} \text{ molecules } H_2O}{\mid 141.9 \text{ g } P_2O_5 \mid 1 \text{ mol } P_2O_5 \mid 1 \text{ mol } H_2O}$$

$$= 4.52 \times 10^{25} \text{ molecules } H_2O$$

With this addition, the flow chart for stoichiometric problems is as follows:

Figure 8C-4

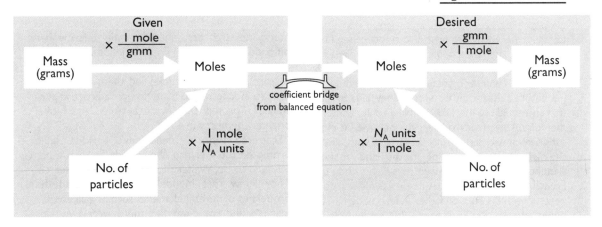

Figure 8C-4 is the completed "stoichiometric roadmap." As on any map, you must first determine where you are and where you want to arrive. The left side of the "map" shows the various starting points of the stoichiometric trip—the given information of the question. All roads lead to the "coefficient bridge," which is determined from the balanced equation. Once across the bridge, you can map your way to the various destinations (moles, mass, or number of particles) using dimensional analysis.

Sample Problem How many grams of sodium chloride must be decomposed to yield 27 grams of chlorine gas?

Solution
First, write a complete balanced equation: $2 \text{ NaCl} \longrightarrow 2 \text{ Na} + \text{Cl}_2$. Next, from Figure 8C-4, you must go from the known side to the unknown side, using the coefficients from the equation as the bridge between the two sides.

$$\frac{27 \text{ g } Cl_2 \mid 1 \text{ mol } Cl_2 \mid 2 \text{ mol NaCl} \mid 58 \text{ g NaCl}}{\mid 71 \text{ g } Cl_2 \mid 1 \text{ mol } Cl_2 \mid 1 \text{ mol NaCl}} = 44 \text{ g NaCl}$$

Chemical Families

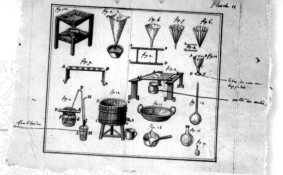

Marie Lavoisiers' sketches

Fourteen-year-old Marie Anne Pierretti sat on the edge of her chair and quietly followed the conversation of the prominent men gathered at her father's house. On this evening in 1769, she was particularly interested in one man, Antoine Laurent Lavoisier. Approximately twenty-five, good-looking, keen-minded, and a good conversationalist, the young scientist also noticed her. Less than a year later, the couple were making plans for marriage. They decided that Marie would study English, Latin, and even science so that she could aid her future husband in his work. They also agreed that she could use her artistic ability to illustrate his memoirs. After their marriage she became his closest collaborator and spent many hours with him in the laboratory, recording notes and making sketches. Eighteen happy years passed, but then the French Revolution came. Antoine Lavoisier's connection with his political father-in-law made him an object of suspicion. In May of 1794 he and his father-in-law were executed at the guillotine. In the following years Marie would marry another scientist named Count Rumford. But the pages of history record that Marie Anne and Antoine Lavoisier were possibly the first husband-and-wife research team.

Perhaps the best-known family research team of all was that of Pierre and Marie Curie. The Curies spent most of their careers discovering radioactive elements. To do this, they laboriously analyzed tons of uranium ore. Eventually they isolated a new element, number 84. They named it polonium after Marie's native land, Poland. A continued probing of the ore finally led to the discovery of radium, an extremely rare element. This research on radioactivity earned Marie and Pierre the Nobel prize in physics in 1903. After Pierre was accidentally killed, Marie continued the search and isolated pure radium. She was awarded the Nobel prize in chemistry in 1911 for the discovery of polonium and radium.

Despite the death of her husband, Marie Curie's work with family members was far from over. Throughout World War I, Marie and her daughter Irene worked on the application of X rays for medical diagnoses. As Irene worked in her mother's laboratory, she met a research assistant who later became her husband. Irene and Jean Frederic Joliot also became a famous husband-and-wife research team. Their research on alpha particles led them to discover how to induce radioactivity in nonradioactive substances. For this they were awarded the Nobel prize in chemistry in 1935.

Scattered gaps in Mendeleev's periodic table prompted the German chemist Ida Tacke and her husband-to-be, Walter Noddack, to begin a search. They knew that Mendeleev had predicted the properties of several elements such as gallium, germanium, and scandium. But there were two gaps for which he made no predictions—those representing elements 43 and 75. Both were located below manganese in the seventh column of the table. Since these elements were transition metals, their properties could not be easily predicted. Early investigators had searched for the missing elements in manganese ores but had failed. In 1922 Ida and Walter began their search. Ores containing molybdenum, tungsten, ruthenium, and osmium were concentrated and studied with X rays. In June of 1925 the Noddack team announced the discovery of element 75. They named the new element rhenium, after the Rhineland. Element 43, on the other hand, never showed itself. To this day it has not been found in nature. Instead, it has been found among the fission products of nuclear reactors. Since this is an artificial means of production, the element has been named technetium, meaning "artificial."

Pierre and Marie Curie in their laboratory

Section Review Questions 8C

1. The rusting of iron is fairly complex but may be written as follows, showing the overall process:

$$4\ Fe\ (s) + 3\ O_2\ (g) \longrightarrow 2\ Fe_2O_3\ (s)$$

   a. How many moles of rust can be produced from 3.2 moles of iron metal?
   b. How many moles of oxygen will be consumed in this process?
   c. What mass of iron metal is needed to form 100.0 grams of rust?
   d. How many molecules of oxygen will be needed to react with 50.0 g iron metal?

Chapter Review

Coming to Terms

chemical equation
balanced chemical equation
reactant
product
precipitate
precipitation
catalyst
combination reaction
synthesis reaction

decomposition reaction
single replacement reaction
activity series
double replacement reaction
ionic equation
spectator ion
net ionic equation
stoichiometry

Review Questions

1. List several evidences that tell whether a chemical reaction has occurred.
2. What three criteria must be met before a chemical equation is acceptable?
3. Explain what each highlighted term means.

$$2\ HgO\ (s) \xrightarrow{\Delta} 2\ Hg\ (l) + O_2\ (g)$$

4. $HCl\ (aq) + H_2O\ (l) \rightleftharpoons Cl^-\ (aq) + H_3O^+\ (aq)$
   a. Why are two opposing arrows shown?
   b. What does the symbol (l) mean?
   c. Why is (aq) written after several of the substances?

5. Determine whether the following equations are balanced. If they are not, balance them.
   a. BaO_2 (s) $\xrightarrow{\Delta}$ BaO (s) + O_2 (g)
   b. Li (s) + H_2O $\xrightarrow{\Delta}$ LiOH (s) + H_2 (g)
   c. Hydrogen peroxide can decompose when exposed to bright sunlight.
   H_2O_2 (l) $\longrightarrow$ H_2O (l) + O_2 (g)
   d. A metallic hydride can form when hydrogen gas is bubbled through molten sodium.
   Na (l) + H_2 (g) $\longrightarrow$ NaH (s)
   e. H_2SO_4 $\xrightarrow{\Delta}$ H_2O (g) + SO_2 (g) + O_2 (g)
   f. Ammonia is produced commercially from nitrogen and hydrogen.
   N_2 (g) + H_2 (g) $\xrightarrow{\Delta,\text{ Fe, high pressure}}$ NH_3 (g)
   g. Ammonium chloride is produced when ammonia and hydrogen chloride vapors mix.
   NH_3 (g) + HCl (g) $\longrightarrow$ NH_4Cl (s)
   h. Carbon black is used in rubber tires and black ink. It is produced by the "cracking" of methane.
   CH_4 (g) $\xrightarrow{\Delta}$ C (s) + H_2 (g)
   i. High-purity silicon is used to produce microcomputer chips. One process for producing chip-grade silicon entails three steps. The first step is to obtain impure silicon from molten sand (SiO_2)
   SiO_2 (l) + C (s) $\longrightarrow$ Si (l) + 2 CO (g)
   j. The second step in the production of pure silicon is to produce silicon tetrachloride from the impure silicon.
   Si (s) + Cl_2 $\xrightarrow{\Delta}$ $SiCl_4$ (l)
   k. The last step is to pass hot silicon tetrachloride vapor and hydrogen gas through a tube. Pure silicon condenses.
   $SiCl_4$ (g) + H_2 (g) $\xrightarrow{\Delta}$ Si (s) + HCl (g)
   l. Ingested barium sulfate causes the intestinal tract to be emphasized in X ray pictures. It may be produced when barium reacts with sulfuric acid.
   Ba (s) + H_2SO_4 (aq) $\longrightarrow$ $BaSO_4$ (s) + H_2 (g)
   m. Milk of magnesia is an aqueous suspension of $Mg(OH)_2$. When ingested, it reduces the amount of hydrochloric acid in the stomach by neutralizing this acid.
   $Mg(OH)_2$ (aq) + HCl (aq) $\longrightarrow$ $MgCl_2$ (aq) + H_2O (l)
6. Tell whether each of the reactions in the previous question is a composition, decomposition, single replacement, or double replacement reaction.
7. Predict whether the following single replacement reactions will occur. Base your answers on the activity series given in the text.

a. $BaSO_4 + Ca \longrightarrow CaSO_4 + Ba$
b. $BaCl_2 + Br_2 \longrightarrow BaBr_2 + Cl_2$
c. $Ni(OH)_2 + Mg \longrightarrow Mg(OH)_2 + Ni$
d. $2\ FeCl_3 + 3\ Mg \longrightarrow 3\ MgCl_2 + 2\ Fe$
e. $Al_2(SO_4)_3 + 2\ Fe \longrightarrow Fe_2(SO_4)_3 + 2\ Al$
f. $Ni + 2\ H_2O \longrightarrow Ni(OH)_2 + H_2$ (at room temperature)

8. Complete ionic equations for double replacement reactions are given below. For each equation, identify the spectator ions and write the net ionic equation.
   a. Balanced equation:
   $AgNO_3\ (aq) + HCl\ (aq) \longrightarrow AgCl\ (s) + HNO_3\ (aq)$
   Complete ionic equation:
   $Ag^+ + NO_3^- + H^+ + Cl^- \longrightarrow AgCl\ (s) + H^+ + NO_3^-$
   b. Balanced equation:
   $CaCl_2\ (aq) + Na_2CO_3\ (aq) \longrightarrow CaCO_3\ (s) + 2\ NaCl\ (aq)$
   Complete ionic equation:
   $Ca^{2+} + 2\ Cl^- + 2\ Na^+ + CO_3^{2-} \longrightarrow CaCO_3\ (s) + 2\ Na^+ + 2\ Cl^-$
   c. Balanced equation:
   $H_2SO_4\ (aq) + 2\ KOH\ (aq) \longrightarrow K_2SO_4\ (aq) + 2\ H_2O\ (l)$
   Complete ionic equation:
   $2\ H^+ + SO_4^{2-} + 2\ K^+ + 2\ OH^- \longrightarrow 2\ K^+ + SO_4^{2-} + 2\ H_2O\ (l)$
   d. Balanced equation:
   $Mg(OH)_2\ (s) + 2\ HCl\ (aq) \longrightarrow MgCl_2\ (aq) + 2\ H_2O\ (l)$
   Complete ionic equation:
   $Mg(OH)_2\ (s) + 2\ H^+ + 2\ Cl^- \longrightarrow Mg^{2+} + 2\ Cl^- + 2\ H_2O\ (l)$
   e. Balanced equation:
   $Ba(NO_2)_2\ (aq) + Na_2SO_4\ (aq) \longrightarrow BaSO_4\ (s) + 2\ NaNO_2\ (aq)$
   Complete ionic equation:
   $Ba^{2+} + 2\ NO_2^- + 2\ Na^+ + SO_4^{2-} \longrightarrow BaSO_4\ (s) + 2\ Na^+ + 2\ NO_2^-$

9. Chlorine was the first of the halogens to be isolated. C. W. Scheele carried out the following reaction in 1774:
   $4\ NaCl\ (aq) + 2\ H_2SO_4\ (aq) + MnO_2\ (s) \longrightarrow$
   $\quad 2\ Na_2SO_4\ (aq) + MnCl_2\ (aq) + 2\ H_2O\ (l) + Cl_2\ (g)$
   a. If you start with 1.00 g of NaCl and an excess of the other reagents, how many grams of Cl_2 will be produced?
   b. If you start with 1.0 g H_2SO_4 and an excess of the other reagents, how many grams of Cl_2 will be produced?
   c. If you start with 1.0 g MnO_2 and an excess of the other reagents, how many grams of Cl_2 will be produced?
   d. How many grams of NaCl must react to produce 1.00 g Cl_2?
   e. How many grams of H_2SO_4 must react to produce 1.00 g Cl_2?

f. How many grams of MnO_2 must react to produce 1.00 g Cl_2?

g. How many grams of Na_2SO_4 will be produced along with 1.00 g Cl_2?

h. How many grams of $MnCl_2$ will be produced along with 1.00 g Cl_2?

i. How many grams of H_2O will be produced along with 1.00 g Cl_2?

j. Show that the law of mass conservation is upheld in the production of 1.00 g Cl_2.

10. Priestley discovered oxygen when he decomposed mercury (II) oxide into oxygen and mercury.

$$2\ HgO\ (s) \xrightarrow{\Delta} 2\ Hg\ (l) + O_2\ (g)$$

a. If 0.440 mole of HgO reacts, how many moles of O_2 will be produced?

b. If 0.580 mole of Hg is produced, how many moles of O_2 are produced?

c. How many grams of HgO must react to produce 3.75 moles of O_2?

d. How many grams of Hg are produced when 14.7 g of HgO react?

e. How many grams of O_2 are produced when 6.20 g of HgO react?

f. How many grams of HgO must react to produce 2.30 g of Hg?

11. Hematite, Fe_2O_3, is converted to molten iron in a blast furnace and is then poured into molds. The balanced equation is

$$Fe_2O_3\ (s) + 3\ CO\ (g) \longrightarrow 2\ Fe\ (l) + 3\ CO_2\ (g)$$

a. How many moles of Fe_2O_3 must react to produce 262 moles of Fe?

b. How many moles of CO_2 are produced by the reaction of 64.0 moles of CO?

c. How many moles of Fe_2O_3 must react to produce 760 kg of Fe?

d. How many grams of CO_2 will be produced when 40.0 moles of CO react?

e. What mass (in kg) of Fe is produced when 299 kg of Fe_2O_3 react?

12. Natural gas consists primarily of methane (CH_4). When natural gas burns in a gas-burning appliance, methane reacts with oxygen to produce carbon dioxide and water.

$$CH_4 + 2\ O_2 \longrightarrow 2\ H_2O + CO_2$$

a. How many moles of CH_4 must react to produce 13.7 moles of H_2O?

b. How many moles of CO_2 will be produced by the reaction of 8.31 moles of O_2?
c. What mass of CH_4 must react to produce 4.00 moles of H_2O?
d. How many grams of CO_2 are produced when 6.00 moles of CH_4 react?
e. How many grams of O_2 will be consumed by the combustion of 2.18 g of CH_4?
f. How many grams of water are produced by the reaction of 139 g of CH_4?

13. Bleach is an aqueous solution of sodium hypochlorite. A bleach solution may be prepared when chlorine gas is bubbled through aqueous sodium hydroxide.
$$2\ NaOH\ (aq) + Cl_2\ (g) \longrightarrow NaCl\ (aq) + NaOCl\ (aq) + H_2O\ (l)$$
    a. How many moles of NaOCl will be produced from 5.73 moles of Cl_2?
    b. How many moles of NaOH must react to produce 13.7 moles of NaCl?
    c. How many grams of H_2O will be produced if 0.750 mole of NaOH react?
    d. How many grams of Cl_2 must react to produce 6.70 moles of NaCl?
    e. What mass of NaOCl will be produced when 65.5 g of NaOH react?
    f. How many grams of NaOH will react with 37.5 g of Cl_2?

14. Without fertilizers farmers could not give their crops enough nitrogen. Urea (NH_2CONH_2) is used as a common nitrogen fertilizer. Urea may be produced from ammonia and carbon dioxide.
$$2\ NH_3 + CO_2 \xrightarrow{\Delta,\ \text{pressure}} NH_2CONH_2 + H_2O$$
    a. How many moles of NH_2CONH_2 are produced when 35.0 moles of CO_2 react?
    b. How many moles of NH_3 must react to produce 75.0 moles of H_2O?
    c. How many grams of CO_2 will react with 5.75 moles of NH_3?
    d. How many moles of NH_2CONH_2 are produced by the reaction of 287 g of CO_2?
    e. How many moles of NH_2CONH_2 are produced by the reaction of 603 g of NH_3?
    f. How many grams of CO_2 must react to produce 454 g of NH_2CONH_2?

9A The Nature of Gases page 227
9B Gas Laws page 234
9C Gases and the Mole page 242
FACETS: Gases: Good and Bad page 232

Gases 9

Molecules on the Move

The word *gas* comes from the Greek word *khaos*, meaning "formless matter." Of the three common states of matter, gas particles are the most chaotic. Unlike solids, whose particles are held firmly in place, or liquids, whose particles are in constant contact with each other, gas particles move freely.

9A The Nature of Gases: What Energetic Particles You Have!

Kinetic Description of Gases

The kinetic theory explains the behavior of matter on the basis of particle motion. According to the kinetic theory,

1. "Gases consist of a vast number of independent particles." These particles have unique sizes and masses.
2. "Particles move at random, with high velocities, in all directions, and at many different speeds." Collisions constantly change the speeds and directions of the molecules.
3. "Particles are separated by great distances." The average distance an oxygen molecule must travel between collisions is about 20,000 times its own diameter!
4. "Particles do not interact except during momentary collisions; therefore, any gravitational, electrical, or chemical forces between the molecules can be ignored." It is important to note that many collisions occur every second. Air molecules at 0°C undergo 5×10^9 collisions every second.

A steel ball is lifted to give it energy (left). When the steel ball is released, the energy is transmitted through elastic collisions (right).

5. "Collisions are elastic." That is, they conserve energy. Two particles leave an impact with the sum of their energies unchanged.

The kinetic theory serves as a remarkably successful basis for explaining the properties of gases.

The Physical Properties of Gases

Low Density. Compared to solids and liquids, gases have hardly any mass per unit of volume. While ice at -25°C has a density of 0.917 gram per milliliter, and liquid water at 25°C has a density of 0.9971 g/mL, steam at 125°C has a density of only 0.000561 g/mL. At 25°C air has a density of 0.001185 g/mL. Some gases, such as hydrogen and helium, have even lower densities. This property is not surprising, since gases have much empty space between their molecules.

Diffusibility. When a gas enters a vacuum or another gas, it spreads out to fill the entire volume with an even concentration. This action is called **diffusion.** Diffusion occurs because the gas molecules are in constant motion.

Permeability. **Permeability,** the ability of a gas to mingle with another porous substance, occurs because the constantly moving particles move into the spaces between other molecules.

Compressibility and Expansibility. Whereas liquids do not easily compress, the volumes of gases have the ability to change to fit their containers. High pressures can squeeze gases into smaller volumes. This property is called **compressibility.** When gases encounter a region of low pressure, they quickly expand to fill the available space. This property is called **expansibility.** Gases can expand without limit, and they always fill their containers. The empty spaces between molecules and the constant motion of the molecules make these properties possible.

Gases such as oxygen and nitrogen can be compressed into containers (top). A demand regulator feeds compressed air to a diver (bottom).

How Gases Cause Pressure

Gas molecules can collide with each other billions of times each second. They can also bang into trees, buildings, people, and the walls of their containers. Although individual collisions are not very forceful, they add up to produce a significant force. **Pressure** is the average force exerted per unit area when molecules collide with a boundary. It is the observable result of molecular collisions.

Pressure is measured in force per unit area. Normal atmospheric pressure at sea level is 14.7 pounds per square inch (or **psi**). When atmospheric pressure is measured with a **barometer**, another unit of pressure can also be used. Mercury barometers allow air pressure to support a column of mercury. As pressure increases, the column of mercury rises. The pressure and the length of the column can be expressed in **millimeters of mercury.** For example, normal atmospheric pressure at sea level can support a column of mercury 760 millimeters high. This pressure is expressed as 760 millimeters of mercury (760 mm Hg). Another common name for millimeters of mercury is the **torr,** named after Evangelista Torricelli, the inventor of the barometer.

Another measure of pressure that is sometimes used in chemistry is the **atmosphere.** One atmosphere (atm) is simply the normal atmospheric pressure at sea level at 45° latitude. Two atmospheres are double the normal pressure at sea level, and $\frac{1}{2}$ atmosphere is half the normal pressure. The SI unit of pressure is the **pascal** (Pa), which is named for Blaise Pascal. One atmosphere is equal to 1.01325×10^5 Pa or 101.3 kPa.

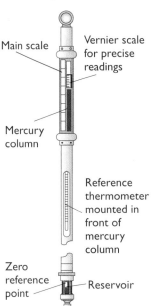

Mercury Barometer

$$1 \text{ atm} = 760 \text{ mm Hg} = 760 \text{ torr} = 14.7 \text{ psi} = 101{,}325 \text{ Pa} = 101.3 \text{ kPa}$$

The facts above justify the following unit-analysis conversion factors and their reciprocals.

$$\frac{1 \text{ atm}}{760 \text{ torr}} \quad \frac{1 \text{ atm}}{14.7 \text{ psi}} \quad \frac{760 \text{ torr}}{14.7 \text{ psi}} \quad \frac{1 \text{ atm}}{101{,}325 \text{ Pa}}$$

Sample Problem Express a pressure of 1.20 atmospheres in torr and psi.

Solution

$$\frac{1.20 \text{ atm} \mid 760 \text{ torr}}{\mid 1 \text{ atm}} = 912 \text{ torr}$$

$$\frac{1.20 \text{ atm} \mid 14.7 \text{ psi}}{\mid 1 \text{ atm}} = 17.6 \text{ psi}$$

Chapter 9A

Pressure, Volume, and Temperature: One Good Change Deserves Another

The volume of a gas depends not only on the number of gas molecules but also on the temperature and pressure of the gas. A gas at a given temperature contains molecules moving at many different speeds. The average kinetic energy of the molecules determines the temperature of the gas.

At 0°C many hydrogen molecules move at velocities near 1500 meters per second. Some move slower; some move faster. The average velocity of 1500 meters per second causes the temperature to be 0°C. At 500°C many hydrogen molecules move faster. The average velocity is thus greater.

This graph describes how gas molecules behave at different temperatures. At higher temperatures gas molecules have a higher average kinetic energy. The molecules move faster, collide more often, and strike with more force. The pressure is bound to increase because of the additional, more forceful collisions. If the gas is confined to a fixed volume, the pressure will increase as the temperature increases.

When external forces act upon a gas, they oppose the effect of all the submicroscopic collisions. When external forces exceed the internal pressure of a gas, they squeeze the gas into a smaller space. Volume therefore decreases when external pressure increases, and it increases when the internal pressure increases. The internal pressure can be increased by increasing the quantity of the gas, as when you inflate a tire with an air pump. If the external pressure decreases, the volume has the chance to increase. Table 9A-2 summarizes the effects of pressure, temperature, and volume on gases.

Table 9A-1
Frequency Distribution of Hydrogen Molecules at Two Different Temperatures

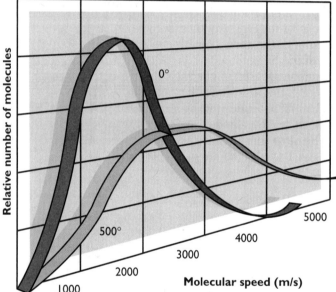

Table 9A-2

Case	Pressure (P)	Temperature (T)	Volume (V)
1.	increase	constant	decrease
2.	constant	increase/decrease	increase/decrease
3.	increase/decrease	increase/decrease	constant

Examples:

Case 1: An inverse relationship. When temperature is constant and pressure increases, the volume decreases.

Case 2: A direct relationship. When pressure is constant and temperature increases, the volume also increases. When pressure is constant and temperature decreases, the volume also decreases.

Case 3: A direct relationship. When volume is constant and temperature increases, the pressure also increases. When volume is constant and temperature decreases, the pressure also decreases.

The Scriptures identify several direct and inverse relationships that affect the lives of Christians. Matthew 24:12 tells of an inverse relationship between iniquity and love for God: "And because iniquity shall abound, the love of many shall wax cold." As iniquity increases, the love for God decreases. On the other hand, an increased love for God causes a decrease in iniquity. Luke 12:48 informs us about a direct relationship between spiritual knowledge and responsibility: "For unto whomsoever much is given, of him shall be much required: and to whom men have committed much, of him they will ask the more." Advantages such as a good family, a sharp mind, and an early salvation increase a person's responsibility toward God.

Section Review Questions 9A

1. Why can you smell a cake baking in the kitchen when you are in your bedroom?
2. Aside from diffusion, why can you smell coffee and bacon at the same time?
3. How are mm Hg, torr, psi, atm, and Pa related to each other?
4. Express a pressure of 0.85 atm in torr, psi, and Pa. Remember significant figures.
5. What will happen to the volume of a gas in a closed container if you increase the external pressure and the temperature remains constant?
6. Explain why a balloon becomes smaller when placed in a freezer.

FACETS OF CHEMISTRY

Gases: Good and Bad

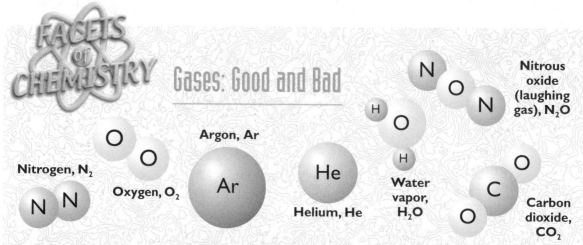

This world is filled with many kinds of gases. Some of these gases are harmful if breathed, while others are not. For the most part, molecules in harmless gases have strong, stable bonds. In some cases the gases consist of atoms that do not participate in any reactions. When gases with stable bonds or nonreactive atoms are breathed, no harmful reactions occur.

Gases We Breathe

Nitrogen, N_2. This colorless, odorless, tasteless, and relatively inactive gas makes up 78 percent (by volume) of the lower atmosphere. This gas is slightly lighter than air, and the molecules are extremely stable.

Oxygen, O_2. This active gas in our atmosphere is colorless, odorless, and tasteless. Oxygen makes up 21 percent of the volume of the lower atmosphere. It is slightly heavier than air. While humans and animals take in a constant supply of oxygen during respiration, plants give off the gas during photosynthesis.

Argon, Ar. The third most abundant component of the earth's atmosphere (1%) is a colorless, odorless, tasteless noble gas. Industries make use of its chemical inactivity. It is used in light bulbs and neon signs to protect the filaments. When welding aluminum and stainless steel, arc welders use argon to momentarily shield the molten metals from the atmosphere. It is also used in medical laser surgeries.

Helium, He. This noble gas is present in the earth but not in the atmosphere. It floats out into space shortly after it escapes from underground deposits or is produced by radioactive processes. Natural gas deposits serve as the primary source of this odorless, tasteless gas. Helium is chemically inert, has an extremely low density, and has the lowest condensation point of all the gases.

Water vapor, H_2O. Normally a liquid, water molecules are a vital part of the air that we breathe. Without some humidity, throat and lung tissues would dry out. As a drawback, water vapor contributes to the rusting of iron and the corrosion of many other metals.

Carbon dioxide, CO_2. This gas is colorless and odorless, and it has a slightly sour taste. Plants get the carbon atoms they need from atmospheric CO_2. The burning of fuels, whether in biological respiration or in a car's engine, produces CO_2. This gas is thought to be responsible for keeping solar heat energy from escaping back into space. One and one-half times as heavy as air, the gas sinks to a tabletop or floor whenever it is released. It is used to extinguish fires and to make carbonated beverages. Some scientists have speculated that recent increases in CO_2 concentrations due to the burning of more and more fuels may cause the earth's temperature to rise in a "greenhouse effect." The "greenhouse effect" theory is debated in the scientific community.

Nitrous oxide (laughing gas), N_2O. The discovery of nitrous oxide and its effects on the human body marked a great step forward in surgical practices. This colorless gas with a somewhat sweet odor and taste induces mild hysteria and an insensitivity to pain. Widely used as an anesthetic when it was first introduced, it now is used only for short operations and in combination with other gases. Long exposures to N_2O can result in death.

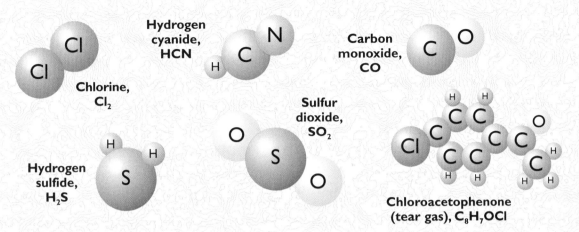

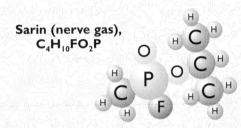

Gases We Should Not Breathe

As a rule of thumb, gases that are chemically reactive should not be breathed. Some should not even be touched. Although some of these gases do not cause harmful reactions, they have such irritating odors that it is wise to avoid them.

Chlorine, Cl_2. Like most harmful gases, chlorine gas is very reactive. This heavy yellowish-green gas has a disagreeable, suffocating odor. Chlorine has the distinction of being one of the first poisonous gases to be used in warfare. Fortunately, the gas can be smelled and detected at levels far below its lethal concentration.

Hydrogen sulfide, H_2S. The decay of organic matter that contains sulfur invariably produces one of the foulest smells known to man: hydrogen sulfide. This colorless, heavy gas has the odor of rotten eggs. Large amounts of it are produced as refineries remove sulfur-containing impurities from petroleum.

Hydrogen cyanide, HCN. A dose of 0.05 gram or 200 to 500 parts per million of this poisonous, colorless gas quickly paralyzes the central nervous system. A fatal dose of hydrogen cyanide inhibits respiration by blocking the ability of the cells to use oxygen. An odor like that of bitter almonds is characteristic of this gas.

Sulfur dioxide, SO_2. Burning sulfur gives off a heavy, colorless gas having the odor of a freshly struck match. This gas readily dissolves in water to form sulfurous acid, H_2SO_3. Sulfur dioxide is extremely irritating to the nose and throat. Manufacturers use it to bleach straw, paper, silk, and wood.

Carbon monoxide, CO. The incomplete combustion of fuels produces a colorless, odorless, tasteless gas that is extremely dangerous to humans. If carbon monoxide is breathed, it will quickly bind to hemoglobin in the blood and prevent the hemoglobin from carrying oxygen. Many tragic deaths have occurred because concentrations of carbon monoxide built up in garages, homes, and cars. One of the leading causes of carbon monoxide poisoning is faulty gas furnaces. For safety, every home should be equipped with a CO detector.

Chloroacetophenone (tear gas), C_8H_7OCl. Police use tear gas to disperse riots and to flush criminals out of buildings. Although the gas has an odor resembling that of apple blossoms, most people who have smelled it remember the strong irritation to their eyes. Exposure brings tears, eye irritation, and uncontrollable coughing.

Sarin (nerve gas), $C_4H_{10}FO_2P$. A one-milligram dose of this poisonous gas paralyzes an entire nervous system in minutes. The gas can be neutralized with applications of water and basic solutions. Sarin was first developed in Germany after World War I.

9B Gas Laws: Mathematical Descriptions of Physical Properties

So far the behavior of gases has been described qualitatively—that is, without numbers. However, a series of gas laws, each based on the qualitative kinetic theory, can make these descriptions quantitative. These gas laws make it possible to calculate volumes, pressures, and temperatures of gases at various conditions.

Standard Conditions: A Reference Point

Because the volume of a gas can change with temperature and pressure, reporting a volume without specifying these conditions would be meaningless. Scientists have defined a standard set of conditions called **standard temperature and pressure** to be used when measuring and comparing gases. Standard temperature is 0°C or 273 K. Standard pressure is 760 torr, 101,325 Pa, or 1 atmosphere (atm). The conditions of standard temperature and pressure are often abbreviated **STP**.

Boyle's Law: Pressures and Volumes

Robert Boyle (1627-91) was the first to measure the effects of pressure on the volume of a gas. He found that increased pressure decreased the volume and that decreased pressure allowed the volume to increase (Case 1 from Table 9A-2). He also found the mathematical relationship that governed the changes.

Twice the pressure yields half the volume.

Three times the pressure yields one-third the volume.

Half the pressure yields twice the volume.

One-third the pressure yields three times the volume.

From data such as this, he formulated **Boyle's law:** *The volume of a dry gas is inversely related to the pressure if the temperature is held constant.* When placed into an equation, this law appears as follows:

(P) Pressure × (V) Volume = k (a constant value)

The product of pressure and volume remains the same, even at different conditions. The equation listed below is a form of Boyle's law and can be used to solve problems concerning changing pressures and volumes.

$$P_1V_1 = k = P_2V_2$$

$P_1V_1 = P_2V_2$
Common form of Boyle's law

Figure 9B-1

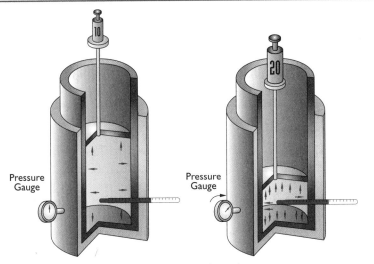

As pressure increases, volume decreases.

Sample Problem A sample of gas occupies 450 mL when it is under a pressure of 780 torr. What volume will it occupy if the pressure is increased to 850 torr?

Solution

$$P_1V_1 = P_2V_2$$

$$(780 \text{ torr})(450 \text{ mL}) = (850 \text{ torr})(V_2)$$

$$\frac{(780 \text{ torr})(450 \text{ mL})}{850 \text{ torr}} = V_2$$

$$V_2 = 410 \text{ mL}$$

Charles's Law: Temperatures and Volumes

Charles's law deals with the relationship between a gas's temperature and its volume (Case 2 from Table 9A-2). Careful measurements of the volume of a gas at different temperatures and constant pressures could yield the following data in Table 9B-1.

Volumes keep shrinking as temperatures decrease. The graph shows that the volume would theoretically become zero at a temperature of -273°C. This extrapolation of gas volumes led to the invention of the Kelvin temperature

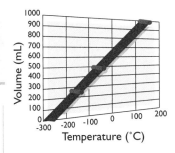

Table 9B-1

Temperature	Volume
127°C	1000 mL
-73°C	500 mL
-173°C	250 mL

scale. On this scale, the lowest temperature that is theoretically possible (-273°C) is labeled 0 K. The two scales are related by the formula K = °C + 273. Table 9B-2 compares Celsius temperatures with Kelvin temperatures at the same volume.

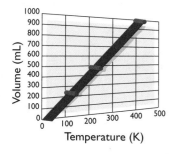

Table 9B-2 Celsius and Kelvin Temperatures Compared

Temperature	Temperature	Volume
127°C	400 K	1000 mL
-73°C	200 K	500 mL
-173°C	100 K	250 mL

When the Kelvin temperature scale is used, the relationship between temperature and volume can easily be seen. Doubling the temperature doubles the volume. **Charles's law** states that *when the pressure on a sample of a dry gas is held constant, the Kelvin temperature and the volume are directly related.* Mathematically, this law is stated as follows:

$$\frac{V}{T} = k$$

If the temperature decreases, the volume must also decrease to keep the constant the same. No matter what changes occur, the ratio between volume and temperature (in the Kelvin scale) will be the same for a given sample of gas at a constant pressure.

$$\frac{V_1}{T_1} = k = \frac{V_2}{T_2}$$

$$\frac{V_1}{T_1} = \frac{V_2}{T_2}$$

Common form of Charles's law

Figure 9B-2

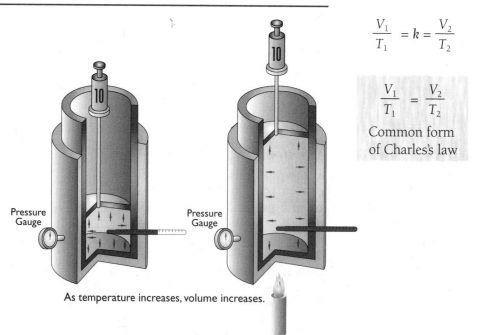

As temperature increases, volume increases.

Sample Problem A sample of gas occupies 430 mL when it is at a temperature of 25°C. What volume will it occupy when it is at standard temperature?

Solution

$$\frac{V_1}{T_1} = \frac{V_2}{T_2}$$

$$\frac{430 \text{ mL}}{25°C + 273 \text{ K} = 298 \text{ K}} = \frac{V_2}{273 \text{ K}}$$

$$\frac{(430 \text{ mL})(273 \cancel{K})}{298 \cancel{K}} = V_2$$

$$V_2 = 390 \text{ mL}$$

A balloon immersed in liquid nitrogen shrinks because the air inside contracts.

Gay-Lussac's Law: Temperatures and Pressures

The pressure in a car's tires increases as the tires heat up on a lengthy trip. A basketball loses its "bounce" when it is taken outside on a cold winter day. Both of these changes are examples of **Gay-Lussac's law:** *pressure is directly proportional to Kelvin temperature for a fixed mass of gas held in a constant volume* (Case 3 from Table 9A-2).

In the first example above, friction with the road raised the temperature of the air in the tires. The gas molecules moved faster, collided more often, and transferred more force to the inner walls of the tires. As a result, pressure increased. In the case of the basketball, the molecules moved more slowly at colder temperatures. The air pressure in the ball decreased accordingly.

The mathematical expression of Gay-Lussac's law is

$$\frac{P}{T} = k$$

As long as the volume is held constant, this equality holds true for many different pressures and temperatures.

Figure 9B-3

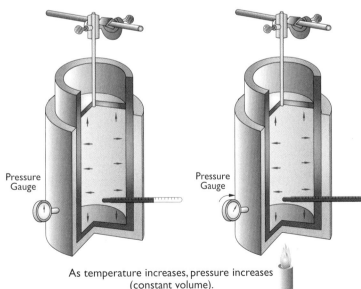

As temperature increases, pressure increases (constant volume).

237

$$\frac{P_1}{T_1} = k = \frac{P_2}{T_2}$$

$P_2 = P_1 \times$ temperature fraction
$T_2 = T_1 \times$ pressure fraction

$$\frac{P_1}{T_1} = \frac{P_2}{T_2}$$

Common form of Gay-Lussac's law

Sample Problem Before Molly Kuhl began a car trip, she measured the air pressure in her car tires and found that it was 32 psi at a temperature of 18°C. After two hours of driving, she found that the pressure had increased to 34 psi. What was the new temperature of the air in her tires?

Solution
In this problem the pressure is increasing. In order for the pressure to increase, the temperature must also increase. Change Celsius temperatures to Kelvin.

$$\frac{P_1}{T_1} = \frac{P_2}{T_2}$$

$$\frac{32 \text{ psi}}{18°C + 273 \text{ K} = 291 \text{ K}} = \frac{34 \text{ psi}}{T_2}$$

$$\frac{(34 \text{ psi})(291 \text{ K})}{32 \text{ psi}} = T_2$$

$$T_2 = 310 \text{ K}$$

Combined Gas Law: Putting It All Together

The gas laws that have been discussed so far have applied to situations in which one quantity stayed the same.

Boyle's law: $PV = k$, where T is constant.
Charles's law: $V/T = k$, where P is constant.
Gay-Lussac's law: $P/T = k$, where V is constant.

When properly combined, the gas laws form a single equation called the **combined gas law**.

$$\frac{P_1 V_1}{T_1} = \frac{P_2 V_2}{T_2}$$

This equation contains the original three equations. If the temperature is held constant ($T_1 = T_2$), the equation becomes Boyle's law. T_1 and T_2 cancel. If the pressure does not change ($P_1 = P_2$), the two pressures cancel each other algebraically, and Charles's law emerges. When the volume does not change ($V_1 = V_2$), Gay-Lussac's law can be seen. The advantage of the combined gas law is that it allows problems in which pressure, volume, and temperature change to be solved in a single step.

Sample Problem A gas has a volume of 3.6 L when it is under a pressure of 1.05 atm and a temperature of -15°C. What will its volume be at STP?

Solution

$$\frac{P_1 V_1}{T_1} = \frac{P_2 V_2}{T_2}$$

$$\frac{(1.05 \text{ atm})(3.6 \text{ L})}{(-15°C + 273 \text{ K}) = 258 \text{ K}} = \frac{(1.00 \text{ atm})(V_2)}{273 \text{ K}}$$

$$\frac{(1.05 \text{ atm})(3.6 \text{ L})(273 \text{ K})}{(258 \text{ K})(1.00 \text{ atm})} = V_2$$

$$V_2 = 4.0 \text{ L}$$

Pay careful attention to the units on pressure, volume, and temperature when solving gas-law problems. The sample problem above demonstrates how the units on pressure and temperature cancel out. The only unit that is left is liters. Simple checks like this can show whether the equation is set up properly.

Dalton's Law of Partial Pressures: Mixtures of Gases

The gases in this world are often mixtures. Even the gases that chemists produce in their laboratories contain some impurities. These mixtures complicate gas law calculations. John Dalton's law of partial pressures describes the behavior of gaseous mixtures. Suppose that 1 liter of oxygen at STP is added to 1 liter of nitrogen gas at the same pressure. If the two gases are held in a 1-liter jar, the pressure will be 1520 torr—the sum of the two pressures. When molecules of gases do not react with each other, they can behave independently of each other. Thus oxygen molecules exert a pressure of 760 torr just as they did before they were mixed with the nitrogen gas. The nitrogen gas exerts the same pressure, so the total pressure from the two gases is 1520 torr. **Dalton's law of partial pressures** puts these observations into general terms. *The total pressure of a mixture of gases equals the sum of the partial pressures.*

A sample of dry air contains 78 percent nitrogen, 21 percent oxygen, and 1 percent argon. If the mixture exerts a pressure of 760 torr, 78 percent of the pressure (592.8 torr) comes from nitrogen. Twenty-one percent of the pressure (159.6 torr) is from oxygen, and 1 percent of the pressure (7.6 torr) is from argon. The sum of all the partial pressures equals the total pressure of the mixture.

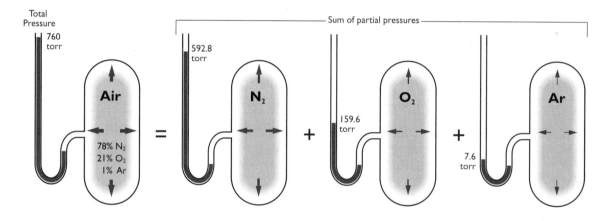

In the laboratory, chemists often collect a sample of a gas by trapping it at the top of a water-filled container. The gas bubbles

Collection of gas over water

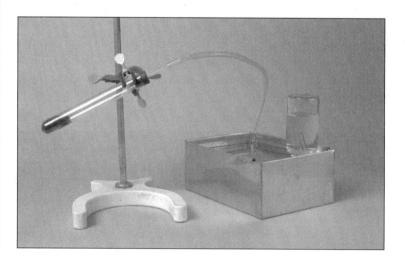

up through the water, collects at the top, and forces the water out the bottom. This technique is called collecting a gas over water, or collection by water displacement.

A gas being collected over water may be pure initially, but that quickly changes. As the gas bubbles through the water, some water evaporates and mixes with the gas being collected. Accurate measurements of the gas cannot be made when it is mixed with water vapor. The extra water molecules exert a pressure called **vapor pressure.** The total pressure, which equals the atmospheric pressure, is made up of pressure from the gas and pressure from the water vapor. To find the pressure due to the gas, the pressure from the water vapor must be subtracted from the atmospheric pressure. The pressure from water vapor depends upon its temperature. Table 9B-3 lists vapor pressures of water at various temperatures.

$$P_{atm} = P_{H_2O} + P_{gas}$$
$$P_{atm} - P_{H_2O} = P_{gas}$$

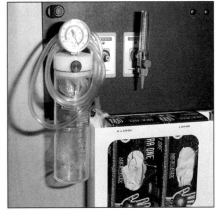

Ambulance equipment humidifies oxygen by bubbling it through water. Without this precaution a patient's mouth, throat, and lungs would become dry.

Table 9B-3 Vapor Pressure of Water at Various Temperatures

Temperature (°C)	Vapor Pressure (torr)	Vapor Pressure (kPa)
0.0	4.579	0.6105
5.0	6.543	0.8723
10.0	9.209	1.228
15.0	12.788	1.705
20.0	17.535	2.338
25.0	23.756	3.167
30.0	31.824	4.243
35.0	42.175	5.623
40.0	55.324	7.376
45.0	71.88	9.583
50.0	92.51	12.33
55.0	118.04	15.74
60.0	149.38	19.92
65.0	187.54	25.00
70.0	233.7	31.16
75.0	289.1	38.54
80.0	355.1	47.34
85.0	433.6	57.81
90.0	525.76	70.10
95.0	633.90	84.51
100.0	760.00	101.3

Sample Problem 46 mL of O_2 gas is collected over water at 25°C when the atmospheric pressure is 102 kPa. What volume of pure oxygen would this be at STP?

Solution
It is not possible to use the combined gas law with the data given. Since the gas was collected over water, water vapor is mixed with the oxygen. The pressure of oxygen gas must be less than 102 kPa. The pressure from water vapor at 25°C can be found in Table 9B-3. Subtract the pressure from water vapor at 25°C from the original pressure. This gives the equation, 102 kPa − 3.167 kPa = 98.8 kPa. Celsius must also be changed to Kelvin.

$$\frac{P_1 V_1}{T_1} = \frac{P_2 V_2}{T_2}$$

$$\frac{(98.8 \text{ kPa})(46 \text{ mL})}{(25°C + 273 \text{ K}) = 298 \text{ K}} = \frac{(101.3 \text{ kPa})(V_2)}{273 \text{ K}}$$

$$\frac{(98.8 \text{ kPa})(46 \text{ mL})(273 \text{ K})}{(298 \text{ K})(101.3 \text{ kPa})} = V_2$$

$$V_2 = 41 \text{ mL}$$

Section Review Questions 9B

1. A gas has a volume of 95 mL at a pressure of 930 torr. What volume will the gas occupy if the pressure is increased to 970 torr?
2. A gas has a volume of 111 mL at a temperature of 32°C. What volume will the gas occupy at standard temperature?
3. A particular sample of pantothenic acid, a B vitamin, gives off 72.6 mL of nitrogen gas at 23°C and 795 torr. What is the volume of the nitrogen at STP?
4. A sample of 40.0 mL of hydrogen is collected by water displacement at a temperature of 20°C. The barometer reads 751 torr. What is the volume of the hydrogen at STP?

9C Gases and the Mole

When 1 mole of water is decomposed into its elements, 16 grams of oxygen and 2 grams of hydrogen are produced. Yet when the gases are collected and their volumes are measured, there is twice as much hydrogen as oxygen. This observation invites further questions about the nature of gases. What is the relationship between the number of molecules, their mass, the volume they occupy, and a gas's density? These questions are the heart of gas stoichiometry.

The Law of Combining Volumes: Gases in Reactions

In addition to his studies of pressure and temperature, Gay-Lussac studied the chemical reactions of gases. In particular, he measured and compared the volumes of gases that reacted with each other. One reaction that he studied was between hydrogen and chlorine ($H_2 + Cl_2 \longrightarrow 2\ HCl$). He found that if the gases had identical pressures and temperatures, 1 liter of hydrogen combined with 1 liter of chlorine formed 2 liters of hydrogen chloride. When he studied the reaction between hydrogen and oxygen ($2\ H_2 + O_2 \longrightarrow 2\ H_2O$), he found different ratios between combining volumes. When another chemist investigated the reaction between nitrogen and hydrogen ($N_2 + 3\ H_2 \longrightarrow 2\ NH_3$), he found yet another set of volume ratios.

Volcanic eruptions involve chemical reactions as well as changes in the temperature, pressure, and volume of gas.

$H_2 + Cl_2 \longrightarrow 2\ HCl$ (mole ratio)
$1\ L\ H_2 + 1\ L\ Cl_2 \longrightarrow 2\ L\ HCl$ (volume ratio)
$2\ H_2 + O_2 \longrightarrow 2\ H_2O$ (mole ratio)
$2\ L\ H_2 + 1\ L\ O_2 \longrightarrow 2\ L\ H_2O$ (volume ratio)
$N_2 + 3\ H_2 \longrightarrow 2\ NH_3$ (mole ratio)
$1\ L\ N_2 + 3\ L\ H_2 \longrightarrow 2\ L\ NH_3$ (volume ratio)

In 1808 Gay-Lussac formulated the **law of combining volumes.** He said that under equivalent conditions, the volumes of reacting gases and their gaseous products are expressed in small whole numbers. Although he did not know it at the time, the ratios of these small whole numbers were the ratios between moles of reactants and moles of products. This law along with the law of definite proportion (which says that the masses of reactants and products are related by ratios of small whole numbers) eventually led chemists to understand the relationship between atoms, molecules, and compounds.

The Molar Volume of a Gas

The law of combining volumes led Amedeo Avogadro to propose a principle. He sought to explain why volumes combined in simple ratios. **Avogadro's principle** soon became a fundamental idea of chemistry.

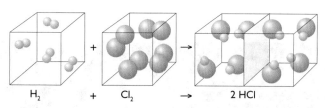

Under equivalent conditions, equal volumes of gases contain the same number of molecules.

Each volume of gas in a reaction contains the same number of molecules. When hydrogen and chlorine react, one hydrogen molecule reacts with one chlorine molecule to form two hydrogen chloride molecules. The one volume of hydrogen contains the same number of molecules as the one volume of chlorine. Since twice as many hydrogen chloride molecules are produced, twice the volume results.

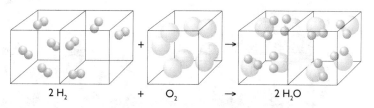

In the formation of water from hydrogen and oxygen, the two volumes of hydrogen contain twice as many molecules as the one volume of oxygen. The two volumes of water vapor contain the same number of molecules as the two volumes of hydrogen and twice the number of molecules as the one volume of oxygen. Likewise, the volumes of gases involved in the formation of ammonia (NH_3) show the relative number of molecules.

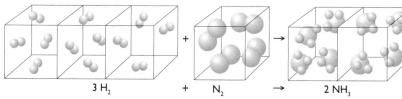

Avogadro's ideas eventually helped establish the idea of a mole. After Avogadro's death, Avogadro's number was determined and named in his honor. Experiments also determined how many molecules were present in a given volume. At STP a volume of 22.4 liters contains 6.022×10^{23} molecules, or 1 mole, of a gas. For this reason, 22.4 liters is called the **molar volume** of a gas. No matter what type of gas is being considered, 1 mole at STP occupies approximately 22.4 liters, or in SI units, 0.0224 m^3.

1 mole H_2

1 mole CH_4

1 mole O_2

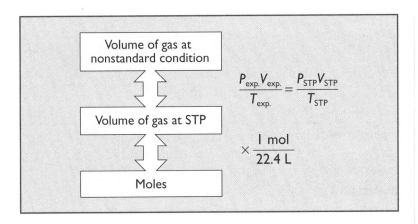

Sample Problem What volume would 4.00 mol of ammonia occupy at STP?

Solution
The fact that 1 mol fills 22.4 L at STP makes the conversion factor 1 mol NH_3/22.4 L valid.

$$\frac{4.00 \text{ mol NH}_3}{} \cdot \frac{22.4 \text{ L}}{1 \text{ mol NH}_3} = 89.6 \text{ L at STP}$$

Sample Problem What volume will 2.50 mol of hydrogen gas occupy at 300. K and at a pressure of 400. torr?

Solution
The conversion factor of 22.4 L/mol changes 2.50 mol of hydrogen to a volume at STP.

$$\frac{2.50 \text{ mol H}_2}{} \cdot \frac{22.4 \text{ L}}{1 \text{ mol H}_2} = 56.0 \text{ L at STP}$$

The combined gas laws can now be used to adjust this volume to the nonstandard conditions.

$$\frac{P_1 V_1}{T_1} = \frac{P_2 V_2}{T_2}$$

$$\frac{(760. \text{ torr})(56.0 \text{ L})}{273 \text{ K}} = \frac{(400. \text{ torr})(V_2)}{300. \text{ K}}$$

$$\frac{(760. \text{torr})(56.0 \text{ L})(300. \text{K})}{(273 \text{ K})(400. \text{torr})} = V_2$$

$$V_2 = 117 \text{ L}$$

Sample Problem A sample of oxygen gas occupies 1.00 L when its temperature is 190. K and its pressure is 129 kPa. How many moles of oxygen are present?

Solution
The given solution must first be corrected to standard conditions.

$$\frac{P_1 V_1}{T_1} = \frac{P_2 V_2}{T_2}$$

$$\frac{(129 \text{ kPa})(1.00 \text{ L})}{190. \text{ K}} = \frac{(101.3 \text{ kPa}) V_2}{273 \text{ K}}$$

$$\frac{(129 \text{ kPa})(1.00 \text{ L})(273 \text{ K})}{(190. \text{ K})(101.3 \text{ kPa})} = V_2$$

$$V_2 = 1.83 \text{ L at STP}$$

Once the gas's volume at STP has been calculated, the number of moles can be found.

$$\frac{1.83 \text{ L}}{} \bigg| \frac{1 \text{ mol } O_2}{22.4 \text{ L}} = 0.0817 \text{ mol } O_2$$

A weather balloon

The Densities of Gases: Why They Are Different

As with all types of matter, the density of a gas is defined as mass per unit volume. The usual units of gas density are grams per liter. Some people assume that all gases have identical, or at least similar, densities. But this is not so. Helium (0.1785 g/L) has a density so low that it can be used to lift weather balloons. Air at 25°C and 1 atmosphere has a density of 1.185 grams per liter. Some gases, such as nitrogen dioxide (1.977 g/L), have densities so high that they immediately sink in air and roll along the ground.

The molar volume concept explains why gases have different densities. One mole of any gas will occupy 22.4 liters at STP. It is reasonable then that the mass of that mole determines the density of the gas. Gases with small molecules and low gram-molecular masses will have low densities. The density of a gas is a function of the gram-molecular mass. Gases with large molecules have greater densities.

$$\text{Density at STP} = \frac{g}{L} = \frac{g/mol}{L/mol} = \frac{\text{Gram-molecular mass}}{\text{Molar volume}}$$

This equation is a powerful tool for calculations involving gases. The theoretical density of a gas can be calculated if the gram-molecular mass of the molecules is known.

Sample Problem What is the density of hydrogen gas (gram-molecular mass = 2.016 g/mol)?

Solution

$$\text{Density of } H_2 = \frac{\text{Gram-molecular mass}}{\text{Molar volume}} =$$

$$\frac{2.016 \text{ g/mol}}{22.4 \text{ L/mol}} = 0.0900 \text{ g/L}$$

Have you ever wondered how science arrived at the figure of 22.4 liters per mole? If the density and gram-molecular mass of a gas are known, the density equation can be used to determine the molar volume.

$$\text{Density} = \frac{\text{Gram-molecular mass}}{\text{Molar volume}}$$

$$\text{Molar volume} = \frac{\text{Gram-molecular mass}}{\text{Density}}$$

Sample Problem Given that the density of oxygen gas at STP is 1.429 g/L and that the gram-molecular mass is 32.00 g/mol, find the molar volume of the gas.

Solution

$$\text{Molar volume} = \frac{\text{Gram-molecular mass}}{\text{Density}} =$$

$$\frac{32.00 \text{ g/mol}}{1.429 \text{ g/L}} = 22.39 \text{ L/mol}$$

Analytical chemists can use this equation to find the gram-molecular mass of an unknown gas. When given an unknown gas, they measure the density and then solve for the gram-molecular mass.

$$\text{Density} = \frac{\text{Gram-molecular mass}}{\text{Molar volume}}$$

$$\text{Gram-molecular mass} = \text{Molar volume} \times \text{Density}$$

Sample Problem An unknown gas has a density of 2.144 g/L at STP. What is its gram-molecular mass?

Solution
$$\begin{aligned}\text{Gram-molecular mass} &= \text{Density} \times \text{Molar volume} \\ &= 2.144 \text{ g/L} \times 22.4 \text{ L/mol} \\ &= 48.0 \text{ g/mol}\end{aligned}$$

Further tests could have proved that this gas is ozone, which has a gram-molecular mass of 48.00 g/mol.

Stoichiometric Conversions with Gases

Chapter 8 showed how the coefficients on balanced chemical equations give the molar ratios between reactants and products. The number of particles and their masses can be calculated once the number of moles is known. The information that 1 mole of a gas at STP occupies 22.4 liters can be added to the flow chart used for stoichiometric problems.

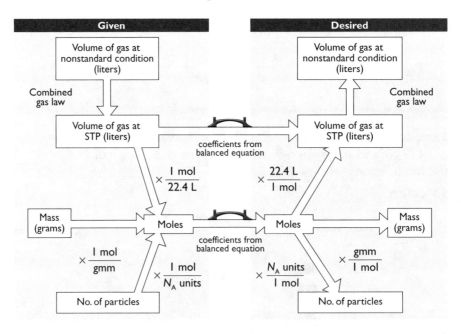

Sample Problem When 2.00 mol of calcium react with water (Ca + 2 H$_2$O $\longrightarrow$ Ca(OH)$_2$ + H$_2$), what volume of hydrogen gas at STP will be produced?

Solution
Moles of calcium can be used to find moles of hydrogen, which can then be converted to the volume of hydrogen.

$$\frac{2.00 \text{ mol Ca}}{} \left| \frac{1 \text{ mol H}_2}{1 \text{ mol Ca}} \right| \frac{22.4 \text{ L}}{1 \text{ mol H}_2} = 44.8 \text{ L H}_2 \text{ at STP}$$

Sample Problem How many grams of water will be produced if 0.500 L of oxygen gas at STP is burned with hydrogen?

Solution
The balanced equation of the reaction is 2 H$_2$ + O$_2$ $\longrightarrow$ 2 H$_2$O. The volume of oxygen must be converted to moles of oxygen. The moles of oxygen can be used to find moles of water, which can then be converted to the mass of water.

$$\frac{0.500 \text{ L O}_2}{22.4 \text{ L O}_2} \left| \frac{1 \text{ mol O}_2}{1 \text{ mol O}_2} \right| \frac{2 \text{ mol H}_2\text{O}}{1 \text{ mol O}_2} \left| \frac{18.016 \text{ g H}_2\text{O}}{1 \text{ mol H}_2\text{O}} \right| = 0.804 \text{ g H}_2\text{O}$$

These conversions apply only to gases at STP. If gases are not at STP, their volumes must be adjusted before they are converted to moles.

Ideal Gases: Nonexistent, but Certainly Useful

Did you know that gas molecules do not have to obey Boyle's law? For that matter, Charles's law holds no authority over gas molecules either. It is important to remember that gas laws do not govern; they describe. Boyle and Charles provided accurate descriptions of volume, pressure, and temperature, but their laws are valid only to the extent that they work. The limitations of these laws point out that science is only man's weak attempt to understand God's creation. Through science man tries to see patterns and to formulate generalizations about the universe. Although sometimes tedious, the endeavor can be challenging, the work can be exciting, and the discoveries can be exhilarating. Yet science must be kept in its proper perspective. Man does not rule the universe. He rather seeks to understand and use what God allows him to know.

An **ideal gas** is a gas that behaves just as the kinetic theory says it should. Like other ideals on this earth, the perfect ideal gas has yet to be found, but some come very close at certain temperatures and pressures. Because the kinetic theory's description of gases is simple, it cannot describe the behavior of most gases at extreme temperatures and pressures.

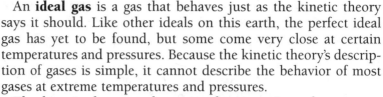

Pressurized gases allow astronauts to breathe in space and aid patients with respiratory difficulties.

The kinetic theory makes several assumptions that are not always accurate.

1. "Gas molecules are extremely small." This statement is true in most cases; however, gases with larger molecules behave differently than expected.

2. "Molecules are located great distances away from each other." This assumption is fine for gases under ordinary conditions. Great pressures or extremely low temperatures can force gases to approach the point of liquefaction. Under these conditions the molecules slow down considerably, and the spaces between them become small.

3. "Forces act on the particles only during collisions." Under normal conditions gas molecules move so fast and are so far apart that intermolecular forces have little chance to act. Yet they always act to some degree. When a gas is near liquefaction, dipole-dipole interactions and other forces increase their effects. Consequently, the gas molecules start "sticking" together and actual pressures and volumes are smaller than expected.

The molar volume of all gases is 22.4 L, right? Not exactly. The volume 22.4 L is only a useful approximation. When gaseous molecules are large, molar volumes decline slightly because intermolecular forces become significant. Table 9C-1 shows the difference in molar volume for some common gases.

Table 9C-1 Standard Molar Volumes for Some Common Gases [0°C]

Gas	Gas Density (g/L)	Molecular Mass (g/mole)	Standard Molar Volume (L)
H_2	0.0899	2.016	22.428
He	0.1785	4.003	22.426
Ne	0.9002	20.18	22.425
N_2	1.251	28.02	22.404
O_2	1.429	32.00	22.394
Ar	1.784	39.95	22.393
CO_2	1.977	44.01	22.256
NH_3	0.7710	17.03	22.094
Cl_2	3.214	70.90	22.063

The Ideal Gas Law

Boyle's law states that volume is inversely proportional to pressure when temperature remains constant. Charles's law states that volume is directly proportional to the Kelvin temperature when pressure is constant. It is also true that the volume is directly proportional to the number of moles of gas (n) in the sample.

$$V \propto 1/P$$
$$V \propto T$$
$$V \propto n$$

When all these proportions are put together, they form the basis for the ideal gas law.

$$V \propto nT/P$$
$$PV \propto nT$$

With the right numerical constant, this proportion can be made into an equality called the **ideal gas law.**

$$PV = nRT$$

R, the **universal gas constant,** relates the units of pressure, volume, temperature, and quantity. Its value and its units depend on the units being used for P, V, n, and T. The ideal gas law can be rearranged to give the value of R.

$$R = \frac{PV}{nT}$$

Since 1 mole of gas at 1 atmosphere of pressure and 273 K occupies approximately 22.4 liters, these values can be substituted into the equation.

$$R = \frac{(1 \text{ atm})(22.4 \text{ L})}{(1 \text{ mol})(273 \text{ K})}$$
$$R = 0.0821 \text{ L} \cdot \text{atm}/(\text{mol} \cdot \text{K})$$

When pressure is measured in torr instead of atmospheres, R has a value of 62.36 L · torr/(mol · K).

In SI units, 1 mole of gas at 101,325 Pa of pressure and 273 K occupies 2.24×10^{-2} m³. Therefore, R in SI units is given by

$$R = \frac{(101{,}325 \text{ Pa})(2.24 \times 10^{-2} \text{m}^3)}{(1 \text{ mol})(273 \text{ K})} = 8.31 \; \frac{\text{m}^3 \cdot \text{Pa}}{\text{mol} \cdot \text{K}}$$

The equation $PV = nRT$ serves chemists well. Like the combined gas law, it relates pressure, volume, and temperature to each other. It also relates the number of gas molecules. Any problem that can be solved with the combined gas law can also be solved with the ideal gas law. As an added benefit, the equation can be used to solve for the number of molecules (measured in moles) in a sample.

Sample Problem How many moles of a gas are present in a 2.4 L sample at 1.25 atm of pressure and 27°C?

Solution

$$PV = nRT$$

$$n = \frac{PV}{RT}$$

Identify the variables, and express them with the correct units. The units in the constant must match the units on the other variables.

$$(1.25 \text{ atm})(2.4 \text{ L}) = n(0.0821 \text{ L} \cdot \text{atm/(mol} \cdot \text{K)})(300. \text{ K})$$

$$\frac{(1.25 \text{ atm})(2.4 \text{ L})}{(0.0821 \text{ L} \cdot \text{atm/(mol} \cdot \text{K)})(300. \text{ K})} = n$$

$$n = 0.12 \text{ mol}$$

If chemists measure the mass, pressure, volume, and temperature of an unknown gas, they can find the gas's gram-molecular mass. They first use the ideal gas law to find the number of moles present. They then form a ratio between the mass of the sample and the number of moles in the sample (n). To complete the problem, they reduce the ratio to find the mass of 1 mole. This mass is the gram-molecular mass of the gas.

$$\frac{\text{Mass of the sample}}{\text{Number of moles in sample }(n)} = \frac{\text{Gram-molecular mass}}{1 \text{ mole}}$$

Sample Problem A 5.04×10^{-4} kg sample of gas occupies 4.57×10^{-4} m³ when under a pressure of 9.63×10^{4} Pa and a temperature of 293 K. What is the gram-molecular mass of this unknown gas?

Solution
The first step is to use the ideal gas law to determine how many moles are present. Since pressure is given in Pa, the constant $R = 8.31$ m³ · Pa/(mol · K) must be used.

$$PV = nRT$$

$$(9.63 \times 10^4 \text{ Pa})(4.57 \times 10^{-4} \text{ m}^3) = n(8.31 \text{ m}^3 \cdot \text{Pa/(mol} \cdot \text{K)})(293 \text{ K})$$

$$\frac{(9.63 \times 10^4 \cancel{\text{Pa}})(4.57 \times 10^{-4} \cancel{\text{m}^3})}{(8.31 \cancel{\text{m}^3} \cdot \cancel{\text{Pa}}/(\text{mol} \cdot \cancel{\text{K}}))(293 \cancel{\text{K}})} = n$$

$$n = 0.0181 \text{ mol}$$

Since 0.0181 mol has a mass of 0.504 g, the mass of 1 mol can be calculated.

$$\frac{0.504 \text{ g}}{0.0181 \text{ mol}} = \frac{27.8 \text{ g}}{1 \text{ mol}}$$

Section Review Questions 9C

1. How many moles of ammonia would produce a volume of 134.4 L at STP?
2. What is the density of chlorine gas, Cl_2, when the gram-molecular mass = 70.91 g/mol?
3. How many mL of propane, C_3H_8, must be burned to form 100.0 mL of CO_2? Use the following equation:
$$C_3H_8 + 5\ O_2 \longrightarrow 3\ CO_2 + 4\ H_2O.$$
4. How many moles of O_2 will occupy 1.00 L at a temperature of -118°C and a pressure of 49.77 atm? Use the ideal gas law to solve the problem.

Chapter Review

Coming to Terms

diffusion
permeability
compressibility
expansibility
pressure
psi
barometer
millimeters of mercury

torr
atmosphere
pascal
standard temperature and pressure (STP)
Boyle's law
Charles's law
Gay-Lussac's law
combined gas law

Dalton's law of partial pressures
vapor pressure
law of combining volumes
Avogadro's principle
molar volume
ideal gas
ideal gas law
universal gas constant

Review Questions

1. Use the kinetic theory of gases to explain why
   a. air has a low density.
   b. on a day with no wind, you can smell a dead skunk that is 500 yards away.
   c. a large volume of air can be pumped into a small bicycle tire.

2. Use your knowledge of the gas laws to predict what will happen to the
   a. pressure in a can of spray paint when the gas is heated (no gas escapes). (case 3)
   b. volume of a balloon as it is warmed (pressure stays the same). (case 2)
   c. volume of a tire tube when it is immersed in cold water (pressure stays the same). (case 2)
   d. pressure in a can of spray paint when some of the paint is released (the can remains at a constant temperature).

3. Convert between pressure units to fill in the blanks.

atm	psi	torr	kPa
		2300	
500.0			50,650
	32.00		
		14.7	1.960

4. After pumping air into a bicycle tire, you notice that the nozzle on the pump is quite hot. Why? What gas law applies to this situation?

5. Boyle's law
   a. A gas at 1.00 atm occupies 5.00 L. Under what pressure will the volume be 10.0 L if the temperature remains constant?
   b. A gas occupying 25.0 L has a pressure of 25.0 lb/in^2. What volume will it occupy if the pressure changes to 35.0 lb/in.^{2} while the temperature remains constant?
   c. A 3.50 L sample of neon gas has a pressure of 0.950 atm at 20°C. What would the volume be if the pressure increased to 1.50 atm and the temperature remained constant?
   d. A gas with a pressure of 845 torr occupies 11.0 L. What will the pressure be for the following volumes if the temperature remains constant?
      1. 2.50 L
      2. 4.00 L
      3. 15.0 L
      4. 22.0 L

6. Charles's law
   a. A sample of nitrogen gas occupies 130 mL at 20°C. What volume will the gas occupy at 45°C if the pressure remains constant?
   b. A particular tank of oxygen gas contains 785 L at 21°C. If the pressure remains constant, what volume will the gas occupy if the temperature is changed to 28°C?
   c. An experiment calls for 5.50 L of sulfur dioxide, SO_2, at 0°C and 1.00 atm. What would the volume of this gas be at 30°C if the pressure remains constant?
   d. At 19°C, a chemical reaction produces 4.30 mL of oxygen gas. How much oxygen gas would be produced at the following temperatures if the pressure remains constant?
      1. 22°C
      2. 32°C
      3. 40°C
      4. 45°C

7. Gay-Lussac's law
   a. A gas occupying 0.500 L at 1200 torr and 16.5°C undergoes a temperature change so that the pressure is now 300 torr. If the volume has remained constant, what is the new temperature?
   b. A reaction requires 1.50 mL of ammonia, NH_3, at 1.65 atm and 23°C. If the temperature is changed to 30°C, what will the new pressure be if the volume remains constant?
   c. A 1.25 mL sample of a colorless, odorless gas had a pressure of 806 torr at 120°C. If the volume remained constant, what would the gas's pressure be if the temperature dropped to 100°C?
   d. A 3.20 mL sample of hydrogen sulfide, H_2S, had a pressure of 970 torr at 55°C. At what temperature would the pressure drop to 900 torr?

8. Combined gas laws
   a. If a gas at 1.2 atm of pressure and 22.0°C occupies 0.350 L, what pressure will hold the same sample of gas in a volume of 0.050 L if the temperature of the gas increases to 25.0°C?
   b. A 15.04 L volume of gas at 700 torr of pressure has a temperature of 35.0°C. What will its temperature be if the pressure increases to 735 torr and the volume increases to 23.8 L?
   c. 40.0 mL of hydrogen are collected by water displacement at a temperature of 20°C. The barometer reads 751 torr. What is the volume of the dry hydrogen at STP?
   d. A sample of nitrogen gas at 16°C and 760 torr has a volume of 2.60 L. What is the volume of the gas at STP?

9. Ideal gas laws
   a. A 500-mL flask is filled with Kr at STP. How many moles of Kr are present? Calculate the density of Kr at STP.
   b. A weather balloon contains 10.0 L of helium at STP. How many moles of He are present? Calculate the density of He at STP.
   c. A 1.00 L flask is filled with Ar by water displacement at 20.0°C. The atmospheric pressure is 750.6 torr. After correcting for the water vapor in the flask, determine how many moles of Ar are present. What is the density of Ar at STP?
   d. A tank of oxygen gas contains 89.0 g O_2. If the volume of the tank is 7.5 L, what is the pressure of the O_2 if the temperature is 21°C?

10. A technician produced H_2 gas by reacting Zn with H_2SO_4 (2 Zn (s) + H_2SO_4 (aq) $\longrightarrow$ Zn_2SO_4 (aq) + H_2). He collected the hydrogen in a flask by water displacement. The atmospheric pressure was 737.2 torr, and room temperature was 25.0°C.
    a. What was the vapor pressure of water in the flask?
    b. What was the pressure of H_2 in the flask?
    c. Suppose that a leak in the gas collection tubes accidentally let a small amount of air into the flask. The unwanted air exerts 25.2 torr of pressure in the flask. How much pressure is the H_2 exerting in the flask?

11. An unknown gas has a density of 0.714 g/L at STP. What is its gram-molecular mass?

12. Carbon dioxide has a density of 1.809 g/L at 25°C and 1 atm. Calculate the gram-molecular mass of CO_2 from this information and calculate a percent error from the actual value. Why should there be an error?

13. What is the gram-molecular mass of a gas whose density is 5.531 g/L at STP?

14. Use the ideal gas law to obtain the answers to the following questions.
    a. How many moles of O_2 will occupy 1.00 L at a temperature of -118°C and a pressure of 49.77 atm?
    b. How many moles of H_2 occupy 5.00 L at 1.70 atm and 35.6°C?
    c. What is the temperature of 0.257 mole of O_2 occupying 6.78 L at 0.856 atm?
    d. What volume does 35.8 g of CO_2 occupy at 1.56 atm and 125.8°C?
    e. What is the pressure of 25.6 g of Cl_2 occupying 15.6 L at 28.6°C?

15. When nitroglycerine (227.1 g/mol) explodes, N_2, CO_2, H_2O, and O_2 gases are released initially. Assume that the gases from the explosion cool to standard conditions without reacting further. 4 $C_3H_5N_3O_9$ (s) $\longrightarrow$ 6 N_2 (g) + 12 CO_2 (g) + 10 H_2O (g) + O_2 (g)
   a. If 16.7 moles of nitroglycerine react, how many liters of N_2 are produced?
   b. What volume of CO_2 (at STP) will be produced when 100.0 g of nitroglycerine reacts?
   c. What is the total volume of gas (at STP) produced when 1.000 kg of nitroglycerine reacts?

16. A reaction between NH_3 and O_2 is the first step in the preparation of nitric acid (HNO_3) on a commercial scale. The products are produced at 1000°C (1273 K) and at atmospheric pressure.
    4 NH_3 (g) + 5 O_2 (g) $\xrightarrow{\text{catalyst}}$ 4 NO (g) + 6 H_2O (l)
   a. What volume of NO is produced in the reaction vessel by the reaction of 0.500 mol O_2?
   b. What mass of H_2O is produced by the reaction of 15.0 L of NH_3?
   c. How many liters of O_2 must react to produce 35.5 L of NO?

17. Copper may be found in an ore named chalcocite, which contains copper(I) sulfide (Cu_2S). Crude copper may be obtained by the roasting of chalcocite (Cu_2S (s) + O_2 $\longrightarrow$ 2 Cu (s) + SO_2). Assume that a smelting plant doing this process is heating some chalcocite to 800°C at an atmospheric pressure of 755 torr. (The O_2 reacts at this temperature and pressure.) Assume also that the SO_2 is being vented from the process at the atmospheric pressure and at 350°C.
   a. What volume of SO_2 is released by the reaction of 454 kg of Cu_2S?
   b. What mass of copper is produced by the reaction of 8.00×10^6 L of O_2?

10A Intermolecular Forces   page 259
10B Solids   page 262
10C Liquids   page 272
FACETS: Diamonds   page 270
FACETS: Cryogenics: It's Really Cold in Here!   page 281

Solids & Liquids 10

Packed and Stacked, Hence More Dense

The kinetic theory of matter ranks as one of the foremost theories of science. Like all good models, the kinetic theory organizes many observations. It explains why gases cause pressure and why evaporating liquids feel cool. With this model scientists can understand phenomena such as thermal expansion and melting. Scientists and students alike appreciate the way this model allows them to visualize microscopic events in solids and liquids.

10A Intermolecular Forces: Weak but Effective

Molecules interact with each other—some more than others. The electrostatic attractions between molecules are called **van der Waals forces.** These bonds are much weaker than the bonds that hold molecules together. If an arbitrary scale were used that measured the strength of the bonds between atoms within a molecule at 100, then the forces between the molecules would range from 0.001 to 15. Nonetheless, these forces play a major role in determining the physical properties of various substances. Why is hydrogen a gas, water a liquid, and sugar a solid at room temperature? The types and strengths of van der Waals forces determine such physical properties. Intermolecular forces can be classified into three groups, and the existence of each type can be explained by the fact that regions of opposite charge on the molecules attract each other.

Figure 10A-1

Figure 10A-2

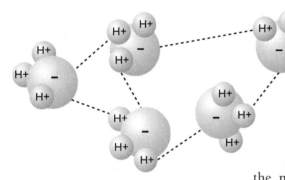

Dipole-dipole interactions. Recall that polar molecules have regions of unevenly distributed electrical charge. The positive areas of one molecule attract the negative areas of other molecules—the stronger the polarity, the stronger the force. These polar molecules then align themselves so that the positive end of one molecule is near the negative end of another molecule. It is this attraction between these negative-to-positive aligned molecules that makes up the dipole-dipole forces. The dipole-dipole forces of polar molecules are similar to the forces in crystals of ionic compounds, but they are much weaker.

Hydrogen bonds. As the name implies, one of the participating atoms in the bond is hydrogen. The other member of the bond is a highly electronegative element. Whenever hydrogen is bonded to a highly electronegative element, the shared electrons shift away from the hydrogen atom. The hydrogen nucleus, which is actually a proton, is left somewhat exposed. Any negatively charged regions of other molecules will quickly interact with the exposed proton. Remember that the hydrogen is the smallest atom, so the molecules can get extremely close together. The combination of the high polarity and the close proximity of the molecules will result in a very strong bond. Hydrogen bonds are possible whenever hydrogen atoms are bonded to nitrogen, oxygen, or fluorine atoms. DNA, the "messenger of life," is composed of two strands of atoms held together by hydrogen bonds in a helical arrangement.

Dispersion (London) forces. Nonpolar substances like wax and gasoline can exist as solids and liquids. This fact shows that dipole-dipole interactions and hydrogen bonds are not the only intermolecular forces. Nonpolar molecules do not have regions of electrical charge, yet they somehow attract each other. This third type of intermolecular force is called a dispersion force. On the average, the electrons in a nonpolar bond spend an equal amount of time around each atom. However, during the course of their movements, the electrons can momentarily concentrate at one end of the molecule. When this occurs, a temporary region of charge forms. Although it lasts for just an instant, this region can pull on a neighboring molecule. Thus, dispersion forces result from the random, unequal dispersion of electrons around the atoms in a molecule. Also, the larger the molecule, the stronger the dispersion force. Can these weak forces have any effect on a substance's physical characteristics? Yes! The cumulative effect of these forces is demonstrated by the fact that methane (gas), octane (free-flowing liquid), octadecane (grease), and paraffin wax (solid) all have the same intermolecular forces and differ only in the number of carbon atoms in each molecule.

Dispersion forces work between all kinds of molecules: polar and nonpolar. Dipole-dipole interactions exist only between polar molecules, and hydrogen bonds exist only between polar molecules that have the necessary atoms (H along with N, O, or F).

Sample Problem List the types of intermolecular forces that act between the molecules of the following compounds:

a. $\ddot{O}=C=\ddot{O}$ b. H:F: c. :Br:Cl:

Solution
a. CO_2 molecules are nonpolar, so only dispersion forces act.
b. HF is polar, and it has the necessary elements for hydrogen bonds to form; so hydrogen, dipole-dipole, and dispersion forces are present.
c. BrCl molecules are polar, so dipole-dipole and dispersion forces act.

Table 10A-1

Structural Units	Forces Between Units	Properties
Ions	Ionic bonds	Very high melting/boiling points; usually soluble in polar solvents; conduct electricity when molten or dissolved, but not in solid state
Nonpolar molecules	Dispersion forces	Low melting/boiling points; molecules soluble only in nonpolar solvents; nonconductors of electricity
Polar molecules	Dispersion, dipole-dipole, possibly hydrogen bonds	Boiling/melting points slightly higher than those of nonpolar molecules; usually soluble in polar solvents; nonconductors of electricity
Cations, mobile electrons (metals)	Metallic bonds	High melting/boiling points; insoluble; electrical conductors in all states

Just as ionic, covalent, and metallic bonds greatly affect the physical properties of the compounds that they form, van der Waals forces also affect these physical properties. Table 10A-1 organizes the cause-effect relationships for the three types of compounds.

Sample Problem Considering the types of chemical bonds and the intermolecular forces that are present, predict the member of each pair that should have the higher boiling point on the basis of the intermolecular forces that are present.

   a. KF, BrF   b. Cl_2, ICl

Solution
a. KF is an ionic compound that has extremely strong ionic forces between its units. Its boiling point should be higher than the covalently bonded BrF.
b. Both Cl_2 and ICl are covalently bonded. ICl molecules are slightly polar, while Cl_2 molecules are not. Since ICl molecules have dipole-dipole interactions, as well as dispersion forces, it is not surprising that their boiling point (97.4°C) is higher than the boiling point of Cl_2 (-34.6°C).

Section Review Questions 10A

1. How are the three types of van der Waals forces similar?
2. What type of van der Waals forces would you expect to find between molecules of H_2O? Of ICl?

10B Solids

Kinetic Description and Basic Properties of Solids

Particles in solids vibrate, but not over great distances. The particles in solids always have some motion. Temperature affects how much the particles in a solid move. At low temperatures the particles barely vibrate. As temperatures increase toward a solid's melting point, the particles have more kinetic energy and the vibrations increase in magnitude. The strength of the attractive forces between molecules in a solid also affects the movements of particles. Strong forces stifle the motions, whereas weak forces allow the particles to move more freely. Table salt exists as a solid at the scorching temperature of 801°C because attractions between the Na$^+$ ions and the Cl$^-$ ions are extremely strong. Nonpolar carbon tetrachloride (CCl_4) molecules, with only their weak van der Waals forces, can move more freely. Consequently, carbon tetrachloride will melt into a liquid at a much lower temperature (-23°C) than sodium chloride.

The kinetic theory explains why solids have the properties they do. Because particles in solids are held close together, they usually have high densities. There is little empty space between them, causing the solids to have fixed shapes and definite volumes. For this reason solids resist compression. Atoms, molecules, or ions would have to be deformed for solid matter to be compressed significantly. Because the particles in solids have little motion, solids have low rates of diffusion, and they are not permeable. If a silver coin and a copper coin were clamped together for several years, not many silver atoms would mix with the copper, and vice versa.

Crystalline and Amorphous Solids: In Line or Jumbled Up

Some solids naturally have orderly shapes. They form regular three-dimensional patterns with distinct edges and sharp angles. When they are shattered, smaller shapes form with similar edges and angles. These solids are called **crystalline solids.** Other solids have no preferred shape. When they are split or shattered, all kinds of fragments result. These solids are called **amorphous solids.**

The differences between crystalline and amorphous solids result from the particular structures of the solids. The particles in crystalline solids such as salt, sugar, and monoclinic sulfur are arranged in well-defined, orderly 3-D patterns. Atoms, ions, or molecules are stacked row upon row, column by column. The patterns, when repeated many times, result in the regular shapes of crystals.

The particles in amorphous solids such as rubber, some plastics, asphalt, paraffin, and amorphous sulfur are not arranged in any particular pattern. Their random, disordered structure results in globular microscopic shapes. Some amorphous solids are called supercooled liquids or glasses. If a liquid can be cooled fast enough, its particles may not have time to get into their preferred crystalline pattern before they stop moving. The molecules become fixed in random positions.

Figure 10B-1

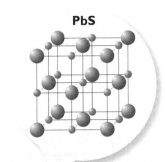

The internal structure of galena (PbS) is reflected in the external shape of its crystals.

Figure 10B-2

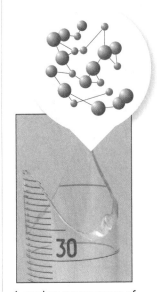

A random arrangement of particles in glass causes glass fragments to have irregular shapes.

Melting and Freezing: Solid, Meet Liquid

Melting and freezing are changes between the solid and liquid states. Melting is the transition from a solid to a liquid; freezing is the reverse. Although people normally think of these terms in relation to water, every substance can melt and freeze. Furthermore, freezing points do not always occur at cold temperatures.

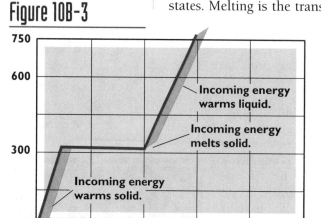

Figure 10B-3

The warming curve of lead shows how temperature changes as heat energy is added.

When a lump of lead is placed in a ladle and heated, the atoms begin to vibrate more vigorously. Temperature, which is a measure of the vibrations, rises. When the melting point of the lead is reached, the temperature ceases to rise. At this point all the atoms are at the brink of liquefaction. Additional heat overcomes the attractive forces and melts the solid. Not until all the lead is melted will the temperature resume its upward climb.

The heat energy that is applied to a substance can be divided into two categories. **Sensible heat** is heat that, when applied to a substance, results in a temperature change in the substance. It is so named because we can "sense" the change in temperature. The second category is called **latent heat.** The latent heat is the heat energy that results in a phase change—the temperature will remain constant. Look at Figure 10B-4 to see where these two categories of energy occur on the warming curve.

The melting of water follows a similar pattern. Ice can be heated until it reaches a temperature of 0°C (sensible heat). Any heat that is added at this point does not raise the temperature—it serves to supply the energy for the breaking of intermolecular bonds (latent heat). Not until the last piece of ice has melted will the temperature of the water rise above 0°C.

A significant amount of heat energy must be added to a solid at its melting point to break intermolecular bonds. This latent heat, called the **heat of**

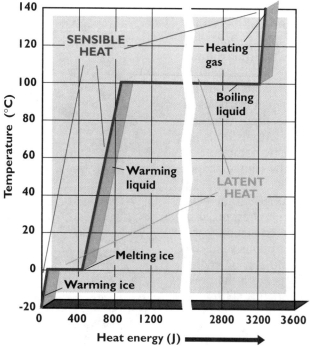

Figure 10B-4

fusion, is different for individual substances. It is defined as the quantity of heat required to change one gram of solid to a liquid with no temperature change. It is usually expressed in units of calories per gram or joules per gram.

Table 10B-1 Heats of Fusion

Substance	(cal/g)	(J/g)
Mercury, Hg	2.8	11
Water, H_2O	79.8	334
Salt, NaCl	124	519
Aluminum, Al	94.5	396
Gold, Au	15.3	64.0
Benzene, C_6H_6	30.5	127

Melting point apparatus. An optical lens allows chemists to observe the point at which crystals on the heating pad begin to melt.

Crystalline substances such as ice and lead have distinct melting points. Warming-curve graphs show clear plateaus that correspond to sharp melting points. An entire sample melts at a clearly defined temperature because all particles are held by nearly identical forces.

Amorphous solids do not have sharp melting points. Their particles are in random positions at different distances from each other. Since the forces vary with distance, not all particles are held together with identical forces. At a specific temperature during the melting process, only some of the forces will be overcome. Amorphous solids gradually soften as some of the attractive forces are overcome. A warming curve of an amorphous solid could look like that shown in Figure 10B-5.

Analytical chemists use melting points to help determine the identity and purity of compounds. Many pure substances are crystalline, and their melting points are listed in reference tables. Chemists can tell how pure a compound is by observing the melting point range. A narrow range indicates a pure sample, and a wide range means that impurities are present.

Figure 10B-5

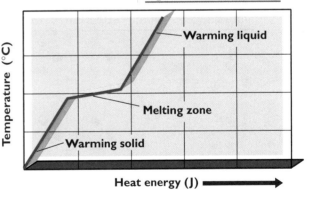

Sublimation: Solid, Meet Gas

Many caterers use dry ice (frozen CO_2) to refrigerate their foods. It has a strong advantage over frozen water because when

Frozen CO_2 sublimes directly into gaseous CO_2.

Iodine sublimes from solid to gas when heated and from gas to solid when cooled.

it warms up, it changes directly into a gas instead of melting into a liquid. This state change bypasses the liquid state and alleviates the problems of messy puddles.

Sublimation is the direct change in state between the solid and gaseous states. Dry ice and many other substances can sublime. It is possible for an occasional molecule to leave the surface of a solid. The smell of naphthalene or paradichlorobenzene mothballs attests to this fact. Substances having many molecules that easily leave the surface sublime readily. Iodine is one of these substances. When heated, the iodine crystals at the bottom of a beaker sublime. The vapor ascends until it hits the cold surface of an ice filled watch glass placed on top of the beaker. Here sublimation occurs again when the vapor changes directly to a solid.

Crystalline Structures: Fourteen Ways to Build a Crystal

The particles in crystals are arranged in orderly, repeating patterns. These patterns vary, as shown by the wide variety of beautiful shapes found in natural crystals. Several factors influence how the particles will be arranged. Solids seek to maximize the distance between similar electrical charges. At the same time they seek to minimize the distance between opposite charges. A crystal's three-dimensional pattern, or **crystal lattice**, depends on the number and kinds of particles, their relative sizes, and their electrical natures. Scientists have found and classified seven basic classes of crystals. (Figure 10B-6)

Figure 10B-6 ▪ Basic Crystal Structures

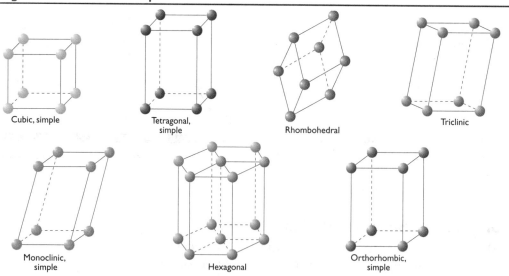

Scientists divide the natural structures of crystals into blocks that contain the fundamental patterns of the lattices. These portions of crystals can be compared to the basic pattern in a piece of wallpaper. The wallpaper pattern can be repeated indefinitely to cover large walls. Unit cells, when repeated many times in three dimensions, form crystals. A **unit cell** is the smallest unit of a crystal that can be used as a building block (Figure 10B-7).

Figure 10B-7

Unit cells are the basic building block for any type of crystal.

Some of these classes can be slightly modified by the addition of particles on the faces or interiors. A body-centered crystal not only has particles at each of the corners, but also has one in the center (body) of the crystal. A face-centered crystal has particles on the corners and on the plane made by one or more of the sides. There is no particle in the center of the crystal. These modifications expand the seven basic classes into fourteen lattices (Figure 10B-8).

Figure 10B-8 ▪ Modifications of Crystal Structure

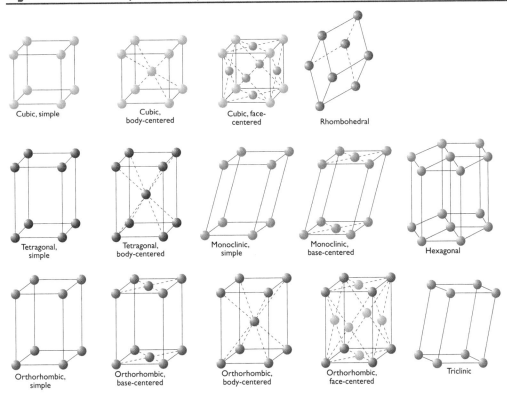

Chapter 10B

Aragonite

Quartz

Hematite

Because crystals are three-dimensional accumulations of unit cells, they often assume the same shape as their unit cells. Salt and sugar crystals appear as tiny cubes because during their formation, their units build upon themselves to form larger cubes with approximately the same number of units along each edge. A sodium chloride unit cell, which contains four Na⁺ ions and four Cl⁻ ions, has a length of 0.56 nanometer (1 nanometer = 10^{-9} m). A salt crystal 0.5 millimeter along each edge has nearly one million (10^6) unit cells along each edge and approximately 10^{18} unit cells in its total structure.

Many crystals found in nature do not have the same shape as their constituent unit cells. Because of varying conditions of temperature, pressure, and other environmental interactions, unit cells may stack in such a way that a different external structure forms. Figure 10B-9 shows some ways in which cubic unit cells may stack. The "steps" on the surfaces of the crystals are only a fraction of a nanometer wide, so they are not visible to the eye. The crystal faces appear to be smooth even though they are ragged on the atomic scale. Despite the fact that all the known minerals have one of the basic crystal structures, mineral crystals can exhibit a great variety of external forms.

Figure 10B-9

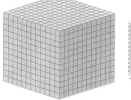 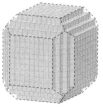 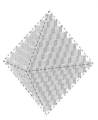 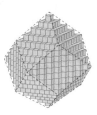

Calcite

Note the difference in structure between calcite crystals and aragonite in the photo above.

Polymorphs and Allotropes: One Substance, Many Faces

Some substances can form more than one type of crystal lattice. Such elements and compounds are said to be **polymorphous*** (pol ee MORE fus). When Ca^{2+} and CO_3^{2-} ions crystallize at a low temperature, they fall into a rhombohedral lattice. Mineralogists call this form of calcium carbonate *calcite*. When these same ions crystallize at a high temperature, they orient themselves in an orthorhombic lattice and form a substance called *aragonite*.

Polymorphous elements are given the special label **allotropic**. The different forms of allotropic elements are called **allotropes**. Sulfur, phosphorus, and arsenic are a few allotropic elements.

---

*polymorphous: poly- (Gk. *polu* – many) + -morphous (Gk. *morphos* – shape)

When sulfur is in the solid state, it exists as ring-shaped S_8 molecules. These rings can be arranged in either the rhombic or the monoclinic lattices. Rapid cooling of liquid sulfur from 300°C can produce yet another allotrope: amorphous sulfur. This globular form changes into the rhombic crystalline form after a couple of days.

Binding Forces in Crystals: What Makes Crystals Strong

Crystals are formed by electrical forces. Oppositely charged regions attract, and similarly charged regions repel each other. The balance between attractions and repulsions determines how tightly the particles in the crystal are bound. If attractions just barely overcome the repulsions, the crystal will have a weak bond. If attractions are much greater than repulsions, the crystal will have strong binding forces.

Sulfur exhibits three crystalline forms: rhombic, monoclinic, and amorphous.

The **lattice energy** of a crystal is the energy that is released when gaseous particles form crystals. It is equal in magnitude but opposite in sign to the energy that must be supplied to pull a crystal apart. The lattice energy shows the difference in energy between the particles when they are in the crystal and when they are free.

What makes a strong crystal? (In other words, what determines the lattice energy?) It has been found that the number of the electrical charges affects the stability of a crystal. Strong charges, like those in ionic compounds, interact strongly and result in strong crystals. The size of the particles also affects the binding forces. Small particles can be more tightly bound than large particles. Finally, the structure of the crystal affects the binding forces.

Binding forces must be overcome whenever a crystal is melted or dissolved. When a crystal melts, thermal energy is used to overcome binding forces. When a crystal dissolves, attractions between the particles of the crystal and the molecules of the solvent are able to break apart the crystal structure.

Section Review Questions 10B

1. Explain the difference between crystalline and amorphous solids.
2. Draw warming curves that would represent a crystalline solid and an amorphous solid. Is there a difference between the two curves? If so, why?
3. How is a unit cell like a formula unit?

FACETS of CHEMISTRY: Diamonds

In January, 1933, a poor prospector and his native helper were searching the diamond field of Pretoria, South Africa. The helper bent down and picked up an earth-encrusted lump. He handed it to the old prospector, Jacobus Jonker. Out of habit, Jonker wiped away the mud from the rather ordinary looking rock. Then he stared at his hand in disbelief. He was holding a rare "blue-white" diamond about the size of a hen's egg. Needless to say, Jonker spent a sleepless night guarding the diamond. He could not relax until he had it safely deposited in the vaults of the Diamond Corporation. Jonker sold his diamond for $315,000—what seemed like a fantastic amount to him. The new owners, however, knew that $315,000 was a small price to pay. With the proper cutting and polishing, the original stone could yield several gems whose combined value would be several times more than that of the rough diamond.

The following year many leading European diamond cutters submitted their plans for cutting the huge diamond. But one lone American expert, Lazare Kaplan, declared that the stone would be ruined if it were cut to the plans that had been submitted. He was one against many, yet the owners finally decided to trust him with their 726-carat treasure. Kaplan planned, measured, and scrutinized for a year and then announced that he was ready. With the help of his son, Leo, he cut a groove on a line of cleavage. Now came the nerve-racking moment: the famous diamond might be shattered into useless fragments, or it might split as planned. Leo held a steel rule in the groove and Kaplan gave it a sharp tap. The Jonker diamond fell apart as planned. The first cleavage yielded a 35-carat chunk. After two more cleavages, the rest of the division was completed by sawing. All of this work was done to increase the value of the rough diamond.

To determine the value of a gem, diamond cutters have to consider the four Cs: carat, clarity, color, and cut. Of these, the weight usually affects the value most. A diamond's weight is measured in carats—a unit of weight equal to 200 mg. Unfortunately, cutting a diamond usually reduces the weight by one-half. Diamond owners are compensated for the weight loss because diamond cutters increase clarity by removing flaws as they work. A diamond cutter looks for flaws

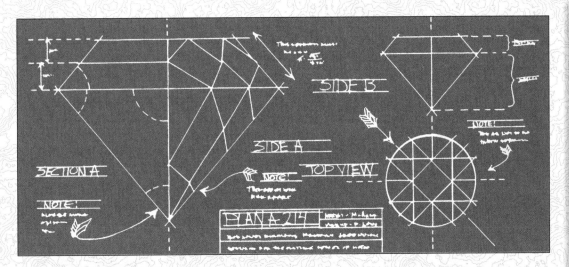

by immersing the diamond in a liquid that bends light rays just as much as diamonds do. This makes the position of any flaws easy to note, and they can be avoided when the diamond is cut. The color is also a factor in the value of a gem. A completely colorless stone, referred to as "blue-white," is the most valuable. Other colors can be valuable only if the color is definite and attractive. Finally, the cut of the diamond influences the value. The shape of the stone should conform to certain proportions, and the facets must be symmetrical, the same size, and well polished.

The cutting of a diamond involves two general steps: dividing and faceting. Although dividing a diamond is only the first step, it is the most important. That is why Kaplan studied the Jonker diamond for a year before attempting to divide it. Two methods can be used to divide diamonds: cleaving and sawing. A diamond can be cleaved in only four directions. To cleave a diamond, cutters form a small scratch in the crystal with a sharp diamond point. They then insert a steel blade into the groove and give it a sharp blow with a mallet. While a diamond will not cleave with ordinary wear, it will cleave cleanly with correct preparation and a hard blow. A diamond can be sawed in only nine directions. The sawing is accomplished with a thin disk of phosphor bronze that revolves 5000 to 6000 times each minute. Initially, a mixture of diamond powder and olive oil serves as the abrasive on the disk. In time, however, diamond dust from the cut diamond replaces the original dust. The bonds in the crystal are so strong that sawing a one-carat diamond may take as long as eight hours.

The final step in cutting the diamond is faceting. The object of faceting is to grind smooth surfaces that will allow the light to enter through the top and be internally reflected as many times as possible. The facets also break up light into its component colors and give the diamond its characteristic fire. Faceting for the common round brilliant gem begins with grinding the stone on a coarse silicon carbide wheel until it is the general size and shape desired. When the stone is satisfactorily shaped, it is closely examined for surface imperfections and chipping. The diamond is then secured in a holder. The intricate pavilion facets are then cut and polished. Again the gem is inspected for chipping. The diamond is then turned around in the holder so that the crown facets can be cut and polished. After the diamond is cleaned, it is ready for sale.

The process of cutting the Jonker diamond produced a total of twelve gems ready for sale. The largest weighed 143 carats. All were of the finest blue-white color, and their total value was $2,000,000 ($2.0 $\times$ 10^6)—quite a change from the original price Mr. Jonker received for his unexpected find.

10C Liquids

If liquids were not so familiar, most people would regard them as amazing substances. Drops of water have been seen beading up into spherical globs so often that this action seems perfectly natural. But is it? People intuitively know that rubbing alcohol will evaporate after it is rubbed onto the skin and that it will feel cold. Although many properties of liquids are well known, they are not necessarily well understood by most people. What makes the surface of a liquid in a test tube curve? What holds an insect up as it walks on the surface of a pond?

A drop of food coloring diffusing through water shows that particles in the liquid state are quite mobile and active.

Kinetic Description of Liquids

Molecules in liquids are held together by intermolecular forces that balance out the kinetic energy of the molecules. They move, but not with the reckless motion of gaseous molecules. They have less energy, and they experience stronger restraining forces than gases do. Yet the attractive forces holding the molecules together do not totally dominate. There is enough freedom for liquid particles to roll and slide over each other. The particles are not fixed in any one position.

Liquids are fluids. They flow and match the shape of a container. In this respect liquids are like gases. Other similar properties are diffusibility and permeability. A drop of food coloring spreading throughout a glass of water, or spilled milk seeping into a paper towel provides evidence that liquids consist of moving particles in unfixed positions.

Liquids have densities that are markedly greater than the densities of gases. This is to be expected, since there is little empty space between the molecules of liquids. Unlike gases, liquids do not expand much. They also are not easily compressed. People rely on this latter property every time they touch the brakes on an automobile. The pressure that is applied on the liquid near the brake pedal is transferred through the brake line to the brake pads or discs. Since the molecules already touch each other, liquids can transfer the incredible pressures of even the most powerful hydraulic systems without being compressed much. The molecules themselves would have to be crushed before a liquid could be compressed to any great degree.

Since liquids cannot be compressed much, hydraulic pumps can force fixed volumes of fluid against great resistances.

Figure 10C-1

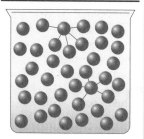

Unbalanced forces act on molecules on the surface of a liquid.

Effects of Intermolecular Attractions

The curved surface of a liquid in a test tube, a drop of oil, and a water spider skating across the surface of a pond are all evidence of intermolecular forces at work. All these examples show the effects of **surface tension.**

Surface tension is the result of molecules forming an elastic "skin" over the surface of liquids. It's not that the intermolecular forces at the surface are stronger than those in the interior of the liquid; instead, the skin forms because the intermolecular forces all point in the same direction at the surface. A molecule within the bulk of a liquid has neighbors on all sides. Each neighbor holds some attraction, so the molecule experiences forces in all directions. A molecule at the surface, however, has no neighboring molecules on one of its sides. All forces are directed toward the interior of the liquid, and the unbalanced forces bind surface molecules together. If the force exerted on the surface of a liquid is less than its surface tension, the object will float. If you are careful, you can float a paper clip on some water in a beaker!

Surface tension causes water droplets to form into spheres (above) and a paper clip to float (below).

Surface tension affects the shapes of liquids. The spherical shape of liquid droplets results when these forces pull the liquid into the shape with the least surface area per unit volume. This action minimizes the unbalanced forces.

Liquids wet surfaces, right? Not always. Wetness is a chemical effect that comes into play in certain situations. For instance, water wets a cotton cloth, but not wax candles. Water molecules are polar, so they readily adhere to sugar, cotton, skin, and any other materials that have polar regions on their surfaces. Nonpolar surfaces cause water to bead up and roll away. On the other hand, nonpolar liquids wet nonpolar surfaces but not polar surfaces. In each situation, the intermolecular attractions determine whether the liquids are "wet."

Sometimes chemists work very hard developing compounds that will reduce surface tension. Detergents are surface-active agents, or **surfactants;** they are able to break down the normal surface tension of water and allow grease and oil to be dissolved. Oil drillers use surfactants to aid in oil recovery after a well has been pumped "dry." The addition of the surfactant allows the 70 percent of the oil that remains in a "dry" well to mix with water that is forced down into the well. The oil and water can be separated later by a different procedure.

A. The surface tension of mercury surpasses mercury's attraction for the glass sides of the dropper. **B.** The strong attraction of water molecules for the glass cause the meniscus of the water to curve upward.

Have you ever observed the curved surface of some liquid in a test tube? This surface, called a **meniscus,** results from intermolecular attractions within the liquid as well as between the liquid and the container. When attractions between the container and the molecules of the liquid exceed the surface tension attractions, the liquid can climb the walls of the container. In narrow glass tubes called capillary tubes, this effect becomes greatly exaggerated. Water rises up narrow capillary tubes easily and exhibits what is called **capillary rise.** Mercury, on the other hand, has little attraction for the glass but has strong internal cohesive forces. These cohesive forces are so

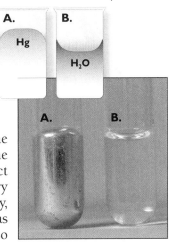

Chapter 10C

Strong intermolecular forces make syrup viscous.

strong that the surface molecules are pulled away from the glass. Thus mercury does not exhibit a capillary rise.

Thickness and *gumminess* are common words for the scientific term **viscosity.** Viscosity is a liquid's ability to resist flowing. It comes about partly because molecules attract each other. The stronger the attractions are, the more viscous the fluid will be. Molasses is viscous, so it can be expected to have some strong attractions between its particles. Thinner, less viscous fluids like water and gasoline pour well and spread out quickly because they have weak intermolecular attractions. Temperature often affects the strength of intermolecular forces. For example, at cold temperatures the forces act more strongly on the slow-moving particles. That is why syrup just taken out of the refrigerator flows so slowly.

Evaporation: The Energetic Ones Come Out on Top

The molecules in a liquid are not all moving at the same speed. Many do move at similar speeds, but random motions rule out complete uniformity. Some molecules are slowed down by a series of head-on collisions. Others are accelerated by a series of favorable shoves. Molecules with above-average speeds can sometimes break away from liquids if they are near the surface. This process, whether it occurs above or below the boiling point, is called **evaporation.** Evaporation can occur at any temperature, but it happens much more readily at higher temperatures.

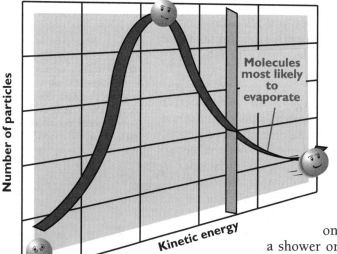

Figure 10C-2

Frequency distribution of water molecules

Evaporation is a cooling process. This is evident when someone shivers before drying off after a shower or a swim. The molecules with the most kinetic energy leave. The ones that remain have a lower average kinetic energy. As a result the unevaporated liquid draws heat from the surrounding environment. The incoming heat does not raise the temperature of the unevaporated liquid; it merely replaces the energy that left with the departed molecules.

The amount of heat required to convert a gram of a liquid at its boiling point to its vapor at the same temperature is called the **heat of vaporization.** Heats of vaporization differ widely because liquids have different intermolecular attractions. For instance, it takes 540 calories to vaporize 1 gram of water at 100°C and 1 atmosphere. A gram of ethyl alcohol at its boiling point can be vaporized with the addition of 205 calories. Table 10C-1 shows the heats of vaporization for several liquids.

Molecules that evaporate can easily re-enter liquids. Suppose that a liquid evaporates in a closed jar. At first the molecules evaporate into the space above the liquid. During their random movements, these molecules can bounce back into the surface of the liquid. When enough molecules have evaporated, the number of molecules that go back into the liquid equals the number of molecules that evaporate. At this point the level of the liquid will remain the same. The volume of the liquid stays constant because the opposite process, condensation, nullifies the effect of evaporation. **Condensation** is the reverse of vaporization—it is the formation of a liquid from a gaseous state. This situation—called a **dynamic equilibrium**—occurs when the two processes of condensation and evaporation oppose each other so that no net effect can be seen.

Table 10C-1 Heats of Vaporization at Boiling Point

Substance	cal/g	J/g
Water	540	2260
Ethyl alcohol	205	854
Ethyl chloride	97.8	409
Diethyl ether	89.6	375

Vapor Pressure: The Impact of Evaporated Molecules

When molecules evaporate, they enter the gaseous state. They then move and collide just like any other gaseous molecules. As temperatures rise, the molecules move faster and will evaporate more easily. The intermolecular attractions within a liquid will also affect the rate of evaporation. Strong attractions restrain par-ticles from evaporating, while weak attractions allow quick evaporation.

When confined in a closed container, the evaporated molecules will exert pressure on the walls of the container. Thus the pressure inside the container is determined by the number of molecules that evaporate. The more molecules that are evaporated, the greater the pressure, and vice versa. The pressure that is exerted by the evaporated molecules is called the **vapor pressure**.

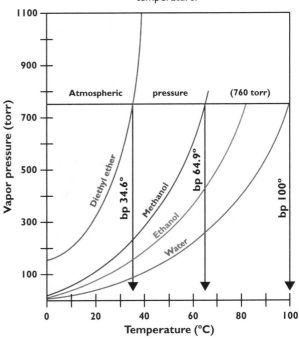

Figure 10C-3

Vapor pressures of liquids increase steadily with temperature.

As expected, temperature and intermolecular forces affect evaporation and vapor pressure similarly. As the temperature is increased, vapor pressure will increase. The vapor pressure will be less for a substance with strong intermolecular forces than for a substance with weaker intermolecular forces.

We can use the calculated vapor pressures of various substances to compare the rate at which they will evaporate. Suppose that some mercury, some water, and some diethyl ether are spilled in a room at 25°C. From Table 10C-2, we can see that the mercury will exert 0.002 torr of vapor pressure; the water, 24 torr; and the ether, 541 torr. These numbers show that the ether will evaporate quickly, the water will take longer, and the mercury will take decades to evaporate.

Table 10C-2 Vapor Pressures (torr)

Substance	0°C	25°C	50°C	100°C
Mercury	0.000185	0.00185	0.0127	0.273
Water	4.58	23.8	92.5	760
Ethanol	12.2	68	234	1663
Methanol	23.7	122	404	—
Diethyl ether	185	541	1216	4859

Boiling:
Full Steam Ahead Despite the Opposition

Boiling is a rapid state change between the liquid and the gaseous states. Vapor escapes from a liquid's surface, as in evaporation, and it also forms internally and collects in bubbles that rise to the surface.

Boiling occurs when the vapor pressure of a liquid equals the atmospheric pressure. Normally, atmospheric pressure is greater than vapor pressure, and it prevents liquids from boiling. Table 10C-2 shows that the vapor pressure of ethanol at 25°C is only 68 torr. The normal atmospheric pressure of 760 torr has no trouble suppressing this weak tendency to vaporize. But when the alcohol is heated to 78.4°C, the story changes. At this temperature the vapor pressure equals normal atmospheric pressure. Now the tendency of ethanol to vaporize wins out, and boiling occurs.

Mountain climbers and users of autoclave sterilizers know from experience that temperature is not the only factor that affects boiling points. Atmospheric pressure affects them too. Climbers at the summit of Mount Everest would find that water boils at 70°C, not at 100°C. At an elevation of 8850 meters

(29,000 ft), the prevailing atmospheric pressure is only 236 torr. This small pressure cannot hold water in its liquid state as long as normal pressure can. If the vapor pressure exceeds 236 torr, the water will boil.

Autoclaves are used in health care facilities to sterilize instruments. An autoclave uses a process that raises boiling points by creating a high pressure "atmosphere" within itself. As the autoclave is heated, a tightly clamped door allows pressure to build up within it to 1036.3 torr, exerting extra pressure on the liquid's surface. At higher pressures, higher-than-normal temperatures are needed to produce matching vapor pressures. At this pressure, the water inside is converted to steam at 232.2°C, rather than at 100°C. This pressure and temperature are then held constant for 30 minutes. That is sufficient time and temperature to achieve complete sterility.

Because boiling points change with pressure, the **boiling point** of a liquid is defined as the temperature at which the vapor pressure equals the applied pressure. The **normal boiling point** is the temperature at which the vapor pressure equals 760 torr.

Hospital autoclave

Sample Problem What is the boiling point of water that is subjected to an atmospheric pressure of 500 torr?

Solution
Using Figure 10C-3, draw a horizontal line from 500 torr on the vertical axis to the curve that represents water's vapor pressure curve. From that intersection, drop a vertical line to the axis representing the temperature. It will intersect at approximately 89°C. At this temperature, the vapor pressure of water will match the given atmospheric pressure, and the water can boil.

Distillation: Separating Liquids

The processes of vaporization and condensation can be linked together and used to separate mixtures. This technique is called **distillation.** Through this method salt water can be purified, and combinations of liquids can be separated into pure samples called fractions.

The mixture to be separated is placed into a distilling flask and is heated. The temperature rises steadily until it reaches the boiling point of the liquid that boils first. The vapor of this liquid enters the condenser and flows over its water-cooled glass walls.

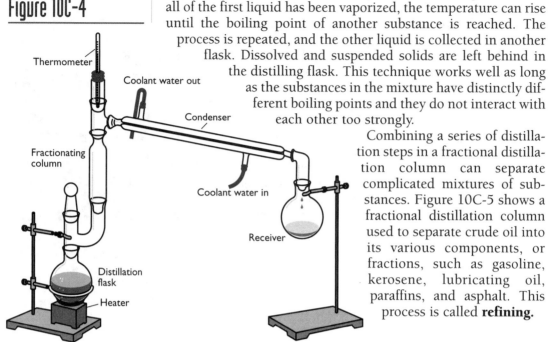

Figure 10C-4

The vapor soon condenses and drips into a collecting flask. Once all of the first liquid has been vaporized, the temperature can rise until the boiling point of another substance is reached. The process is repeated, and the other liquid is collected in another flask. Dissolved and suspended solids are left behind in the distilling flask. This technique works well as long as the substances in the mixture have distinctly different boiling points and they do not interact with each other too strongly.

Combining a series of distillation steps in a fractional distillation column can separate complicated mixtures of substances. Figure 10C-5 shows a fractional distillation column used to separate crude oil into its various components, or fractions, such as gasoline, kerosene, lubricating oil, paraffins, and asphalt. This process is called **refining.**

Critical Values: Temperature and Pressure

The combustion of 1,444,000 liters of liquid hydrogen and 530,000 liters of liquid oxygen thrusts space shuttles into their orbits. Blood banks can store frozen blood cells for years at the super-cold temperatures of liquid nitrogen. Physicists use liquefied helium to refrigerate metals and to cause them to become superconductors.

Oxygen, hydrogen, nitrogen, and helium are usually thought of as gases because their boiling points are well below normal temperatures. Yet these gases can be liquids under the proper conditions. Low temperatures and high pressures can condense and even solidify any gas. The low temperatures slow the molecules down, and the high pressures pack them together.

Scientists have found that high pressures alone cannot liquefy gases. Their temperatures must be lowered past a certain point. This value, called the **critical temperature,** is the highest temperature at which a gas can be liquefied. Each gas has its own characteristic critical temperature.

Hydrogen gas at 35 K cannot be liquefied, even with tremendous pressures. The molecules are moving too quickly. If the temperature is lowered to 33 K, it can be squeezed into a liquid if enough pressure is applied. Gases that have critical temperatures above room temperatures can be liquefied at room temperature by pressure alone. Other gases require a combination of refrigeration and compression.

The pressure that is required to liquefy a gas at its critical temperature is called the **critical pressure**. Hydrogen gas at its critical temperature can be liquefied under a pressure of 12.8 atmospheres. If the gas is colder than its critical temperature, less pressure will be required to liquefy it.

Table 10C-3

Substance	Critical Temperature (K)	Critical Pressure (atm)
Hydrogen, H_2	33.1	12.8
Nitrogen, N_2	126	33.5
Oxygen, O_2	155	50.1
Carbon dioxide, CO_2	304	72.9
Ammonia, NH_3	406	112
Chlorine, Cl_2	417	76.1
Sulfur dioxide, SO_2	431	77.7
Water, H_2O	647	218

Specific Heats: Keeping Your Cool

If an aluminum pan full of water were placed on a stove and the burner were turned on high, after two minutes the pan would be painfully hot, but the water would still be relatively cool. Why? Although the pan does receive the heat first, there is a more significant reason that it becomes hotter more quickly. Not all substances heat at the same rate. The amount of heat energy that raises 1 gram of water 1°C will raise the temperature of 1 gram of aluminum 4.54°C. Given the same amount of heat, the aluminum will get hotter.

These differences arise because substances have characteristic specific heats. The **specific heat** of a substance is the amount of heat required to raise the temperature of 1 gram of the substance 1°C. The usual unit for this quantity is calories per gram · °C (cal/(g · °C)) or joules per gram · °C (J/(g · °C)).

One gram of water can be raised 1°C with the input of 1 calorie, thus the specific heat of water is 1 calorie per gram · °C. A gram of aluminum requires only 0.22 calorie for the same rise in temperature. The specific heat of aluminum is 0.22 calorie per gram · °C.

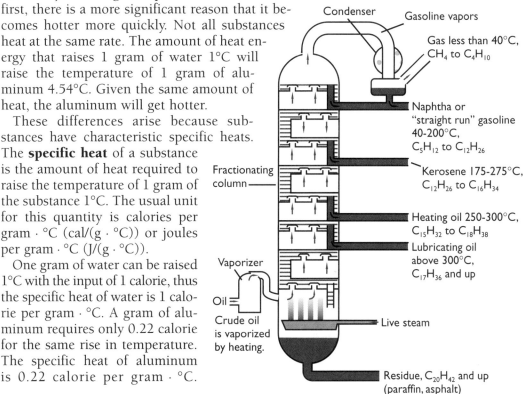

Figure 10C-5

Condenser

Gasoline vapors

Gas less than 40°C, CH_4 to C_4H_{10}

Naphtha or "straight run" gasoline 40-200°C, C_5H_{12} to $C_{12}H_{26}$

Kerosene 175-275°C, $C_{12}H_{26}$ to $C_{16}H_{34}$

Heating oil 250-300°C, $C_{15}H_{32}$ to $C_{18}H_{38}$

Lubricating oil above 300°C, $C_{17}H_{36}$ and up

Fractionating column

Vaporizer

Oil

Crude oil is vaporized by heating.

Live steam

Residue, $C_{20}H_{42}$ and up (paraffin, asphalt)

Specific heats of some common substances are listed in Table 10C-4. Those substances with high specific heats require large amounts of heat for a given temperature change. Low specific heats indicate that the substance will change temperature rapidly.

Table 10C-4

Substance	Specific Heat	
	cal/(g · °C)	J/(g · °C)
Mercury	0.033	0.14
Lead	0.038	0.16
Silver	0.056	0.23
Copper	0.092	0.39
Chlorine	0.11	0.46
Carbon	0.12	0.50
Aluminum	0.22	0.92
Oxygen	0.22	0.92
Benzene	0.42	1.8
Acetic acid	0.49	2.1
Ethyl alcohol	0.58	2.4
Water	1.00	4.19

Like heats of vaporization, specific heats serve as rough indicators of intermolecular attractions. Molecules that are strongly attracted to each other require extra thermal energy to increase their kinetic energies.

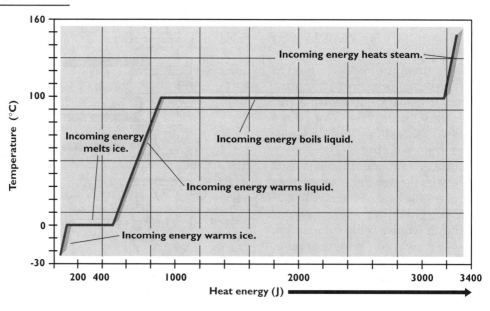

Figure 10C-6

Cryogenics: It's Really Cold in Here!

In the bizarre world of cryogenics, rubber balls shatter on impact and bananas are so strong that they can be used to pound nails into wood. Cryogenics is the science of the supercold. This word *cryogenics* comes from the Greek word *kryo*, meaning "frost." Cryogenics does not deal with the "moderately" cold temperatures of a freezer, but with the intensely cold world of -150°C and below. Applications of this fascinating area of study are found in such wide-ranging fields as space, food preservation, and medicine.

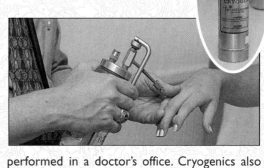

Supercooled gases are used extensively in the space industry. Liquid hydrogen and oxygen are used together as a powerful fuel. Life support systems and space refrigeration systems also make ample use of cryogenic techniques.

Another application of cryogenics has been the preservation of food, especially during shipment. Instead of a mechanical refrigerator, liquid nitrogen can be used to cool food. No moving parts are needed, and exceedingly low temperatures are attainable. Nitrogen has the added advantage of being more chemically inert than the oxygen in air. Since it will not react with the food, even perishable foods such as lettuce and strawberries arrive at their destination in perfect condition.

Cryosurgery, the use of a freezing probe in place of a surgeon's scalpel, has been used with good success for removing warts, tonsils, and cataracts. Cryosurgery offers advantages of less pain and a freedom from hemorrhaging. Many operations that previously required hospitalization have been replaced by procedures that can be performed in a doctor's office. Cryogenics also allows cells and tissues to be frozen for later use.

One area involving irresponsible experimentation is cryonics. This is the process of freezing and storing a human body in hope that the individual can be brought back to life at some future date. Some men hope that people with incurable diseases could be stored in a deep freeze until a cure is found. This extravagant dream has serious ethical and moral implications. Freezing people before they die involves either the sin of murder or of suicide. If the freezing is done after death, there is no real hope of reviving the person, since, as far as science has been able to determine, the biological processes of death are irreversible. Leading authorities also seriously doubt that an entire human body could ever be frozen intact. When more than a few cells are frozen at a time, scientists find it impossible to maintain a fast, uniform rate of freezing. Consequently, destructive ice crystals form that destroy the cell membranes. Despite these facts, people still seek immortality, and there are unscrupulous "freezatorium" operators who continue to offer families false hopes and collect huge sums of money for their futile efforts.

Isn't it amazing that people will spend huge sums of money to extend their mortal life? Even if it were possible, what kind of world would they come back to? The question is not *whether* you can attain immortality, but rather *where* it will be spent. Jesus Christ is the answer, for if you accept Him as your personal Savior, you can be assured that you will spend eternity with Him.

The warming curve of a substance reveals the relative specific heats. If the temperature rises quickly with the addition of heat, the specific heat must be low. In this case, a small amount of heat results in a large change in temperature. A substance that has a gentle slope on its warming curve has a high specific heat because added heat does not change the temperature very much. The warming curve of water illustrates this well. Look back at Figure 10C-6. The specific heat of ice is approximately 0.5 calorie per gram · °C at 0°. Ice heats up twice as quickly as liquid water. The slope of the warming curve that corresponds to the heating of ice is steeper than the portion that corresponds to the heating of the liquid.

Sample Problem If 225 cal of heat energy is added to 50.0 g of water at 25.1°C, what is the final temperature of the water?

Solution
You must first determine the temperature change that occurred by adding the 225 cal of heat energy to the 50.0 g of water using the specific heat of water as the conversion factor as follows:

$$\frac{225 \text{ cal}}{50.0 \text{ g}} \cdot \frac{1.00 \text{ g} \cdot °C}{1.00 \text{ cal}} = 4.50°C$$

Then add this value to the starting temperature of 25.1°C.
$$25.1° + 4.5° = 29.6°C$$

Sample Problem A 28-g mass of silver is heated from 15°C to 85°C. How many joules were added to the sample?

Solution
First determine the rise in temperature: 85° − 15° = 70.°C
Next, use the specific heat determined from Table 10C-4 as the conversion factor. Remember that the units for specific heat is grams · °C, not grams per °C. So, the next step is to multiply the specific heat of silver (from Table 10C-4) times the grams of silver times the temperature change of the silver.

$$\frac{28 \text{ g} \cdot 70.°C \cdot 0.23 \text{ J}}{1.00 \text{ g} \cdot °C} = 450 \text{ J}$$

There has been a tremendous amount of information given—heats of fusion and vaporization, specific heats—what does it all mean? How can this information be used? Suppose you have an ice cube with a mass of 5.00 g at a temperature of 273 K. If that ice cube is heated until it is completely converted to steam at 373 K, how many joules of heat must be added? Remember from the previous sections regarding latent and sensible heat—latent heat is that heat added to make a phase change with no change in temperature, and sensible heat is that heat added with a noticeable temperature change. In this example, ice is being converted to water and then to steam. The first step is to gather all of the needed values from the various tables in this chapter.

1. Heat of fusion of water (ice $\longrightarrow$ water) 334 J/g from Table 10B-1
2. Heat of vaporization (water $\longrightarrow$ steam) 2.26×10^3 J/g from Table 10C-1
3. Specific heat of water 4.19 J/(g · °C) from Table 10C-4

The second step is to determine the latent heat energy (J) needed to melt 5.00 g of ice:

$$5.00 \text{ g} \times 334 \text{ J/g} = 1.67 \times 10^3 \text{ J}$$

Next, determine the sensible energy needed to raise the temperature from 273 K to 373 K:

The temperature change is 100. K, and since the Kelvin and the Celsius scales use equal values, the 100 can be changed to 100.°C. Remember that the specific heat is that heat required to raise 1 g of a substance 1°C, so

$$\frac{5.00 \text{ g}}{} \bigg| \frac{4.19 \text{ J}}{1.00 \text{ g} \cdot °C} \bigg| \frac{100.°C}{} = 2.10 \times 10^3 \text{ J}$$

Then, the latent heat of vaporization as the water is converted to steam is

$$5.00 \text{ g}(2.26 \times 10^3 \text{ J/g}) = 1.13 \times 10^4 \text{ J}$$

Finally, add all of the previously determined energy amounts to obtain the total energy needed to completely convert the 5.00 g of ice to steam:

$$1.67 \times 10^3 \text{ J} + 2.10 \times 10^3 \text{ J} + 1.13 \times 10^4 \text{ J} = 1.51 \times 10^4 \text{ J}$$

Sample problem If you heat 10.0 g of copper to 1083°C and then drop it into 10.0 g water at 10.0°C, what will be the final temperature of both the copper and the water?

Solution

First, since this problem deals with temperature changes, we need to examine the table of specific heats. From Table 10C-4, we can see that the specific heats for copper and water are 0.092 cal/(g · °C) and 1.00 cal/(g · °C). Next, recall that when a warmer object is placed into a colder substance, heat will be transferred from the warmer to the colder object until the temperatures are equal. Thus, the copper will cool down an unknown amount (1083°C − x) and the water will warm up (x − 10.0°C). The final temperature will be the same for both the copper and the water.

The heat lost by the copper will be

$$\frac{10.0 \text{ g} \mid 0.092 \text{ cal/g} \mid (1083°C - x)}{\mid 1.00 \text{ g} \cdot °C \mid}$$

The heat gained by the water will be

$$\frac{10.0 \text{ g} \mid 1.00 \text{ cal} \mid (x - 10.0°C)}{\mid 1.00 \text{ g} \cdot °C \mid}$$

Since the heat lost by the copper will be equal to the heat gained by the water, the final equation would be

$$\frac{10.0 \cancel{g} \mid 0.092 \text{ cal} \mid (1083°C - x)}{\mid 1.00 \cancel{g} \cdot °C \mid} = \frac{10.0 \cancel{g} \mid 1.00 \text{ cal} \mid (x - 10.0°C)}{\mid 1.00 \cancel{g} \cdot °C \mid}$$

The 10.0 will cancel from both sides of the equation, leaving

$$\frac{0.092 \text{ cal} \mid (1083°C - x)}{1.00°C \mid} = \frac{1.00 \text{ cal} \mid (x - 10.0°C)}{1.00°C \mid}$$

Now, solve for the unknown.

$$99.6 \text{ cal} - \left(\frac{0.092 \text{ cal}}{1.00°C}\right)x = \left(\frac{1.00 \text{ cal}}{1.00°C}\right)x - 10.0 \text{ cal}$$

Add 10.0 cal to both sides, as well as (0.092 cal/°C)x:

$$109.6 \text{ cal} = (1.092 \text{ cal/°C})x$$

Solve for x.

$$x = \frac{109.6 \text{ cal}}{1.092 \text{ cal/°C}} = 100.0°C$$

Therefore, the copper will cool down to 100.0°C, and the water will heat up to its boiling point.

Section Review Questions 10C

1. In the previous section, we saw that a paper clip could float on the surface of a beaker of water. What do you think would happen if some detergent were added to the water? Explain.
2. For the boiling point to be considered a unique characteristic of a particular substance, it is always the temperature at 1 atm. Why?
3. How does the specific heat of a substance relate to its ability to change temperature?

Chapter Review

Coming to Terms

van der Waals forces
dipole-dipole interaction
hydrogen bond
dispersion forces
crystalline solid
amorphous solid
sensible heat
latent heat
heat of fusion
sublimation
crystal lattice
unit cell
polymorphous
allotropic
allotrope
lattice energy
surface tension
surfactant
meniscus
capillary rise
viscosity
evaporation
heat of vaporization
condensation
dynamic equilibrium
vapor pressure
boiling
boiling point
normal boiling point
distillation
refining
critical temperature
critical pressure
specific heat

Chapter Review

1. Predict the types of intermolecular forces that may act between the molecules in these substances.
   a. CO_2
   b. NH_3
   c. HCl
   d. C_3H_8
2. Based on your understanding of intermolecular forces, which substance of the pairs should have the higher boiling point?
   a. NF_3, NH_3
   b. $NaCl$, HCl
   c. CF_4, CHF_3
   d. Cl_2, C_2H_5Cl

3. Use the kinetic theory to explain each observation given.
   a. Wax melts near the flame of a burning candle.
   b. Liquid water may be converted into ice cubes in a freezer.
   c. Ginger ale flows to match the shape of the glass.
   d. Water gradually evaporates from a swimming pool.
   e. Water vapor condenses inside house windows on cold days.
   f. Alcohol boils when heated strongly.
   g. Snow gradually disappears, even when the temperature remains below freezing.
   h. Solids and liquids cannot be compressed as much as gases.
4. What is the major difference between a crystal and an amorphous solid?
5. Assume that the particles in table salt (NaCl) vibrate just as forcefully as the particles in Pb. Can you explain why NaCl remains a solid at 500°C while Pb exists as a liquid?
6. When heat is being removed from a liquid, why does the temperature of the liquid at its freezing point remain constant until all of the liquid freezes?
7. What would happen to the temperature and physical state of a gram of liquid diethyl ether at its standard boiling point if its heat of vaporization (375 J/g) were added?
8. What is wrong with the statement, "As a substance freezes, it absorbs energy equal in amount to its heat of fusion"?
9. After a jar of liquid has been sealed, the level of the liquid decreases slightly because of evaporation. After a slight decrease, the level of the liquid ceases to change. Why?
10. Name three factors that determine the structure of a crystal.
11. Name three factors that determine the strength of a crystal.
12. Predict which member of the following pairs of crystals has the stronger binding forces. Explain the reasons for your prediction.
    a. NaCl, I_2 b. KBr, NaBr   c. CaI_2, KI
13. In terms of attractive forces and kinetic energies of particles, explain what happens during the following phase changes. Example: melting—the particles gain enough kinetic energy to break away from the attractive forces that hold them in fixed positions.
    a. boiling
    b. evaporation
    c. freezing
    d. condensation
    e. sublimation
14. A white powder contains tiny, cube-shaped grains and melts at a temperature between 141.6° and 142.2°C. Is this solid most likely to be a crystalline solid or an amorphous solid?

15. Fill in the following chart, which summarizes the properties of solids, liquids, and gases.

State	Compressible	Fluid	Density
Solid	_____	_____	relatively high
Liquid	_____	yes	_____
Gas	yes	_____	_____

16. What causes surface tension?
17. Water rolls off a duck's back but thoroughly wets a head of human hair. What do these observations reveal about the chemical nature of these two surfaces?
18. Why do raindrops not assume triangular or cubic shapes?
19. Human bodies sweat when they overheat. Why does perspiration on the skin cool a person's body?
20. Why does the surface of water in a glass test tube curve upward at the edges?
21. Automobile engines require low-viscosity oils in extremely cold weather. What causes low-viscosity oil to work better at low temperatures than high-viscosity oils? What can you conclude about the intermolecular attractions in low-viscosity oil?
22. What is the essential difference between boiling and evaporation?
23. Refer to Figure 10C-3 to determine
    a. the boiling point of ethanol when it is at normal atmospheric pressure.
    b. the boiling point of methanol at 720 torr.
    c. the atmospheric pressure at which diethyl ether boils at 20°C.
24. Water in a truck's radiator can get hotter than 100°C when the radiator is sealed tightly. How is it possible for water to exist as a liquid at temperatures above its normal boiling point?
25. Crude oil is purified through a process called fractional distillation. In this process the different components of oil (lubricating oil, grease, gasoline, natural gas, etc.) are separated by distillation. What do you conclude about the physical properties of the different components?
26. A scientist claims to have a cooling apparatus kept at -100°C by liquid nitrogen. Is this possible? Why or why not?
27. An ice cube at 0.00°C with a mass of 18.0 g is heated until it is converted to steam at 100.0°C. How many calories of heat have been added to the sample?
28. If you put an 18.0 g ice cube at 0°C in 500 g of water at 20°C, what will be the final temperature of the water after the ice cube completely melts? Hint: you may use the latent heat of fusion value from problem 27.

11A The Water Molecule   page 289
11B The Reactions of Water   page 298
11C Water in Compounds   page 304
FACETS: Water Use and Purification   page 296
FACETS: An F⁻ That Made the Grade   page 303

Water 11

Small Bonds Have Big Effects

What could be so spectacular about water that an entire chapter should be devoted to its qualities? It is such an ordinary compound. It has no color, no taste, and no odor. The compound is so harmless that it serves as a home for creatures ranging from amoebas to blue whales. It makes up approximately 65 percent of the human body. Water is seen everywhere—oceans, lakes, rivers, ice, and clouds. All in all, it has been estimated that there are 1358 million cubic kilometers (1.358×10^{21} L) of water on the earth. That amounts to 7.546×10^{22} moles of water.

Only an amazing compound could be so ordinary. Water must have unique properties that allow it to do so many things in the earth's environment. Indeed, when compared to other compounds that should be similar, water emerges as unique, fascinating, and intriguing.

11A The Water Molecule

Structure

Water owes its unique properties to its molecular structure. The atoms and the shape and polarity of the molecules cause the compound to act the way it does. A water molecule is made of two hydrogen atoms that are covalently bonded to an oxygen atom.

Before bonding, hydrogen atoms have one valence electron (one short of a full outer energy level), and oxygen atoms have six valence electrons (two short of an octet). Two single bonds allow all the atoms to achieve stable electronic configurations.

$$H{:}\ddot{\underset{\times\times}{O}}{:} \\ \overset{..}{H}$$

According to the valence bond theory, the orbitals of the oxygen atom change before they bond. The one 2s and the three 2p orbitals hybridize to form four similar orbitals. Two of the new orbitals contain two electrons, but the other two are only partially filled. The hydrogen atoms bond covalently by sharing electrons in the two unfilled hybridized orbitals.

Figure 11A-1

Because the hybridized orbitals are oriented in the shape of a tetrahedron, the two bonds should form a 109.5° angle. The molecule does have a bent shape, but the angle is 104.5° instead of the expected 109.5°. The difference between the theoretical and the actual bond angle results from the fact that the unbonded orbitals take up more space than the bonded orbitals. Electrostatic repulsions between the electrons in unbonded orbitals are greater than the electrostatic repulsions in bonded orbitals according to the VSEPR theory.

Hydrogen Bonding

An oxygen atom has a high electronegativity (3.5 on the Pauling scale). A hydrogen atom has a lower electronegativity—2.1. The combination of oxygen and hydrogen atoms forms a polar bond, and electrons are shifted toward the oxygen atom. The positive charge on each hydrogen nucleus is left partially exposed while the shared electrons are pulled close to the oxygen atom. The polar bonds and the unsymmetrical shape make the water molecule polar. The region of the molecule that contains the exposed hydrogen nuclei is partially positive, and the region that contains the oxygen is partially negative.

Figure 11A-2

Bond lengths and angles in water molecules

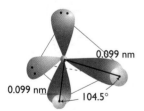

Hydrogen bonds form between the hydrogen atoms of one water molecule and the oxygen atom of another water molecule. These bonds, like other intermolecular forces, are not as strong as the covalent bonds that form molecules. While 68.3 kilocalories of energy would be required to break the covalent bonds in a mole of water molecules, only 4 or 5 kilocalories of energy are required to break apart a mole of hydrogen bonds. The average hydrogen-bond length between water molecules is 0.177 nm (compared to 0.099 nm for the O-H covalent bond). Despite the apparent weakness of these bonds, they have a great effect on the physical properties of water.

Figure 11A-3
Hydrogen bonding between water molecules

The effects of the hydrogen bonding can be seen in a demonstration where water appears to defy the laws of gravity. Water will not flow from an inverted bottle if the opening of the bottle is small enough in size. If the opening of the bottle is 1.6 cm or less, it can be filled with water, covered with a plastic plate, inverted, and the plate removed horizontally. The water will remain in the bottle. This behavior is a result of the hydrogen bonds effectively counteracting the force of gravity.

Physical Properties

Hydrogen bonds raise melting and boiling points. It may seem perfectly natural for water to melt (or freeze) at 0°C and to boil at 100°C, but these facts are actually rather surprising. Look at the series of compounds in which two hydrogen atoms are bonded to other elements in the VIA family.

Hydrogen bonds can effectively counteract gravity.

Table 11A-1

Compound	Molecular Mass	Melting Point °C	Boiling Point °C
H_2Te	129.6	-49.0	-2.0
H_2Se	80.98	-60.4	-41.5
H_2S	34.07	-85.5	-60.7
H_2O	18.02	0.0	100.0

The large mass of each hydrogen telluride (H_2Te) molecule causes the comparatively high melting point of -49°C and boiling point of -2°C. Hydrogen selenide (H_2Se) and hydrogen sulfide (H_2S) have lower melting points because their molecules have less mass. One would expect that water would follow this

Figure 11A-4

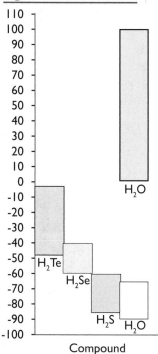

The temperatures at which water exists as a liquid are much higher than the temperatures at which related compounds exist as liquids.

trend of decreasing boiling and melting points, but it does not. It is the hydrogen bonds in water that raise the boiling point despite the low mass of each molecule. These attractions hold water molecules in the solid state at temperatures far above those at which the other compounds vaporize. The hydrogen bonds also hold the water molecules in the liquid state at the surprisingly high temperature of 100 °C. Figure 11A-4 shows where water would exist if it followed the pattern based on decreasing molecular mass, as shown by the empty box.

Water changes from a liquid to a gas at 100 °C when the atmospheric pressure is 760 torr. At higher pressures, higher temperatures are needed to boil water. Lower-than-normal pressures cause lower boiling temperatures. Figure 11A-5 shows the boiling points at different pressures (dashed line). Note that water can boil even at temperatures near 0 °C if the atmospheric pressure is low enough.

Is it merely coincidental that atmospheric conditions on this planet allow water to exist in the liquid state? Hardly. The Lord providentially engineered the sun's thermal output, the dimensions of the solar system, and the earth's cloud cover to provide conditions that would allow water to exist as a liquid. Other planets and moons do not have appreciable amounts of water. Those that do have water have surface temperatures and pressures that will not allow it to exist as a liquid.

Water's normal melting point (or freezing point) is 0 °C. Differences in atmospheric pressure affect the melting point, but not nearly as much as they affect the boiling point. As the pressure rises, the melting point decreases slightly. The almost vertical line in Figure 11A-5 shows the melting points at different pressures.

A **phase diagram** is a graphical way to summarize the conditions under which a substance exists as a solid, liquid, or gas. Figure 11A-6 is called a phase diagram of water because it represents the conditions of pressure and temperature at which water exists as a solid, liquid, or gas. All of the points along the liquid-solid curve (AB) are melting points, all of the points along the liquid-gas curve (AC) are boiling points, and all of the points along the solid-gas curve (AD) are sublimation points.

Figure 11A-5

The boiling and freezing points of water at different pressures

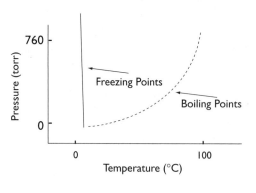

Sample Problem Determine the state of water at the following conditions using Figure 11A-6.
  a. 60°C, 400 torr
  b. -10°C, 1000 torr
  c. 80°C, 200 torr

Solution
  a. The point that corresponds to 60°C and 400 torr is between the line of melting points and the line of boiling points. This region represents the conditions at which water is in the liquid state.
  b. The point corresponding to -10°C and 1000 torr is to the left of the line of melting points. Water is a solid under these conditions.
  c. The point corresponding to 80°C and 200 torr lies beyond the line of boiling points. Water will exist as a gas under these conditions.

In what phase would water exist at a temperature of 0.01°C and a pressure of 4.6 torr? The line of boiling points and the line of melting points intersect at this point (A). Under this condition water can exist in all three states: solid, liquid, and gas. The pressure is low enough that some liquid water can change to a gaseous state. At the same time, the temperature is low enough that some liquid water can begin to freeze into a solid. This specific set of conditions is called the **triple point** of water.

Normally the density of water is considered to be 1 gram per milliliter (cm³ or cc—all are equivalent units) but that density changes slightly with temperature. The kinetic theory states that particles in matter vibrate less violently at lower temperatures. It logically follows that particles that are vibrating less will occupy less space than other particles. Because of this, most matter "shrinks" as its temperature is lowered. Since the same mass occupies less space, the density increases with lower temperatures.

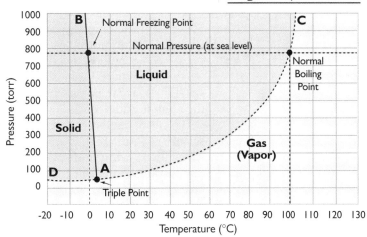

Figure 11A-6 ▪ Phase Diagram of Water

Figure 11A-7 • Density of Water vs. Temperature

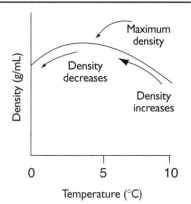

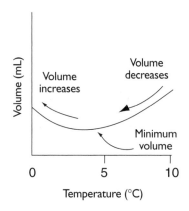

The volume of water changes in accordance with the kinetic theory when it is cooled from 100°C to 3.98°C. As the molecules lose kinetic energy, they vibrate less and take up less space. But at 3.98°C (approximately 4°C) something surprising happens: water begins to expand. It continues to expand as the temperature falls down to 0°C. When water freezes, it suddenly expands even more. This characteristic is completely unlike the characteristics of most matter because most matter shrinks when it freezes.

Density is inversely related to volume. As volume increases, density decreases. Conversely, if volume decreases, density increases. This means that water's density is greatest at 4°C rather than at 0°C when water is frozen. Since the density of an object determines whether it will float or sink in water, ice always floats on the top of liquid water. God's wisdom in designing this earth becomes evident. If water did not expand when it froze, ice would sink to the bottom of oceans, lakes, and rivers instead of forming a layer on the surface. Long, cold winters would cause small bodies of water to freeze solid, killing off many fish and plants and disrupting a vital ecological food chain.

Why does water exhibit this novel behavior? Hydrogen bonds and molecular shapes are the causes. When water freezes, molecules fall into an open, hexagonal crystalline lattice. Hydrogen bonds stabilize the structure and hold molecules in place. Note the empty spaces in the structure (Figure 11A-8). If the molecules were free to roll and slide, this space would be filled. But since the hydrogen bonds dictate where the molecules are when they are in the solid state, the space remains empty. It is this empty space that increases the volume and decreases the density of water when it freezes.

The hexagonal structure is seen mostly in solid ice, but some scientists think that it is present to some degree in very cold water. Even while molecules are rolling over one another, they may arrange into the hexagonal form. When the temperature falls below 4°C, the thermal movement decreases and the hydrogen bonds begin to orient the water molecules into the orderly but spacious crystal system.

The extraordinary surface tension of water further illustrates water's strong intermolecular attractions. Of all the common liquids, only hydrazine (N_2H_4), hydrogen peroxide (H_2O_2), and mercury (Hg) have greater surface tensions.

The Water Molecule

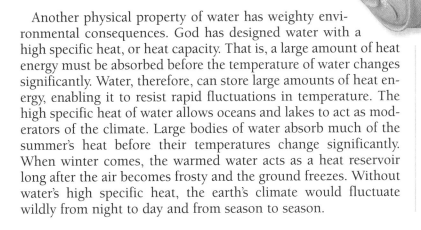

Another physical property of water has weighty environmental consequences. God has designed water with a high specific heat, or heat capacity. That is, a large amount of heat energy must be absorbed before the temperature of water changes significantly. Water, therefore, can store large amounts of heat energy, enabling it to resist rapid fluctuations in temperature. The high specific heat of water allows oceans and lakes to act as moderators of the climate. Large bodies of water absorb much of the summer's heat before their temperatures change significantly. When winter comes, the warmed water acts as a heat reservoir long after the air becomes frosty and the ground freezes. Without water's high specific heat, the earth's climate would fluctuate wildly from night to day and from season to season.

Since water expands as it freezes, icebergs float and sealed cans burst.

Figure 11A-8

When ice forms, hydrogen bonds hold water molecules in an open hexagonal crystalline structure.

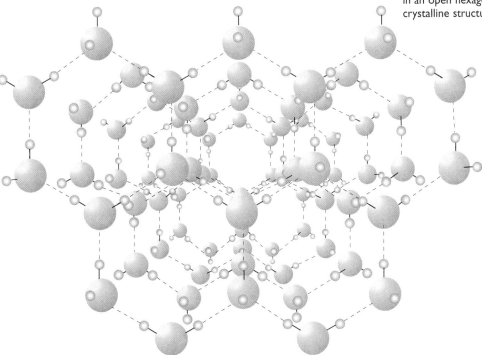

295

Water Use and Purification

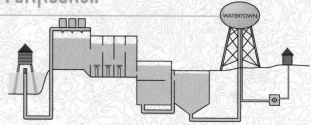

Water is one of the most essential resources on the earth. Our bodies and our society need a constant supply of water in order to function. Each of us should drink at least 2 liters of water per day to replace body losses. In fact, you can live only 5 to 10 days without water. In most of the industrialized countries, people take clean drinking water for granted. However, it is estimated that more than two-thirds of all residents of the world's poorest countries have no access to safe water, and 80 percent subsist on inadequate sanitation facilities.

Are we in danger of running out of water? That would appear to be an illogical question since two-thirds of the earth is covered in water. However, over 99 percent of that water is either in the oceans or in polar ice caps and glaciers. As more and more fresh water is used to claim previously nonarable land by irrigation throughout the world, problems have developed in some areas. Some major rivers now run dry for a portion of the year, including the Yellow River in China and the Indus River in Pakistan. In the western and southwestern United States, where 85 percent of water is used for irrigation, the Colorado River sometimes runs dry before it reaches its mouth at the Gulf of California. In contrast, households use only about 7 percent of the available water, mostly for washing—tub bath (35 gal/bath), shower (5 gal/min), washing machine (45 gal/load), dish washer (12 gal/load)—rather than for drinking. In fact, Americans are unique in the world in that they consume more soft drinks than water!

Pure water is a vital commodity in today's society. Consequently, the purification of water is an important industrial process. A gigantic natural distillation process called the water cycle purifies the earth's water. Water continuously moves from the continents to the oceans. Evaporation takes place from both the continents and the oceans, leaving impurities behind. Water vapor rising to colder layers of air condenses to liquid droplets that form clouds. The cycle is completed when the clouds pour the purified water back onto the earth's surface as rain. This divinely engineered system continuously purifies water on a global scale.

Early settlers in the United States were able to meet their water requirements by drilling wells. Ground water is available almost anywhere if a person drills to a sufficient depth. Residents of many rural and resort areas still obtain their water from wells. In more densely populated areas, however, it is difficult, if not impossible, to keep sewage and other contaminants out of the ground water. Since it would be impractical for each home owner to have his own water-treatment plant, well water as a source for homes has largely given way to community water supplies with large-scale treatment plants.

Water generally goes through three types of purification: (1) coagulation and settling, (2) filtration, and (3) disinfection. In the first step, technicians add alum (aluminum sulfate), which reacts with the water to form aluminum hydroxide, a gelatinous precipitate that slowly settles to the bottom of the container. As it settles, it traps bacteria and impurities that color the water. After the precipitate settles, the water is drawn off and passed through a filter, which often consists of a bed of sand on top of a bed of gravel. Finally, the water is disinfected by the addition of chlorine. In some localities where taste or odor is a particular problem, the water is also aerated (mixed with air) or treated with activated charcoal. Heavy metal ions are difficult to remove and represent a significant health risk. Lead (Pb) from lead pipes

and solder joints is of a great concern to the Environmental Protection Agency (EPA), a government agency that sets the safe drinking water standards. Mercury (Hg) in its organic form, the methylmercury cation (CH_3Hg^+), is increasingly finding its way into the water supply from the soil and from mercury-contaminated rain.

Many regions of the world are now seeking additional sources of drinking water. Researchers continue to improve the efficiency of desalinating seawater (less than 3 percent salt content), and brackish ground water. Distillation is the most common and the most straightforward method to purify water containing dissolved solids; however, large amounts of energy are required to boil the water. Water may be boiled at a lower temperature than the normal boiling point if the pressure is reduced in a process called vacuum distillation. Even though less energy is used in the heating process, energy is still required to produce the reduced vacuum. Because of the high energy costs, alternatives to distillation have been sought.

In one alternate process called reverse osmosis, salt water is forced under pressure against a semipermeable membrane. Water molecules squeeze through the membrane, but the salt stays behind. In another process salt water is cooled to the freezing point. Crystals of pure ice form while the salt stays in solution. The ice is washed free of salt, then melted for use. Both of these alternate methods work better for brackish water than for seawater, since brackish water contains a lower concentration of impurities. Because of the cost, most of the seawater purification processes are economically feasible only for producing drinking water, not the huge amounts of water required for irrigation. An additional problem for each of these processes is what to do with the heavily concentrated brine from the desalination plants. Pumping it back into the ocean could cause environmental problems.

Are there other alternatives? At one time, government planners in southern California considered a proposal to build a freshwater pipeline from Alaska to California to supplement the water supply for the area. However, the cost was prohibitive. Another idea that has been considered is to transport fresh water from Alaska to California in tanker ships similar to those used in the oil shipping industry. The proponents of this idea contend that seagoing water tankers could deliver more water of higher quality at less cost and with less damage to the environment than the desalination plants. Until the problem is solved, researchers and government planners worldwide will continue to seek ways to supply everyone with clean water.

Section Review Questions 11A

1. According to the VSEPR theory, what accounts for water's bond angle of 104.5°?
2. What characteristics of water's crystalline structure contribute to its unique physical characteristics at freezing?
3. Using Figure 11A-6, determine the state of water at the following conditions. More than one condition may exist.
   a. 500 torr, 90°C
   b. 200 torr, 45°C
   c. 900 torr, 100°C
4. List two characteristics of water that have environmental significance.

11B The Reactions of Water

Water molecules are stable. Large releases of energy usually accompany their formation. Once water molecules form, it is extremely difficult to break them apart with heat alone (temperatures around 2700°C are required). This great stability makes water an energetically favored product of many reactions. On the other hand, water is also a reactant in many chemical processes. Many reactive substances can react with it despite its stability.

Formation of Water

It may come as a surprise, but one of the chemicals emitted from an automobile's exhaust pipe is plain water. The combustion of gasoline yields water as a product. In fact, the combustion of any hydrocarbon produces water.

Ethane: $2\ C_2H_6 + 7\ O_2 \longrightarrow 4\ CO_2 + 6\ H_2O$
Propane: $C_3H_8 + 5\ O_2 \longrightarrow 3\ CO_2 + 4\ H_2O$
Octane: $2\ C_8H_{18} + 25\ O_2 \longrightarrow 16\ CO_2 + 18\ H_2O$

The combustion of hydrogen also produces water. This combustion reaction releases energy with an explosive force like that from the combustion of hydrocarbons. Because this reaction sidesteps problems associated with the burning of hydrocarbons (such as high cost of fuel and the unwanted polluting by-products), scientists have considered it as an alternative source of energy.

$2\ H_2 + O_2 \longrightarrow 2\ H_2O$

Hydrogen-powered vehicles may soon become a reality for consumers.

Reactions between acids and bases also produce water. These two types of compounds can neutralize each other and produce water in the process.

$$HCl + NaOH \longrightarrow NaCl + H_2O$$
$$acid + base \longrightarrow salt + water$$

Decomposition of Water

The heat and the explosive force that result from the formation of water by combustion attest to the great amount of energy that is given off when hydrogen and oxygen bond. Once water molecules form, either a large amount of energy or a catalyst is needed to break them apart. Nickel metal has long been used as a catalyst in the steam-reforming process in which steam and hydrocarbons from petroleum or natural gas react at high pressures to form hydrogen and carbon monoxide, as in the following example:

$$C_3H_8\ (g) + 3\ H_2O\ (g) \xrightarrow{Ni} 3\ CO\ (g) + 7\ H_2\ (g)$$

Another thermal means of decomposing water has been the reaction of coal with steam. In this reaction, steam is passed over red-hot coal:

$$C\ (s) + H_2O\ (g) \longrightarrow CO\ (g) + H_2\ (g)$$

Hydrogen is used extensively in the production of ammonia, in petroleum refining, and as a fuel for NASA's space vehicles. Researchers have sought to overcome the large thermal energy requirements, especially in view of the high cost of petroleum and natural gas. One result of that research has been to utilize a dissociation scheme using a type of blue-green algae, *Anabaena cylindrica*, to convert sunlight and water into hydrogen and oxygen. Using this concept to supply a fuel cell with oxygen and hydrogen, electricity could be produced during daylight hours. Solar energy has also been proposed to convert water directly into oxygen and hydrogen. If a process could be developed that would split water more

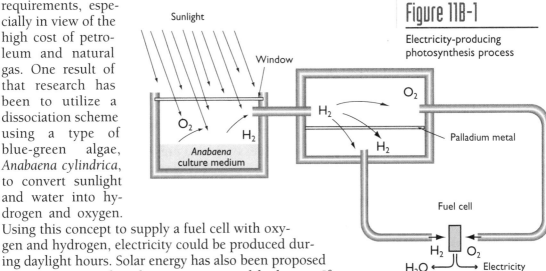

Figure 11B-1

Electricity-producing photosynthesis process

economically than the present techniques, the economic benefits would be enormous.

Electrolysis is another popular method of producing pure oxygen and hydrogen from water by using electricity. An electrical voltage of at least 1.23 volts is required to pull the water molecules apart. This voltage can be supplied by a dry cell battery, a lead-storage battery, or any other source of direct current. The voltage is applied to two chemically inert electrodes dipped in the water sample. The electrode connected to the negative terminal (the cathode) adds electrons to water molecules to produce hydrogen gas.

$$\text{Cathode: } 4\ H_2O + 4\ \text{electrons} \longrightarrow 2\ H_2 + 4\ OH^-$$

The electrode connected to the positive terminal (the anode) pulls electrons from water molecules to form oxygen.

$$\text{Anode: } 2\ H_2O \longrightarrow O_2 + 4\ H^+ + 4\ \text{electrons}$$

The total reaction is the combination of the reactions at the anode and the cathode:

$$6\ H_2O + 4\ \text{electrons} \longrightarrow 2\ H_2 + O_2 + 4\ OH^- + 4\ H^+ + 4\ \text{electrons}$$

The 4 electrons on each side cancel out. The 4 OH⁻ and 4 H⁺ on the right side combine to form 4 H_2O, and when subtracted from the 6 H_2O on the right side, leave 2 H_2O remaining, and a final equation that looks as follows:

$$2\ H_2O \longrightarrow 2\ H_2 + O_2$$

Figure 11B-2 shows an apparatus that is often used to separate and collect the gases that are formed. A small amount of sulfuric acid, hydrochloric acid, sodium chloride, or potassium hydroxide is added to the water being decomposed. These substances do not participate in the electrolysis except to carry the electrical current between the two electrodes. Pure water cannot conduct electricity, so substances that ionize must be present. Moving ions can shuttle electrical current. Substances

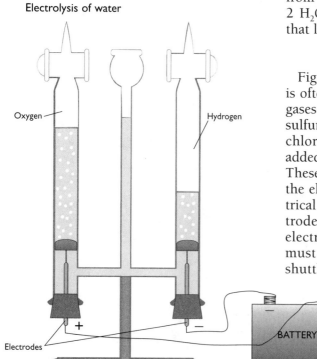

Figure 11B-2

Electrolysis of water

that dissociate in water and provide ions that can carry an electrical current are called **electrolytes.** All acids, bases, and salts are electrolytes.

Reactions with Elements

Many of the more active metals can react with water at room temperature. In essence these reactions are single replacement reactions (discussed in Chapter 8) in which the metals replace hydrogen in the water molecule. The reaction between sodium and water proceeds quickly and explosively.

$$2\ Na + 2\ H_2O \longrightarrow 2\ NaOH + H_2$$

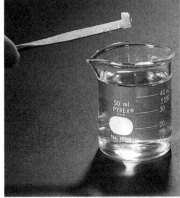

Sodium hydroxide (NaOH) splits into Na^+ and OH^- ions in the water solution, and hydrogen gas is released. Potassium undergoes a similar reaction except that it forms potassium hydroxide (KOH) and reacts even faster and more violently than sodium. Calcium metal also reacts with water at room temperatures.

$$Ca + 2\ H_2O \longrightarrow Ca(OH)_2 + H_2$$

Since magnesium is not as reactive as calcium, it requires boiling water to react. As in the other reactions, a metal hydroxide and hydrogen gas are the products.

Moderately reactive metals such as iron, zinc, and aluminum will react with water only when the water is heated into steam.

$$2\ Fe + 6\ H_2O\ (g) \longrightarrow 2\ Fe(OH)_3 + 3\ H_2$$
$$Zn + 2\ H_2O\ (g) \longrightarrow Zn(OH)_2 + H_2$$
$$2\ Al + 6\ H_2O\ (g) \longrightarrow 2\ Al(OH)_3 + 3\ H_2$$

The most significant water-and-nonmetal reactions involve the halogens. Iodine, bromine, and chlorine can react to form acids: the halogen and water molecules are split. The reaction for chlorine illustrates the general form:

$$\underset{\underset{H}{|}}{H-O} + Cl-Cl \longrightarrow Cl-OH + H-Cl$$

Sodium reacts violently with water.

The extremely reactive fluorine molecule is the only halogen that displaces oxygen from a water molecule.

$$2\ F_2 + 6\ H_2O \longrightarrow 4\ H_3O^+ + 4\ F^- + O_2$$

Reactions with Compounds

Compounds that can react with water in composition reactions are called **anhydrides*** (an HI drides). Two types of anhydrides will be considered: metal oxides and nonmetal oxides. The oxides of very active metals such as sodium, potassium, barium, and calcium react to form compounds with basic properties. The properties of acids, bases, and salts will be discussed in Chapter 14.

$$Na_2O + H_2O \longrightarrow 2\ NaOH$$
$$CaO + H_2O \longrightarrow Ca(OH)_2$$

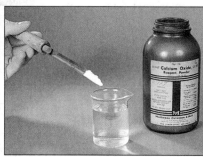

$CaO + H_2O \longrightarrow Ca(OH)_2$ Phenolphthalein in the water is colorless before the reaction occurs. After the reaction it turns pink, signifying the presence of the base $Ca(OH)_2$.

Potassium oxide and barium oxide undergo similar reactions. Because the products of reactions like these have the properties of bases, metal oxides are called **basic anhydrides.**

Oxides of nonmetals such as sulfur, carbon, and phosphorus react with water to form oxyacids. Because of these reactions, oxides of nonmetals are given the name **acidic anhydrides.**

$$CO_2 + H_2O \longrightarrow H_2CO_3$$
$$SO_3 + H_2O \longrightarrow H_2SO_4$$
$$P_4O_{10} + 6\ H_2O \longrightarrow 4\ H_3PO_4$$

Section Review Questions 11B

1. Give two reasons some researchers consider H_2 an ideal fuel.
2. Complete and balance the following reactions.
    a. HI + KOH $\longrightarrow$
    b. Ca + H_2O $\longrightarrow$
    c. BaO + H_2O $\longrightarrow$

$SO_3 + H_2O \longrightarrow H_2SO_4$
Burning sulfur produces SO_3 gas, which mixes with the water. The color of the indicator changes to show that an acid has formed.

*anhydride: an- (Gk. prefix – without) + -hydride (Gk. *hydros* – water)

Facets of Chemistry: An F⁻ That Made the Grade

In the early 1900s Dr. Frederick McKay noticed that some of his patients had discolored and mottled (spotted) tooth enamel. To his surprise, these patients seemed to have less dental decay than his patients with tooth enamel that appeared normal. He observed this condition of mottled enamel only in individuals who had lived in certain geographical areas during their childhood. Adults who moved into these areas did not exhibit any effect to their teeth. One locality in particular that came to Dr. McKay's attention was Bauxite, Arkansas. His intuition told him that something must be in the drinking water, but what? In 1931, ALCOA (Aluminum Company of America) detected elevated levels of fluoride (F^-) ions in the water. The cause of the tooth enamel's abnormal condition had finally been discovered. The fluoride in the water was incorporated into the teeth during the mineralization process, causing the abnormal appearance, which was later termed dental fluorosis.

Later in the 1930s the U.S. Public Health Service made a study of selected localities that had either unusually high or unusually low levels of fluoride. On the basis of these investigations, authorities determined that a level of 1 part per million (ppm) dramatically reduced dental decay without causing mottling of the enamel. Several of the investigators suggested that it would be beneficial to raise the fluoride concentration to 1 ppm in regions where the natural fluoride level was low. As a result, community fluoridation was born.

In the mid-1940s three cities were selected for pilot fluoridation projects—Newburgh, New York; Grand Rapids, Michigan; and Brantford, Ontario. The test results were amazing: the occurrence of dental cavities was reduced 41 to 63 percent. A fourth study, undertaken in Evanston, Illinois in 1947, resulted in reductions ranging from 49 to 75 percent. At first, the researchers were puzzled at the wide range of results. Further study determined that the amount of benefit from the fluoridated water depends strongly on the age of the individual—the younger the person, the greater the benefit. From these communities, the practice of community fluoridation quickly spread to other areas. Soon foreign countries were adopting water fluoridation. Today more than half of the United States enjoys the benefits of water fluoridation, which has been one of the greatest public health successes on record.

What happens if community fluoridation is discontinued? The dental decay rate will surely increase. Antigo, Wisconsin, started fluoridation in 1949 and ceased adding fluoride to the water in 1960. Within six years, second-grade children had over 200 percent more decay, fourth graders 70 percent, and sixth graders 91 percent. A more recent study (1979) in Scotland showed similar results after discontinuing community water fluoridation.

Many people drink bottled water and have home water filtration systems because of the concern over contaminated water. It is important to note that many of the bottled waters on the market do not contain the optimal levels of fluoride. Also, many of the home filtration systems that remove contaminants from the water may remove the fluoride as well. It is important to read the labels of these products to determine their effect on fluoridated water.

In addition to studying the benefits of fluoridation, researchers have studied the risks of high concentrations, and none can be documented at the level of 1 ppm. It is generally agreed that fluoridation at the 0.7 to 1.0 ppm level is as safe as it is effective.

11c Water in Compounds

*Water, water, everywhere,
Nor any drop to drink.*

These lines from Samuel Coleridge's "Rime of the Ancient Mariner" were written to show the plight of sailors on a ship stranded in a salty sea. These lines could also describe the presence of water in chemical compounds. Water molecules can be found in many compounds and in surprising numbers. Though not always easily obtained, water is nearly everywhere.

Hydrates

Hydrates are chemical compounds that include water in their crystal structures. Water molecules occupy empty space in some crystals. Electrically charged particles in the crystal structure can interact with the regions of charge in water molecules to hold them in place. In one sense hydrates are like a chemical sponge: they hold water molecules. Yet hydrates are different from sponges in two important aspects: they hold a set amount of water, and they have crystal structures.

The majority of hydrates are salts. Copper (II) sulfate ($CuSO_4$) crystals can bind 5 moles of water molecules for every mole of copper sulfate. The water that is incorporated into the crystal structure is called the **water of hydration.**

Many other compounds naturally hold water in their crystal structures. Recall that hydrated compounds are named according to the same rules that were used for other compounds. The word *hydrate* follows a Greek prefix.

$CuSO_4 \cdot 5\ H_2O$: copper (II) sulfate pentahydrate
$Ba(OH)_2 \cdot 8\ H_2O$: barium hydroxide octahydrate
$Na_2CO_3 \cdot 10\ H_2O$: sodium carbonate decahydrate

Water molecules can be driven out of hydrates with high temperatures or low pressures. Copper (II) sulfate pentahydrate can be dehydrated if it is heated with a burner.

$$CuSO_4 \cdot 5\ H_2O \xrightarrow{\Delta} CuSO_4 + 5\ H_2O$$

The dehydrated substance is called anhydrous copper (II) sulfate to distinguish it from the hydrated compound.

Hygroscopic, Deliquescent, and Efflorescent Compounds

Some substances have such a strong tendency to attract water into their structures that they can remove water from the air.

$CuSO_4 \cdot 5\ H_2O$ changes color from dark blue to white as it is dehydrated.

These compounds are called **hygroscopic*** (hi gruh SCAHP ic) compounds. Sodium hydroxide is hygroscopic. If placed on a balance, its mass can be observed to increase as it pulls water from the air.

The hygroscopic properties of some compounds must be dealt with in the laboratory. If the compounds are left exposed to the air, their masses will change because of the water they collect. Measurements will not be accurate if precautions are not taken. Hygroscopic compounds must be kept in airtight containers called **desiccators.** Yet even this precaution is sometimes not enough. Powerfully hygroscopic compounds, called drying agents, are stored with the compound being protected in a desiccator. Any moisture in the desiccator is captured by the drying agent (usually $CaCl_2$) rather than by the compound being stored. Many commercially prepared containers of medicine, especially vitamin tablets, have small packets in them that look like tea bags or small perforated plastic cylinders. They contain hygroscopic compounds (usually silica gel) to prevent moisture from degrading the medication.

A desiccator

What happens when a compound is highly hygroscopic and also very soluble in water? It dissolves in the water that it captures from the air. The process of attracting enough water to completely dissolve is called **deliquescence.** Compounds that do this become moist. Later the crystals become completely dissolved in a small puddle of water. Some common drying agents, or **dessicants,** that are also deliquescent include $CaCl_2$, $Mg(ClO_4)_2$, and P_2O_5. The nondeliquescent category includes $CaSO_4$, CaO, silica gel, and Al_2O_3.

Some hydrates lose water when they are exposed to air. This process is called **efflorescence.** Sodium sulfate decahydrate ($Na_2SO_4 \cdot 10\ H_2O$) is one compound that effloresces easily. Copper sulfate pentahydrate effloresces if the humidity is sufficiently low.

Dry NaOH pellets gain moisture and mass when they are exposed to air.

Section Review Questions 11c

1. What can be done to prevent the hydration of a hygroscopic substance?
2. Compound A and Compound B are both hygroscopic. Compound B is also deliquescent. Which would you choose to protect your valuables from water/humidity damage?
3. You ordered 100 grams of a desiccant for your laboratory desiccator. When it arrived, you noticed that it weighed 110 grams. What could have happened? Does it need to be returned to the chemical supply house?

*hygroscopic: hygro- (Gk. *hygro* – moisture) + -scopic (Gk. *skopion* – to see)

Chapter Review

Coming to Terms

triple point
phase diagram
electrolysis
electrolyte
anhydride

basic anhydride
acidic anhydride
water of hydration
hydrate
hygroscopic

desiccator
desiccant
deliquescence
efflorescence

Review Questions

1. Use your knowledge of the structure and properties of water to explain the following observations.
   a. H_2O molecules are polar.
   b. Of all the group VIA nonmetal hydrides, H_2O has the highest melting and boiling points.
   c. Icebergs float.
   d. Northern coastal cities have warmer winters than do cities located inland at the same latitude.
   e. Snowflakes have crystalline structures.
   f. H_2O molecules are bent.
   g. The angle at which an H_2O molecule is bent is smaller than the angle expected from a tetrahedrally hybridized oxygen.

2. What effect do hydrogen bonds have on the following?
   a. the melting point of water
   b. the boiling point of water
   c. the density of ice
   d. the specific heat of water

3. List three types of reactions that produce water.

4. With the aid of the phase diagram in Figure 11A-6, tell what state water exists in under the set of conditions given. If water can exist at more than one state at a set of given conditions, list all possible states.

	Pressure (torr)	Temperature (°C)	State of Water
a.	780	-25	
b.	740	40	
c.	500	125	
d.	760	0	
e.	760	100	
f.	300	125	
g.	4.60	0.01	

5. Complete and balance the following reactions.
   a. $C_5H_{12} + O_2 \longrightarrow$
   b. $H_2 + O_2 \longrightarrow$
   c. $HBr + KOH \longrightarrow$
   d. $Ba(OH)_2 + HNO_3 \longrightarrow$
   e. $Ba(OH)_2 \cdot 8\ H_2O \xrightarrow{\Delta}$

6. If reaction (e) of the previous question occurred spontaneously without heat, the substance would be said to be _____, and the process would be called _____.

7. List four methods by which water can be decomposed.

8. Complete and balance the following reactions.
   a. $Li + H_2O \longrightarrow$
   b. $Mg + H_2O \longrightarrow$
   c. $F_2 + H_2O \longrightarrow$
   d. $Br_2 + H_2O \longrightarrow$
   e. $MgO + H_2O \longrightarrow$
   f. $K_2O + H_2O \longrightarrow$
   g. $SO_2 + H_2O \longrightarrow$

9. What kind of anhydride is
   a. K_2O (see reaction (f) of the previous question)?
   b. SO_2 (see reaction (g) of the previous question)?

10. A 2-lb. box of washing soda ($Na_2CO_3 \cdot 10\ H_2O$) manufactured and packaged in Chicago was found to contain only 28 oz. when it was opened in Tempe, Arizona. Assuming that the manufacturer put in a full 2 lb. of washing soda and that the box did not leak, can you explain why the box weighs 4 oz. less?

11. Calculate the gram-formula mass of $CuSO_4 \cdot 5\ H_2O$.

12. Knowing that H_2O is a stable compound, what can you infer about the strength of its chemical bonds?

13. How could you keep a hygroscopic compound anhydrous during several months of storage?

14. If a hygroscopic compound had been left exposed to humid air during storage, what could you do to make it anhydrous?

12A The Dissolving Process page 309
12B Measures of Concentration page 317
12C Colligative Properties page 322
12D Colloids page 328
FACETS: An In-Depth Look at the Dead Sea page 316

Solutions 12

Consistently Mixed Up

Shampoos, soft drinks, and perfumes are mixtures of many ingredients. Their ingredients do not clump together, separate from each other, or fall to the bottom of the container. These products are not heterogeneous mixtures, since their compositions are uniform. They are not compounds, because no reaction has bound them together.

The substances listed above are **solutions.** In other words, they are homogeneous mixtures of variable composition. The major substance in a solution is called the **solvent.** In shampoo and soft drinks the solvent is water; in perfume, the solvent is often some type of alcohol. The substances that are dissolved are called **solutes.** Solutes in a solvent make a solution.

12A The Dissolving Process

People use solutions whenever they add antifreeze to a radiator, salt an icy sidewalk, use sterling silver, disinfect a cut with iodine tincture, or clean windows with ammonia water. They benefit from a myriad of products manufactured with and from solutions. Almost every chemical reaction takes place in solution. People breathe a solution called air, and most of the human body is a water solution. The study of chemistry is saturated with solutions.

Types of Solutions

Most of the common solutions that people come into contact with each day are liquid solutions. The most abundant part of

Oxygen dissolved in aquarium water is essential for fish to live.

the solution—the solvent—is a liquid. Liquid solvents can dissolve solids, other liquids, and even gases. When sugar is dissolved in tea, a liquid-solid solution is created. Rubbing alcohol, a mixture of 30 percent water dissolved in 70 percent isopropyl alcohol, is a common liquid-liquid solution. When two liquids mix to form a homogeneous mixture, they are said to be **miscible*** (MISS uh bul). Some chemicals form a special kind of solution called an **azeotrope*** (uh ZEE uh trope). An azeotrope is a mixture of liquids that has a constant boiling point and thus cannot be separated by normal distillation techniques. For example, when ethanol (boiling point 78.5°C) and water (boiling point 100°C) are mixed, they form an azeotrope with a boiling point of 78.2°C and a composition of 95 percent ethanol. If you tried to separate these two liquids by fractional distillation, the liquid would boil at 78.2°C until both solvent and solute were gone, resulting in a product of 95 percent ethanol. This problem must be overcome by special distillation techniques.

When liquids refuse to remain mixed, they are said to be **immiscible** (im MISS uh bul). For example, the oil and vinegar in Italian salad dressing are a common pair of immiscible liquids. No matter how hard oil-and-vinegar salad dressing is shaken, the oil and vinegar will separate into two distinct layers. Likewise, the fact that gases can dissolve in liquids is all-important to fish. Fish "breathe" by removing dissolved oxygen from water with the use of thin, membranous gills. Since fish are cold blooded, the amount of oxygen they remove from the water is determined by their body temperature and by the rate at which they move and eat. During the summer months when competition for dissolved oxygen may be severe, some streams may experience large fish kills in which hundreds of fish suffocate.

Solids can also act as solvents—liquids, gases, and even other solids can dissolve in solid solvents. The most common solid-solid solutions are metal alloys. Brass, a copper and zinc alloy, blends the two metals into a uniform mixture. Gold and mercury can form a solid-liquid solution that is sometimes undesirable—liquid mercury will dissolve in gold and will permanently discolor gold jewelry. Hydrogen gas can dissolve in palladium metal and produce a solid-gas solution. Scientists have used this process to purify hydrogen by allowing it to dissolve into the metal and then forcing it out with heat.

---

*miscible: *miscere* (L. – to mix)
*azeotrope: *a* (Gk. prefix – without) + *zeo* (Gk. *zein* – to boil) + *trope*
 (Gk. *tropous* – turning or changing in a specified manner)

A third general type of solution has a gas as the solvent. Although some scientists will argue the point, the only permanent homogeneous mixtures that occur at ordinary pressures with gas solvents are those in which a gas is also the solute. The solution of gases forming the air we breathe consists of oxygen, carbon dioxide, and trace amounts of other gases dissolved in approximately 78 percent nitrogen. Gases cannot serve as solvents for liquid or solid particles, because gases cannot support solid or liquid particles for indefinite periods of time.

Dissolving Mechanisms: Small-Scale Interactions

The water in coffee will eventually dissolve a sugar cube. Although stirring will speed up the process, it is not necessary. The sugar will become evenly distributed throughout the liquid without any outside help. How does this happen? The process of dissolving occurs through molecular interactions called the *dissolving mechanism*.

For a substance to be dissolved, the attractive forces between the particles must be overcome. **Dissociation** is the first step in the dissolving process. In dissociation, the attractive forces between the solute and the solvent must be overcome. In a sugar-water solution, the solute particles (sugar molecules) must be pulled away from each other, and the attractions between solvent particles (water) must also be overcome. Both of these steps absorb heat; therefore, heat energy must be supplied before these intermolecular forces can be broken.

The next step in the dissolving mechanism is called solvation. **Solvation** is the process in which the solvent particles surround and interact with the solute particles. In a sugar-water solution, the positive portions of the water molecules attach to the negative portions of the sugar molecules. The water molecules then pull sugar molecules away from their neighbors. Negative portions of water molecules attract positive parts of sugar molecules in the same manner. Hydrogen bonds form and dispersion forces take effect. As the solute particles are dissolved, energy is released as heat. When the solvent is water, the process of solvation is called **hydration**.

The total input or output of heat energy during the dissolving process depends on the amount of heat energy required for dissociation and the amount of heat given off in solvation. When dissociation absorbs more heat than solvation releases, the energy change shows up as a decrease in the temperature of the solution—an endothermic process. When the process of solvation releases more energy than dissociation absorbs, heat energy is released.

Figure 12A-1

When sugar dissolves, energy is required to break water-water and sugar-sugar intermolecular forces. Energy is given off when hydrogen bonds form between water and sugar molecules.

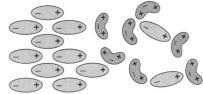

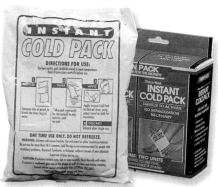

Endothermic reactions make cold packs feel cold.

The motions of the particles increase, and the temperature of the solution rises—an exothermic process.

After solvation, molecular motions carry the dissolved particles throughout the solution. This molecular motion is called **diffusion**. The end result of the dissolving process is an even concentration of solute particles that will remain in solution.

Solvent Selectivity: Like Dissolves Like

Acetic acid mixes with water. Oil does not. Why? What determines which substances can mix or dissolve in others? One general rule is that *like dissolves like*. In chemical terms, polar solvents dissolve polar solutes, and nonpolar solvents dissolve nonpolar solutes.

Ionic compounds and covalent molecules with dipole moments are polar because they have regions of electrical charge. The charges can interact with neighboring ions or polar portions of molecules. Water (H_2O), acetic acid ($HC_2H_3O_2$), ammonia (NH_3), and hydrogen chloride (HCl) are all polar substances. They will mix because forces exist between their molecules. Forces can also exist between ions and molecules, as in a solution of salt water. Positive sodium ions are strongly attracted to the negative portions of water molecules, while negatively charged chloride ions are attracted to the positive portions.

Nonpolar substances such as hexane, pentane, and petroleum ether cannot mix with polar substances. They lack dipoles, and they cannot form hydrogen bonds. Consequently, they are "squeezed out" of polar solvents, forming distinct layers in the container. If you were to mix water and hexane, they would be immiscible and show two layers.

Table 12A-1
Miscible Combinations

	Polar	Nonpolar
Polar	Miscible	Immiscible
Nonpolar	Immiscible	Miscible

Figure 12A-2

As sodium chloride dissolves, negative portions of water molecules attach to Na^+ ions. Positive portions of water molecules interact with Cl^- ions.

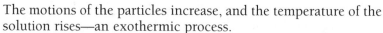

Water and hexane are immiscible.

Sample Problem Considering the electrical natures of methanol, hexane, and aluminum chloride, which of these substances can dissolve in water?

methanol — H—C(H)(H)—OH

hexane — H—C(H)(H)—C(H)(H)—C(H)(H)—C(H)(H)—C(H)(H)—C(H)(H)—H

aluminum chloride — $[Al^{3+}][Cl^-]_3$

Solution
Polar water molecules can dissolve polar methanol molecules and ionic aluminum chloride, but they cannot dissolve nonpolar hexane molecules.

"Like dissolves like" is only a general rule; there are several exceptions. Table salt dissolves in water, but chalk ($CaCO_3$) does not. Both compounds are ionic, so it seems that both should be soluble in a polar solvent like water. Why is there a difference? It is the strength of the ionic attractions within compounds that helps account for differences in solubility. The process of dissociation must overcome the lattice energy of the crystalline structures. It is possible for ions to be bound together so tightly that the solvent cannot break them apart. The solute is then described as being **insoluble.**

Solution Equilibria: More Than Meets the Eye

When a glass is filled half with cold water and half with salt, water molecules immediately begin colliding with the salt, dissociating and hydrating the Na^+ and Cl^- ions in the process. As ions are carried away, additional water molecules move in to pull more of the crystals apart. At this stage when more solute can still be dissolved, the solution is said to be **unsaturated.** The heat energy required for the dissociation of salt is greater than the heat given off in hydration. Heat is therefore absorbed in the solution process, and the solution becomes slightly cooler.

$$H_2O\ (l) + NaCl\ (s) + Heat \longrightarrow Na^+\ (aq) + Cl^-\ (aq)$$

Eventually, however, no more salt appears to dissolve. The temperature and the amount of salt at the bottom of the glass remain constant. When a solution contains the maximum amount of solute at a given temperature, it is **saturated.** But this condition does not stop the solution process. The water molecules are still moving, dissociating, and hydrating ions. The reason that the amount of salt at the bottom of the glass remains constant is that a reverse process also occurs.

$$Na^+\ (aq) + Cl^-\ (aq) \longrightarrow NaCl\ (s) + Heat + H_2O\ (l)$$

When both processes occur at the same rate, no noticeable changes occur. This condition is called a **dynamic equilibrium.**

$$H_2O\ (l) + NaCl\ (s) + Heat \rightleftharpoons Na^+\ (aq) + Cl^-\ (aq)$$

Supersaturation occurs when the solution contains more dissolved solute than normally occurs at equilibrium. For example, the solubility of sodium thiosulfate, $Na_2S_2O_3$, is about 50 g/100 mL at room temperature. However, if you raise the temperature of the solution to 100°C, the solubility increases to 231 g/100 mL—thus more sodium thiosulfate can be dissolved. If the solution is then slowly cooled back to room temperature, the excess sodium thiosulfate will not crystallize out, making 231 g of sodium thiosulfate in 100 mL of water—more than four times the expected amount!

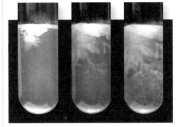

Crystallization begins almost immediately when a crystal of sodium thiosulfate is added to a supersaturated solution (left). Within seconds crystals grow throughout the solution (center and right).

Supersaturated solutions are not in equilibrium, and the supersaturated condition can be destroyed almost immediately by the addition of a small crystal of sodium thiosulfate.

Equilibria can also exist for liquid-gas solutions. Carbonated drinks are actually solutions of carbon dioxide. When a bottle of carbonated beverage is opened, carbon dioxide gas escapes as a "fizz." When the cap is replaced, the escaping gas is again confined in the bottle. As the pressure in the bottle builds up, the rate at which carbon dioxide escapes from the solution decreases. Eventually, the rate at which carbon dioxide re-enters the solution will match the rate at which it leaves the solution. Because the two processes oppose each other, the drink does not lose all of its carbonation.

Rates of Solution

The rate at which something dissolves depends on two factors. The first is the inherent solubility of the solute. Substances that dissolve well dissolve quickly. The second major factor is the number of collisions that occur between solvent molecules and solute molecules: the more collisions, the faster the process.

The temperature of a solution can affect the number of molecular collisions. Consider a glass of warm, unsweetened tea. Which should be put into the tea first—sugar or ice? If the dissolving process is to occur quickly, the sugar should be put in first. The active molecules in warm tea will dissolve sugar faster than the more sluggish molecules in iced tea.

Stirring a solvent will increase the number of collisions. A moving spoon more rapidly brings the solvent molecules into contact with the sugar at the bottom of the glass.

The number of collisions also depends on the surface area exposed to solvent action. Which dissolves more quickly in tea—granulated sugar or sugar cubes? The granulated sugar does. In a sugar cube, many sugar molecules are shielded from the solvent. In granulated sugar, many more sugar molecules are exposed. Increased surface area speeds up the rate of solution.

Factors That Affect Solubility: Temperature and Pressure

Will the solubility of a substance increase or decrease as temperature increases? In general, higher temperatures help solids and liquids to dissolve by increasing the number of particle collisions. Increased temperature also increases solubility because most solids and liquids require heat to dissolve. Suppose that a saturated solution of potassium chlorate ($KClO_3$) is at equilibrium.

$$KClO_3 + H_2O + Heat \rightleftharpoons Solution$$

Adding heat is equivalent to adding more of a reactant. This heat fuels the forward reaction. With the addition of more heat, more potassium chlorate will dissolve until a new equilibrium is established at the new temperature. Figure 12A-3 shows how much $KClO_3$ can be dissolved in 100 grams of water at various temperatures.

However, when gases dissolve, the opposite usually occurs—heat is released.

$$CO_2 + H_2O \rightleftharpoons \text{Solution} + \text{Heat}$$

In the above reaction, adding heat is equivalent to increasing the amount of products. This heat fuels the reverse reaction and causes carbon dioxide to escape from the solution until a new equilibrium is established. Figure 12A-4 shows how much carbon dioxide dissolves in 100 grams of water at various temperatures. Since the increase in heat drives gases out of a solution, this is another reason that fish kills due to suffocation can occur in the summer months. Because the warmer water is not able to dissolve as much oxygen, there is greater competition between the fish and plant life for the available oxygen. Fish cannot live where dissolved oxygen is less than 0.004 g per 1000 g of solution.

Pressure also affects solubility. The effect of pressure on the solubility of solid and liquid solutes is minimal. However, its effect on the solubility of gases can be drastic. **Henry's law** (Chapter 9) states that the solubility of gases increases with the partial pressures of the gases above the solutions. This law is illustrated vividly when a bottle of carbonated soft drink is opened. When the pressure above the beverage is removed, the solubility of carbon dioxide in the beverage decreases, and the excess gas escapes.

Figure 12A-3

Solubility curve of $KClO_3$

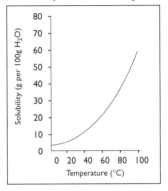

Figure 12A-4

Solubility curve of CO_2

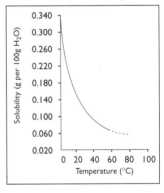

Figure 12A-5 ▪ Solubilities of Common Compounds

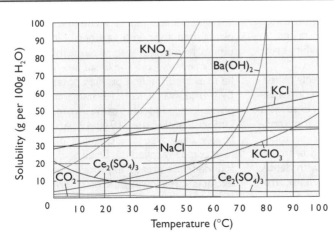

Henry's law in action

Facets of Chemistry: An In-Depth Look at the Dead Sea

It is a landlocked lake, yet people call it a sea. Certain microscopic bacteria seem to thrive in its waters, yet people call it "dead." Together, these misnomers form the name of a most unusual body of water: the Dead Sea.

Nestled in the hills of Judea, the Dead Sea lies 1296 feet below sea level. It covers an area 48 miles long and 10 miles wide. A peninsula divides the sea into two unequal basins. The large northern basin averages about 1300 feet deep. The smaller southern basin has an average depth of only 20 feet.

The Dead Sea receives its water from the Jordan River and four minor streams. The water from these sources picks up many chemicals as it mixes with hot sulfur springs and flows through the region's salty soil. Unlike many other lakes, the Dead Sea has no outlets. Scorching temperatures keep the water level constant by evaporating as much as six million tons of water per day. Sometimes so much water evaporates that heavy clouds form above the surface of the water. While evaporation removes water, it leaves many dissolved minerals behind. Over the centuries, minerals have accumulated and increased in concentration. Dissolved minerals now account for nearly 30 percent of the water's weight. This is seven and one-half times the concentration of salts in the oceans. The brine is so concentrated that it leaves a feeling of nausea if it is tasted; an oily, slippery feeling if touched; and a white crust of chemicals after it dries.

The water of the Dead Sea is not evenly mixed. Different areas of water contain different minerals. The upper layer, for example, is rich in sulfates and bicarbonates, whereas the lower zone contains hydrogen sulfide and strong concentrations of magnesium, potassium, chlorine, and bromine. The concentration of salts increases with depth. Concentrated salt solutions are more dense than pure water, so they sink toward the bottom. Sodium chloride saturates the deep waters of the Dead Sea so much that it remains permanently on the bottom. The floor of the sea is covered with salt crystals.

If the waters of the Dead Sea could speak, they could add many exciting details to biblical history. Much history lies beneath the murky waters of the southern basin. Here the cities of Sodom and Gomorrah once existed. When their wickedness became too great for Jehovah to tolerate, He rained down fire and brimstone (burning sulfur) upon the cities. Today the waters of the southern basin cover the sites of these ancient cities.

The many valuable chemicals in the Dead Sea's waters have attracted several mining operations. Today the area near historical Sodom is the site of a large chemical industry. The first chemical plant established at Sodom was begun in 1929. At that time a Jewish engineer secured the rights to extract minerals from the Dead Sea. Two years later he marketed the first purified potassium chloride. By World War II, half of Britain's annual supply of potassium chloride came from this source. Unfortunately, the 1948-49 Arab-Israeli war largely destroyed the chemical plant. In 1952 a new company called the Dead Sea Works, Ltd. was founded. In a greatly expanded operation, it now extracts large quantities of chlorides, bromides, chlorates, and bromates, as well as many other chemicals from the waters of the Dead Sea.

Section Review Questions 12A

1. Explain why the dissolving process for one substance may be endothermic while for a different substance, it may be exothermic.
2. The chief engineer at a soft drink bottling plant wants to increase the amount of carbonation in the sodas. What can he do to increase the amount of gas dissolved in the sodas?

12B Measures of Concentration

What is the best way to express the concentration of a solution? Terms such as *diluted* and *concentrated* refer to small and large amounts of solute in a solvent. These terms are easy to understand, but they do not precisely describe concentrations. Quantitative measurements of solutes are much more useful in chemistry. Many methods can be used, but the most common are percent by mass, molarity, normality, and molality. Each expression has its own advantages and unique uses.

Percent by Mass

One of the most common methods of expressing concentrations compares the mass of the solute to the mass of the solution. **Percent by mass** is used on the ingredient labels of many household products. Vinegar is approximately 5 percent acetic acid by mass. That means that acetic acid contributes 5 grams for every 100 grams of solution.

$$\% \text{ by mass} = \frac{\text{mass of solute}}{\text{mass of solution}} \times 100\% = \frac{5 \text{ g}}{100 \text{ g}} = 5\%$$

Sample Problem Concentrated hydrochloric acid is 37.2 percent hydrogen chloride by mass. How much hydrogen chloride is in 500.0 mL of the acid (the solution has a density of 1.19 g/mL)?

Solution
The given volume of solution can be converted to mass, since the density is known.

$$\frac{500.0 \text{ mL HCl}}{} \cdot \frac{1.19 \text{ g HCl}}{1 \text{ mL HCl}} = 595 \text{ g HCl}$$

The amount of hydrogen chloride (solute) in the acid can now be calculated.

$$\frac{\text{mass of solute}}{\text{mass of solution}} \times 100\% = \% \text{ by mass}$$

$$\text{mass of solute} = \frac{(\% \text{ by mass})(\text{mass of solution})}{100\%}$$

$$\text{mass of solute} = \frac{(37.2\%)(595 \text{ g HCl})}{100\%}$$

$$\text{mass of solute} = 221 \text{ g HCl}$$

Unit analysis can be used to calculate the same answer.

$$\frac{500 \text{ mL HCl}}{} \left| \frac{1.19 \text{ g HCl}}{1 \text{ mL HCl}} \right| \frac{37.2 \text{ g HCl}}{100 \text{ g HCl}} = 221 \text{ g HCl}$$

Sample Problem A chemical reaction requires 40.0 g of hydrogen chloride. What volume of the hydrochloric acid in the previous sample problem should be added?

Solution

$$\frac{40.0 \text{ g HCl}}{} \left| \frac{100 \text{ g HCl}}{37.2 \text{ g HCl}} \right| \frac{1 \text{ mL HCl}}{1.19 \text{ g HCl}} = 90.4 \text{ mL HCl}$$

Molarity

The most common measure of concentration deals not with mass but with moles. The **molarity** of a solution is the number of moles of solute per liter of solution. The word molar is abbreviated with an M, as in a 6 M HCl solution.

Sample Problem What is the molarity of a solution that contains 3.40 mol of solute in 245 mL of solution?

Solution
After expressing the volume in units of liters, you can write the ratio 3.40 mol/0.245 L, which will give you the units required for molarity (moles of solute/liter of solution). This ratio can be divided by the denominator to yield the number of moles in 1 L.

$$\frac{3.40 \text{ mol}}{0.245 \text{ L}} = \frac{13.9 \text{ mol}}{1 \text{ L}} = 13.9 \, M$$

Solutions measured by molarity are convenient to use in chemical reactions. They make it easy for a chemist to measure out a precise number of atoms, molecules, or ions.

Sample Problem A chemical reaction requires 0.180 mol of silver nitrate ($AgNO_3$). How many milliliters of a 0.800 M solution should be added to the reaction vessel to provide this amount?

Solution

$$\frac{0.180 \text{ mol } AgNO_3}{} \left| \frac{1 \text{ L solution}}{0.800 \text{ mol } AgNO_3} \right| \frac{1000 \text{ mL}}{1 \text{ L}} = 225 \text{ mL solution}$$

Equivalents and Normality

Another measure of concentration is called normality. Before this second measure of concentration can be understood, a more fundamental concept called *chemical equivalents* must be learned. In chemistry the term **equivalent** refers to the combining capacity of a substance in a specific reaction. There are several similar ways to define the term, depending on the type of reaction being considered. When reactions transfer electrons, an equivalent is defined as the amount of substance that gains or releases 1 mole of electrons.

By this definition in the equation below, sodium has 1 equivalent per mole when it reacts with chlorine to form sodium chloride.

$$Na + Cl \longrightarrow NaCl$$

Each mole of sodium loses 1 mole of electrons. The **gram-equivalent mass** of a substance is the mass of an equivalent expressed in grams. The gram-equivalent mass of sodium in this reaction is easily calculated with unit analysis.

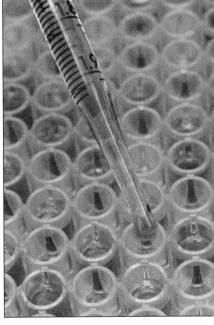

Laboratory solutions are often measured in units of molarity and normality.

$$\frac{22.99 \text{ g Na}}{1 \text{ mol Na}} \left| \frac{1 \text{ mol Na}}{1 \text{ eq Na}} \right. = \frac{22.99 \text{ g Na}}{1 \text{ eq Na}}$$

When copper (II) sulfate reacts with zinc metal, each mole of copper (II) sulfate releases 2 moles of electrons.

$$CuSO_4 \text{ (aq)} + Zn \longrightarrow ZnSO_4 \text{ (aq)} + Cu$$

Since 1 mole of electrons equals 1 equivalent, copper sulfate contains 2 equivalents since it released 2 moles of electrons.

Thus, the gram-equivalent mass of copper (II) sulfate is one-half its gram-formula mass.

$$\frac{159.61 \text{ g CuSO}_4}{1 \text{ mol CuSO}_4} \cdot \frac{1 \text{ mol CuSO}_4}{2 \text{ eq CuSO}_4} = \frac{79.805 \text{ g CuSO}_4}{1 \text{ eq CuSO}_4}$$

In acid-base reactions, an equivalent is defined as the quantity that donates or accepts 1 mole of hydrogen ions. Acids donate hydrogen ions and bases accept hydrogen ions. When hydrogen chloride is dissolved in water, each hydrogen chloride molecule splits into an H^+ ion and a Cl^- ion.

$$HCl \longrightarrow H^+ + Cl^-$$

One mole of hydrochloric acid gives up a mole of hydrogen ions. The gram-equivalent mass of this substance is identical to its gram-molecular mass. For sulfuric acid each molecule gives away two hydrogen ions when it dissociates completely in water.

$$H_2SO_4 \longrightarrow SO_4^{2-} + 2\ H^+$$

Therefore, one mole of sulfuric acid gives up 2 moles of hydrogen ions. The gram-equivalent mass of sulfuric acid can be determined as follows:

$$\frac{98.08 \text{ g H}_2\text{SO}_4}{1 \text{ mol H}_2\text{SO}_4} \cdot \frac{1 \text{ mol H}_2\text{SO}_4}{2 \text{ eq H}_2\text{SO}_4} = \frac{49.04 \text{ g H}_2\text{SO}_4}{1 \text{ eq H}_2\text{SO}_4}$$

Since the 1 mole of sulfuric acid gives up 2 moles of hydrogen ions, the gram-equivalent mass is $\frac{1}{2}$ the gram-molecular weight of sulfuric acid.

The **normality** of a solution is the number of equivalents of solute per liter of solution. The word *normal* is abbreviated with an N, as in a 6 N HCl solution. Recall that molarity is the number of moles per liter of solution. As you can see from the diagram, these two are related by the mole to equivalent ratio.

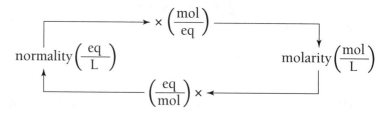

If you know the molarity, you can calculate the normality by multiplying the molarity by the number of equivalents per mole. Conversely, if the normality is known, multiply it by the number of moles per equivalent to calculate the molarity.

Sample Problem
a. Calculate the normality of a 0.75 M solution of H_2SO_4.
b. Calculate the molarity of a 0.5 N solution of HCl.

Solution
a. Using unit analysis,

$$\frac{0.75 \text{ mol } H_2SO_4}{1 \text{ L}} \left| \frac{2 \text{ eq } H_2SO_4}{1 \text{ mol } H_2SO_4} \right. = \frac{1.5 \text{ eq } H_2SO_4}{1 \text{ L}} = 1.5 \text{ N solution } H_2SO_4$$

b. Using unit analysis,

$$\frac{0.5 \text{ eq HCl}}{1 \text{ L}} \left| \frac{1 \text{ mol HCl}}{1 \text{ eq HCl}} \right. = \frac{0.5 \text{ mol HCl}}{1 \text{ L}} = 0.5 \text{ M solution HCl}$$

Sample Problem What is the normality of an 875 mL sulfuric acid solution that contains 126 g of sulfuric acid and reacts with sodium hydroxide? The reaction is as shown:

$$H_2SO_4 + 2\ NaOH \longrightarrow Na_2SO_4 + 2\ H_2O$$

Solution
In this reaction each mole of sulfuric acid donates 2 moles of hydrogen ions. The conversion factor of 2 equivalents per 1 mole sulfuric acid must be used.

$$\frac{126 \text{ g } H_2SO_4}{875 \text{ mL}} \left| \frac{1000 \text{ mL}}{1 \text{ L}} \right| \frac{1 \text{ mol } H_2SO_4}{98.08 \text{ g } H_2SO_4} \left| \frac{2 \text{ eq } H_2SO_4}{1 \text{ mol } H_2SO_4} \right.$$

$$= \frac{2.936 \text{ eq } H_2SO_4}{1 \text{ L solution}} = 2.94 \text{ N } H_2SO_4 \text{ solution}$$

Molality

The **molality** of a solution is defined as the number of moles of solute per kilogram of solvent. It is different from the molarity of the solution, and it is used in different instances. The word *molal* is abbreviated with an *m*, as in a 6 *m* NaCl solution.

Sample Problem What is the molality of a solution composed of 30.0 g of sodium nitrate ($NaNO_3$) and 400. g of water? The gram-formula mass of sodium nitrate is 85.00 g.

Solution

The problem can be solved by setting up an equation using unit analysis with the information given. Since the definition for molality is moles of solute/kg of solvent, the 400. g of water must be converted to 0.400 kg of water. The equation is as follows:

$$\frac{30.0 \text{ g NaNO}_3}{0.400 \text{ kg H}_2\text{O}} \times \frac{1 \text{ mol NaNO}_3}{85.00 \text{ g NaNO}_3} = 0.882 \, m \text{ solution}$$

Section Review Questions 12B

1. Calculate the percent by mass of a solution containing 0.21 moles of $C_{12}H_{22}O_{11}$ (gram-molecular mass = 342 g/mole) in 250 g of water.
2. What mass of K_2SO_4 would you measure out to prepare 550 mL of a 0.76 M solution?
3. A radiator is filled with a mixture of 3.25 kg ethylene glycol ($C_2H_6O_2$) in 7.75 kg of water. Calculate the molality of this solution.
4. How many grams of H_2SO_4 are contained in 4.5 L of 0.30 N sulfuric acid?
5. Calculate the molarity of the following solutions.
   a. 5.0 N H_3PO_4 b. 0.75 N HBr
6. Calculate the normality of the following solutions.
   a. 5 M HCl   b. 10.5 M $HC_2H_3O_2$

12C Colligative Properties: The Effects of Solutes

The presence of solutes causes solutions to behave differently from their pure solvents. Freezing points, boiling points, vapor pressures, and osmotic pressures all change. It makes no difference what types of solutes are present; the new properties, called **colligative properties**, depend on the *number*, not the type, of particles in solution.

Decreased Vapor Pressures

At the surface of a solution, solute particles fill positions that are normally occupied by solvent molecules. Since fewer solvent molecules are exposed to the surface, fewer have the chance to evaporate. The result of this is that the vapor pressure of the solution is lower than the vapor pressure of the pure solvent.

Pure water has a set of vapor pressures that increase with temperature. If a mole of sugar were added to a liter of water, the vapor pressures would be lower than those of pure water. If a mole of sodium chloride were added to a liter of water, the vapor pressures would be even lower. Sodium chloride units dissociate into Na⁺ and Cl⁻ ions, and these extra particles lower the vapor pressures. If aluminum chloride ($AlCl_3$) were dissolved in a liter of water, 4 moles of particles for every 1 mole of $AlCl_3$ would be released, decreasing the vapor pressures even more.

These effects can be summarized in a single statement called **Raoult's Law:** The lowering of the vapor pressure of a solvent is directly proportional to the number of solute particles. Therefore, a 1.0 molal solution of $AlCl_3$ has a **particle molality** of 4.0. It is this particle molality that is directly proportional to the decrease in vapor pressure.

Boiling Point Elevation

When solute particles lower vapor pressures, they prevent the solvent from boiling at its expected boiling point. Higher-than-normal temperatures are then necessary to raise vapor pressures up to atmospheric pressures. The effect that solute particles have on the boiling points of substances is called **boiling point elevation.**

A 1 molal solution of sugar in water boils at 100.512°C—an increase of 0.512°C. The boiling point elevation is represented by the symbol ΔT_{bp}. Δ is the capital Greek letter delta. It stands for the words *change in*. The change in boiling temperature is the difference between the new boiling point and the original boiling point.

$$\Delta T_{bp} = bp_{new} - bp_{original}$$

It has been found that 1 mole of particles—molecules, atoms, or ions—in 1 kilogram of water elevates the boiling point of water 0.512°C. This value is called the **molal boiling point elevation constant (K_{bp})** for water. Other solvents have their own unique molal boiling point elevation constants. Note that concentration is always measured in molality for these constants.

The more concentrated the solution is, the greater the boiling point elevation will be. Exactly how much higher depends on the precise concentration and the molal boiling point elevation constant of the solvent.

$$\Delta T_{bp} = K_{bp}m$$

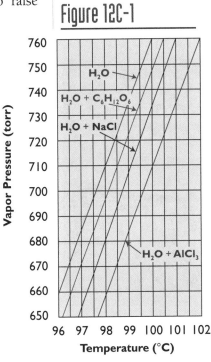

Figure 12C-1

Table 12C-1 Boiling Point Elevations of Solvents

Solvent	Normal Boiling Point (°C)	Molal Boiling Point Elevation Constant (K_{bp}) (°C/molal)
Acetic acid	117.9	3.07
Acetone	56.2	1.71
Benzene	80.15	2.53
Carbon tetrachloride	76.50	5.03
Ethanol	78.26	1.22
Ether	34.42	2.02
Phenol	181.8	3.56
Water	100.0	0.512

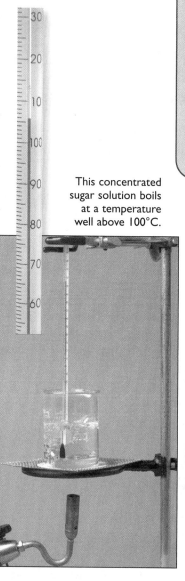

This concentrated sugar solution boils at a temperature well above 100°C.

Sample Problem What is the boiling point of acetone when 0.500 mol of naphthalene ($C_{10}H_8$) are added to 1.00 kg of acetone? The molal boiling point elevation constant of acetone is 1.71°C/m.

Solution
Remember that the molality is moles of solute/kg of solvent, so a solution of 0.500 mol naphthalene per kilogram of acetone constitutes a 0.500 molal solution.

$$\Delta T_{bp} = K_{bp}m$$
$$\Delta T_{bp} = (1.71°C/m)(0.500\ m)$$
$$\Delta T_{bp} = 0.855°C$$

To find the new boiling point, add the temperature change to the normal boiling point of acetone (see Table 12C-1).

$$56.2°C + 0.885°C = 57.1°C$$

Recall that ionizable solutes increase the concentration of particles in solution. This fact must be considered when working colligative property problems.

Sample Problem If 425 g of magnesium chloride ($MgCl_2$) are added to 675 g of hot water, what is the boiling point of the solution?

Solution
The first step is to determine the molality of the solution. Using

unit analysis, it can be determined that a 6.61 m solution will be obtained.

$$\frac{425 \text{ g MgCl}_2}{675 \text{ g H}_2\text{O}} \bigg| \frac{1000 \text{ g}}{1 \text{ kg}} \bigg| \frac{1 \text{ mol MgCl}_2}{95.21 \text{ g MgCl}_2} = \frac{6.61 \text{ mol MgCl}_2}{1 \text{ kg H}_2\text{O}} = 6.61 \ m$$

Because each mole of magnesium chloride ionizes to form 1 mole of Mg^{2+} ions and 2 moles of Cl^- ions, the particle molality of the solution is 3 × 6.61, or 19.8 moles per kilogram. See Table 12C-1 for the K_{bp} of water.

$$\Delta T_{bp} = K_{bp} m$$
$$\Delta T_{bp} = (0.512°C/m)(19.8 \ m)$$
$$\Delta T_{bp} = 10.1°C$$

The elevation in boiling point (10.1°C) is then added to the normal boiling (100.0°C) to obtain the new boiling point of 110.1°C.

Freezing Point Depression

Solutions freeze at lower temperatures than their pure solvents do. A 1 molal solution of sugar in water freezes at -1.86°C, or 1.86°C lower than the normal freezing point of water. The change in freezing point is called the **freezing point depression** and is represented by the symbol ΔT_{fp}. As with molal boiling point elevation constants, each solvent has its own characteristic **molal freezing point depression constant** (K_{fp}).

Table 12C-2 Freezing Point Depressions of Solvents

Solvent	Normal Freezing Point (°C)	Molal Freezing Point Depression Constant (K_{fp}) (°C/molal)
Acetic acid	16.6	3.90
Benzene	5.48	5.12
Ether	-116.3	1.79
Phenol	40.9	7.40
Water	0.00	1.86

Sample Problem How much will the freezing point of phenol change if 4.00 mol of naphthalene are dissolved in 2.00 kg of phenol?

Solution

Four moles of naphthalene in 2.00 kg of phenol is equivalent to 2.00 mol/kg, or a 2.00 m solution. The molal freezing point depression constant of phenol is 7.40°C/m.

$$\Delta T_{fp} = K_{fp}m$$
$$\Delta T_{fp} = (7.40°C/m)(2.00\ m)$$
$$\Delta T_{fp} = 14.8°C$$

The freezing point will be 14.8°C lower than the freezing point of pure phenol.

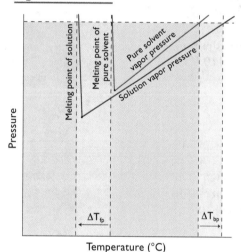

Figure 12C-2

Antifreeze is added to engine coolant systems for this very reason. Even though it is called "antifreeze," it affects both the freezing point and the boiling point of the water in the system. A 25 percent by volume solution of ethylene glycol (antifreeze) will not only depress the freezing point by about 12°C, but also elevate the boiling point. If antifreeze were not added, the water in the cooling system would freeze in the winter, causing hoses and pipes to burst. In the summer months, a higher boiling point keeps the water from vaporizing completely and keeps the engine from overheating.

Calcium chloride, rather than sodium chloride, is the salt used on icy roads. The reason calcium chloride is used lies partly in the fact that it lowers the freezing point of water more and therefore melts ice at lower temperatures. It does this because it releases more ions upon dissolution than sodium chloride does. When a mole of calcium chloride dissociates, 3 moles of ions can be released into solution. The effective concentration of the solution is tripled, rather than doubled, so there is a greater change in freezing point.

Sample Problem Compare the freezing points of a 3.00 molal calcium chloride solution and a 3.00 molal sodium chloride solution.

Solution

The chemical formula for calcium chloride is $CaCl_2$. In solution, it dissolves into 2 moles of Ca^{2+} ions and 1 mole of Cl^- ions. Sodium chloride when dissolved in solution will yield 1 mole of Na^+ ions and 1 mole of Cl^- ions. Therefore, the concentration of the calcium chloride solution will be tripled (3 × 3.00 m = 9 mol/kg), whereas the concentration of the sodium chloride solution will only be doubled (2 × 3.00 m = 6 mol/kg).

$\Delta T_{fp} = K_{fp}m$
$\Delta T_{fp} = (1.86°C/m)(9.00\ m)$
$\Delta T_{fp} = 16.7°C$ for $CaCl_2$

The freezing point of the calcium chloride solution is 16.7°C lower than the normal freezing point of water, making the new freezing point -16.7°C.

$\Delta T_{fp} = K_{fp}m$
$\Delta T_{fp} = (1.86°C/m)(6.00\ m)$
$\Delta T_{fp} = 11.2°C$ for NaCl

The sodium chloride solution freezes at -11.2°C. Thus, the calcium chloride solution will freeze at a temperature 5.5°C *colder* than a solution of sodium chloride.

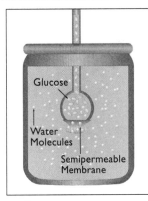

Figure 12C-3

Osmotic Pressure

Wrap the end of a thistle tube with a sheet of semipermeable membrane—plastic wrap will do. Pour a concentrated sugar solution into this tube, and quickly dip the tube into a sample of pure water.

Osmosis, the process in which a solvent (in this case, water) moves across a semipermeable membrane (the plastic wrap), begins immediately. The semipermeable membrane allows the small water molecules to pass through it, but it retains the larger sugar molecules. Only a few water molecules leave the sugar solution. Many more enter the solution from the pure water outside. As a result, the level of liquid in the tube gradually rises. Eventually the level of the solution ceases to rise because the weight of the liquid in the column pushes water molecules out of the tube as fast as they enter.

The **osmotic pressure** of a solution is the amount of pressure required to prevent osmosis from occurring. The more solute particles there are in a solution, the higher the osmotic pressure will be and the higher the column will rise. The osmotic pressure of a solution is a colligative property because it depends on the number of particles in solution.

Osmosis is important not only in chemistry, but also in biology. For example, a cell could be thought of as a small amount of an aqueous solution enclosed by a semipermeable membrane. The solution surrounding the cell should be kept at

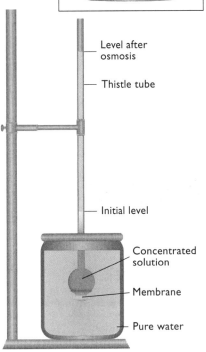

the same osmotic pressure as the solution inside the cell. If this equilibrium is disturbed, water could leave the cell in sufficient quantities to cause dehydration or enter the cell and cause the membrane to burst! The concentration of solutes inside a cell is approximately 0.3 M. If a cell is placed in a solution that is more concentrated—hypertonic—than the inside of the cell, water will flow out of the cell into the surrounding solution. If the cell is placed in a solution that is less concentrated—hypotonic—than the inside of the cell, water will flow into the cell, causing it to swell and possibly burst. When health care providers are administering intravenous fluids, the concentration of the solution must be considered to prevent harm to the patient.

Section Review Questions 12c

1. What is the determining factor in various colligative properties?
2. What are the freezing point and boiling point of an aqueous solution of 15.5 g glucose ($C_6H_{12}O_6$) dissolved in 150 g of water?

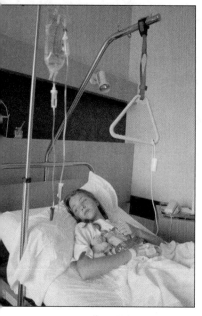

Healthcare workers must ensure proper IV fluid management.

12D Colloids: Almost Solutions, But Not Quite

Some mixtures cannot be classified as solutions or suspensions. The mixture of water in air that is called fog appears to have a uniform consistency. Is it a solution? Not really. The water molecules have not been completely separated from each other. Groups of molecules float through the air without becoming dissociated. Is fog a suspension? Again the answer is no. Fog has a uniform consistency and does not readily separate like suspensions do. A class of mixtures called colloids lies between solutions and suspensions.

What Are Colloids?

Colloids are mixtures that contain small particles dispersed in a medium. Fog is a colloid: it consists of water droplets (particles) in air (medium). Since colloids are different from solutions, terms such as *solute*, *dissolved*, and *solvent* do not apply.

The colloidal particles that are dispersed in the medium usually exceed 1 nanometer (1 nm = 10^{-9} meters), which is the approximate size limit on dissolved particles. The size of particles in a colloid cannot exceed 1000 nm, since particles of this

Table 12D-1

Mixture	Particle Size
Solution	< 1 nm
Colloid	1-1000 nm
Suspension	> 1000 nm

size would quickly settle out from their medium. Intermolecular collisions can buffet particles smaller than 1000 nm with sufficient force to counteract the constant tug of gravity.

Colloids are commonly found in everyday surroundings. Colloids can be solids, liquids, or gases. Eight of the nine possible combinations of solids, liquids, and gases can form colloids, but gas-gas mixtures are always solutions.

Table 120-2 Types of Colloids

Particles	Medium	Common Name	Examples
Liquid	Gas	Aerosol	Fog, clouds, mist
Solid	Gas	Aerosol	Smoke, dust in the air
Gas	Liquid	Foam	Shaving cream, whipped cream
Liquid	Liquid	Emulsion	Mayonnaise (oil dispersed in water)
Solid	Liquid	Sol, gel	Paint, pudding
Gas	Solid	Solid foam	Plastic foam, marshmallows, pumice
Liquid	Solid	Solid emulsion	Cheese, butter
Solid	Solid	Solid sol	Ruby glass (glass with dispersed metal)

The Properties of Colloids

In 1869 a Scottish physicist named John Tyndall demonstrated that the particles in colloids were large enough to scatter light waves. This effect, called the **Tyndall effect** in his honor, is often used to distinguish colloids from solutions. A beam of light will pass through a solution without being seen. The particles that are small enough to be dissolved are too small to disturb light waves. When a beam of light goes through a colloid, the beam's outline shows up distinctly. This effect can be seen when automobile headlights shine through air into a patch of fog. The outline of the beams does not show up in the air, because air is a solution. However, the beams will show up clearly in fog, since it is a colloid.

Headlights in fog demonstrate the Tyndall effect.

A British botanist named Robert Brown discovered a major proof of the kinetic theory by observing a colloid. In 1827 he watched pollen particles that were dispersed in water. He saw that the pollen particles moved slightly, as if they were being jostled by many small collisions. This movement, called **Brownian movement,** results from colloidal particles being buffeted by molecular collisions. This observation led scientists to conclude that matter is made of moving particles.

The intermediate size of colloidal particles allows them to pass through most filtering systems. Most membranes can catch suspended particles, but special membranes are needed to catch particles smaller than 1000 nm.

The chemistry of colloids is largely determined by the electrical charges on the surfaces of the particles. What is on the surface of the particles is important because colloidal particles have a great amount of surface area compared to their volume. It is the repulsive forces between the charged particles that keep them suspended in solution. **Flocculating agents** are compounds that form ions that can effectively screen the repulsive forces between the colloidal particles. When this occurs, the particles will settle out, a process called **coagulation.** The opposite of coagulation is **peptization,** a process whereby the particles revert to the colloidal state. A widely used flocculating agent is alum, $KAl(SO_4)_2 \cdot 12\ H_2O$. Alum can also be used as an astringent, that is, it has the ability to coagulate proteins and can be used to stop bleeding from minor cuts.

Mayonnaise is an emulsion of oil and water (from milk). The oil in mayonnaise particles does not naturally combine with the water medium. Makers of mayonnaise must use a chemical principle called adsorption to keep the oil and water together. **Adsorption** is the attachment of charged particles to the particles in a colloid. In mayonnaise, oil and water are "tied together" by an emulsifying agent in egg yolks. A protein from the egg covers the oil particles and makes them compatible with the aqueous medium. Without the emulsifier in the eggs, mayonnaise could not be a stable emulsion.

Section Review Questions 12D

1. What determines whether a mixture is a solution, colloid, or suspension?
2. What is the Tyndall effect, and how is it used?

Chapter Review

Coming to Terms

solution
solvent
solute
miscible
azeotrope
immiscible
dissociation
solvation
hydration
diffusion
insoluble
unsaturated
saturated
dynamic equilibrium
supersaturated
Henry's law
dilute
concentrated
percent by mass
molarity
molality
equivalent
gram-equivalent mass
normality
colligative property
particle molality
Raoult's law
boiling point elevation
molal boiling point
   elevation constant
freezing point depression
molal freezing point
   depression constant
osmosis
osmotic pressure
colloid
Tyndall effect
Brownian movement
flocculating agent
coagulation
peptization
adsorption

Review Questions

1. Identify the solvent and solute in the following solutions.
   a. salt water used for gargling
   b. carbonated water
   c. air
   d. carbon steel (1% Mn, 0.9% C, 98.1% Fe)
2. Predict whether each of the following pairs of liquids is miscible or immiscible.
   a. popcorn oil and vegetable oil
   b. gasoline and water
   c. iced tea and salt water
   d. popcorn oil and tea

3. Will the following substances dissolve in water or in an oil?
   a. carbon tetrachloride (CCl_4)
   b. LiCl
   c. NaBr
   d. methanol (CH_3OH)
4. Using the kinetic theory, explain each of the following observations.
   a. Sugar dissolves more quickly in hot coffee than in cold coffee.
   b. Finely ground salt dissolves more quickly than large chunks of salt.
   c. Powders dissolve more quickly when stirred.
   d. Popcorn oil and vegetable oil mix more quickly than do vegetable oil and sugar.
5. Will a carbonated drink go "flat" faster if it is heated? Why?
6. Thimerosal is an organic compound (gram-molecular mass = 404.8 g/mol) used to inhibit bacterial growth. A 0.100% solution of thimerosal is sometimes used as an antiseptic. Assume that this solution has a density of 1.00 g/mL.
   a. How many grams of thimerosal are found in 0.500 g of 0.100% solution?
   b. How many grams of thimerosal are required to make 50.0 g of 0.100% solution?
   c. What mass of thimerosal must be mixed with 25.0 g of water to make a 0.100% solution?
   d. How many grams of thimerosal are required to make 75.0 mL of 0.100% solution?
   e. What volume of 0.100% solution contains 0.750 g of thimerosal?
7. A 10.0% aqueous solution of sulfuric acid has a density of 1.0661 g/mL at 20°C.
   a. How many grams of H_2SO_4 are found in 50.0 g of this solution?
   b. How many grams of water are found in 50.0 g of this solution?
   c. What mass of H_2SO_4 must be mixed with 50.0 g of water to make this solution?
   d. How many grams of H_2SO_4 are required to make 50.0 mL of this solution?
   e. What volume of this solution contains 50.0 g of H_2SO_4?
   f. What is the molarity of this solution?
   g. Assuming a complete dissociation of H_2SO_4 in water ($H_2SO_4 \longrightarrow 2H^+ + SO_4^{2-}$), what is the normality of the solution?
   h. What is the gram-equivalent mass of H_2SO_4 given in the dissociation in (g)?

i. How many equivalents of H_2SO_4 are in 50.0 g of this solution?
   j. What is the molality of this solution?
   k. What is the particle molality of this solution?
   l. What is the boiling point of this solution?
   m. What is the freezing point of this solution?
   n. What is the osmotic pressure of this solution. Hint: use information from (f).

8. A 9.168 M aqueous solution of H_2SO_4 has a density of 1.4987 g/mL at 20°C.
   a. How many grams of H_2SO_4 are in 50.0 g of this solution?
   b. What is the percent by mass of this solution?
   c. From the answer to (b) we see that H_2SO_4 is the solvent and H_2O is the solute. What is the molarity of water in this solution?
   d. What is the molality of this solution?
   e. What is the osmotic pressure of this solution?

9. An aqueous solution of sodium borate (201.2 g/mol) is sometimes used to fire-proof wood.
   a. What is the molarity of 2.50 L of solution that contain 1.85 moles of sodium borate?
   b. What is the molarity of 45.0 L of solution that contains 6.78 kg of sodium borate?
   c. How many moles of sodium borate are in 600.0 mL of a 1.57 M sodium borate solution?
   d. A chemist needs 50.8 g of sodium borate for a reaction. How many milliliters of 1.87 M solution contain this mass?

10. What is the molality of the following aqueous sucrose solutions? Sucrose, $C_{12}H_{22}O_{11}$, has a gram-molecular mass of 342.30 g/mol.
    a. A 0.100 M solution with a density of 1.0119 g/mL.
    b. A 0.100 N solution with a density of 1.0119 g/mL.
    c. A 0.100% by mass solution.
    d. A solution with a $\Delta T_{bp} = 0.100°C$.
    e. A solution with a $\Delta T_{fp} = 0.100°C$.

11. Sodium monofluorophosphate (144.0 g/mol) is the fluoride component in some modern fluoride-containing toothpastes. What is the molality of a solution containing
    a. 1.85 kg of H_2O and 1.00 mol of sodium monofluorophosphate?
    b. 125.0 g of H_2O and 0.356 g of sodium monofluorophosphate?
    c. 500.0 g of solution and 12.0 g of sodium monofluorophosphate?

Chapter 12 Review

12. Consider the following reaction: $2\,H_2 + O_2 \longrightarrow 2\,H_2O$.
    a. What is the gram-equivalent mass of hydrogen?
    b. What is the gram-equivalent mass of oxygen?
    c. What is the gram-equivalent mass of water?

13. Consider a reaction in which sulfuric acid (H_2SO_4) (98.08 g/mol) donates two H^+ ions to two NaOH units.
    $$H_2SO_4 + 2\,NaOH \longrightarrow Na_2SO_4 + 2\,H_2O$$
    a. How many equivalents per mole of H_2SO_4 are in this reaction?
    b. What is the gram-equivalent mass of H_2SO_4 in this reaction?
    c. What is the normality of a 365 mL solution that contains 76.8 g of H_2SO_4?

14. From each set choose the solution that has the lowest vapor pressure.
    a. 1.8 m CH_3OH, 0.7 m CH_3OH, 2.9 m CH_3OH, 0.2 m CH_3OH
    b. 0.5 m Na_3PO_4, 0.5 m $MgCl_2$, 0.5 m $NaCl$, 0.5 m $Al_2(SO_4)_3$

15. A solution of an unknown compound boils at 101.00°C (760 torr).
    a. What is the freezing point of this solution?
    b. What is the molality of this solution?
    c. What is the gram-molecular mass of this unknown compound if 7.80 g of this solute was added to 100.0 g of water in order to make this solution?

16. A housewife reads that eggs will boil faster if salt is added to the hot water.
    a. Suggest an explanation for this fact.
    b. What is the boiling point of 1.00 L of water in which 10.0 g of NaCl (58.44 g/mol) have been dissolved?
    c. How many grams of NaCl must be added to a liter of water to raise its boiling point to 105.00°C?
    d. In light of the answers to questions (b) and (c), do you think that adding salt to water is a convenient method for decreasing the cooking time of eggs?

17. Automobile antifreeze consists primarily of ethylene glycol ($C_2H_6O_2$) (62.07 g/mol). Small amounts of dye and anti-corrosion substances are also added.
    a. The coldest temperature of the year in Augusta, Maine, is expected to be -26.0°C. What must the molality of the radiator fluid be to prevent it from freezing?
    b. How many grams of ethylene glycol must be added to 4.00 L of water to attain this molality?

c. How many quarts of antifreeze is this? Assume that the density of antifreeze is 1.11 g/mL and ignore the fact that there are small amounts of other substances present.

d. Why is it important to have a solution of antifreeze in the radiator during the summer?

18. At what temperature will a solution consisting of 100.0 mL of H_2O and 2.50 g of sucrose ($C_{12}H_{22}O_{11}$) (342.3 g/mol) freeze?

19. Explain why
    a. red blood cells that are placed in pure water absorb water until they explode but are fine when they are surrounded by blood plasma.
    b. red blood cells shrink when they are in a very concentrated salt solution.

20. How can you determine whether a substance is a solution or colloid? Why does this method of determination work?

13A Thermodynamics page 337
13B Kinetics page 351
FACETS: Spontaneous Combustion page 354

Thermodynamics & Kinetics 13

Determining whether a reaction can occur is crucial in the study of chemistry. **Thermodynamics,** the branch of science which studies the transformation of energy from one form to another, is the tool that chemists use for this purpose. A reaction that involves a favorable energy change can be spontaneous. It may not be spectacular when the reactants are mixed, but there is a good chance that it will proceed on its own. If the energy changes required for a reaction are unfavorable, the reaction cannot occur on its own. It is therefore nonspontaneous.

The fact that a reaction can occur does not mean that it will occur at an observable rate when the reactants combine. Questions about the speed at which reactions occur and the way in which they get started are answered in a field of chemistry called kinetics. While thermodynamics answers the basic question "Can it react?" kinetics answers the broader questions "Will it react?" and "How fast will it react?" Therefore, **kinetics** is the study of the rates of reactions and the steps by which they occur.

13A Thermodynamics: Can Things React?

Chemical Energy: Stored in Bonds

Chemical bonds possess stored energy. As long as the bonds do not break or change, the energy that they contain cannot be

Chapter 13A

Though it is not immediately observable, there is more energy stored in the pile of gunpowder on the left than in the pile of sand on the right.

observed. When the bonds are broken, however, their energy takes a more observable form. Chemical bonds can be broken in three ways. The atoms can be vibrated so strongly that the attraction holding them together breaks; they can be rotated so vigorously that they fly apart; or the electrons can be moved out of the orbitals that are involved in bonding.

In chemistry the most common, and usually most important, energy transformations are between chemical bonds and heat. Although some reactions involve significant amounts of light energy, electrical energy, or physical work, these transformations are minimal when compared to the heat energy transformations. Therefore, thermodynamics usually concentrates on transformations of heat energy.

The principles of thermodynamics are of fundamental importance. The laws of thermodynamics, discovered in the nineteenth century, govern the nature of all thermodynamic processes and place limits on them.

The First Law of Thermodynamics: Energy Conservation

After six days of Creation, God saw that the universe was good and pronounced it complete. With the exception of miracles, the total energy content of the universe has been preserved. The replenishment of oil and meal for the widow of Zarephath (I Kings 17) and Christ's feeding of the multitudes (Mark 6, 8) are examples of these miraculous exceptions.

Scientists have found that energy is conserved in every natural process. This observation has been expressed as the **first law of thermodynamics:** *energy can be neither created nor destroyed, only changed in form.*

In thermodynamics both exothermic and endothermic reactions occur. The launch of a space shuttle requires a series of reactions to thrust the space shuttle into the air. These reactions produce a great deal of heat. All reactions that release heat energy are exothermic reactions. Do these reactions violate the first law? No, they do not. The energy present before the reaction equals the energy present after the reaction; however, the energy after the reaction is in a much more visible form. The following equation shows a possible reaction that could occur at a space shuttle launch.

$$6NH_4ClO_4 + 10Al \longrightarrow 5Al_2O_3 + 6HCl + 3N_2 + 9H_2O + Heat$$

High energy content $\longrightarrow$ Low energy content + Energy

The launch of a space shuttle is fueled by an exothermic reaction.

Endothermic reactions absorb energy from their surroundings. Photosynthesis is an example of an endothermic reaction since it requires the input of light energy. A photosynthetic organism converts the energy in sunlight to chemical forms of energy for the synthesis of organic compounds. Again the amount of energy present before the reaction matches the amount of energy present after the reaction—the first law of thermodynamics holds true. The following formula is for photosynthesis.

$$6CO_2 + 6H_2O + Light \longrightarrow C_6H_{12}O_6 + 6O_2$$

Low energy content + Energy $\longrightarrow$ High energy content

Enthalpy: Heat Content of Compounds

All substances contain energy. **Enthalpy (H)** is a thermodynamic quantity that describes the energy of a substance at constant pressure. In this text enthalpy will be expressed in kilocalories. Scientists sometimes call the change in enthalpy that occurs during a reaction **the heat of reaction,** and they represent it by the symbol ΔH. The enthalpy change of a reaction is the difference between the enthalpy of the products and the enthalpy of the reactants.

$$\Delta H = H_{products} - H_{reactants}$$

An exothermic reaction releases energy; the products have less enthalpy than the reactants. The formation of water from its two elements illustrates this process. When 1 mole of hydrogen gas is ignited in the presence of ½ mole of oxygen gas, an explosive reaction that forms 1 mole of water occurs. Careful measurements reveal that 68.3 kilocalories of heat are released in the explosion.

$$H_2 + ½O_2 \longrightarrow H_2O + 68.3 \text{ kcal}$$

An exothermic reaction has a negative ΔH. When the large enthalpy of the reactants is subtracted from the small enthalpy of

Figure 13A-1

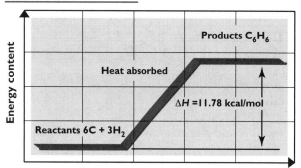

Exothermic reaction progress

The products of exothermic reactions have less enthalpy than the reactants do.

Figure 13A-2

Endothermic reaction progress

Laboratory calorimeters measure heat exchanges during reactions.

the products, the difference is negative. Figure 13A-1 illustrates the energy changes that are common for exothermic reactions. The products have less enthalpy than the reactants, and the difference is released as heat.

Products of endothermic reactions have more enthalpy than the reactants. The formation of a mole of benzene (C_6H_6) from its elements requires an input of 11.78 kilocalories.

$$6\ C + 3\ H_2 + 11.78\ kcal \longrightarrow C_6H_6$$

An endothermic reaction has a positive ΔH. Figure 13A-2 illustrates the difference in energies between the reactants and the products of an endothermic reaction.

If the ΔH of a reaction is known, the ΔH of the reverse reaction can be easily calculated. The ΔH of the reverse reaction has the same magnitude as the ΔH of the forward reaction, but it has the opposite sign. For example, the reaction that formed water had a ΔH of -68.3 kilocalories. The breakdown of water has a ΔH of 68.3 kilocalories.

Calculating Enthalpy Changes

Chemists could measure the change in enthalpy for thousands of known reactions. This would be difficult, and the data would fill several volumes of books. Therefore, a quick technique for calculating changes in enthalpy for reactions was devised. A table that lists enthalpies of formation for common compounds is used. A compound's **enthalpy of formation,** also called the heat of formation, is the change in enthalpy that occurs when 1 mole of the compound is formed from its elements. The enthalpy of formation of water is -68.3 kilocalories.

$$H_2 + \tfrac{1}{2}O_2 \longrightarrow H_2O;\ \Delta H = -68.3\ kcal$$

The amount of enthalpy in a substance varies with temperature and pressure, so these conditions must be specified. In thermodynamics the **standard state** is defined to be 25°C, or 298 K, and 1 atmosphere of pressure. The **standard molar enthalpy of formation,** $\Delta H°_f$, is defined as the enthalpy change for the reaction that produces 1 mole of a compound in its standard state from its elements in their standard states. The degree symbol in $\Delta H°_f$ signifies that this ΔH refers to standard conditions. Table 13A-1 lists several standard molar enthalpies of formation. Table 13A-5 contains a more extensive listing.

As will be seen later in Table 13A-5, most of the $\Delta H°_f$s are negative. Thus the reactions that produce most of the common compounds are exothermic. Compounds with large negative numbers release large amounts of energy when they form. Molecules such as Cl_2, H_2, and O_2 have no enthalpies of formation because these substances are the reference point to which the enthalpies of compounds are compared. This is true for all elements in their naturally occurring form.

Table 13A-1 Standard Molar Enthalpies of Formation

Compound	$\Delta H°_f$ (kcal/mol)
NH_4NO_3	-87.3
NO	21.6
NO_2	8.1
$PbCl_2$	-85.9

Sample Problem What is the $\Delta H°$ for $N_2 + 2\ O_2 \longrightarrow 2\ NO_2$?

Solution

Table 13A-1 shows that the standard molar enthalpy of formation of nitrogen dioxide is 8.1 kcal/mol. The given reaction forms 2 moles of nitrogen dioxide.

$$\frac{2\ \text{mol } NO_2 \mid 8.1\ \text{kcal}}{\mid 1\ \text{mol } NO_2} = 16\ \text{kcal}$$

A reaction has two theoretical steps. The first step involves the breaking down of the reactants. The second step is the formation of the products. The following reaction between lead oxide and carbon monoxide illustrates these two steps.

$$PbO + CO \longrightarrow Pb + CO_2;\ \Delta H° = -15.6\ \text{kcal}$$

The first step in the reaction is the breakdown of lead oxide and carbon monoxide. The $\Delta H°_f$ for the formation of lead oxide at standard conditions is -52.1 kilocalories per mole. The breakdown of lead oxide, which is the reverse reaction, requires 52.1 kilocalories.

$$PbO \underset{-52.1\ \text{kcal}}{\overset{52.1\ \text{kcal}}{\rightleftarrows}} Pb + \tfrac{1}{2} O_2$$

The $\Delta H°_f$ for the breakdown of carbon monoxide is 26.4 kilocalories per mole. The $\Delta H°$ for the first part of the reaction is 78.5 kilocalories.

$$CO + O_2 \underset{94.1\ \text{kcal}}{\overset{-94.1\ \text{kcal}}{\rightleftarrows}} CO_2$$

The second part of the reaction is the formation of products. The $\Delta H°$s for these changes can be easily found in the table of

ΔH°_fs. Since lead is an element, its ΔH°_f is zero. The ΔH°_f for the formation of 1 mole of carbon dioxide from its elements at standard conditions is -94.1 kilocalories. The total ΔH° for the formation of products is -94.1 kilocalories.

The ΔH° for the reaction is -15.6 kilocalories. This result shows that the reaction is exothermic.

In the previous problem, calculating the ΔH° of a reaction involved combining the ΔH°s of two partial reactions. A law of thermochemistry called Hess's law justifies this method of calculation. **Hess's law** states that the enthalpy change of a reaction equals the sum of the enthalpy changes for each step of the reaction. The example shows how one reaction could be thought of as the sum of two other reactions.

$$\begin{array}{ll} PbO + CO \longrightarrow Pb + O_2 + C & \Delta H^\circ = 78.5 \text{ kcal} \\ Pb + O_2 + C \longrightarrow Pb + CO_2 & \Delta H^\circ = -94.1 \text{ kcal} \\ \hline PbO + CO \longrightarrow Pb + CO_2 & \Delta H^\circ = -15.6 \text{ kcal} \end{array}$$

The enthalpy change of the total reaction equals the sum of the two enthalpy changes. The mathematical version of this statement is as follows:

$$\Delta H^\circ_{\text{reaction}} = \Delta H^\circ_{\text{product formation}} + \Delta H^\circ_{\text{reactant breakdown}}$$

Because the table refers to the information of compounds, the values for reactant-breakdown enthalpies will be opposite in sign to those listed. The equation can be changed to a form that uses the data in Table 13A-5 without changing the signs of the numbers. The symbol Σ in this equation is the Greek letter *sigma*. It signifies a summation of numbers.

$$\Delta H^\circ_{\text{reaction}} = \Sigma \Delta H^\circ_{f \text{ products}} - \Sigma \Delta H^\circ_{f \text{ reactants}}$$

Sample Problem Estimate the ΔH° for the combustion of methane, using standard enthalpies of formation.

$$CH_4 + 2\, O_2 \longrightarrow CO_2 + 2\, H_2O$$

Solution
The equation requires the use of the following enthalpies of formation from Table 13A-5.

$$\begin{array}{lcl} \Delta H^\circ_f \text{ of } CO_2 & = & -94.1 \text{ kcal/mol} \\ \Delta H^\circ_f \text{ of } H_2O & = & -68.3 \text{ kcal/mol} \\ \Delta H^\circ_f \text{ of } CH_4 & = & -17.9 \text{ kcal/mol} \\ \Delta H^\circ_f \text{ of } O_2 & = & 0 \end{array}$$

Two moles of water and 2 moles of oxygen participate in the reaction. This fact must be dealt with when calculating the enthalpy change of the entire reaction.

$$\Delta H°_{reaction} = \Sigma \Delta H°_{f\ products} - \Sigma \Delta H°_{f\ reactants}$$

$$\begin{aligned}\Delta H°\text{reaction} &= [(1)(-94.1) + (2)(-68.3)] - [(1)(-17.9) + (2)(0)] \\ &= (-94.1 + -136.6) - (-17.9) \\ &= (-230.7) + (17.9) \\ &= -213 \text{ kcal}\end{aligned}$$

This reaction is highly exothermic.

Bond Enthalpies: The Stronger the Better

The **enthalpy of bond formation**, or bond enthalpy, is defined as the $\Delta H°$ that occurs when a mole of bonds in a gaseous compound is broken. For example, an input of 46 kilocalories is necessary to break apart a mole of Br-Br bonds.

$$Br_2 + 46 \text{ kcal} \longrightarrow Br + Br; \quad \Delta H° = 46 \text{ kcal}$$

Because bonded atoms are more stable than unbonded atoms, energy is always required to break bonds. Consequently, all bond enthalpies are positive numbers. Conversely, the formation of bonds always releases energy and is thus exothermic.

Bond enthalpies can help explain why some reactions give off energy and others require energy. Strong bonds are stable bonds; large amounts of energy are necessary to break them. The bond enthalpies of strong bonds are thus quite high. Strong bonds also give off large amounts of energy when they form. Weak bonds can be broken with small amounts of energy. Predictably, they give off small amounts of energy when they form. The bond enthalpies of weak bonds are relatively low.

A reaction that breaks strong bonds and forms weak bonds will require energy. More energy is required to break the strong bonds than is released when the weak bonds form. As a result, reactions that produce compounds with weaker, less stable bonds are endothermic.

Stronger bonds $\longrightarrow$ Weaker bonds; $\quad \Delta H > 0$, Endothermic

A reaction that forms stable, low-energy compounds from compounds with high-energy, weak bonds releases energy. The amount of energy that must be used to break the weak bonds is

Figure 13A-3

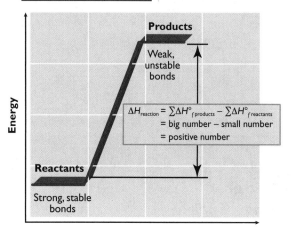
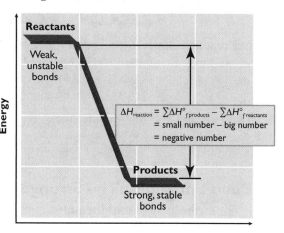

small compared to all the energy that is given off when the strong bonds form.

Weaker bonds ⟶ Stronger bonds; $\Delta H < 0$, Exothermic

Exothermic reactions form stable, low-energy bonds. Endothermic reactions form unstable, high-energy bonds.

The Second Law of Thermodynamics

In most cases exothermic processes are thermodynamically favorable, and endothermic ones are not. But this rule does not always hold true. More than a negative ΔH is necessary in order for a process to be favorable. An example of an endothermic process that tends to proceed is the releasing of compressed gas into the atmosphere. As gas under high pressure is released, it gets colder. Something besides changes in enthalpy influences spontaneity.

Two natural tendencies govern all thermodynamically favorable processes. The first is that matter seeks stable, low-energy states. The second is that natural processes decrease the order of the universe. The latter is a statement of the **second law of thermodynamics.** The second law of thermodynamics is also known as the law of increasing entropy.

The Bible and the Laws of Thermodynamics

Is it likely that the universe formed by itself? The answer according to Scripture is no, and the laws of thermodynamics support this truth of Scripture as well. The first law of thermodynamics shows that the universe could not have begun by itself, because energy cannot be created or destroyed. The fundamental structure of the universe is one of conservation, not the innovation necessary for evolutionary theory. The second law of thermodynamics forbids the spontaneous evolution of the universe because the orderliness of this world could not have arisen on its own. Not only could this orderliness fail to arise on its own, but the present orderliness is constantly decreasing. This principle is seen in Scripture in Isaiah 51:6, Hebrews 1:10-11, and Psalm 102:25-26. The laws of thermodynamics are excellent scientific tools against the theory of evolution.

Entropy: The Chaos Factor

Entropy (S) is a measure of randomness (disorder) or lack of orderliness in a system. Disordered substances have high entropies; very ordered substances have low entropies. Therefore, the greater the randomness, the greater the entropy. Gases, with their free-flying molecules, have high entropies. Crystalline solids have particles arranged in definite, repeating, organized systems; therefore, they possess low entropies. However, even crystals which appear to have increased order will eventually decay.

The entropy of substances increases with temperature. As the temperature becomes greater, the increased kinetic energy gives the particles greater motion. For example, when ice (a crystalline solid) melts to the state of water, entropy has increased. Because the amount of entropy in a substance varies with temperature as well as pressure, these conditions must be specified. Table 13A-2 lists the entropies per mole of several common substances at standard conditions. Although elements are not listed for enthalpies in the extensive table, they are included for entropies because all substances have some entropy (see Table 13A-5).

A. Highly organized crystalline structure
B. Ice melting to water

Table 13A-2 Standard Molar Entropies

Compound	Entropy (cal/(mol · K))
HBr (g)	47.44
HCl (g)	44.6
NH_3 (g)	46.0
NH_4Cl (s)	22.6

Calculating Entropy Changes

From the data in Table 13A-2 the entropy changes of many reactions can be calculated. An entropy change is the difference between the entropy of the products and the entropy of the reactants.

$\Delta S°$ = Total entropy in products − Total entropy in reactants

$$\Delta S° = \sum S°_{products} - \sum S°_{reactants}$$

A reaction with a positive $\Delta S°$ increases entropy. The products have more entropy than the reactants. Negative $\Delta S°$ signifies a decrease in entropy.

Sample Problem Does the reaction $NH_3 + HCl \longrightarrow NH_4Cl$ increase or decrease entropy?

Solution

$\Delta S°$ of NH_3 = 46.0 cal/(mol·K)

$\Delta S°$ of HCl = 44.6 cal/(mol·K)

$\Delta S°$ of NH_4Cl = 22.6 cal/(mol·K)

$\Delta S° = \sum S°_{products} - \sum S°_{reactants}$

$\Delta S° = [(1)(22.6)] - [(1)(46.0) + (1)(44.6)]$

$= 22.6 - 90.6$

$= -68.0$ cal/(mol·K)

Since the $\Delta S°$ value is negative, the amount of entropy has decreased. The product (a solid) is more ordered than the reactants (gases).

Natural processes always increase the entropy of the universe. In fact, all processes, whether natural or unnatural, increase the entropy of the universe.

A positive $\Delta S°$ aids spontaneity. A reaction with a negative $\Delta S°$ goes against this tendency and attempts to decrease entropy. What can force a reaction to violate the trend toward disorder?

The Driving Force of Reactions: Free-Energy Change

Two tendencies drive reactions. The first is the tendency to decrease enthalpy, and the second is the tendency to increase entropy. Reactions can be either exothermic or endothermic, and they can either increase or decrease entropy. Four possible enthalpy/entropy combinations exist.

1. Exothermic and increasing entropy ($-\Delta H$, $+\Delta S$)
2. Exothermic and decreasing entropy ($-\Delta H$, $-\Delta S$)
3. Endothermic and increasing entropy ($+\Delta H$, $+\Delta S$)
4. Endothermic and decreasing entropy ($+\Delta H$, $-\Delta S$)

Reactions are driven by the combined effects of ΔH and ΔS. A single expression that relates ΔH and ΔS is needed to determine how they combine. J. Willard Gibbs, an American physicist and mathematician of the 1800s, formulated a single criterion of favorability. He combined the change in enthalpy and the change in entropy into a single term called **free energy,** which is sometimes called Gibbs free energy (*G*) in his honor. The **free-energy**

change, ΔG, for a reaction can be calculated by the following equation, where T is temperature in kelvins:

$$\Delta G = \Delta H - T\Delta S$$

J. Willard Gibbs

ΔG is the change in free energy. It is the difference between the free energy of the products and the free energy of the reactants. A negative ΔG indicates a decline in free energy and signifies that a reaction may occur naturally. A positive ΔG indicates a net increase in free energy and signifies that a reaction is not favorable.

The free-energy change can be negative under several conditions. A negative ΔH always contributes to a negative ΔG; so does a positive ΔS. When the two tendencies oppose each other, the temperature often determines which one will "win." Table 13A-3 shows when ΔG is negative.

Case 1 shows that exothermic reactions that increase entropy ($-\Delta H$, $+\Delta S$) are always favorable. They release energy and they increase disorder. ΔG for such reactions is always negative.

Case 2 shows that exothermic reactions that decrease entropy ($-\Delta H$, $-\Delta S$) may or may not be favorable. The two tendencies oppose each other. In cases like this, the temperature determines whether the reaction is favorable. The entropy change hinders the reaction less at low temperatures than it does at high temperatures, so ΔG is negative at low temperatures.

Table 13A-3 When ΔG is Negative

Case	ΔH	ΔS	ΔG
1.	−	+	$-\Delta H - T(+\Delta S) = -\Delta G$ at all Ts
2.	−	−	$-\Delta H - T(-\Delta S) = -\Delta G$ at low Ts
3.	+	+	$+\Delta H - T(+\Delta S) = -\Delta G$ at high Ts
4.	+	−	$+\Delta H - T(-\Delta S) = +\Delta G$ at all Ts

Case 3 represents endothermic reactions that increase entropy ($+\Delta H$, $+\Delta S$). These reactions can be favorable if the entropy change is greater than the change in enthalpy. High temperatures magnify the effect of the entropy change and make ΔG negative. As a result, the reaction can proceed.

Case 4 shows that endothermic reactions that decrease entropy ($+\Delta H$, $-\Delta S$) do not occur naturally. To proceed naturally, these reactions would have to store energy in chemical bonds and increase order. Reactions that exhibit positive enthalpy changes can be forced if other types of energy are used. Otherwise, they are unfavorable at any temperature.

Examples of each case using information from Table 13A-5 can be seen in Table 13A-4 on the next page.

The ΔG of a reaction can be calculated from the values for ΔH, ΔS, and T by plugging them into the free-energy equation. For practice, calculate the free-energy change for the reaction $Mg(OH)_2 \longrightarrow MgO + H_2O(l)$. According to Table 13A-5, once the change in free energy is known, the probability of the reaction can be determined at 25°C.

$$\Delta H = (-143.8 - 68.3) - (-221.0) = 8.9 \text{ kcal/mol}$$
$$\Delta S = (6.4 + 16.7) - (15.09) = 8.0 \text{ cal/(mol·K)}$$

Note that the two values are not expressed in similar units. Converting the ΔS value from 8.0 cal/(mol·K) to 0.0080 cal/(mol·K) can remedy the problem.

$$\Delta G = \Delta H - T\Delta S$$
$$= 8.9 - 298 (0.0080)$$
$$= 6.5 \text{ kcal/mol}$$

Calculating ΔG from Table 13A-5 gives a result very close to the same value.

$$\Delta G = (-136.1 - 56.7) - (-199.2) = 6.4 \text{ kcal/mol}$$

Since the ΔG is positive, the reaction cannot be favorable at 298 K. At a higher temperature the favorable ΔS has a greater effect. This effect could be seen by a calculation of ΔG at 1000 K if the appropriate values for ΔH_f° and S°, or ΔG at 1000°C were available.

Table 13A-4 Examples of the Four Cases

Case 1: $2Ag + S \rightarrow Ag_2S$

$\Delta H^\circ_f = -7.79$ kcal/mol
$\Delta S = [34.4] - [(2)(10.17) + 7.6] = 6.5$ cal/(mol · K)
$\Delta G = \Delta H - T\Delta S = -7.79 - (298)(0.0065) = -9.7$ kcal/mol

Case 2: $2Fe + \frac{3}{2}O_2 \rightarrow Fe_2O_3$

$\Delta H^\circ_f = -195.0$ kcal/mol
$\Delta S = [20.9] - [(2)(6.52) + (\frac{3}{2})(49.0)] = -65.6$ cal/(mol · K)
$\Delta G = \Delta H - T\Delta S = -195.0 - (298)(-0.0656) = -175.5$ kcal/mol

Case 3: $2C + H_2 \rightarrow C_2H_2$

$\Delta H^\circ_f = 54.2$ kcal/mol
$\Delta S = [48.0] - [(2)(1.4) + 31.2] = 14.0$ cal/(mol · K)
$\Delta G = \Delta H - T\Delta S = 54.2 - (298)(0.0140) = 50.0$ kcal/mol

Case 4: $2C + 2H_2 \rightarrow C_2H_4$

$\Delta H^\circ_f = 12.5$ kcal/mol
$\Delta S = [52.5] - [(2)(1.4) + (2)(31.2)] = -12.7$ cal/(mol · K)
$\Delta G = \Delta H - T\Delta S = 12.5 - (298)(-0.0127) = 16.3$ kcal/mol

Table 13A-5 Standard Thermodynamic Property Values

Compound	$\Delta H°_f$	$S°$	ΔG	Compound	$\Delta H°_f$	$S°$	ΔG
Ag (s)	0	10.17	0	H_2SO_4 (l)	-193.9	37.5	-164.9
AgBr (s)	-23.8	25.6	-23.2	Hg (l)	0	18.5	0
AgCl (s)	-30.4	23.0	-26.2	HgO (s)	-27.1	17.2	-14.0
AgI (s)	-14.9	27.3	-15.8	HgS (s)	-13.9	18.6	-12.1
Ag_2O (s)	-7.42	29.1	-2.68	I_2 (s)	0	27.9	0
Ag_2S (s)	-7.79	34.4	-9.72	K (s)	0	15.2	0
Al (s)	0	6.77	0	KBr (s)	-93.7	23.1	-91.0
Al_2O_3 (s)	-399.1	12.2	-378.2	KCl (s)	-104.2	19.8	-97.8
Ba (s)	0	16.0	0	$KClO_3$ (s)	-93.5	34.2	-70.8
$BaCl_2$ (s)	-205.6	30.0	-193.7	KF (s)	-135.6	15.9	-128.5
$BaCO_3$ (s)	-291.3	26.8	-271.9	KOH (s)	-101.8	18.9	-90.6
$BaSO_4$	-350.2	31.6	-325.6	Mg (s)	0	7.8	0
Br_2	0	36.4	0	$MgCl_2$ (s)	-153.4	21.4	-141.5
C (s)	0	1.4	0	$MgCO_3$ (s)	-261.9	15.7	-241.9
Ca (s)	0	10.0	0	MgO (s)	-143.8	6.4	-136.1
$CaCl_2$ (s)	-190.0	27.2	-178.8	$Mg(OH)_2$ (s)	-221.0	15.09	-199.2
$CaCO_3$ (s)	-288.5	22.2	-269.8	$MgSO_4$ (s)	-305.5	21.9	-279.8
CaO (s)	-151.9	9.5	-144.4	Mn (s)	0	7.6	0
$Ca(OH)_2$ (s)	-235.8	18.2	-214.8	MnO (s)	-92.0	14.4	-86.7
$CaSO_4$ (s)	-342.4	25.5	-315.9	MnO_2 (s)	-124.3	12.7	-111.2
CCl_4 (l)	-33.3	51.3	-15.6	N_2 (g)	0	45.8	0
CH_4 (g)	-17.9	44.5	-12.1	Na (s)	0	12.2	0
C_2H_2 (g)	54.2	48.0	50.0	NaBr (s)	-86.0	20.8	-83.4
C_2H_4 (g)	12.5	52.5	16.3	NaCl (s)	-98.2	17.3	-91.8
C_2H_6 (g)	-20.24	54.85	-7.86	NaF (s)	-136.0	14.0	-129.9
C_3H_8 (g)	-24.8	64.5	-5.61	NaI (s)	-68.8	23.5	-68.4
C_6H_6 (l)	19.8	64.3	31.0	NaOH (s)	-102.0	15.4	-90.7
CH_3OH (l)	-57.0	30.3	-39.8	NH_3 (g)	-11.0	46.0	-3.94
C_2H_5OH (l)	-66.37	38.4	-41.8	NH_4Cl (s)	-75.4	22.6	-48.5
Cl_2 (l)	0	53.3	0	NH_4NO_3 (s)	-87.3	36.1	-44.0
CO (l)	-26.4	47.3	-32.8	NO (g)	21.6	50.3	20.7
CO_2	-94.1	51.1	-94.3	NO_2 (g)	7.93	57.4	12.3
Co (s)	0	6.8	0	O_2 (g)	0	49.0	0
Cr	0	5.7	0	O_3 (g)	34.1	56.8	39.0
Cr_2O_3 (s)	-272.4	19.4	-252.9	Pb (s)	0	15.5	0
Cu (s)	0	8.0	0	$PbBr_2$ (s)	-66.6	38.6	-62.6
CuO (s)	-37.6	10.4	-31.0	$PbCl_2$ (s)	-85.9	32.6	-76.0
Cu_2O (s)	-40.3	24.1	-34.9	PbO (s)	-52.1	16.2	44.9
CuS (s)	-12.7	15.9	-12.8	PbO_2 (s)	-66.3	18.3	-52.0
$CuSO_4$ (s)	-184.4	27.1	-158.2	Pb_3O_4 (s)	-171.7	50.5	-143.7
F_2 (g)	0	48.6	0	PCl_3 (g)	-73.2	74.5	-64.0
Fe (s)	0	6.52	0	S (s)	0	7.6	0
Fe_2O_3 (s)	-195.0	20.9	-177.4	Si (s)	0	4.5	0
Fe_3O_4 (s)	-267.3	35.0	-242.7	SiO_2 (s)	-217.2	10.0	-204.75
H_2 (s)	0	31.2	0	Sn (s)	0	12.3	0
HBr (s)	-8.7	47.44	-12.8	$SnCl_4$ (l)	-122.2	61.8	-105.2
HCl (s)	-22.1	44.6	-22.8	SnO (s)	-68.3	13.5	-61.4
HF (g)	-64.8	41.5	-65.3	SnO_2 (g)	-138.8	12.5	-124.2
HI (g)	6.33	49.3	0.41	SO_2 (g)	-70.9	59.4	-71.7
HNO_3 (l)	-41.6	37.2	-19.3	SO_3 (g)	-94.5	61.2	-88.7
H_2O (g)	-57.8	45.1	-54.6	Zn (s)	0	10.0	0
H_2O (l)	-68.3	16.7	-56.7	ZnO (s)	-83.2	10.5	-76.1
H_2S (g)	-4.93	49.2	-8.02	ZnS (s)	-48.5	13.8	-48.1

$\Delta H°_f$ and ΔG are in units of kcal/mol. $S°$ is in units of cal/(mol·K).

Sample Problem Determine whether the reaction between ammonia and hydrogen chloride ($NH_3 + HCl \longrightarrow NH_4Cl$) is probable at 298 K and whether it is probable at 1000 K. According to Table 13A-5

$$\Delta H = -75.4 - (-11.0 + -22.1) = -42.3 \text{ kcal/mol}$$
$$\Delta S = 22.6 - (46.0 + 44.6) = -68.0 \text{ cal/(mole} \cdot \text{K)}$$

Solution

To find the ΔG at 298 K, plug the appropriate values into the free-energy equation. Remember to convert the units of ΔS from cal/(mol·K) to kcal/(mol·K).

$$\Delta G = \Delta H - T\Delta S$$
$$= -42.3 \text{ kcal/mol} - [(298 \text{ K})(-0.0680 \text{ kcal/(mol} \cdot \text{K)})]$$
$$= -22.0 \text{ kcal/mol}$$

Since the ΔG is negative, the reaction is favorable at this temperature. Even though the values of $\Delta H_f°$ and $S°$ at 1000 K are not known for this reaction, the value of ΔG can be estimated by using the 298 K figures in the free energy equation and solving for ΔG.

$$\Delta G = \Delta H - T\Delta S$$
$$= -42.3 \text{ kcal/mol} - [(1000 \text{ K})(-0.0680 \text{ kcal/(mol} \cdot \text{K)})]$$
$$= 25.7 \text{ kcal/mol}$$

The positive ΔG shows that the reaction is not favorable.

Section Review Questions 13A

1. Evaluate why the laws of thermodynamics are excellent scientific tools against the theory of evolution.
2. Explain the difference between enthalpy and entropy.
3. Arrange the following examples in order of increasing entropy. (Begin with the most ordered and work to the least ordered.)
   a. 1 mole of H_2O (g) at 125°C
   b. 1 mole of H_2O (g) at 140°C
   c. 1 mole of H_2O (s)
   d. 1 mole of H_2O (l)
4. Define the following symbols:
   a. H b. $\Delta H°_f$ c. ΔH d. S
   e. $\Delta S°$ f. T g. G h. ΔG

Use the following equations to answer questions 5-7:
   a. N_2 (g) + 3 H_2 (g) $\longrightarrow$ 2 NH_3 (g)
   b. $Ca(OH)_2$ (s) $\longrightarrow$ CaO (s) + H_2O (g)
   c. 2 $KClO_3$ (s) $\longrightarrow$ 2 KCl (s) + 3 O_2 (g)
   d. SnO_2 (s) + 2 H_2 (g) $\longrightarrow$ Sn (s) + 2 H_2O (l)

5. Calculate the standard change in enthalpy (ΔH) for each reaction and tell whether the reactions are endothermic or exothermic.
6. Calculate the standard entropy change (ΔS) for each reaction.
7. Calculate the change in free energy (ΔG) for each reaction at 25°C, and tell whether the reactions are favorable or not favorable at this temperature.

13B Kinetics: How Fast Will Things React?

The fact that a reaction is thermodynamically favorable does not mean that it will proceed automatically. Some reactions proceed at an extremely slow rate, some need a push to get started, and some go on their own. Consider three reactions. Each of them has a negative ΔG, so they are all favorable under the proper conditions of temperature and pressure.

1. The oxidation of a diamond: ΔG = -94.7 kilocalories
2. The burning of methane: ΔG = -138.9 kilocalories
3. The mixing of Ba(OH)$_2$ and H$_2$SO$_4$:   ΔG = -31.3 kilocalories

The first reaction does not proceed at a significant rate. The second reaction does not start unless it is given an energetic push (from a lighted match, for example). The third reaction proceeds as soon as the reactants are mixed together. Something besides thermodynamics must be used to explain why reactions have such a wide variety of rates. This "something" is kinetics.

A building is imploded when stored energy is released in a chemical reaction.

Energy Diagrams: Mapping Energy Changes

Thermodynamics relates only to the starting and ending points of a reaction and is therefore path-independent. Kinetics is path-dependent and is concerned with how a reaction proceeds from point A to point B. Thus, kineticists strive to determine what happens between the start and finish of a reaction. Are the energy changes during a reaction simple and direct, or must the reactants gain some energy before the reaction can proceed?

Figure 13B-1 ▪ Possible Reaction Pathways

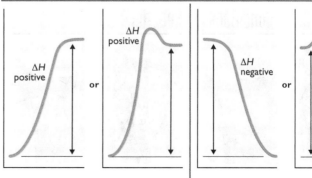

Possible Endothermic Pathways | Possible Exothermic Pathways

351

Diagrams of energy changes during the course of reaction are often used to illustrate the energetic aspects of the reaction and are commonly used in kinetics to illustrate complex ideas. The vertical scale represents the energy of a particular chemical entity. Usually the energy scale does not have any specific units. The horizontal scale represents the progress of the reaction being considered. The highest point on the energy diagram corresponds to the energy of the rate-limiting step in the reaction. The rate-limiting step is the slowest step in the reaction—the one that limits the overall rate of the reaction. For example, compare the rate-limiting step with the principle of an assembly line. Suppose that one job on an assembly line requires the worker to perform a task that is much more difficult than that of any other worker. The product can be produced only as fast as the worker with the difficult task completes his part.

Collision Theory: Prerequisites for Reactions

A hydrogen gas molecule, H_2, is combined with a gaseous iodine molecule, I_2. A forceful collision disrupts the electron orbitals of these molecules. For a moment, four atoms hang in one highly energized group; a reaction then occurs to make hydrogen iodide molecules, HI, according to the equation $H_2 + I_2 = 2HI$.

Three conditions must be met in order for any reaction to occur. The first is a fundamental principle of kinetics: (1) things must collide before they react. However, a collision alone does not guarantee a reaction. The collision must also be (2) properly oriented for the necessary rearrangement of atoms and electrons and (3) forceful enough— possessing enough energy to form products. Reactions follow only forceful, properly oriented collisions.

The **collision theory** explains why reactions go at greater rates, depending on reaction conditions. Any factor that increases the number of effective collisions increases the rate of a reaction. Therefore, the rate at which a chemical reaction

A. Proper orientation but not enough force
B. Enough force but not properly oriented
C. Properly oriented and enough force

Figure 13B-2 ▪ Conditions for Reactions

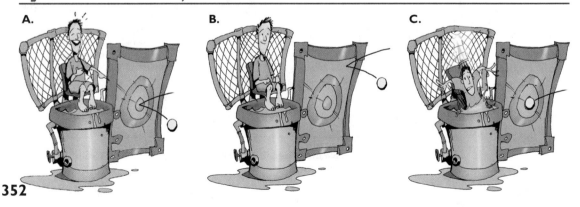

proceeds is equal to the frequency of effective collisions. Fast reactions occur when the three criteria are easy to meet. However, if even one criterion is difficult, the reaction will occur slowly.

Activated Complex: Over the Hump!

Despite a negative ΔG, indicating a favorable reaction, a reaction will not proceed rapidly until the initial energy barrier is overcome.

Recall that a reaction can be pictured as a two-step process: (1) the breaking of bonds, and (2) the formation of bonds. Breaking bonds in molecules requires energy. Not until this step is completed can the resulting atoms form bonds and release energy.

Molecules need kinetic energy before their collisions can be forceful enough to cause reactions. This energy, called the **activation energy** (E_a), is the minimum amount of kinetic energy that must be possessed by the colliding molecules before they can react. As a rule, the lower the activation energy, the faster the reaction will occur. There is a specific activation energy for each reaction.

Colliding reactants can form a transitional structure between reactants and products. This structure, called an **activated complex,** is an intermediate group of reactants. Its high energy content makes it extremely unstable. The activated complex can break up to form the products of the reaction, or it can revert to the separate reactants. If the activated complex goes on to complete the reaction, a large amount of energy will be released. In exothermic reactions, this energy activates more reactants, and their reaction rate may accelerate.

In conclusion, the activation energy plays a crucial role in determining how fast a reaction will proceed under certain conditions of temperature and pressure. A large activation energy can prevent a spontaneous reaction from proceeding quickly. Figure 13B-5 shows energy diagrams for several reactions. Which reaction is most likely to proceed rapidly?

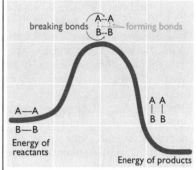

Figure 13B-3

Reactions must often overcome an initial energy barrier before they can proceed.

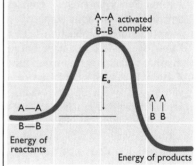

Figure 13B-4

An activated complex is a transitional structure.

Figure 13B-5 ▪ Reaction Pathways

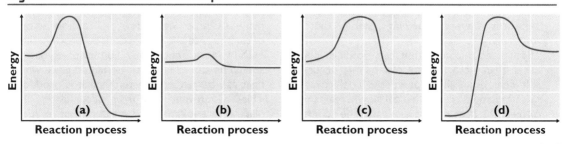

Facets of Chemistry: Spontaneous Combustion

It had been a hot day in North Arlington, New Jersey. Swimmers seeking relief from the oppressive heat swarmed a beach along a river bank. Between them and the road was a storehouse full of pyroxylin—a highly combustible material used in lacquers, plastics, and artificial leather. Across the road was the Atlantic Pyroxylin Waste Company. Here workers sorted pyroxylin scraps collected from surrounding factories so that they could be used again. At 9:12 P.M. lingering swimmers noticed fire coming from the roof of the factory. With a puff, both ends of the building blew out, and a vast flame swept over the surrounding area. The flames crossed the road and set fire to the small storehouse beside the beach. Another puff occurred, and flames engulfed the beach. The burning of the Atlantic Pyroxylin Waste Company is a classic example of spontaneous combustion.

Spontaneous combustion fires start without a flame or spark. They start because of the role that heat plays in speeding up reaction rates. The Atlantic Pyroxylin Waste Company stored pyroxylin scraps swept up from factory floors. These scraps often had machine oil on them. The heat necessary for the fire to start came from the reaction between this oil and atmospheric oxygen. The reaction does not normally generate great amounts of heat, because it occurs slowly. Whatever heat is produced is released gradually. However, if the heat from the oxidation cannot escape, it accumulates and the temperature increases. The increased temperature causes the reaction to proceed faster and to produce more heat. This vicious cycle repeats itself until the reaction proceeds fast enough to generate temperatures above the kindling point of the surrounding material.

Spontaneous combustion can occur in places other than factories. For example, it occurs in soft coal, especially if powdered coal covers a mass of lump coal. As the coal slowly reacts with oxygen, the heat cannot dissipate. The temperature increases until the coal ignites. Soft coal should be spread over a large area to allow the heat to escape.

Farmers must also be aware of the dangers of spontaneous combustion. If they pack freshly cut hay into a barn before it is properly dried, oxidation will continue in the confined space. When a certain temperature is reached, the hay will burst into flames. Hay should be stored in a cool, dry, well-ventilated barn. Compost piles containing leaves and other organic material can also lead to spontaneous combustion if not properly ventilated.

Oily, paint-saturated rags pose a threat to many homes and workshops. As paint dries, the oil it contains reacts with oxygen and forms an elastic solid. This reaction produces heat as the paint dries. A pile of oily rags provides the perfect conditions for a cycle of higher temperatures and faster reaction rates to be set up. Again the temperature increases until the rags catch fire. Paint rags should be hung outdoors where there is good air circulation, and they should then be stored in a metal can.

Preventing spontaneous combustion fires is a matter of controlling the rates of chemical reactions. Caution should be taken with combustible materials in powdered forms. The large amount of surface area allows reactions to proceed more quickly. More importantly, combustible material should be stored in a well-ventilated place where heat can be removed faster than it is produced. This will prevent heat from accumulating in a place where it can increase the rate of reaction.

Reaction (d) requires a large initial input and a net gain in energy; it is doubtful that such a reaction will proceed rapidly on its own. The other reactions release energy but must overcome the activation-energy barrier to do so. Reaction (b) has the smallest barrier to overcome, so it is the most likely to proceed rapidly as soon as reactants are mixed.

Rates of Reactions

During a reaction the concentrations of reactants and products change constantly. At the beginning of a reaction, the reactants are packed together in high concentrations. As the reaction progresses, product concentrations increase as reactants are consumed. Consider the reaction between iodine chloride and hydrogen: $2 \text{ ICl } (g) + H_2 (g) \longrightarrow I_2 (g) + 2 \text{ HCl } (g)$. Figure 13B-6 shows how the concentrations of the substances in this reaction vary with time.

Reaction rates tell how fast reactants change into products. They can describe how fast the reactants disappear or how fast the products appear. Although reaction rates may be measured in many different units, they usually tell how fast concentrations change with time. Units such as molarity per second or moles per liter per hour are common.

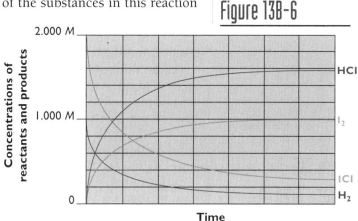

Figure 13B-6

Factors That Affect Reaction Rates

The most obvious factor that controls the rate of a reaction is the chemical nature of the reactants. Logically enough, reactive substances react quickly. The chemical nature of reactants cannot be changed, but other factors can. The rate of a given reaction can be modified by changes in concentration, temperature, the amount of surface area, and the presence of catalysts.

Concentration. The typical test for the presence of oxygen gas relies on the fact that high concentrations of reactants improve the rate of a reaction. A glowing wooden splint is thrust into a gas-collection bottle. If the bottle contains pure or relatively pure oxygen gas, the glowing splint will burst into flames. The atmosphere contains only

Glowing splint test for oxygen gas

21 percent oxygen. This small amount of oxygen supports the slow smoldering while the splint is glowing. An atmosphere of 100 percent oxygen increases the reaction rate; the combustion thus proceeds more quickly.

The effect of concentration on the reaction rate varies from reaction to reaction. Doubling the concentration of a reactant in one reaction might not affect the rate at all. Doubling the concentration of a reactant in a second reaction could double or even quadruple its rate. No rule can predict the effect of concentration changes on reaction rates. Thus, reactions must be studied individually.

Temperature. One rule of thumb in kinetics states that reaction rates double for every 10°C rise in temperature. Although the rule has many exceptions, it illustrates the vital role that temperature plays in determining the rates of reactions. Higher temperatures increase reaction rates in two ways. First, they increase the number of collisions between reactants. Second, they increase the force with which collisions occur. In the final analysis, higher temperatures increase the number of effective collisions.

The decomposition of lactose (a sugar in dairy products) into lactic acid turns fresh milk into a foul-smelling fluid. The lactic acid that is produced gives old milk its distinctive sour taste. Can this reaction be stopped? Not totally; but it can be greatly slowed down if the milk is kept cool in a refrigerator. The lower temperature in the refrigerator extends the usable period of milk by lowering the reaction rate.

Surface Area. Small particles have greater surface area per unit volume than do large particles. The amount of surface area exposed affects the rate of a reaction. If a substance is broken up into pieces rather than being in one piece, more surface area is exposed and the reaction will occur faster. Increased rate of reaction is due to an increase in collision rate. For example, if two antacid tablets, one whole and one broken, are placed in separate cups and the same amount of water is added to each cup, then the tablet that is broken into pieces will dissolve faster than the one that is left whole. Figure 13B-7 illustrates this example.

Figure 13B-7 ▪ Surface Area and Reaction

● Water molecules

● Antacid tablet

Catalysts. A **catalyst** is a substance that changes a reaction rate without being permanently changed by the reaction. It is present during the reaction, but it is neither a reactant nor a product. A catalyst provides an alternate route from reactants to products that is easier—possessing a smaller activation energy. A catalyst however, does not effect the equilibrium of a reaction, only the rate at which the equilibrium is obtained. The catalyst for a given reaction accelerates both the forward and reverse reactions equally.

How do catalysts work? Theoretically, catalysts hold reactants in just the right positions for favorable collisions. Catalysts always lower the activation energy of a reaction. For example, a mixture of hydrogen and oxygen gas does not react to form water at room temperature. A catalyst of powdered platinum must first be introduced. The platinum causes an explosion as its surface becomes covered with absorbed oxygen. The platinum atoms stretch and weaken the bonds of the O_2 molecules, thereby lowering the activation energy. The oxygen atoms can now react rapidly with the hydogen molecules to form water.

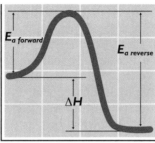
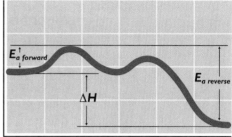

Figure 13B-8

There are several types of catalysts. A **homogeneous catalyst** is in the same phase as a reactant or in solution with a reactant. Homogeneous catalysts combine with one of the reactants to form an intermediate compound that will react more readily with the other reactants.

A catalyst that is in a separate phase from the reactants is said to be a **heterogeneous catalyst.** Heterogeneous catalysts are materials capable of absorbing molecules of gases or liquids onto their surfaces. Platinum, nickel, palladium, and other finely divided metals and metalloids are examples of heterogeneous catalysts.

A special and distinct class of catalysts comprises naturally occurring biological substances known as **enzymes.** Enzymes are responsible for many essential biochemical reactions. More than a thousand enzymes have been identified, and each one is specific to a chemical reaction occurring within a living organism.

Figure 13B-9 ■
Model of Enzymes

Reactants enter the active site of the enzyme.

The reaction occurs.

One product leaves the active site.

A water molecule leaves the active site.

Some catalysts, such as those which promote spoiling of food, are undesirable. **Inhibitors** are used to reduce the catalyst's undesirable effects. Inhibitors work by bonding to the catalysts and slowing the increase in rate of reaction.

Mechanisms: One Step at a Time

Chemists can easily identify the substances that go into reactions. They can also routinely analyze the compounds that emerge. The fragments of molecules that exist during reactions are much more difficult to pinpoint. For chemists, observing reactions is like looking at a factory from the outside. They can see the raw materials going in and the finished products coming out, but they cannot see the individual steps within the reactions. The series of steps that make up a reaction is called a **reaction mechanism.**

The reaction between hydrogen and iodine gases was one of the first to be studied in the field of kinetics. Scientists knew that $H_2 + I_2 \longrightarrow 2$ HI. What they did not know was what actually happened during the reaction. Did the two molecules collide to form an activated complex that split into two hydrogen iodide molecules? Did the atoms in both molecules split and then rearrange themselves, or did the iodine molecule split, surround the hydrogen molecule, and cleave it? Theoretically, any one of these proposed mechanisms could produce hydrogen iodide from hydrogen and iodine. Mechanisms are temperature dependent; therefore, temperature is the deciding factor in determining which mechanism will occur.

Figure 13B-10 ▪ Possible Mechanisms for $H_2 + I_2 \longrightarrow 2HI$

Rate Laws: Equations that Describe Reaction Rates

As stated before, the concentrations of reactants influence reaction rates. The effect of concentration varies from reaction to reaction. For example, the rate of $H_2 + I_2 \longrightarrow 2$ HI is found by experiment to be directly proportional to the concentration of hydrogen gas in the reaction vessel. The concentration of reactants in terms of moles per liter (molarity) is often signified by square brackets. Doubling the concentration of hydrogen, $[H_2]$, doubles the reaction rate. Tripling the concentration triples the reaction rate. The rate of this reaction is also found by experiment to be directly proportional to the concentration of iodine gas.

$$\text{Rate} \propto [H_2][I_2]$$

This expression, called a rate law, becomes an equality when the numerical constant k is inserted. A **rate law** is an equation that mathematically describes how fast a reaction occurs. Each reaction has its own rate constant (k).

$$\text{Rate} = k[H_2][I_2]$$

Table 13B-1 gives several examples of rate laws. It is important to note that rate laws must be determined experimentally, not by using the balanced equation coefficients as exponents. Reactions often have several possible mechanisms, and an equation alone cannot reveal which mechanism was used. Chemical equations show only the reactants and the products they formed, not the steps the reactants underwent to form those products. If there are several possible reaction mechanisms, the rate law can sometimes be used to eliminate one or more of the possibilities. This would be the case if the experimentally determined rate law did not match the theoretical rate law derived from the proposed mechanism.

Table 13B-1 Rate Laws for Several Reactions

Reaction	Rate Law
$2 H_2 + 2 NO \longrightarrow N_2 + 2 H_2O$	Rate = $k[H_2][NO]^2$
$2 NO + Br_2 \longrightarrow 2 NOBr$	Rate = $k[NO]^2[Br_2]$
$O_3 + NO \longrightarrow NO_2 + O_2$	Rate = $k[O_3][NO]$
$2 NO + O_2 \longrightarrow 2 NO_2$	Rate = $k[NO]^2[O_2]$
$C_4H_9Br + 2 H_2O \longrightarrow C_4H_9OH + Br^- + H_3O^+$	Rate = $k[C_4H_9Br]$

The reaction between nitrogen dioxide and carbon monoxide at low temperatures has a completely different rate law.

$$NO_2 + CO \longrightarrow NO + CO_2; \quad \text{Rate} = k[NO_2]^2$$

Changing the concentration of carbon monoxide has no effect on the rate of the reaction. Therefore, the concentration of carbon monoxide does not appear in the rate law. The exponent on the [NO_2] term indicates that the concentration of nitrogen dioxide plays the crucial role in determining how fast the reaction goes. If the concentration of nitrogen dioxide is doubled, the rate quadruples. If the concentration of nitrogen dioxide is tripled, the rate speeds up by a factor of nine.

Section Review Questions 13B

1. Give two reasons that a collision between two reactive molecules might not result in a reaction.
2. Draw energy diagrams for both an endothermic and exothermic reaction. Include the following labels in your diagram: reactants, products, activated complex, activation energy, and ΔH.
3. There is usually one step in a reaction that is rate determining. What are the characteristics of a rate-determining step? Explain.
4. What are the four variables that can determine the rate of reaction?
5. How does a catalyst speed up a reaction rate?
6. Does a catalyst affect the activation energy of a reaction? the enthalpy? Explain.
7. Why is it generally impossible to use only a chemical equation to predict a rate law for a particular reaction?

Chapter Review

Coming to Terms

thermodynamics
kinetics
first law of thermodynamics
enthalpy
heat of reaction
enthalpy of formation
standard state
standard molar enthalpy of formation
Hess's law
enthalpy of bond formation
second law of thermodynamics
entropy
free energy (Gibbs free energy)
free energy change
collision theory
activation energy
activated complex
reaction rate
catalyst
homogeneous catalyst
heterogeneous catalyst
enzyme
inhibitor
reaction mechanism
rate law

Review Questions

1. What two tendencies influence all chemical reactions?
2. Which of the following situations is/are possible according to the laws of thermodynamics?
   a. Insect larvae automatically emerging from rotting meat
   b. The human body converting the energy in food to other forms of energy
   c. The energy and matter in the universe coming into being from nothing without any intervention from God
   d. The invention of an automobile engine that is 100 percent efficient
3. For each of these reactions, give the enthalpy of reaction and tell whether the reaction is endothermic or exothermic.
   a. $C\ (s) + O_2\ (g) \longrightarrow CO_2\ (g)$
   b. $CO_2\ (g) \longrightarrow C\ (s) + O_2\ (g)$
   c. $2\ C\ (s) + 2\ H_2\ (g) \longrightarrow C_2H_4\ (g)$
   d. $C_2H_4\ (g) \longrightarrow 2\ C\ (s) + 2\ H_2\ (g)$

For questions 4-8, refer to these reactions:
   a. $4\ NH_3\ (g) + 5\ O_2\ (g) \longrightarrow 4\ NO\ (g) + 6\ H_2O\ (g)$
   b. $3\ NO_2\ (g) + H_2O\ (l) \longrightarrow 2\ HNO_3\ (l) + NO\ (g)$
   c. $2\ NH_4NO_3\ (s) \longrightarrow 2\ N_2\ (g) + O_2\ (g) + 4\ H_2O\ (g)$
   d. $CH_4\ (g) + 2\ H_2O\ (g) \longrightarrow CO_2\ (g) + 4\ H_2\ (g)$

4. Calculate the standard change in enthalpy ($\Delta H°$) for each reaction, and tell whether the reactions are endothermic or exothermic.
5. Calculate the standard entropy change ($\Delta S°$) for each reaction.
6. Calculate the change in free energy (ΔG) for each reaction at 25.0°C, and tell whether the reactions are spontaneous or nonspontaneous at this temperature.
7. Calculate ΔG for each reaction at 1000 K, and tell whether the reaction is spontaneous or nonspontaneous at this temperature.
8. Calculate the temperature (if any) at which the reactions change from being spontaneous to nonspontaneous (the temperature at which ΔG equals zero).
9. Based on your calculations of ΔG at 298 K and 1000 K for the four reactions, graph ΔG versus temperature. Graph ΔG on the y-axis and temperature on the x-axis. Based on the trend from the graph, state whether the reaction is more or less spontaneous at higher temperatures.

10. The reactants for a spontaneous reaction are mixed, but no reaction appears to occur. Suggest an explanation for the nonreaction.

11. According to the collision theory,
    a. why does increased temperature increase the reaction rate?
    b. why does a greater concentration of reactants increase the reaction rate?
    c. why do powders react more quickly than crystals?
    d. why will a reaction involving the collision of three molecules proceed more slowly than one involving the collision of two molecules (all other things being equal)?

12. Sugar needs temperatures much higher than 98.6°F in order to burn. Yet sugar can be "burned" in your digestive tract at this temperature. What is responsible for the ability of your body to burn sugar at this low temperature? Draw energy diagrams that illustrate the difference between the two situations.

13. This diagram shows how fast an unnamed substance (A) reacts. Curves I, II, and III correspond to the reaction under different conditions.

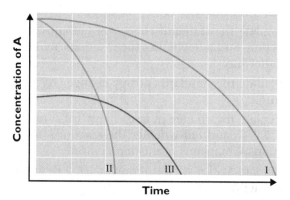

   a. Under which set of conditions does the reaction proceed at the highest rate?
   b. Of reaction conditions I and III, which starts with the highest reactant concentration?
   c. If reaction conditions I and II occur without a catalyst, which one probably occurs at the higher temperature?
   d. If the temperatures for I and II are identical, which occurs in the presence of a catalyst?

14. The two curves in the following energy diagram represent the reaction 2 KClO$_3$ (s) $\longrightarrow$ 2 KCl (s) + 3 O$_2$ (g) occurring under two different sets of reaction conditions.

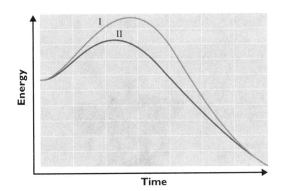

   a. Which set of conditions requires the lower activation energy?
   b. Which reaction pathway would more likely include a catalyst?
   c. Which reaction pathway requires higher temperatures?

15. The reaction 2 NO (g) + 2 H$_2$ (g) $\longrightarrow$ N$_2$ (g) + H$_2$O (l) occurs in two steps, which could be either
   A. (1) 2 NO (g) + H$_2$ (g) $\longrightarrow$ N$_2$ (g) + H$_2$O$_2$ (l)
       (2) H$_2$O$_2$ (l) + H$_2$ (g) $\longrightarrow$ 2 H$_2$O (l)
   or
   B. (1) 2 NO (g) + H$_2$ (g) $\longrightarrow$ N$_2$O (g) + H$_2$O (l)
       (2) N$_2$O (g) + H$_2$ (g) $\longrightarrow$ N$_2$ (g) + H$_2$O (l).

   a. Processes A and B are called two possible _____ of the reaction.
   b. Scientists can decide whether A or B is correct by _____.

16. Propose a possible two-step mechanism for the decomposition of HgO according to the equation
   2 HgO (s) $\longrightarrow$ 2 Hg (l) + O$_2$ (g).

17. A scientist postulates that two possible rate laws for the reaction given in question 20 are as follows:
   Rate = $k[NO]^2[H_2]$
   Rate = $k[NO]^2[H_2]^2$
   a. What is k called?
   b. How could you tell which rate law is correct?

363

14A Theories of Chemical Equilibrium page 365
14B Applications of Equilibrium Chemistry page 377
FACETS: Scientists and Their Faith page 375

Chemical Equilibrium 14

Constant Exchange Without Any Change

Chemical reactions transform reactants into products. Warning: This simple statement may be misleading! When most students think about chemical reactions, they think about reactions that go to completion—that is, reactions in which the reactant or reactants are completely changed to produce one or more products. Students often think that all reactions are **irreversible** reactions. In reality, many natural processes are **reversible** reactions. While products form, a portion of those products simultaneously revert into reactants. Having forward and reverse reactions going at the same time means that

1. reactants and products will always be mixed together;
2. the ratio in which reactants and products are mixed depends on the speed of each reaction;
3. changing the rate of one reaction changes the ratio of products and reactants in the reaction mixture.

14A Theories of Chemical Equilibrium

A chemical equilibrium results when forward and reverse reactions proceed simultaneously. This idea ranks as one of the premier concepts in chemistry. Like the kinetic theory and the atomic theory, it helps to explain many observations. While the study of thermodynamics tries to answer the question "Can the reaction occur?" and kinetics seeks to determine "How fast will the reaction occur?" the study of equilibria* seeks to answer "How far will the reaction go?"

Equilibria is the plural form of *equilibrium*.

Dynamic Equilibrium: Opposing Processes

The term *equilibrium* often evokes images of fixed or static situations such as the balance of forces in a stalemated tug-of-war or a balanced scale. Yet these examples describe only one type of equilibrium: the static equilibrium of balanced physical forces. In chemistry the term *equilibrium* refers to balanced changes, not balanced forces.

A **chemical equilibrium** exists when two opposing reactions occur simultaneously at the same rate. Unlike the previous examples of physical equilibria, chemical equilibria are *dynamic*. Particles constantly move and react. Microscopic activity continues at a hectic pace even though macroscopic changes have ceased.

Unequal forces have upset this physical equilibrium.

Consider the reaction $A + B \longrightarrow C + D$. A, B, C, and D represent particles of the reactants and products. As particles of A and B mix, they collide and produce some C and D particles. As particles of C and D accumulate, the chances that they will collide with each other increase. Some of these collisions result in the re-formation of A and B particles. As more C and D particles accumulate, more A and B particles re-form.

$A + B \longrightarrow C + D$; the formation of C and D particles
$C + D \longrightarrow A + B$; the formation of A and B particles

Eventually the two reactions will proceed at the same rate. To an observer it will seem that the amounts of A, B, C, and D remain constant; yet at the molecular level, the two opposing processes continue. The total system is designated by an equation with a special equilibrium sign.

$$A + B \rightleftharpoons C + D$$

Dynamic equilibria have been presented in earlier chapters. Water molecules in a sealed jar continue to evaporate even after the air is completely saturated. The vapor molecules condense at the surface of the water at the same rate that others evaporate.

$$\text{Liquid} \rightleftharpoons \text{Vapor}$$

Dissolving processes also reach dynamic equilibria at their saturation points. Particles leave and rejoin an undissolved solid at the bottom of the solution at equal but opposing rates.

$$\text{Solid particles} \rightleftharpoons \text{Dissolved particles}$$

Chemical equilibria are vitally important. Many industrial processes are made profitable through the control of reversible reactions. However, before discussing applications of equilibrium chemistry, we must study the way in which chemical equilibria are measured and influenced by reaction conditions.

Shifted Equilibria: Which Way Did that Reaction Go?

Some, but not many, equilibria contain a 50:50 mixture of reactants and products. Ratios of 60:40, 98:2, 1:99, or even 0.001:99.999 are possible. The ratios between products and reactants reflect the speed at which the forward and reverse reactions can go. A story illustration may help explain how equal but opposing rates can allow the substances on one side of an equilibrium to outnumber substances on the other side.

Bob and Betty, two of Ace Catering Service's finest employees, had received the distasteful assignment of serving punch at Mrs. Tisdale's garden party. The party itself was not drudgery; it was the 95° weather that made the job so miserable.

The scorching sun soon made both servers wish for a cool spot in the shade. Suddenly Bob speculated that if he emptied his punch bowl into Betty's bowl, he would be free for a long-desired rest. He grabbed a large ladle and began scooping his punch into Betty's bowl. Realizing what Bob was doing, Betty began to scoop her punch into Bob's bowl. Unfortunately all Betty had was a serving spoon. Soon Bob's greater rate of transfer allowed him to scoop most of his punch into Betty's bowl. Betty worked furiously to keep up, but her rate of transfer was much slower than Bob's.

Just as Bob was feeling confident of victory, he noticed that his large scoop was not working as well as it had before. He could not get a full scoop when there was only a small amount of punch in the bottom of his bowl. Soon Bob could scoop up only what Betty gave him with her serving spoon. At that point, their two rates of transfer were equal. Realizing that they had set up a *dynamic equilibrium*, the two called it a draw and went back to serving punch for Mrs. Tisdale.

Equal but opposing rates resulted in a dynamic equilibrium that favored the transfer of punch to Betty's bowl.

Figure 14A-1

Punch bowl equilibrium

This equilibrium is said to favor the forward (to the right) reaction.

$$\text{Punch in Bob's bowl} \rightleftharpoons \text{Punch in Betty's bowl}$$

A reaction that favors the formation of products is said to be shifted toward the right and is represented in the chemical equation with a long half-arrow pointing to the right. Conversely, a reaction that favors the formation of reactants is shifted to the left and is represented by a long half-arrow pointing to the left.

A spiritual "equilibrium" exists between the old nature and the new nature in a Christian (old nature $\rightleftharpoons$ new nature). Each Christian possesses an old and a new nature. The strengths of the two natures constantly change throughout a Christian's walk with God. The equilibrium can shift to the old nature if the Christian practices sin and neglects to fellowship with God. The equilibrium can shift to the new nature if the Christian stays in fellowship with God and avoids sin.

Equilibrium Constants: Describing Mixtures with Numbers

Chemists must often describe an equilibrium in terms of the ratio of products to reactants. To do this, they use a number called an equilibrium constant. Consider the reaction $A + B \rightleftharpoons C + D$. Suppose that experimentation shows that the rate law for the forward reaction (formation of products) is $R_f = k_f[A][B]$ and that the rate law for the reverse reaction (formation of reactants) is $R_r = k_r[C][D]$. Square brackets signify concentration in units of moles per liter, or molarity. By definition, chemical equilibrium is attained when the forward reaction rate matches the reverse reaction rate ($R_f = R_r$). As a result, the two rate laws can be equated.

$$R_f = R_r$$
$$k_f[A][B] = k_r[C][D]$$

A few algebraic operations can rearrange this equation to a new form.

$$\frac{k_f}{k_r} = \frac{[C][D]}{[A][B]}$$

Both k_f and k_r are constants. Therefore, their quotient is also a constant. This quotient of rate constants is called an **equilibrium constant** and is denoted by an uppercase K.

$$\frac{k_f}{k_r} = K = \frac{[C][D]}{[A][B]}$$

An equilibrium constant tells the ratio of products to reactants. If the value of K is large, the concentrations of products (in the numerator) are greater than the concentrations of the reactants (in the denominator). If the value of K is small, the reverse reaction predominates and the concentration of products is small compared to the concentration of reactants.

The form of an equilibrium constant is determined from the stoichiometry of the reaction. The concentration of each substance is raised to a power that matches the coefficient of that substance in a balanced equation.

$$aA + bB \rightleftharpoons cC + dD$$

$$K = \frac{[C]^c[D]^d}{[A]^a[B]^b}$$

In the equation $I_2(g) + H_2(g) \rightleftharpoons 2\,HI$, the relationship for the equilibrium constant would be written as follows:

$$K = \frac{[HI]^2}{[I_2][H_2]}$$

Sample Problem Write the equilibrium constant for the formation of ammonia from its elements. The reaction is

$$3\,H_2 + N_2 \rightleftharpoons 2\,NH_3$$

Solution

$$K = \frac{[NH_3]^2}{[H_2]^3[N_2]}$$

Writing equilibrium constants for reactions that involve solids or solvents requires an additional bit of knowledge. The concentrations of solids and solvents are defined to be 1; therefore, these substances do not affect equilibrium constants. For instance, the equilibrium between solid silver chloride and its dissolved ions in a saturated solution is not affected by the amount of undissolved solid present.

$$AgCl\,(s) \rightleftharpoons Ag^+\,(aq) + Cl^-\,(aq)$$

The equilibrium constant for this reaction would normally include the AgCl in the denominator, but since the concentration of the solid, [AgCl], is defined to be 1, it drops out of the expression. This seemingly arbitrary mathematics makes sense because the amount of dissolved silver chloride that remains at the bottom of a saturated solution does not affect the equilibrium. The concentration of dissolved ions would remain the same even if an entire handful of silver chloride were added.

$$K = \frac{[Ag^+][Cl^-]}{[AgCl]}$$

$$K = [Ag^+][Cl^-]$$

When acetic acid molecules mix with water, some of them release hydrogen ions. The reaction adds hydrogen ions to some water molecules. Since only an insignificant amount of water is used up in the reaction, the concentration of water molecules is defined to be 1. Although the equilibrium constant could include the concentration of water, it should be left out.

$$HC_2H_3O_2 \,(l) + H_2O \rightleftharpoons C_2H_3O_2^- \,(aq) + H_3O^+ \,(aq)$$

$$K = \frac{[C_2H_3O_2^-][H_3O^+]}{[HC_2H_3O]}$$

Sample Problem Write the equation for the equilibrium constant for the precipitation of zinc hydroxide.

$$Zn^{2+} \,(aq) + 2\, OH^- \rightleftharpoons Zn(OH)_2 \,(s)$$

Solution
Since zinc hydroxide is a solid, its concentration is 1 in the equation for the equilibrium constant.

$$K = \frac{1}{[Zn^{2+}][OH^-]^2}$$

The numerical value of an equilibrium constant must be determined experimentally. Suppose that sulfur dioxide and oxygen are placed in a container and allowed to react to form sulfur trioxide. After the reactants reach equilibrium with the sulfur trioxide, the various concentrations are measured. It is found that 1 mole of sulfur trioxide, 4 moles of sulfur dioxide, and 3.5 moles of oxygen are present.

$$2\, SO_2 + O_2 \rightleftharpoons 2\, SO_3$$

$$K = \frac{[SO_3]^2}{[SO_2]^2[O_2]}$$

$$K = \frac{[1]^2}{[4]^2[3.5]}$$

$$K = 0.02$$

The equilibrium constant 0.02 describes the ratio between products and reactants under a specified temperature and pressure.

Henri Le Châtelier

Sample Problem Calculate the numerical value of K for the equilibrium involved in the formation of ammonia from nitrogen and hydrogen. When the mixture of the gases reaches equilibrium, $[H_2] = 0.05\ M$, $[N_2] = 0.05\ M$, and $[NH_3] = 50\ M$.

$$N_2\ (g) + 3\ H_2\ (g) \rightleftharpoons 2\ NH_3\ (g)$$

Solution
Substitute the given concentration values into the equilibrium constant equation, and solve for K.

$$K = \frac{[NH_3]^2}{[H_2]^3[N_2]}$$

$$K = \frac{[50]^2}{[0.05]^3[0.05]}$$

$$K = 4 \times 10^8$$

Le Châtelier's Principle: How Equilibria Handle Stress

In 1884 a French chemist named Henri (awn REE) Le Châtelier (luh shah tel YAY) questioned how equilibria behave when they are disturbed by an external change. The answer that he found has served the field of chemistry so well that it has been named **Le Châtelier's principle** in his honor. This principle says that when a reversible process is disturbed, it will proceed in the direction that relieves the stress.

Le Châtelier's Principle can be applied to some real life situations, such as the teenage "clean-up-your-room" syndrome:

Clean Room $\rightleftharpoons$ Messy Room

What are the observable properties of the usual teenager's room? One quantitative measure of messiness would be the ratio

Figure 14A-2

Sebastiano del Piombo, *Christ Bearing the Cross*, The Bob Jones University Collection

of the mass of clothes on the floor to the mass still hanging in the closet. Stress can be placed on this system by parents in the terms of an ultimatum: "Clean up your room or you are grounded." This will shift the equilibrium far to the left.

Most people operate according to Le Châtelier's principle—that is, when a stress is placed on us, we attempt to go in a direction that will relieve the stress. Although this is basic human nature, it often is not God's will. Writing about the heroes of the faith, the author of Hebrews in chapter 11, verse 35, tells of those who did not "relieve the stress," but continued to follow in God's will. What if Christ had acted according to Le Châtelier's principle? He would never have suffered the shame nor the humiliation that He endured. Yet, Christ prayed, "nevertheless not my will, but thine, be done," and He willingly went to the cross for you and for me (Luke 22:42).

The Effect of Concentration

Before applying Le Châtelier's principle to chemical reactions, consider how it would work in the story of the two punch bowls. Go back to the point where Bob had put most of the punch into Betty's bowl and had set up the dynamic equilibrium.

$$\text{Punch in Bob's bowl} \rightleftharpoons \text{Punch in Betty's bowl}$$

This equilibrium could be disturbed in several ways. First, if someone added a large amount of punch to Bob's bowl, Bob's rate of transfer could increase. The large scoop would quickly convey the extra punch over to Betty's bowl until an equilibrium was reestablished. Second, if Betty disturbed the equilibrium by doubling the number of times she scooped up punch, the equilibrium would temporarily shift to the left.

Figure 14A-3

Original equilibrium

Equilibrium disturbed

Equilibrium reestablished

Additional SO_2 causes the equilibrium to shift forward.

Changes in concentration definitely affect chemical equilibria. Return to the $2\ SO_2 + O_2 \rightleftharpoons 2\ SO_3$ equilibrium. At one set of conditions, the K has been measured to be 0.02.

$$K = \frac{[SO_3]^2}{[SO_2]^2[O_2]} = 0.02$$

Suppose that more sulfur dioxide were added to the container. The system would be temporarily disturbed from equilibrium. (Just as the addition of more punch to Bob's bowl temporarily disturbed that equilibrium.) Le Châtelier's principle predicts that the forward reaction would proceed to reduce the extra sulfur dioxide—to reduce the stress of the additional sulfur dioxide. This reaction would increase the concentration of sulfur trioxide but decrease the concentration of oxygen. Despite all the changes, the original constant would remain unchanged.

If sulfur trioxide were added, the reverse reaction would work to dissipate the added substance. If some sulfur trioxide were removed, the forward reaction would work to restore the lost amount. Of course, the concentrations of sulfur dioxide and oxygen would decrease when the forward reaction predominated. This action would keep the value of the equilibrium constant the same.

The Effect of Pressure

Some, but not all, gaseous reactions are affected by changes in pressure. Recall that 1 mole of a gaseous substance occupies 22.4 liters at STP. In the equilibrium $2\ SO_2 + O_2 \rightleftharpoons 2\ SO_3$, 2 moles of sulfur trioxide are produced for every 3 moles of gaseous reactants. Since there are fewer gas molecules in the product than in the reactants, the forward reaction has the effect of decreasing the pressure.

Suppose that some additional pressures were placed on the equilibrium. Le Châtelier's principle predicts that the reaction that relieves the stress will predominate. The forward reaction would lower the pressure by producing 2 moles of sulfur trioxide from 2 moles of sulfur dioxide and 1 mole of oxygen. If the equilibrium were depressurized, the reverse reaction would act to fill the void—more sulfur dioxide and oxygen would be produced.

Some gaseous reactions are not affected by pressure changes. $H_2 + I_2 \rightleftharpoons 2\ HI$ has two moles of gas in the products for every two moles of gas in the reactants. The pressure of the entire system does not change with either the forward or the reverse reaction. Extra pressure on this equilibrium will have no effect on the direction of the reaction.

Figure 14A-4

$3\ H_2 + N_2 \rightleftharpoons 2\ NH_3$.
Additional pressure shifts the equilibrium forward.

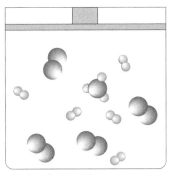

Equilibrium at low pressure

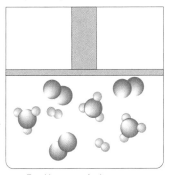

Equilibrium at high pressure

Sample Problem Will the formation of ammonia be helped or hindered by high pressures?

$$3 H_2 (g) + N_2 (g) \rightleftharpoons 2 NH_3 (g)$$

Solution
The balanced equation shows that four moles of gaseous reactants combine to form two moles of gaseous products. High pressures cause the reaction that produces the smallest volume to predominate. Thus high pressures aid the formation of ammonia gas.

Figure 14A-5

$3 H_2 + N_2 \rightleftharpoons 2 NH_3$. Additional heat causes a shift to the left.

Equilibrium at low temperature

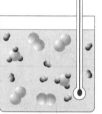

Equilibrium at high temperature

The Effect of Temperature

Heat may be thought of as a participant in chemical reactions. The exothermic reaction between sulfur dioxide and oxygen produces sulfur trioxide and heat. If heat is applied to this reaction, the reaction will shift to the left.

$$2 SO_2 + O_2 \rightleftharpoons SO_3 + \text{Heat}$$

Add heat

Le Châtelier's principle predicts how temperatures affect each of these equilibria. When heat acts as a reactant in an endothermic reaction, it helps the forward reaction. The equilibrium of hydrogen, iodine, and hydrogen iodide will shift to produce more hydrogen iodide if heat is added to the reaction.

$$H_2 + I_2 + \text{Heat} \rightleftharpoons 2 HI$$

Add heat

Sample Problem The formation of ammonia is exothermic. Will high temperatures help or hinder the forward reaction?

$$3 H_2 + N_2 \rightleftharpoons 2 NH_3 + \text{Heat}$$

Solution
Since heat is one of the products, additional heat would encourage the reverse reaction and would hinder the forward reaction. Again, when heat is added or removed, the value of the equilibrium constant will change. This is why the temperature and pressure of the system must always be specified when the value of K is considered.

FACETS of CHEMISTRY: Scientists and Their Faith

Preparing the Mars probe

The popular image of scientists as totally objective, unbiased information processors is a myth. All scientists, no matter how objective or unbiased they are, can be influenced by something that has nothing to do with scientific evidences. What could this influence be? The beliefs that scientists have about the universe, God, and man will invariably affect their work. Simply put, personal philosophy guides scientific work.

Pagan men who worship temperamental, whimsical spirits have little reason to study science. Their beliefs serve as handy explanations for natural events that they do not understand. When pagans observe planets that "wander" through the sky, see comets and meteors that arrive unannounced, and endure natural calamities that seem to strike without warning, they simply assume that evil spirits are sending omens.

The lives of the men who founded modern science illustrate the link between personal beliefs and scientific work. Many of the early "giants of science" were scientists who believed in God. Many, in fact, were Christians. Francis Bacon, the seventeenth-century philosopher who pioneered the scientific method, knew that the Bible was the main source of knowledge for the Christian. "There are two books laid before us to prevent our falling into error," he stated; "first, the volume of the Scriptures, which reveal the will of God; then the volume of the Creatures, which express His power." Robert Boyle, often called the Father of Chemistry, was a humble Christian and a diligent student of the Bible. Lord Kelvin, noted among other things for establishing the Kelvin temperature scale, was a dedicated Christian and a staunch creationist who frequently did battle with the evolutionists of his day. "If you think strongly enough," he said, "you will be forced by science to the belief in God."

Bacon

Boyle

Kelvin

A faith in God showed these men that the universe must be orderly and knowable—like its Creator. Their faith convinced them that it would be possible to find out how the universe worked. They believed that if they looked hard enough, gathered the right data, and arranged their information correctly, they could understand the universe. Many of today's scientists no longer believe that the universe was created. Do these men perform their scientific investigations without being influenced by their beliefs? Hardly. Many present-day scientists have substituted a faith in evolution, humanism, or science for a faith in God.

An example of how evolutionary beliefs guide scientific activities is seen in the ongoing searches of outer space for signs of intelligent life. In 1982 Congress authorized 1.5 million dollars for studies that would explore new worlds and seek out new civilizations. At this moment, space probes search for evolving life forms on other planets, and radio telescopes listen for signals from supercivilizations in other parts of the galaxy. Throughout the 1990s, probes traveled to Venus, Saturn, and Jupiter. Hundreds of millions of dollars have been spent to seek out evidences of life on Mars. By traveling to other planets and exploring space, scientists hope to unlock the secrets of the origin of the universe and evidences of the development of man. The men who conduct such searches are guided by their faith. Their faith in evolution leads them to search for evidences of evolution.

People who do not see the link between personal beliefs and scientific practice may interpret the antagonism of many scientists toward God as a conflict between science and the Bible. People begin to wonder whether science and the Bible conflict. Regrettably, most scientists oppose the truth of God. But the battle between scientists and the Bible stems from personal beliefs, not scientific evidence. True science and the Bible are always in perfect agreement.

The Effect of Catalysts

As discussed in Chapter 8, catalysts are substances that change the rate of a reaction without undergoing any change themselves. A catalyst can affect either the forward or the reverse reaction, causing the equilibrium to be reached sooner than if no catalyst were added. Although it causes equilibrium to be reached more quickly, the addition of a catalyst has no effect on equilibrium concentrations, or on the value of K.

Table 14A-1 summarizes the effects on equilibrium reactions and K by changes in pressure, concentration, temperature, and addition of catalysts.

Table 14A-1 Effects of Stress on Equilibrium Concentrations and K

Stress	Shift	Effect	Value of K
Add catalyst	none	none	no change
Concentration increase			
of reactant	toward right	more products form	no change
of product	toward left	more reactants form	no change
Concentration decrease			
of reactant	toward left	more reactants form	no change
of product	toward right	more products form	no change
Pressure change (only for unequal numbers of moles of gaseous products and reactants)			
increase	toward side having smaller number of gaseous molecules	---	no change
decrease	toward side having larger number of gaseous molecules	---	no change
Temperature increase			
exothermic	toward left	more reactants form	decreases
endothermic	toward right	more products form	increases
Temperature decrease			
exothermic	toward right	more products form	increases
endothermic	toward left	more reactants form	decreases

Section Review Questions 14A

1. What generalizations about an equilibrium constant can be made if the value for K is large? If the value is small?
2. How can the chemical manufacturing industry utilize Le Châtelier's principle?

14B Applications of Equilibrium Chemistry

Chemical industries supply society with such diverse products as medicines, pesticides, fertilizers, paints, textiles, detergents, cosmetics, plastics, fuels, and building materials. Because the molecular transformations involved are usually reversible, the laws of equilibrium chemistry govern the processes. Methods of describing and controlling equilibria become indispensable tools in the fast, efficient production of consumer goods.

Equilibrium chemistry plays a vital role in manufacturing.

Equilibria Between Molecules: Milking Reactions for All They Are Worth

The characteristic flavors and odors of fruits are the results of naturally occurring compounds called *esters*. Manufacturers of artificial flavorings seek to form esters by reacting alcohols and organic acids.

$$\text{Alcohol} + \text{Organic Acid} \rightleftharpoons \text{Ester} + \text{Water}$$

The reaction works well, but it is very easily reversed. Neither the forward nor the reverse reaction dominates. This situation presents several obstacles to the efficient production of esters. Instead of going to completion, the reaction converts only a part of the reactants into the products. Rather than accept the reduced yields, manufacturers produce more esters per batch of reactants by forcing the forward reaction to prevail.

Manufacturers can shift the equilibrium by continually adding a reactant, removing a product, or both. If either the acid or the alcohol is inexpensive, huge quantities can be added to drive the equilibrium forward. Sometimes manufacturers can easily remove a product by boiling it off as it forms. This also results in a shifted equilibrium. Regardless of the technique that is used, the result is increased ester production.

Huge quantities of ammonia are used every year in fertilizers, cleaning compounds, and explosives. Until World War I, industrial countries obtained nitrogen atoms by importing mined saltpeter (KNO_3 and $NaNO_3$). All countries had access to nitrogen molecules in the atmosphere, but no one knew how to liberate the atoms from the stable molecules. A way was sought to form ammonia from atmospheric nitrogen.

Finding the appropriate reaction was no problem: $N_2 + 3\ H_2 \rightleftharpoons 2\ NH_3$ seemed perfect. The huge equilibrium constant of this reaction (4.0×10^8 at 25°C) indicated that the production of ammonia was the favored direction of the reaction. The problem was that nitrogen, hydrogen, and ammonia took a very long time to reach equilibrium. A few years before World War I, a German chemist named Fritz Haber (1868-1934) found a way to manipulate the equilibrium to overcome this difficulty. He invented the **Haber** (HAH ber) **process** that supplied Germany's war machine

Figure 14B-1 ▪ The Haber Process

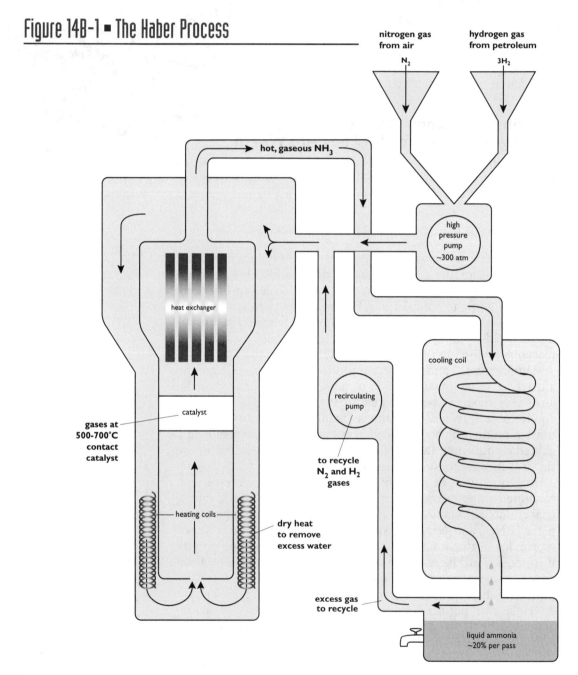

Applications of Equilibrium Chemistry

Nitrogen-based fertilizers are vital to increased farm production.

with ammonia-based explosives. Today the Haber process provides ammonia for use in numerous industries.

$$N_2 + 3\,H_2 \rightleftharpoons 2\,NH_3 + \text{Heat}$$

Because 2 moles of gas are formed from 4 moles of gas, the products take up less space than the reactants. Le Châtelier's principle predicts that high pressures favor the production of ammonia. Because the forward reaction is exothermic, it proceeds best at low temperatures. Thus, a process that utilized high pressure and low temperatures was needed to facilitate the manufacturing of ammonia.

In the Haber process the reaction is performed under high pressures (600-1000 atm), at high temperatures (450-600°C), and with a catalyst. High temperatures shift the equilibrium slightly away from the formation of ammonia. To compensate for this, the Haber process manipulates other variables that affect the reaction to increase the production of ammonia. A catalyst (mostly Fe_3O_4 with traces of K_2O and Al_2O_3) helps to speed up the reaction. To further increase the yield, ammonia is removed by liquefaction as it forms. This continual removal keeps the forward reaction going.

Equilibria Between Ions: Writing a K_{sp}

Saturated solutions have equilibria between undissolved solids and aqueous ions. Some salts, such as calcium sulfate, are slightly soluble. Others, such as lead phosphate, dissolve so little that for all practical purposes they are insoluble.

$$CaSO_4\,(s) \rightleftharpoons Ca^{2+}\,(aq) + SO_4^{2-}\,(aq)$$
$$Pb_3(PO_4)_2\,(s) \rightleftharpoons 3\,Pb^{2+}\,(aq) + 2\,PO_4^{3-}\,(aq)$$

The forward reaction is dissolution, and the reverse reaction is precipitation. As with all equilibria, the rate of the forward reaction matches the rate of the reverse reaction. The relative amounts of dissolved ions and undissolved solids can be expressed with equilibrium constants. The equilibrium constant for the dissolution of $Pb_3(PO_4)_2$ is expressed as

$$K = [Pb^{2+}]^3[PO_4^{3-}]^2$$

As in other equilibrium expressions, solids are left out; thus the $[Pb_3(PO_4)_2]$ term is omitted in the expression of the equilibrium constant. The resulting solubility equilibrium constant is given the special name **solubility product constant (K_{sp})**. The concentrations of the ions are raised to the appropriate powers as in other equilibrium constant expressions.

$$K_{sp} = [Pb^{2+}]^3[PO_4^{3-}]^2$$

Table 14B-1 K_{sp} for Minimally Soluble Substances

Salt	Product	K_{sp} at 25°C	g/formula unit	Solubility g/L
Silver chloride, AgCl	$[Ag^+][Cl^-]$	1.8×10^{-10}	143.32	1.9×10^{-3}
Silver carbonate, Ag_2CO_3	$[Ag^+]^2[CO_3^{2-}]$	8.2×10^{-12}	275.75	3.5×10^{-2}
Silver phosphate, Ag_3PO_4	$[Ag^+]^3[PO_4^{3-}]$	1.3×10^{-20}	418.58	2.0×10^{-3}
Silver sulfide, Ag_2S	$[Ag^+]^2[S^{2-}]$	6.0×10^{-50}	247.80	6.1×10^{-15}
Aluminum hydroxide, $Al(OH)_3$	$[Al^{3+}][OH^-]^3$	2.0×10^{-32}	78.00	4.1×10^{-7}
Barium carbonate, $BaCO_3$	$[Ba^{2+}][CO_3^{2-}]$	5.1×10^{-9}	197.35	1.4×10^{-2}
Barium sulfate, $BaSO_4$	$[Ba^{2+}][SO_4^{2-}]$	1.3×10^{-10}	233.40	2.7×10^{-3}
Bismuth sulfide, Bi_2S_3	$[Bi^{3+}]^2[S^{2-}]^3$	1.0×10^{-97}	514.15	8.2×10^{-18}
Calcium carbonate, $CaCO_3$	$[Ca^{2+}][CO_3^{2-}]$	4.8×10^{-9}	100.09	6.9×10^{-3}
Calcium sulfate, $CaSO_4$	$[Ca^{2+}][SO_4^{2-}]$	1.9×10^{-4}	136.14	1.9
Cobalt (II) sulfide, CoS	$[Co^{2+}][S^{2-}]$	5.0×10^{-22}	91.00	2.0×10^{-9}
Copper (II) sulfide, CuS	$[Cu^{2+}][S^{2-}]$	6.0×10^{-36}	95.60	2.3×10^{-16}
Iron (III) hydroxide, $Fe(OH)_3$	$[Fe^{3+}][OH^-]^3$	4.0×10^{-38}	106.87	2.1×10^{-8}
Iron (II) hydroxide, $Fe(OH)_2$	$[Fe^{2+}][OH^-]^2$	8.0×10^{-16}	89.86	5.3×10^{-4}
Iron (II) sulfide, FeS	$[Fe^{2+}][S^{2-}]$	6.0×10^{-18}	87.91	2.2×10^{-7}
Magnesium carbonate, $MgCO_3$	$[Mg^{2+}][CO_3^{2-}]$	1.0×10^{-5}	84.32	2.7×10^{-1}
Magnesium hydroxide, $Mg(OH)_2$	$[Mg^{2+}][OH^-]^2$	1.8×10^{-11}	58.33	9.6×10^{-3}
Lead (II) carbonate, $PbCO_3$	$[Pb^{2+}][CO_3^{2-}]$	3.3×10^{-14}	267.20	4.9×10^{-5}
Lead (II) phosphate, $Pb_3(PO_4)_2$	$[Pb^{2+}]^3[PO_4^{3-}]^2$	3.0×10^{-44}	811.51	6.3×10^{-7}
Lead (II) sulfide, PbS	$[Pb^{2+}][S^{2-}]$	1.0×10^{-28}	239.25	2.4×10^{-12}
Lead (II) sulfate, $PbSO_4$	$[Pb^{2+}][SO_4^{2-}]$	1.6×10^{-8}	303.25	3.8×10^{-2}
Tin (II) sulfide, SnS	$[Sn^{2+}][S^{2-}]$	1.0×10^{-25}	150.75	4.8×10^{-11}
Zinc hydroxide, $Zn(OH)_2$	$[Zn^{2+}][OH^-]^2$	1.2×10^{-17}	99.38	1.4×10^{-4}
Zinc sulfide, ZnS	$[Zn^{2+}][S^{2-}]$	2.0×10^{-24}	97.43	1.4×10^{-10}

Solubilities and Equilibria: K_{sp} Is a Guide

The K_{sp} provides a description of salt solubilities. Large numbers of ions in solution will cause the solubility product to be relatively large (much greater than 1). Small numbers of ions in solution cause the K_{sp} to be small (less than 1). Thus, the larger the K_{sp}, the more soluble the salt, and conversely, the smaller the K_{sp}, the less soluble the salt. The K_{sp} of some slightly soluble salts are listed in Table 14B-1.

K_{sp} values can be determined from measured solubilities. Calcium sulfate has a solubility of 1.4×10^{-2} moles/liter. When calcium sulfate dissolves, each formula unit releases 1 Ca^{2+} ion and 1 SO_4^{2-} ion.

$$1.4 \times 10^{-2} \text{ mol } CaSO_4 \rightleftharpoons 1.4 \times 10^{-2} \text{ mol } Ca^{2+} + 1.4 \times 10^{-2} \text{ mol } SO_4^{2-}$$

$$K_{sp} = [Ca^{2+}][SO_4^{2-}]$$
$$K_{sp} = (1.4 \times 10^{-2})(1.4 \times 10^{-2})$$
$$K_{sp} = 2.0 \times 10^{-4}$$

If the K_{sp} of a substance is known, it can be used to determine the solubility in units of moles per liter. The K_{sp} of lead (II) phosphate is known to be 3.0×10^{-44}. The equilibrium for its dissociation can be written as follows:

$$Pb_3(PO_4)_2 \rightleftharpoons 3\ Pb^{2+} + 2\ PO_4^{3-}$$

Thus,

$$K_{sp} = [Pb^{2+}]^3[PO_4^{3-}]^2$$
$$3.0 \times 10^{-44} = [Pb^{2+}]^3[PO_4^{3-}]^2$$

Notice that when a mole of lead (II) phosphate dissolves, 3 moles of Pb^{2+} ions and 2 moles of PO_4^{3-} ions are released. The number of Pb^{2+} ions in a saturated solution must be three times the amount of $Pb_3(PO_4)_2$ units that were dissolved. Likewise, there must be twice as many PO_4^{3-} ions as the number of dissolved formula units. The concentration of the Pb^{2+} ions can be expressed as $3S$, where S stands for the solubility of the salt, in this case Pb_3PO_4. The concentration of PO_4^{3-} ions can be expressed as $2S$. The concentrations of the ions can now be expressed in terms of the solubility of the salt.

$$K_{sp} = [Pb^{2+}]^3[PO_4^{3-}]^2$$
$$K_{sp} = (3S)^3(2S)^2$$

To find the solubility of the salt, substitute the value of the K_{sp} and solve for S.

$$3.0 \times 10^{-44} = (3S)^3(2S)^2$$
$$3.0 \times 10^{-44} = 27S^3 \times 4S^2$$
$$3.0 \times 10^{-44} = 108S^5$$

Take the fifth root of both sides to obtain $7.7 \times 10^{-10} = S$

The solubility of lead (II) phosphate is 7.7×10^{-10} moles/liter. This amount is so small that the salt is, for all practical purposes, insoluble.

Sample Problem Given that the K_{sp} of silver chloride is 1.8×10^{-10}, determine the solubility (mol/L) of silver chloride.

Solution
$$AgCl \rightleftharpoons Ag^+ + Cl^-$$
$$K_{sp} = 1.8 \times 10^{-10}$$
$$K_{sp} = [Ag^+][Cl^-]$$

Since each silver chloride unit releases one Ag^+ ion and one Cl^- ion, the $[Ag^+]$ and the $[Cl^-]$ equal the solubility of the salt (S) in a saturated solution.

$$K_{sp} = [S][S]$$
$$1.8 \times 10^{-10} = S^2$$
$$1.3 \times 10^{-5} = S$$

Thus, 1.3×10^{-5} mol of silver chloride can be dissolved in 1 L of water.

The K_{sp} provides a quick way to see whether a solution is unsaturated, saturated, or supersaturated. If the product of ion concentrations equals the K_{sp}, the solution must be saturated. If the product of ion concentrations is less than the K_{sp}, the solution is unsaturated. A product greater than the K_{sp} indicates a supersaturated solution.

Sample Problem Suppose that 2.0×10^{-6} mol of zinc hydroxide, $Zn(OH)_2$, are dissolved in a mole of water. Use the K_{sp} of zinc hydroxide to determine whether the solution is unsaturated, saturated, or supersaturated.

Solution
The equilibrium is $Zn(OH)_2 \rightleftharpoons Zn^{2+} + 2\ OH^-$. When dissolved, 2.0×10^{-6} mol of zinc hydroxide produces 2×10^{-6} mol of Zn^{2+} ions and $2 (2 \times 10^{-6})$ mol of OH^- ions. The product of these concentrations is

$$K_{sp} = [Zn^{2+}][OH^-]^2$$
$$= [2.0 \times 10^{-6}][2(2 \times 10^{-6})]^2$$
$$= 3.2 \times 10^{-17}$$

This product is greater than the K_{sp} of 1.2×10^{-17}, so the solution must be supersaturated.

Common-Ion Effect: An Overpopulation Problem

Picture an unsaturated silver chloride solution. Less than the allowed 1.34×10^{-5} mol/L has dissolved, and the product of the silver- and chloride-ion concentrations is less than the K_{sp}. Now imagine what happens when a handful of sodium chloride is added to the solution.

Sodium chloride is very soluble (it does not appear on the list of slightly soluble salts). It dissolves and dissociates into sodium and chloride ions. The sodium ions have little or no effect on the original solution, but the chloride ions do. The chloride ion is called the common ion because it is contained in both silver chloride and sodium chloride. The additional chloride ions from the sodium chloride cause a shift to the left in the AgCl (s) $\rightleftharpoons$ Ag$^+$ (aq) + Cl$^-$ (aq) equilibrium to produce more solid silver chloride. The increased concentration of chloride ions raises the solubility product ($[Ag^+][Cl^-]$) above the K_{sp} value. The result of the **common-ion effect** is that the less soluble salt will precipitate out of the solution.

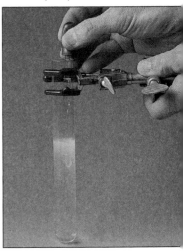

The addition of concentrated HCl to a supersaturated salt solution will cause NaCl to precipitate.

K_{sp} and Precipitation Reactions: Some Ions Just Cannot Stay Apart

Mixing two solutions can sometimes cause a precipitate to form. For instance, suppose that hydrochloric acid (HCl) and silver nitrate (AgNO$_3$) were mixed together. The resulting solution would contain H$^+$, Cl$^-$, Ag$^+$, and NO$_3^-$ ions. Hydrogen chloride and silver nitrate are relatively soluble. However, an alternate combination of ions, Ag$^+$Cl$^-$, is not. A silver chloride precipitate will form if enough of these ions are in the solution. How much is enough? That depends on the K_{sp} of silver chloride. If the value of $[Ag^+][Cl^-]$ in the solution exceeds the K_{sp} for silver chloride, a precipitate will form. How much precipitate will be deposited? Enough to lower the value of $[Ag^+][Cl^-]$ to the K_{sp} for silver chloride. Thus, the K_{sp} of a salt can be used to predict whether precipitation (a double replacement reaction) will occur, and how much precipitate will occur.

Sample Problem Will a precipitate form when 5.0×10^{-5} M barium nitrate—$Ba(NO_3)_2$—is mixed with 1.0×10^{-3} M sodium carbonate—Na_2CO_3?

Solution

Both barium nitrate and sodium carbonate are soluble salts. At this low concentration, they will both be completely dissolved. The first step is to calculate the theoretical concentration of each ion after both salts completely dissolve.

$$Ba(NO_3)_2 \longrightarrow Ba^{2+} + 2\ NO_3^-$$
$$[Ba^{2+}] = 5.0 \times 10^{-5}\ M$$
$$[NO_3^-] = 2(5.0 \times 10^{-5})\ M$$
$$Na_2CO_3 \longrightarrow 2\ Na^+ + CO_3^{2-}$$
$$[Na^+] = 2(1.0 \times 10^{-3})\ M$$
$$[CO_3^{2-}] = 1.0 \times 10^{-3}\ M$$

The next step is to check Table 14B-1 to determine which combination of ions is most likely to precipitate. Both original combinations are very soluble, and so is sodium nitrate. Barium carbonate is the only salt formed by a combination of the ions in this solution that appears in Table 14B-1 ($K_{sp} = 5.1 \times 10^{-9}$). The final step is to determine whether the value of $[Ba^{2+}][CO_3^{2-}]$ exceeds the given K_{sp}. In the mixture,

$$[Ba^{2+}][CO_3^{2-}] = (5.0 \times 10^{-5})(1.0 \times 10^{-3})$$
$$= 5.0 \times 10^{-8}$$

Since 5.0×10^{-8} is larger than the K_{sp} of 5.1×10^{-9}, a precipitate will form.

Section Review Questions 14B

1. An experimental solution has a K_{sp} value of 2.5×10^{-12}. That same solution has a known K_{sp} of 3.2×10^{-10}. Is the experimental solution saturated, unsaturated, or supersaturated? Explain your answer.
2. The solubility of copper (I) chloride is 1.08×10^{-2} g/mL solution.
    a. Write the K_{sp} expression for CuCl.
    b. Find the K_{sp} value.

Chapter Review

Coming to Terms

irreversible reaction
reversible reaction
chemical equilibrium
equilibrium constant
Le Châtelier's principle
Haber process
solubility product constant
common ion effect

Review Questions

1. What is the difference between a dynamic equilibrium and a static one?
2. When a bottle of soft drink is shaken violently, CO_2 gas escapes from solution and exerts increased pressure on the interior of the bottle. If the bottle is opened in this condition, the beverage will spurt out of the bottle. If the bottle is allowed to sit for a short amount of time, the interior pressure subsides. Why does the pressure decrease with time?
3. Write the equilibrium constant for each reaction.
   a. Laughing gas can decompose into nitrogen and oxygen:
      $2\ N_2O\ (g) \rightleftharpoons 2\ N_2\ (g) + O_2\ (g)$.
   b. Carbon monoxide can be converted into methane in the process of converting coal into a gas:
      $CO\ (g) + 3\ H_2\ (g) \rightleftharpoons CH_4\ (g) + H_2O\ (g)$.
   c. Methanol can be synthesized from carbon monoxide:
      $CO\ (g) + 2\ H_2\ (g) \rightleftharpoons CH_3OH\ (l)$.
   d. Baking soda can extinguish fires because it can decompose to produce water and carbon dioxide, both of which smother combustion: $2\ NaHCO_3\ (s) \rightleftharpoons Na_2CO_3\ (s) + H_2O\ (l) + CO_2\ (g)$.
4. A scientist does experiments to determine the equilibrium constant for the Haber process at 450°C.
   a. After permitting a reaction mixture of N_2 and H_2 to reach equilibrium, he finds that $[N_2]$ = 0.100, $[H_2]$ = 0.0300, and $[NH_3]$ = 0.000200. Calculate K at this temperature.

385

b. The scientist repeats the experiment under the same conditions with different amounts of gas in the reaction vessel. The scientist finds that $[H_2]$ = 0.0375 and $[NH_3]$ = 0.000318. Use the value of K already determined to calculate $[N_2]$.

5. What is the value of the equilibrium constant for the following reaction: $2\ NO\ (g) + O_2\ (g) \rightleftharpoons 2\ NO_2\ (g)$ if
   a. $[NO]$ = 0.890, $[O_2]$ = 0.250, and $[NO_2]$ = 0.0320?
   b. If under the same conditions $[NO]$ = 1.00 and $[O_2]$ = 0.500, what will be the concentration of NO_2 in a reaction vessel?

6. Acetic acid ($HC_2H_3O_2$) is the compound that gives vinegar its distinctive smell and taste. When dissolved in water, it can ionize.
$$HC_2H_3O_2\ (aq) + H_2O \rightleftharpoons C_2H_3O_2^- + H_3O^+\ (aq)$$
The equilibrium constant for this reaction is
$$K = \frac{[C_2H_3O_2^-][H_3O^+]}{[HC_2H_3O_2]}$$

   a. A chemist working in a clinical research lab dissolves some acetic acid in water (at 25°C) and finds that $[H_3O^+]$ = 1.01 × 10⁻⁵, $[C_2H_3O_2^-]$ = 1.01 × 10⁻⁵, and $[HC_2H_3O_2]$ = 5.67 × 10⁻⁶. What is the value of K?
   b. If $[H_3O^+]$ = 3.6 × 10⁻³ and $[C_2H_3O_2^-]$ = 3.6 × 10⁻³, what is the concentration of un-ionized acetic acid?
   c. What equation gives the equilibrium constant for the reverse reaction?
   d. What is the value of the equilibrium constant for the reverse reaction?

7. Oxalic acid, $H_2C_2O_4$, experiences a two-step dissociation in an aqueous solution: $H_2C_2O_4 \rightleftharpoons H^+ + HC_2O_4^-$ where K = 5.90 × 10⁻² at 25°C, and $HC_2O_4^- \rightleftharpoons H^+ + C_2O_4^{2-}$ where K = 6.40 × 10⁻⁵ at 25°C. If the equilibrium concentration of $[H^+]$ = 1.95 × 10⁻³ M and $[HC_2O_4^-]$ = 1.95 × 10⁻³, what are the equilibrium concentrations of $H_2C_2O_4$ and $C_2O_4^{2-}$? What is the equilibrium constant for the total reaction: $H_2C_2O_4 \rightleftharpoons 2\ H^+ + C_2O_4^{2-}$?

8. Name three factors that can shift equilibria.

9. Predict the effect of increasing the pressure on each of the four equilibria given in problem 3.

10. Lithium carbonate (Li_2CO_3) is less soluble in hot water than in cold.
    a. Do you think the dissolution of Li_2CO_3 is exothermic or endothermic? Explain.
    b. What will happen if a saturated solution of Li_2CO_3 at 25°C is heated to 100°C? Explain.

11. Write K_{sp} expressions for the dissolution of the following:
    a. $BaSO_4$
    b. MgF_2
    c. $Al(OH)_3$
    d. $Pb_3(PO_4)_2$
    e. $Mg(OH)_2$

12. A scientist measures the solubility of $Mg(OH)_2$. The concentration of the Mg^{2+} ions in a saturated solution at 18°C is 1.44×10^{-4} M, and the concentration of OH^- in the same solution is 2.88×10^{-4} M. What is the value of the K_{sp} at this temperature?

13. Using the solubilities and gram-formula masses in Table 14B-1, calculate the K_{sp} for the following:
    a. Ag_2CO_3
    b. $BaCO_3$
    c. Bi_2S_3
    d. $Fe(OH)_3$
    e. $PbSO_4$

    If the calculations are performed correctly, you should obtain close to the same K_{sp} value listed in the table.

14. A saturated solution of Li_2CO_3 has both Li^+ and CO_3^{2-} ions in it. What will happen if some solid LiCl is added to the solution? (LiCl is much more soluble than Li_2CO_3.)

15. Will a precipitate form if the following solutions are mixed? If so, what substances will precipitate? (Hint: First determine whether each substance is soluble or only minimally soluble. If a salt is minimally soluble, refer to its K_{sp}.)
    a. 0.05 M NaCl and 0.05 M LiCl
    b. 6.3×10^{-4} M $MgCO_3$ and 1.0×10^{-4} M $Mg(OH)_2$
    c. 0.5 M NaOH and 2.0×10^{-5} M $Mg(OH)_2$

16. Barium sulfate ($BaSO_4$) is administered to people when X rays of their digestive systems are taken, despite the fact that the Ba^{2+} ion is toxic to humans. Suggest the method by which $BaSO_4$ could be administered without the danger from the Ba^{2+} ions. Base your method on the common-ion effect.

17. The K_{sp} of AgCl at 50°C is 1.32×10^{-9}. Predict whether AgCl will precipitate if $AgNO_3$ and NaCl solutions of the following concentrations are mixed.
    a. 1.00×10^{-5} M $AgNO_3$ and 1.00×10^{-5} M NaCl
    b. 1.00×10^{-5} M $AgNO_3$ and 1.32×10^{-4} M NaCl
    c. 1.00×10^{-5} M $AgNO_3$ and 4.00×10^{-4} M NaCl

18. Calculate the molar solubilities of each of the following:
    a. barium fluoride, BaF_2, $K_{sp} = 1.7 \times 10^{-6}$
    b. cadmium carbonate, $CdCO_3$, $K_{sp} = 2.5 \times 10^{-14}$
    c. lanthium iodate, $La(IO_3)_3$, $K_{sp} = 6.0 \times 10^{-10}$
    d. silver arsenate, Ag_3AsO_4, $K_{sp} = 1.0 \times 10^{-22}$

15A Definitions and Descriptions page 389
15B Equilibria, Acids, and Bases page 393
15C Neutralization page 406
FACETS: When the Weather Turns Sour—Acid Rain page 395

Acids, Bases, & Salts 15

Ions Make the Difference

Chemists endeavor to classify or group chemicals into categories of common behavior, such as organics, halides, or gases. The classification scheme for acids, bases, and salts is particularly useful for understanding a large number of chemical reactions. The three classification schemes considered in this chapter—Arrhenius, Brönsted-Lowry, and Lewis—have proved to be the most useful in predicting chemical behavior and reactivity.

15A Definitions and Descriptions

Arrhenius Definitions: H⁺ and OH⁻ Ions

What determines whether a compound is an acid, a base, a salt, or none of the above? Acids and bases are often regarded as fuming, caustic, and generally nasty. This impression holds true in many cases, but it is wrong in many others. Scientists focus on chemical behaviors as the basis for definitions. Table 15A-1 lists some acids, bases, and salts that are used in the U.S. economy.

The earliest of the modern acid-base definitions was proposed in the 1880s by the Swedish chemist Svante Arrhenius (uh REE ne us). **Arrhenius acids** release hydrogen ions (H^+) into aqueous solutions. Note that a hydrogen ion is just a proton, or a hydrogen nucleus. **Arrhenius bases** are substances that release hydroxide ions (OH^-) into aqueous solutions. According to the Arrhenius definitions, HCl, HCOOH, and H_2SO_4 are acids and NaOH, $Mg(OH)_2$, and $Al(OH)_3$ are bases.

Hydrogen-containing compounds that do not donate protons are not acids. Methane (CH_4) has four hydrogen atoms, but none of them is bonded in such a way that it can be easily released. Since no protons are released from methane under ordinary circumstances, methane is not considered an acid. Likewise, compounds with OH groups in their formulas are not always bases. Methanol (CH_3OH) does not normally release its OH group.

Svante Arrhenius

The Arrhenius definitions deal only with compounds in aqueous solutions. Normally this is not a great limitation, but modern chemistry has started exploring some reactions that take place in nonaqueous solutions and some that take place without the aid of any solvents at all. The Arrhenius definition of bases does not recognize compounds such as ammonia as being bases, because these compounds do not have an OH group. Nevertheless, ammonia exhibits properties that are generally thought of as being basic. Although the Arrhenius definitions find some use, they are restrictive and are not used as often as other definitions. We honor Arrhenius today because he was the first to recognize that the dissociation of acids, bases, and salts into ions played a key role in their behavior.

Table 15A-1 Some Common Industrial Acids, Bases, and Salts

Name and Formula	Classification	Acid/Base Strength	Uses
Sulfuric Acid, H_2SO_4	Acid	Strong	Chemical fertilizers, automobile batteries
Lime, CaO or $Ca(OH)_2$	Base	Strong	Steel making, air pollution control
Ammonia, NH_3 or NH_4OH (aq)	Base	Weak	Fertilizer production, refrigerant
Sodium hydroxide, NaOH	Base	Very strong	Paper and soap production
Phosphoric acid, H_3PO_4	Acid	Moderately strong	Fertilizer, detergent, and food industry
Ammonium nitrate, NH_4NO_3	Salt	Weakly acidic	Fertilizer, explosives
Carbon dioxide, CO_2	Acid	Weak	Carbonated beverages
Aluminum sulfate, $Al_2(SO_4)_3$	Salt	Weakly acidic	Water treatment, cosmetics
Acetic acid, $HC_2H_3O_2$	Acid	Weak	Vinegar, synthetic fiber production

Table 15A-2 Conjugate Pairs of Brönsted-Lowry Acids and Bases

Name of Conjugate Acid	Conjugate Acid	Conjugate Base	Name of Conjugate Base
Acetic acid	$HC_2H_3O_2$	$C_2H_3O_2^-$	Acetate ion
Hydrochloric acid	HCl	Cl^-	Chloride ion
Perchloric acid	$HClO_4$	ClO_4^-	Perchlorate ion
Water	H_2O	OH^-	Hydroxide ion
Sulfuric acid	H_2SO_4	HSO_4^-	Hydrogen sulfate ion
Hydrogen sulfate ion	HSO_4^-	SO_4^{2-}	Sulfate ion

Brönsted-Lowry Definitions: Trading Protons

In 1923 a Danish chemist named J. N. Brönsted and a British chemist named I. M. Lowry proposed new definitions of acids and bases. A **Brönsted-Lowry acid** is a substance that donates protons, and a **Brönsted-Lowry base** is a substance that accepts protons. A released proton does not float freely in water. Instead, it immediately joins a water molecule to make it a hydronium ion (H_3O^+). The process of losing a proton is called **deprotonation**, and the process of gaining a proton is called **protonation**. This equilibrium is established upon mixing acetic acid with water:

$$HC_2H_3O_2 + H_2O \rightleftharpoons H_3O^+ + C_2H_3O_2^-$$

Proton Donor (Acid) + Proton Acceptor (Base) ⇌ Proton Donor (Acid) + Proton Acceptor (Base)

Brönsted-Lowry acids and bases exist in pairs called **conjugate pairs**. An acid that loses a proton forms a substance capable of accepting a proton. This base is called the **conjugate base** of the acid. The **conjugate acid** of a base is the substance formed by the protonation of the base. Examples of Brönsted-Lowry acids and bases with their conjugates are shown in Table 15A-2.

The Brönsted-Lowry definitions of acids and bases encompass all Arrhenius acids and bases plus many others. Its definition of acids is essentially the same as the Arrhenius definition, but its definition of bases differs greatly. The Brönsted-Lowry definition greatly expands the number of substances called bases because it includes many substances without OH groups.

J. N. Brönsted & I. M. Lowry

Chapter 15A

Gilbert N. Lewis

Figure 15A-1

Ammonia is a Lewis base because it can donate a pair of electrons in the formation of a covalent bond.

Lewis Definitions: Electron Pairs

The Lewis theory of acids and bases is named after Gilbert N. Lewis, an American chemist who also published his ideas in 1923. A **Lewis acid** is any substance that can accept a pair of electrons, and a **Lewis base** is a substance that can donate a pair of electrons. Another way of saying this is that an acid has at least one empty orbital and a base has at least one lone (unbonded) pair of electrons. The formation of an ammonium (NH_4^+) ion entails a reaction between a Lewis acid and a Lewis base. A hydrogen ion acts as a Lewis acid by accepting an electron pair. The electron pair comes from the ammonia molecule, which acts as a Lewis base. When Lewis acids and bases combine, covalent bonds form. One covalent bond forms in the ammonium ion when the nitrogen of the ammonia molecule supplies both electrons in the bond.

Lewis acids include many more substances than do Arrhenius or Brönsted-Lowry acids. Despite this fact, the Lewis definitions are not used as commonly as the Brönsted-Lowry definitions.

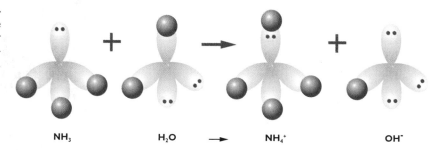

NH_3 H_2O → NH_4^+ OH^-

Observable Properties of Acids and Bases

Aqueous solutions of acids and bases have distinctive properties. Arrhenius and Brönsted-Lowry acids release hydrogen ions that impart several unique characteristics to acids. Citric acid in fruit juices has a tart, sour taste; so do other acids. The putrid, sour taste of lactic acid identifies sour milk. Acetic acid is responsible for vinegar's sharp, sour taste.

Acids react with active metals to produce hydrogen gas and a salt. If the metal is above hydrogen in the activity series, it will replace the hydrogen ion of the acid. The reaction between magnesium and hydrochloric acid illustrates this characteristic.

$$Mg\ (s) + 2\ HCl\ (aq) \longrightarrow Mg^{2+}\ (aq) + 2\ Cl^-\ (aq) + H_2\ (g)$$

Magnesium reacts with hydrochloric acid, producing hydrogen gas and magnesium chloride.

Acids can also be identified by their reactions with compounds that change colors. One such compound is litmus, which turns red in the presence of an acid. A final characteristic of acids is that they neutralize bases. A **neutralization reaction** between an acid and a base is one that produces a salt and water. Probably one of the most familiar is the reaction of hydrochloric acid with sodium hydroxide:

$$HCl\ (aq) + NaOH\ (aq) \longrightarrow NaCl\ (aq) + H_2O$$

Accidental tastes of soap confirm the fact that basic solutions tend to be bitter. The ions also contribute to that slippery sensation that is felt when bases are touched. Bases react with red litmus to turn it blue, and they react with acids to neutralize them.

Figure 15A-2

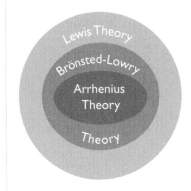

The Lewis definitions of acids and bases include more substances than do the other definitions.

Figure 15A-3

Acids turn litmus red; bases turn litmus blue.

Section Review Questions 15A

1. Define the following.
    a. Arrhenius acid/base
    b. Brönsted-Lowry acid/base
    c. Lewis acid/base
2. What is a neutralization reaction?
3. For each of the following reactions, label which substance is the acid and which is the base. Also, indicate which substances are conjugates.
    a. $H_2C_2O_4 + ClO^- \rightleftharpoons HC_2O_4^- + HClO$
    b. $HPO_4^{2-} + NH_4^+ \rightleftharpoons NH_3 + H_2PO_4^-$
    c. $SO_4^{2-} + H_2O \rightleftharpoons HSO_4^- + OH^-$

15B Equilibria, Acids, and Bases

The deprotonation of acids and the protonation of bases are reversible processes. Equilibrium constants describe the extent to which an acid gives up a proton and how readily a base acquires a proton. Equilibrium chemistry explains why it is all right to use boric acid as an eyewash but why it is unsafe to touch the sulfuric acid in the battery of a car. The equilibrium constants of acid-base reactions provide the theoretical basis for the pH system.

The Autoprotolysis of Water: H_2O Splits Up

In addition to all its unique physical properties, water has one intriguing chemical property: it can react with itself. The reaction is an acid-base reaction in which one water molecule donates a proton to another water molecule. One molecule acts as an acid, and the other acts as a base. This reaction is called the **autoprotolysis** of water. It is also called the auto-ionization, or self-ionization, of water.

$$H_2O + H_2O \rightleftharpoons OH^- + H_3O^+$$

The reaction is an equilibrium in which the reverse reaction predominates. Only a few molecules produce protons. As with all reversible reactions, this one can be described with an equilibrium constant.

$$K = \frac{[H_3O^+][OH^-]}{[H_2O]^2} = [H_3O^+][OH^-]$$

The concentration of water is eliminated from the equation. The resulting constant is given the special name **autoprotolysis constant of water** and the special symbol K_w. Experimental evidence has shown that the concentrations of hydronium and hydroxide ions are both 1.0×10^{-7} mole per liter in pure water at 25°C. Only 1 out of 555 million water molecules is dissociated at any given instant in time. The product of these two concentrations makes the value of the constant 1.0×10^{-14}.

$$K_w = [H_3O^+][OH^-]$$
$$= (1.0 \times 10^{-7})(1.0 \times 10^{-7})$$
$$= 1.0 \times 10^{-14}$$

Despite its small size, 1.0×10^{-14} is a very important number. Whether a solution is acidic, basic, or neutral, the concentration of hydronium ions times the concentration of hydroxide ions always equals 1.0×10^{-14} at 25°C.

Sample Problem The concentration of hydronium ions in a mild acid is found to be 5×10^{-7} mol/L. What is the concentration (molarity) of hydroxide ions?

Solution
Given that $K_w = [H_3O^+][OH^-]$, solve for $[OH^-]$

$$\frac{K_w}{[H_3O^+]} = [OH^-]$$

Substitute the known values.

$$\frac{1 \times 10^{-14}}{5 \times 10^{-7}} = [OH^-] \qquad 2 \times 10^{-8} \, M = [OH^-]$$

FACETS OF CHEMISTRY

When the Weather Turns Sour—Acid Rain

Much of the industrial world has a problem with acid rain. Statues and buildings that have stood for centuries, such as the Parthenon in Greece, are suddenly beginning to corrode because of the acid attack on limestone, marble, and concrete. The reaction with limestone is:

$$H_2SO_4 \,(aq) + CaCO_3 \,(s) \longrightarrow CaSO_4 \,(s) + H_2O + CO_2 \,(g)$$

Additionally, fish have disappeared from acidic lakes, crops tend to grow more slowly in acidic soil, and forests die out.

Although rain is naturally acidic (pH ≅ 5.6) because of dissolved carbon dioxide, the oxides of sulfur and nitrogen from automobiles and coal-burning power plants react with rain water to form more and stronger acids.

$$H_2O \,(l) + SO_2 \,(g) \longrightarrow H_2SO_3 \,(aq) \text{ sulfurous acid}$$

$$H_2O \,(l) + SO_3 \,(g) \longrightarrow H_2SO_4 \,(aq) \text{ sulfuric acid}$$

$$H_2O \,(l) + 2\,NO_2 \,(g) \longrightarrow HNO_3 \,(aq) \text{ nitric acid} + HNO_2 \text{ nitrous acid}$$

At times in the northeastern United States, there has been enough acid in the rain to lower the pH to 3.0.

There is evidence that acid rain removes vital nutrients, such as calcium, from the soil. Additionally, it has been demonstrated that acidic groundwater liberates aluminum from the soil. Aluminum is poisonous to the root hairs of plants, crippling their ability to extract water from the soil. Reforestation efforts in the acid-rain regions are failing because of the unhealthy state of the soil.

New England and Scandinavia are among the regions with the greatest acid rain problem. The acid rain in the New England states may originate in the Midwest. Acid rain that percolates into the groundwater eventually flows into streams and lakes. Drinking water has been contaminated in some areas, and the populations of certain species of fish have reportedly been decimated.

One localized solution that has been successfully used in the Northeast is the addition of large quantities of lime to some of the lakes. However, this is only a temporary solution.

A truly effective solution to the acid rain problem must attack the source—the industrial plants. In this country, most of the pollution seems to come from coal-fired power plants. One option would be to change over to nuclear energy to fuel the power plants. However, the building of nuclear power plants is extremely costly and time consuming, and the question of safe disposal of nuclear waste has not been satisfactorily addressed to all concerned. In addition, since the nuclear power accident at Chernobyl near Kiev on April 26, 1986, which resulted in the evacuation of 135,000 people, long-term environmental damage, and health effects that are still being realized, few communities are willing to assume these risks. Another option is the addition of "scrubbers" to the existing power plants. These tall towers mix the escaping gases with water and lime so that the gases that emerge are "clean." A third approach to the problem is to place marble-sized pieces of limestone and burning coal in a stream of compressed air at the bottom of the furnace. When sulfur dioxide forms, it immediately reacts with the limestone.

The problem of acid rain does not appear to have a simple solution, yet it cannot be ignored—research must continue to find a solution to this problem. After God created the earth, He entrusted it to man's keeping (Gen. 1:28). While man should use the resources God has supplied, he should not recklessly abuse them and spoil the environment.

S. P. L. Sorensen

The pH Scale

The concentrations of hydronium and hydroxide ions in aqueous solutions frequently range between 1 and 1×10^{-14} mole per liter. The repetitive expression of H_3O^+ concentration in scientific notation becomes tedious. Soren P. L. Sorensen, a Danish chemist, proposed the pH scale in 1909 to provide a clear, concise, and convenient way of describing the concentrations of these ions. The **pH** of a solution is the negative logarithm of the hydronium ion molar concentration.

$$pH = -\log [H_3O^+]$$

A pure sample of water has a hydrogen ion concentration of 1.0×10^{-7} moles per liter. The logarithm of 1.0×10^{-7} is -7.0. Since $-(-7.0)$ is $+7.0$, the pH of pure water is 7.0. Table 15B-1 shows hydronium ion concentrations of several solutions, the logarithms of these concentrations, and the pH values.

Table 15B-1

$[H_3O^+]$	$\log [H_3O^+]$	pH, or $-\log [H_3O^+]$
1.0×10^{0}	0.0	0.0
5.0×10^{-3}	-2.3	2.3
1.0×10^{-7}	-7.0	7.0
5.0×10^{-10}	-9.3	9.3
1.0×10^{-14}	-14.0	14.0

Sample Problem The hydronium ion concentration in a shampoo is 2.0×10^{-5} mol/L. What is the pH of this shampoo?

Solution

$$\begin{align} pH &= -\log [H_3O^+] \\ &= -\log (2.0 \times 10^{-5}) \\ &= -(-4.7) \\ &= 4.7 \end{align}$$

If you are using a scientific calculator, enter the $[H_3O^+]$, which is 2.0×10^{-5}. Press the Log key, and then press the +/- key to change the signs.

Just as the pH system describes the hydronium ion concentration, a pOH system describes the hydroxide ion concentration. The **pOH** of a solution is the negative logarithm of the hydroxide

Table 15B-2

[OH⁻]	log [OH⁻]	pOH, or −log [OH⁻]
1.0×10^{-14}	−14.0	14.0
2.0×10^{-12}	−11.7	11.7
1.0×10^{-7}	−7.0	7.0
2.0×10^{-5}	−4.7	4.7
1.0×10^{0}	0.0	0.0

ion concentration. Table 15B-2 shows the hydroxide ion concentrations of several solutions, the logarithms of these concentrations, and the pOH values.

Because the hydronium and hydroxide ion concentrations are related, pH and pOH are also related. The pH plus the pOH equals 14 at 25°C. We will always assume a temperature of 25°C unless otherwise stated.

Table 15B-3

[H₃O⁺]	pH	pOH	[OH⁻]
1.0×10^{0}	0.0	14.0	1.0×10^{-14}
5.0×10^{-3}	2.3	11.7	2.0×10^{-12}
1.0×10^{-7}	7.0	7.0	1.0×10^{-7}
5.0×10^{-10}	9.3	4.7	2.0×10^{-5}
1.0×10^{-14}	14.0	0.0	1.0×10^{0}

A pH meter calculates the pH of a solution.

Many commonly encountered substances contain acids or bases. Figure 15B-1 lists some of these solutions and their pHs.

All solutions with a pH less than 7 are called **acidic solutions** because the hydronium ion concentration is greater than the hydroxide ion concentration. All solutions with a pH greater than 7 are called **basic solutions** because the hydroxide ions outnumber the hydronium ions. When a solution has a pH of 7, it is called a **neutral solution.**

Figure 15B-1

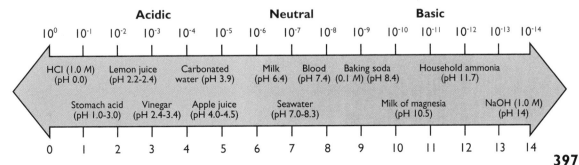

pH and pOH Calculations

The relationships between [H₃O⁺], [OH⁻], pH, and pOH are used in many calculations. Figure 15B-2 is a flow chart that shows how the various conversions can be made.

Household ammonia solutions used for cleaning are aqueous solutions of ammonia (NH₃). In water, ammonia produces ammonium hydroxide, which makes the solution basic (pH = 11.7). If the pH of a solution is known, the concentration of hydronium ions can be calculated with the use of antilogs. Recall that pH is a negative exponent; thus, you can find the [H₃O⁺] by raising 10 to a power given by -pH:

$$[H_3O^+] = 10^{-pH}$$

Using the 10^x button on the calculator, you can determine the [H₃O⁺] in a household ammonia solution. Enter the pH value (11.7), change the sign with the +/- key, and push the 10^x key.

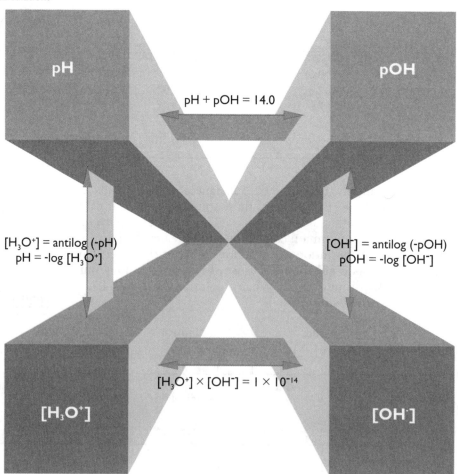

Figure 15B-2

Flow chart for pH and pOH calculations

The result is 2.0×10^{-12}; thus, a pH of 11.7 corresponds to a hydronium concentration of 2.0×10^{-12} M. This process is known as finding the antilog of -11.7.

The pOH and the hydroxide ion concentration of the solution can be calculated directly from the pH, since pH + pOH = 14.0.

$$\text{pOH} = 14.0 - \text{pH}$$
$$= 14.0 - 11.7$$
$$= 2.3$$

The hydroxide ion concentration equals the antilog of -2.3.

$$[OH^-] = 10^{-\text{pOH}}$$
$$= 10^{-2.3}$$
$$= 5.0 \times 10^{-3} \text{ M}$$

Sample Problem A 0.10 molar hydrochloric acid solution has a hydronium ion concentration of 1.0×10^{-1} mol/L. Calculate the pH, the hydroxide ion concentration, and the pOH.

Solution
The pH can be found first.

$$\text{pH} = -\log(1.0 \times 10^{-1})$$
$$= -(-1.0)$$
$$= 1.0$$

Next, find the pOH.

$$\text{pOH} = 14.0 - \text{pH}$$
$$= 14.0 - 1.0$$
$$= 13.0$$

The hydroxide ion concentration can now be calculated from the pOH.

$$[OH^-] = 10^{-\text{pOH}}$$
$$= 10^{-13}$$
$$= 1.0 \times 10^{-13}$$

Describing Acid-Base Strengths: Equilibrium Constants Again

The strength of an acid depends on how easily it releases protons. **Strong acids** give up protons easily and ionize completely. Surprisingly, only six common acids may be classified as strong acids: $HClO_4$, HI, HBr, HCl, H_2SO_4, and HNO_3. **Weak acids** do not ionize completely. Only a portion of their molecules lose protons. Acetic acid, in which only 1 out of 24 molecules loses a proton, is a familiar weak acid.

The strength of a base depends on how easily it accepts protons. Whereas **weak bases** (such as the Cl⁻ ion) are poor proton acceptors, **strong bases** (such as the OH⁻ ion) accept protons readily. The hydroxides of active metals—LiOH, NaOH, KOH, RbOH, CsOH, $Mg(OH)_2$, $Ca(OH)_2$, $Sr(OH)_2$, and $Ba(OH)_2$—are the best sources of the hydroxide ion, so they are commonly called strong bases.

The conjugate base of a strong acid is always weak; it does not readily accept protons to form a strong base. Perchloric acid ($HClO_4$) is a strong acid that readily gives up a hydrogen ion. Once the hydrogen and the chloride ions are apart, they do not readily rejoin to form perchloric acid. Therefore, the perchlorate ion is a weak conjugate base. On the other end of the acid scale are the weak acids. Their conjugate bases are strong.

Acid strengths are described by dissociation equilibrium constants. The **acid dissociation constant** (K_a) describes the extent of the forward reaction in the equilibrium.

$$Acid + H_2O \rightleftharpoons H_3O^+ + Base$$

$$K_a = \frac{[H_3O^+][Base]}{[Acid]}$$

Stronger acids have higher K_as because the forward reaction dominates and many hydrogen ions are released. Table 15B-4 lists acids, dissociation constants, and their deprotonation reactions in order from strong to weak. Because the stronger acids have weaker conjugate bases, the bases at the top of the column are the weakest.

Sample Problem On the basis of dissociation constants, which acid is stronger, hydrofluoric or hydriodic acid?

Solution
By looking at Table 15B-4, you can see that the dissociation constant of hydriodic acid is too large to be measured, whereas the constant for hydrofluoric acid is 3.5×10^{-4}. Therefore, hydriodic acid is the stronger acid.

Boric acid, which has a low dissociation constant, is a weak acid. Since it does not release many hydrogen ions, the diluted eyewash solutions that contain boric acid will not cause burns. On the other hand, sulfuric acid in car batteries releases enough hydrogen ions to blister the skin.

Table 15B-4 Relative Strengths of Some Acids and Bases

Acid	K_a	Equilibrium	Base
Perchloric acid	very large	$HClO_4 \longrightarrow H^+ + ClO_4^-$	perchlorate ion
Hydriodic acid	very large	$HI \longrightarrow H^+ + I^-$	iodide ion
Hydrobromic acid	very large	$HBr \longrightarrow H^+ + Br^-$	bromide ion
Hydrochloric acid	very large	$HCl \longrightarrow H^+ + Cl^-$	chloride ion
Sulfuric acid	very large	$H_2SO_4 \longrightarrow H^+ + HSO_4^-$	hydrogen sulfate ion
Nitric acid	very large	$HNO_3 \longrightarrow H^+ + NO_3^-$	nitrate ion
Sulfurous acid	1.7×10^{-2}	$H_2SO_3 \rightleftharpoons H^+ + HSO_3^-$	bisulfate ion
Hydrogen sulfate ion	1.2×10^{-2}	$HSO_4^- \rightleftharpoons H^+ + SO_4^{2-}$	sulfate ion
Phosphoric acid	7.5×10^{-3}	$H_3PO_4 \rightleftharpoons H^+ + H_2PO_4^-$	dihydrogen phosphate ion
Hydrofluoric acid	3.5×10^{-4}	$HF \rightleftharpoons H^+ + F^-$	fluoride ion
Formic acid	1.8×10^{-4}	$HCHO_2 \rightleftharpoons H^+ + CHO_2^-$	formate ion
Acetic acid	1.8×10^{-5}	$HC_2H_3O_2 \rightleftharpoons H^+ + C_2H_3O_2^-$	acetate ion
Carbonic acid	4.2×10^{-7}	$H_2CO_3 \rightleftharpoons H^+ + HCO_3^-$	hydrogen carbonate ion
Hypochlorous acid	3.2×10^{-8}	$HClO \rightleftharpoons H^+ + ClO^-$	hypochlorite ion
Boric acid	5.8×10^{-10}	$H_3BO_3 \rightleftharpoons H^+ + H_2BO_3^-$	dihydrogen borate ion
Ammonium ion	5.7×10^{-10}	$NH_4^+ \rightleftharpoons H^+ + NH_3$	ammonia
Hydrocyanic acid	4.0×10^{-10}	$HCN \rightleftharpoons H^+ + CN^-$	cyanide ion
Hydrogen carbonate ion	4.8×10^{-11}	$HCO_3^- \rightleftharpoons H^+ + CO_3^{2-}$	carbonate ion
Hydrogen peroxide	2.6×10^{-12}	$H_2O_2 \rightleftharpoons H^+ + HO_2^-$	hydroperoxide ion
Water	1.0×10^{-14}	$H_2O \rightleftharpoons H^+ + OH^-$	hydroxide ion

Causes of Acid-Base Strengths: Unhand That Proton!

Dissociation constants *describe* acid strengths; they do not determine them. The factors that cause one acid to be strong and another to be weak are electrical in nature. The location of shared electron clouds and the charge on an acid determine how easily a proton (H⁺ ion) can be released.

To be easily ionizable, the nucleus of a hydrogen atom must be exposed and open to attack. The negatively charged electron cloud of the covalent bond must be shifted away from the hydrogen and toward a more electronegative atom in the center of the molecule. The series of compounds shown on the next page illustrates how a highly electronegative element can make acids strong by shifting electrons and exposing hydrogen nuclei. The hydrogen atoms in the methane molecule are virtually surrounded by negative charges. The highly electronegative fluorine atom pulls shared electrons to itself. Consequently, hydrofluoric acid is the strongest acid of this series.

Table 15B-5

Compound	Electronegativity of Central Atom	
CH_4	2.4	
NH_3	3.0	
H_2O	3.5	
HF	4.0	

Besides electron-cloud location, electrical charges determine the strengths of acids. The sulfuric acid molecule (H_2SO_4) has a neutral charge and is a strong acid. The hydrogen sulfate (HSO_4^-) ion has an ionizable proton, but this negative ion has little tendency to release another hydrogen ion. Similarly, H_3PO_4, $H_2PO_4^-$, and HPO_4^{2-} have decreasing acidic strengths. Anions hold protons more strongly than neutral molecules.

The same factors that make acids strong make bases weak. Conversely, factors that weaken acids make bases stronger. Strong acids release protons easily; strong bases accept protons easily.

Strong acids	Strong bases
1. have exposed hydrogen nuclei.	1. have exposed electron pairs.
2. have a minimum of negative charges.	2. may have strong negative charges.

It's an Acid! It's a Base! It's Amphiprotic!

Amphiprotic substances can act as both Brönsted-Lowry acids and bases. Whether an amphiprotic substance behaves as an acid or a base depends on the reaction conditions. For example, the hydrogen carbonate ion accepts a proton when an acid is added to the solution. Under those circumstances, therefore, the hydrogen carbonate ion is a base.

$$HCO_3^- + H_3O^+ \rightleftharpoons H_2CO_3 + H_2O$$

When a base is added to the solution, the hydrogen carbonate ion acts as an acid by donating a proton.

$$HCO_3^- + OH^- \rightleftharpoons CO_3^{2-} + H_2O$$

Water is also an amphiprotic substance, as is demonstrated by its reactions with perchloric acid and ammonia.

$$H_2O + HClO_4 \rightleftharpoons H_3O^+ + ClO_4^-$$
$$H_2O + NH_3 \rightleftharpoons NH_4^+ + OH^-$$

In the first reaction water acts as a Brönsted-Lowry base by receiving a proton from perchloric acid. It acts as an acid in the second reaction when it donates a proton to ammonia.

Polyprotic Acids: Protons Galore

Monoprotic Brönsted-Lowry acids can donate only one proton. **Polyprotic** acids can donate more than one proton. Each ionization is characterized by a different equilibrium constant. Carbonic acid (H_2CO_3) is an example of a **diprotic** acid, which can donate two protons. The K_a for the first ionization is called K_{a1} and that for the second ionization is called K_{a2}.

$$H_2CO_3 + H_2O \rightleftharpoons HCO_3^- + H_3O^+$$

$$K_{a1} = \frac{[HCO_3^-][H_3O^+]}{[H_2CO_3]}$$

$$= 4.2 \times 10^{-7}$$

$$HCO_3^- + H_2O \rightleftharpoons CO_3^{2-} + H_3O^+$$

$$K_{a2} = \frac{[CO_3^{2-}][H_3O^+]}{[HCO_3^-]}$$

$$= 4.8 \times 10^{-11}$$

All polyprotic acids have at least one amphiprotic substance in their series of ionizations. In carbonic acid the hydrogen carbonate ion is amphiprotic. Furthermore, the K_{a2} is always smaller than the K_{a1}. The first ionization involves separation of a proton from an uncharged carbonic acid (H_2CO_3) molecule, and the second involves separation of a proton from the negatively charged hydrogen carbonate (HCO_3^-) ion. Because the strength of attraction between opposite charges depends on the magnitude of the charges, the second ionization is less favorable than

Table 15B-6

Polyprotic Acids	Acid-Base Reaction	K_a
Chromic acid	$H_2CrO_4 \rightleftharpoons H^+ + HCrO_4^-$	1.8×10^{-1}
	$HCrO_4^- \rightleftharpoons H^+ + CrO_4^{2-}$	3.2×10^{-7}
Hydrosulfuric acid	$H_2S \rightleftharpoons H^+ + HS^-$	1.1×10^{-7}
	$HS^- \rightleftharpoons H^+ + S^{2-}$	1.0×10^{-14}
Phosphoric acid	$H_3PO_4 \rightleftharpoons H^+ + H_2PO_4^-$	7.5×10^{-3}
	$H_2PO_4^- \rightleftharpoons H^+ + HPO_4^{2-}$	6.2×10^{-8}
	$HPO_4^{2-} \rightleftharpoons H^+ + PO_4^{3-}$	4.8×10^{-13}
Sulfurous acid	$H_2SO_3 \rightleftharpoons H^+ + HSO_3^-$	1.3×10^{-2}
	$HSO_3^- \rightleftharpoons H^+ + SO_3^{2-}$	6.3×10^{-8}

Universal indicator paper can measure a wide range of pHs. More specific indicators are used to test the pH of swimming pools.

the first. In general, $K_{a1} > K_{a2} > K_{a3}$. Acids with three ionizable protons are called **triprotic** acids.

Several more examples of polyprotic acids are shown in Table 15B-6. The values of K_a are for the reactions at 25°C.

Indicators: Color-Coded Chemicals

Indicators are substances that change colors when the pH of a solution changes. Indicators are usually weak acids or bases whose conjugates have different colors. Some indicators change colors at low pHs, some at high pHs. Some are polyprotic, so they may exhibit more than one color change. Examples of indicators, their colors, and the pH range over which their colors change are given in Table 15B-7.

The first synthetic indicator (phenolphthalein) was introduced in 1877. Prior to that time, chemists used natural plant juices as acid-base indicators, some of which are shown in Table 15B-8.

Indicators give rough estimates of pHs. For instance, if a solution is yellow with both bromthymol blue and methyl orange, then the pH must be between 4.4 and 6.0. Indicators usually show pH changes over a narrow range. Congo red changes color when a solution has a pH between 3 and 5. Once the pH is past 5, the color remains essentially constant. The congo red indicator tells nothing about how basic the solution is. This problem may be overcome by the use of carefully chosen combinations of indicators called universal indicators. More accurate values can be obtained when large numbers of indicators are used, but this is tedious. Instruments called pH meters give accurate

Table 15B-7 Colors of Common Indicators

Indicator	1	2	3	4	5	6	7	8	9	10	11	12	13	14
Bromthymol blue (<6.0 yellow; >7.5 blue)														
Methyl orange (<3.1 red; >4.4 yellow)														
Methyl red (<4.4 red; >6.2 yellow)														
Phenolphthalein (<8.5 colorless; >9.0 deep red)														
Alizarin yellow R (<10.2 yellow; >12.0 red)														
Litmus (<4.5 red; >8.3 blue)														
Congo red (<3.0 blue violet; >5.0 red)														

pH measurements of solutions by measuring electrical properties that depend on the number of ions present.

Table 15B-8 Natural Chemical Indicators

Color changes as a function of pH

Indicator	pH 2	3	4	5	6	7	8	9	10	11	12
Red apple skin											
Beets											
Blueberries											
Red cabbage*											
Cherries											
Grape juice											
Red onion											
Yellow onion											
Peach skin											
Pear skin											
Plum skin											
Radish skin											
Rhubarb skin											
Tomato											
Turnip skin											

*Yellow at pH 12 and above

Section Review Questions 15B

1. Find the pH of a solution whose $[H_3O^+] = 9.5 \times 10^{-8}$.
2. What is the $[H_3O^+]$ of a solution with a pOH = 5.45?
3. Find the pOH of a solution whose pH is 3.52.
4. Using Table 15B-4, indicate the stronger acid in each pair.
   a. HI, H_2SO_4
   b. HF, HCN
   c. H_2O_2, $HCHO_2$
   d. H_3BO_4, H_3PO_4
   e. HCl, $HClO_4$
5. A rainwater sample shows a yellow color when tested with both methyl red and bromthymol blue. What is the approximate pH of the sample? Is it acidic, basic, or neutral?
6. Calculate the pH of a solution with $[OH^-] = 1.0 \times 10^{-6}$.

15C Neutralization: When Acids and Bases Mix

The pain and discomfort of heartburn are usually associated with meals that were either too large or too spicy. The burning sensation under the breastbone comes not from oregano or green peppers, but from excess hydrochloric acid. This acid, which is normally confined to the stomach, irritates the tissues of the esophagus if some splashes past the upper part of the stomach. To relieve this pain, a person can neutralize the acid with an antacid (a base). The mixing of two active chemicals results in a set of harmless products.

Salts: Products of Neutralizations

Antacids that contain magnesium hydroxide react to neutralize stomach acid: $Mg(OH)_2 + 2\ HCl \longrightarrow MgCl_2 + 2\ H_2O$. Sodium hydroxide is much too corrosive to be used as an antacid, but it too neutralizes acids: $2\ NaOH + H_2SO_4 \longrightarrow Na_2SO_4 + 2\ H_2O$. According to the Arrhenius definitions, the heart of every neutralization reaction is the formation of water from hydronium and hydroxide ions. The hydronium ions of acids and the hydroxide ions of bases combine to form two water molecules.

$$H_3O^+ + OH^- \longrightarrow 2\ H_2O$$

Metal cations and nonmetal anions are present, but they are not involved in the neutralization reaction. If the resulting solution is evaporated, these ions will crystallize into a salt.

A **salt** is a substance formed when the anion of an acid and the cation of a base combine. In the antacid reaction Cl^- (the acid's anion) combines with Mg^{2+} (the base's cation) to form $MgCl_2$ (the salt). Many different salts can be produced in neutralization reactions. Sodium chloride, zinc bromide, and potassium fluoride are all examples of salts. Most salts, like these three, are composed of a metal and a nonmetal, but this is not always true. Ammonium chloride (NH_4Cl), potassium sulfate (K_2SO_4), and aluminum phosphate ($AlPO_4$) contain polyatomic ions.

Sample Problem Write a neutralization reaction that could produce the salt zinc chloride ($ZnCl_2$).
Solution
Salts are made from the anions of acids and the cations of bases. In this case, the most common acid that releases the chloride ion is hydrochloric acid. Zinc hydroxide is a likely source for the zinc ion.
$$2\ HCl + Zn(OH)_2 \longrightarrow ZnCl_2 + 2\ H_2O$$

Sample Problem What salt results from the neutralization of barium hydroxide by acetic acid ($HC_2H_3O_2$)?
Solution
$$2\ HC_2H_3O_2 + Ba(OH)_2 \longrightarrow Ba(C_2H_3O_2)_2 + 2\ H_2O$$
Barium acetate forms from the barium and acetate ions.

Salts are divided into three general classes: neutral, acidic, and basic. **Neutral salts** result from the reaction between strong acids and strong bases, or evenly matched weak acids and bases. **Acid salts** are produced by the neutralization of weak bases with strong acids. Conversely, **basic salts** are derived from the reactions of strong bases with weak acids. See Table 15C-1 for examples.

Table 15C-1 Salt Producers

Acid	plus	Base	yields	Salt
H_2CO_3 (weak)		NaOH (strong)		Na_2CO_3 (basic salt)
HNO_3 (strong)		NH_4OH (weak)		NH_4NO_3 (acid salt)
H_2SO_4 (strong)		NH_4OH (weak)		$(NH_4)_2SO_4$ (acid salt)
HCl (strong)		KOH (strong)		KCl (neutral salt)
$HC_2H_3O_2$ (weak)		NH_4OH (weak)		$NH_4C_2H_3O_2$ (neutral salt)
H_2SiO_3 (weak)		NaOH (strong)		Na_2SiO_3 (basic salt)

Figure 15C-1

Chemical indicators denote a range of values. The endpoint is a specific value.

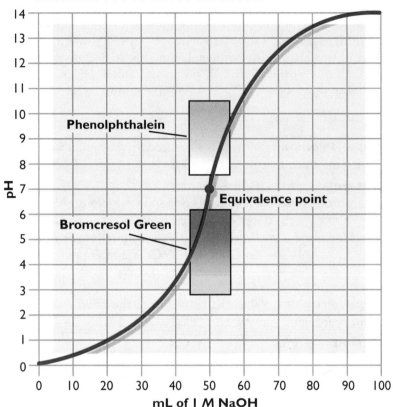

Acid-Base Titrations: Finding Unknown Concentrations

Acid-base **titrations** are controlled reactions in which scientists determine the unknown concentration of a solution by measuring its capacity to react with a solution of known concentration. Suppose that a chemist has 100.0 mL of a hydrochloric acid solution, but he does not know its exact concentration. He can find the concentration by adding small volumes of a precisely measured 1.00 M sodium hydroxide solution until the pH of the solution rises to 7.0. A graph called a **titration curve** shows how pH changes when an acid or base is added to a solution. Figure 15C-1 shows how the pH of a solution of unknown concentration changes with the addition of 1.00 M sodium hydroxide.

Fifty milliliters of the sodium hydroxide raises the pH of the acid to 7; so 50.0 mL of 1.00 M sodium hydroxide is chemically equivalent to 100 mL of the acid. With this information, the chemist knows that the 1.00 M sodium hydroxide is twice as concentrated as the hydrochloric acid. He can then calculate that the hydrochloric acid has a concentration of 0.50 M.

The **equivalence point** is the point in a titration at which an equivalent of titrant is added; that is, the amount of H_3O^+ ions now equals the amount of OH^- ions. The eqivalence point should not be confused with the end point of a titration. The **end point** is the point at which some change in a property of the solution is detected. For example, if a 0.1 N HCl solution is titrated with a 0.1 N NaOH solution and phenolphthalein indicator is added, the equivalence point will occur at pH 7.0, but the endpoint will be detected at pH 8.5—the first noticeable pink tinge.

The concepts of chemical equivalents and normality are used extensively in acid-base titrations. Remember that normality indicates concentration in gram-equivalents per liter of solution. By way of review, the gram-equivalent mass of an acid or base is the amount of substance that accepts or releases 1 mole of protons. The gram-equivalent masses of sodium hydroxide and hydrochloric acid are the same as their gram-molecular masses because 1 mole of sodium hydroxide can accept 1 mole of protons, and 1 mole of hydrochloric acid releases 1 mole of protons. The gram-equivalent mass of sulfuric acid (H_2SO_4) is one-half its gram-molecular mass because each $\frac{1}{2}$ mole of acid can release a full mole of protons.

Sample Problem Calculate the normality of 1.00 L of solution made with 30.0 g of phosphoric acid (H_3PO_4).

Solution
Set up the information as in other concentration calculations. Start with the given information (amount of solute/amount of solution). Use unit analysis to arrive at the normality (equivalents/liter of solution). Since each mole of phosphoric acid can release three protons, 1 mole equals 3 equivalents.

$$\frac{30.0 \text{ g } H_3PO_4}{1.00 \text{ L solution}} \left| \frac{1 \text{ mol } H_3PO_4}{98.00 \text{ g } H_3PO_4} \right| \frac{3 \text{ eq}}{1 \text{ mol } H_3PO_4}$$

$$= 0.918 \text{ N solution } H_3PO_4$$

A simple mathematical equation relates the normalities of solutions and the volumes involved in titrations. The subscript k stands for "known," and the subscript u stands for "unknown."

$$(\text{Normality}_k)(\text{Volume}_k) = (\text{Normality}_u)(\text{Volume}_u)$$

$$(N_k)(V_k) = (N_u)(V_u)$$

The above equation can be transformed to solve for an unknown normality.

$$N_u = \frac{(N_k)(V_k)}{V_u}$$

Putting the values from the titration example for the volumes and the normality of the base into this equation gives the normality of the acid.

$$N_u = \frac{(1.00\ N)(50.0\ mL)}{100.0\ mL}$$

$$= 0.500\ N$$

Sample Problem The addition of 108 mL of 5.00×10^{-3} N sodium hydroxide solution can neutralize 36.0 mL of a nitric acid solution. What is the concentration of the nitric acid?

Solution

$$N_u = \frac{(N_k)(V_k)}{V_u}$$

$$= \frac{(5.00 \times 10^{-3}\ N\ NaOH)(108\ mL\ NaOH)}{36.0\ mL\ HNO_3}$$

$$= 0.0150\ N\ \text{solution of } HNO_3$$

Buffers: Solutions That Are Anti-acid and Anti-base

During the course of a single day, most people probably eat substances with remarkably different pHs. A glass of orange juice has a pH of 4, while the milk of magnesia that follows a late night pizza has an approximate pH of 10.5. It is obvious that the human stomach can handle a wide range of pHs. The bloodstream, however, cannot tolerate this wide pH range. Its pH must stay between 7.35 and 7.45. God designed the bloodstream with an ingenious protective feature: a buffer system.

Buffers are solutions that resist pH changes despite small additions of hydronium or hydroxide ions. Water is not a buffer; a small amount of acid or base changes its pH dramatically. An addition of 0.01 mole of hydrochloric acid or sodium hydroxide to a liter of pure water changes the pH by 5 units. The same substances added to a liter of blood change the pH only 0.1 unit.

Buffer systems usually consist of a weak acid and its conjugate base or a weak base and its conjugate acid. For example, blood is buffered by a mixture of carbonic acid (H_2CO_3), which is a weak acid, and hydrogen carbonate ions (HCO_3^-), its conjugate base. This combination of solutes keeps the pH relatively constant by reacting with hydronium or hydroxide ions.

When an acid intrudes on this blood buffer system, hydrogen carbonate ions snare the hydronium ions.

$$H_3O^+ + HCO_3^- \longrightarrow H_2CO_3 + H_2O$$

When a base disturbs the equilibrium of the same buffer system, carbonic acid molecules spring into action and remove the hydroxide ions.

$$H_2CO_3 + OH^- \longrightarrow H_2O + HCO_3^-$$

Blood has a second buffer system that utilizes hydrogen phosphate ions (HPO_4^{2-}) and dihydrogen phosphate ions ($H_2PO_4^-$). If excess acid enters the bloodstream, it reacts with HPO_4^{2-} to form $H_2PO_4^-$. Bases react with $H_2PO_4^-$ to form HPO_4^{2-}; thus the pH is stabilized in either direction.

By properly selecting the acid-base pair, buffers can be used in almost any pH range. A given buffer acts to maintain the pH relatively constant within its working range. You can see several examples of buffers and their usable ranges in Table 15C-2.

Buffers are most effective in regulating slight pH changes. If an enormous amount of acid or base is added to a buffered solution, the buffer will be depleted and the pH will change drastically. Keep in mind that the buffer does not keep the solution neutral (pH of 7.0), but rather maintains a constant pH for the given solution. For example, suppose that the pH of a given solution containing a buffer is 4.7. As base is added to this solution, the buffer will maintain the pH at 4.7 until it has been depleted. Only then will the pH begin to rise with the addition of more base solution.

Table 15C-2 Common Buffer Systems

Components	Usable pH Range
Formic acid + sodium formate	2.6-4.8
Citric acid + sodium citrate	3.0-6.2
Acetic acid + sodium acetate	3.4-5.9
Sodium hydrogen carbonate + sodium carbonate	9.2-10.6
Sodium hydrogen carbonate + sodium hydroxide	9.6-11.0

Section Review Questions 15c

1. In titrations, what is the difference between the equivalence point and the endpoint?
2. In a titration reaction, 45.6 mL of 0.047 N NaOH was needed to neutralize 30.1 mL of H_2SO_4. Calculate the following measures of concentration.
   a. The normality of the acid solution
   b. The molarity of the acid solution
3. The addition of 37 mL of 4.2×10^{-2} N of HCl solution can neutralize 65 mL of NaOH solution. What is the concentration of NaOH?
4. A solution has a buffer that is effective in the 3.0-6.2 pH range. If the system is overloaded with acid, what will be the effect on the buffer system?

Chapter Review

Coming to Terms

Arrhenius acid
Arrhenius base
Brönsted-Lowry acid
Brönsted-Lowry base
deprotonation
protonation
conjugate pair
conjugate base
conjugate acid
Lewis acid
Lewis base
neutralization reaction
autoprotolysis
autoprotolysis constant of water
pH
pOH
acidic solution
basic solution
neutral solution
strong acid
weak acid
weak base
strong base
acid dissociation constant
amphiprotic
monoprotic
polyprotic
diprotic
triprotic
indicator
salt
neutral salt
acid salt
basic salt
titration
titration curve
equivalence point
end point
buffer

Review Questions

1. Fill in the blanks.
   a. Al(OH)$_3$ is an Arrhenius _____ and therefore must be a Brönsted-Lowry _____.
   b. H$_2$SO$_4$ is an Arrhenius _____ and therefore must be a Brönsted-Lowry _____.
   c. NH$_3$ can be classified as a _____ base and a _____ base, but not as a _____ base.
   d. The Cl$^-$ ion can be classified as a _____ base or a _____ base, but not as a _____ base.

2. True or False
   a. All Lewis acids donate protons.
   b. Brönsted-Lowry acids accept protons.
   c. Arrhenius bases donate protons.
   d. All Brönsted-Lowry acids are Lewis acids.
   e. All Arrhenius acids are Lewis acids.

3. Formic acid (HCO$_2$H) ionizes in water to form a formate ion and a hydronium ion.
   $$HCO_2H\ (aq) + H_2O\ (l) \rightleftharpoons H_3O^+\ (aq) + CO_2H^-\ (aq)$$
   a. What is the conjugate acid in the forward reaction?
   b. What is the conjugate base in the forward reaction?
   c. What is the name of the process by which formic acid loses a proton?

4. Two reactions describe the stepwise ionization of H$_2$SO$_4$:
   $$H_2SO_4\ (l) + H_2O\ (l) \rightleftharpoons HSO_4^-\ (aq) + H_3O^+\ (aq)$$
   $$HSO_4^-\ (aq) + H_2O\ (l) \rightleftharpoons SO_4^{2-}\ (aq) + H_3O^+\ (aq)$$
   a. What is the conjugate acid in the first reaction?
   b. What is the conjugate base in the first reaction?
   c. What is the conjugate acid in the second reaction?
   d. What is the conjugate base in the second reaction?

5. Fill in the blanks.
   a. Aqueous solutions of Arrhenius acids taste _____.
   b. Aqueous solutions of Arrhenius _____ are slippery.
   c. Arrhenius acids react with active metals to produce _____ gas.
   d. An aqueous solution of an Arrhenius base tastes _____.

6. What ion gives H$_2$SO$_4$ the ability to react strongly with many substances? What ion is responsible for the corrosiveness of strong bases like NaOH?

7. List four common properties of acids and four common properties of bases.

8. What particles, other than H_2O molecules, are always present in pure water?
9. Write the equation for the equilibrium constant that describes the reaction of water with itself. What is the numerical value of this constant?
10. Can water molecules act like acids, bases, both, or neither?
11. A young chemist measures $[H_3O^+]$ and $[OH^-]$ in an aqueous solution at 25°C. He reports that $[H_3O^+] = 1 \times 10^{-8}$ and $[OH^-] = 1 \times 10^{-8}$. His lab instructor tells him to go back to the lab and make the measurements again. Why?
12. What advantage does the pH scale offer that scientific notation does not?
13. All the following substances can undergo deprotonation reactions. Write the equations for their dissociation constants.
    a. $HClO_4$ b. H_2CO_3 c. H_3BO_3
14. Calculate the $[H_3O^+]$ and the pH of each solution.
    a. 1×10^{-5} M HCl   c. 1×10^{-1} M H_2SO_4
    b. 5×10^{-3} M HCl   d. 1×10^{-5} M H_2SO_4
15. Calculate the $[OH^-]$ and the pOH of each solution in problem 14.
16. What is $[H_3O^+]$ in each of the following solutions?
    a. orange juice, pH = 4.0
    b. black coffee, pH = 5.0
    c. pure water, pH = 7.0
    d. phosphate detergent solution, pH = 9.5
    e. seawater, pH = 8.0
17. Classify each of the solutions in the previous question as acidic, neutral, or basic.
18. Choose the correct answer.
    a. (Strong, Weak) acids ionize incompletely.
    b. (Strong, Weak) acids have large K_as.
    c. Strong (acids, bases) accept protons easily.
    d. The conjugate base of a strong acid is a (strong, weak) base.
19. Referring to Table 15B-4, tell which acid in each of the following pairs is the stronger.
    a. $HClO_4$, H_3PO_4
    b. formic acid, acetic acid
    c. hydrocyanic acid, formic acid
    d. NH_4^+, HSO_4^-
    e. NH_4^+, H_2O

20. Choose the member of each pair that should be the stronger acid and then write a brief justification for your choice.
    a. $HBrO_4$, HBr
    b. PH_3, H_2S

21. Is the HCO_3^- ion an acid, a base, both, or neither? Explain.

22. Write the equation that shows how the ions in acids act to neutralize the ions in bases during a neutralization reaction.

23. A NaOH solution contains 1.000 M NaOH solution. If 25.00 mL of this solution neutralizes 27.00 mL of a HCl solution of unknown concentration,
    a. what is the normality of the HCl?
    b. what is the molarity of the HCl?

24. How many milliliters of 2.00 M H_2SO_4 will be required to neutralize 45.0 mL of 3.00 N KOH?

25. A careless laboratory assistant set out to determine the concentration of a perchloric acid ($HClO_4$) solution. He found that 50.00 mL of 1.000 N NaOH neutralized 0.5000 L of the $HClO_4$ solution. His calculations were as follows:

$$N_u = \frac{(50.00 \text{ mL})(1.000 \text{ N})}{0.5000 \text{ L}} = 100.0 \text{ N } HClO_4$$

Another laboratory assistant insists that the calculation is incorrect. Why? What is the correct answer?

26. A salad dressing manufacturer desires to make tangy salad dressing by using vinegar with an acetic acid concentration of at least 1 N. The quality control department examines a sample of vinegar and determines that 300.0 mL of 0.100 N sodium hydroxide neutralizes the acetic acid in 25.00 mL of the vinegar. Does this vinegar meet the manufacturer's requirements?

27. You wish to determine the point at which a chemical reaction raises the pH of a solution past 8.8. Exactly how could you do this?

28. What happens to excess H_3O^+ ions when an acid is added to a solution buffered by a combination of acetic acid and sodium acetate?

29. What happens to excess OH^- ions when a base is added to a solution buffered by a combination of acetic acid and sodium acetate?

16A Redox Reactions page 417
16B Electrochemical Reactions page 431
FACETS: The Early Atmosphere page 422
FACETS: The Battle Against Corrosion page 437

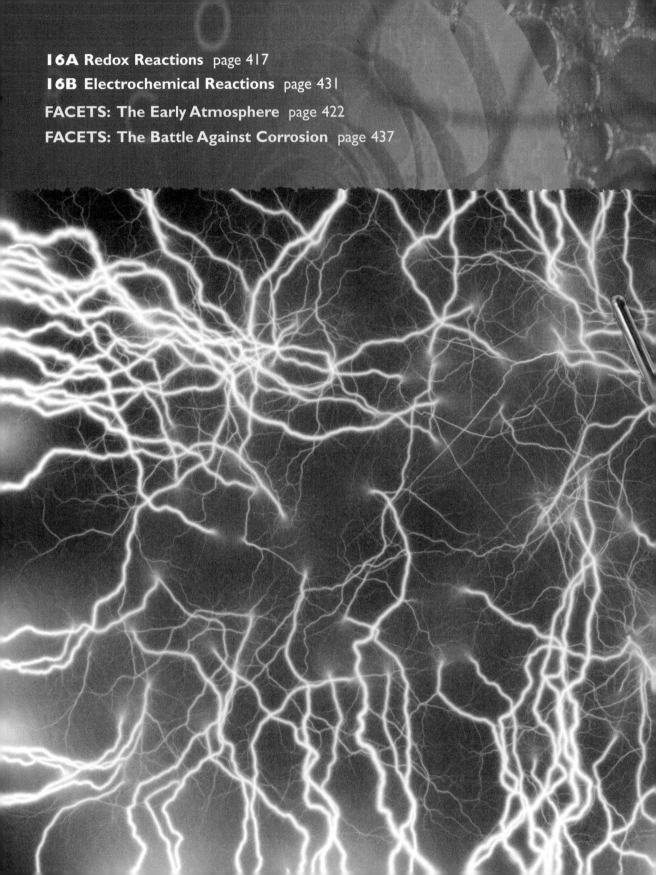

Oxidation-Reduction 16

Electrons on the Move

Reactions that involve transfers or shifts of electrons occur in every branch of chemistry. In living cells and test tubes, the substances that gain and lose electrons are mixed together. In other cases the substances are separated, and the electrons travel through wires, flow out of batteries, split compounds, or help plate metal onto objects.

16A Redox Reactions

How do the chemicals on photographic film record the presence of light? At the heart of the film-developing process is the transfer of electrons between different compounds. Some compounds undergo the process of reduction, and some go through oxidation. The result of these processes is a set of compounds with different colors arranged to record images on film. Because electrons move around in so many reactions, the study of these movements has become an important part of chemistry.

Oxidation: Loss of Electrons

The general name **oxidation-reduction reactions** is given to reactions in which electrons are transferred from one atom or molecule to another. The shortened form of this name is **redox reactions**. The formation of magnesium oxide as magnesium burns involves electron transfers between atoms. In this case, oxygen atoms strip two electrons from each magnesium atom.

Oxidation causes apples to turn brown when they are exposed to air.

Even reactions between covalent compounds often end up shifting electrons toward or away from atoms, depending on their electronegativities. When sulfur dioxide forms from its elements, oxygen attracts the shared electrons closer to itself than sulfur does, so, in a sense, oxygen gains electrons.

Oxidation entails the loss of electrons from an atom. In the reaction between magnesium and oxygen, magnesium atoms are oxidized.

$$Mg: \longrightarrow Mg^{2+} + 2\ e^-$$

Lost electrons (e^-) are shown as reaction products, while gained electrons are treated as reactants. Because electrons with their negative charges are lost, the oxidation numbers of atoms that are oxidized always become more positive. The oxidation number of magnesium is 0 before it reacts and +2 after it reacts. The oxidation number of magnesium increased by two in this reaction because each atom of magnesium lost two electrons.

$$\overset{0}{2\ Mg} + O_2 \longrightarrow \overset{+2}{2\ MgO}$$

At this point it may seem logical to assume that all oxidation reactions involve oxygen. While oxygen usually causes atoms to lose electrons because of its high electronegativity, other atoms can do the same. The reaction between sodium and chlorine provides a good example of an oxidation reaction that does not involve oxygen. Chlorine causes the sodium atoms to be oxidized.

$$\overset{0}{2\ Na} + \overset{0}{Cl_2} \longrightarrow \overset{+1}{2\ Na^+} + \overset{-1}{2\ Cl^-} \longrightarrow \overset{+1\ -1}{2\ NaCl}$$

Reduction: Gain of Electrons

The electrons lost by atoms in oxidation reactions must go somewhere. They are gained by other atoms. An atom that gains electrons in a reaction is said to be reduced, and the process of gaining electrons is called **reduction.** In the reaction between magnesium and oxygen, each oxygen atom is reduced because it gains two electrons. Reduction causes oxidation numbers to become smaller (that is, it *reduces* the oxidation number) as more negative charges join atoms. The oxidation number of oxygen reduces from 0 to -2.

Black-and-white photography is based on the reduction of silver from Ag^+ to Ag. Photographic film contains grains of a silver halide such as silver bromide (AgBr). When exposed to light, silver ions in the silver bromide grains are sensitized. These sensitized ions are more prone to be reduced during the developing process than those in areas not struck by light. The film is developed in a solution containing substances that cause the sensitized silver ions to gain electrons.

$$\overset{+1}{Ag^+} + e^- \longrightarrow \overset{0}{Ag}$$

The silver appears as black areas on the developed film. This stage of development produces a negative on which light objects in the photographed scene appear dark. Conversely, dark objects in the scene appear light on the negative. The film is subjected to a fixing procedure that prevents further blackening. To produce a positive, or print, that looks like the scene that was photographed, a light is projected through the negative to produce an image on paper treated with a silver halide. The paper is then developed with the same procedure used on the film. Although color photography is more difficult to explain, it is also based on reduction processes.

The photographic industry uses silver salts in large quantities, especially in the production of film used by radiography services in health care facilities. The price of silver usually makes its recovery from the discarded film and developing solutions a worthwhile project. The recovery of these discarded silver compounds is accomplished by first burning the film and then reducing the silver ions in the ash by heating it with inexpensive electropositive metals such as steel wool:

$$Fe\ (s) + 2\ Ag^+ \longrightarrow Fe^{2+} + 2\ Ag\ (s)$$

or electrochemically by using a pure silver metal cathode:

$$Ag^+\ (s) + e^- \longrightarrow Ag\ (s)$$

Figure 16A-1

Production of a photographic print

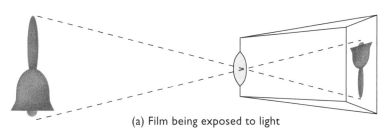
(a) Film being exposed to light

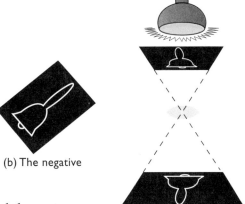
(b) The negative

(c) Light shown through negative onto a second piece of film

(d) The finished, enlarged print

Is the reaction that forms magnesium oxide a reduction process? Is it an oxidation process? Actually, it is both. Oxygen atoms gain electrons and are reduced. But the magnesium atoms lose electrons and are oxidized. Oxidation and reduction **ALWAYS** occur simultaneously. When electrons leave one atom, they must join another. As they shift away from one atom, they must migrate toward another.

Identifying Redox Reactions: Oxidation Numbers Are the Key

The only reactions that qualify as redox reactions are those in which the oxidation numbers of the atoms involved change. The absence of change indicates that neither oxidation nor reduction has occurred.

Chlorine is a potent oxidizing agent often used in public water disinfection.

Sample Problem Is either of the following reactions a redox reaction?
a. $HCl + H_2O \rightleftharpoons H_3O^+ + Cl^-$
b. $2\ KClO_3 \longrightarrow 2\ KCl + 3\ O_2$

Solution
a. After assigning oxidation numbers to each atom, you can see that the oxidation numbers of all the atoms remain the same before and after the reaction. The reaction is not a redox reaction.

$$\overset{+1\ -1}{HCl} + \overset{+1\ -2}{H_2O} \rightleftharpoons \overset{+1\ -2}{H_3O^+} + \overset{-1}{Cl^-}$$

b. A determination of the oxidation numbers before and after the reaction shows that both oxygen and chlorine atoms change their numbers. This reaction involves both oxidation and reduction.

$$\overset{+1\ +5\ -2}{2\ KClO_3} \longrightarrow \overset{+1\ -1}{2\ KCl} + \overset{0}{3\ O_2}$$

Oxidizing and Reducing Agents

A **reducing agent** is a substance that causes another substance to be reduced. Recall that the development steps in photography involve the use of one or more reducing agents. They cause silver ions to be reduced to silver atoms. Because reduction cannot occur without oxidation, reducing agents contain atoms that donate electrons. Because magnesium donates electrons, it is a reducing agent in the reaction between magnesium and oxygen.

$$2\ Mg + O_2 \longrightarrow 2\ MgO$$

Remember that oxidation and reduction always occur simultaneously. If magnesium acts as a reducing agent, it donates electrons. Therefore, it is oxidized (**L**oss of **E**lectrons is **O**xidation). Reducing agents are always oxidized in redox reactions. There are no exceptions.

Oxidizing agents are substances that cause other atoms to be oxidized. Many oxidizing agents are commonly used as cleaners and disinfectants. Chlorine, for instance, is an oxidizing agent frequently added to swimming pools to kill bacteria, algae, and fungi. Chlorine acts as an oxidizing agent by taking electrons from these microbes. In the process, chlorine atoms are reduced to chloride ions.

$$Cl_2 + 2\ e^- \longrightarrow 2\ Cl^-$$

Because chlorine gains electrons in this reaction, it is reduced (**G**ain of **E**lectrons is **R**eduction). Oxidizing agents are always reduced in redox reactions.

Figure 16A-2

"LEO the lion says GER"—a memory aid for redox reactions

Sample Problem Tell which substance in the following reaction is a reducing agent and which is an oxidizing agent.

$$2\ Na + 2\ H_2O \longrightarrow 2\ NaOH + H_2$$

Solution
The oxidation number of sodium goes from 0 in sodium to +1 in sodium hydroxide. The oxidation number of hydrogen goes from +1 in water to 0 in hydrogen gas. Sodium was oxidized (lost electrons), and hydrogen was reduced (gained electrons). The hydrogen in water, therefore, is acting as an oxidizing agent, and sodium is acting as a reducing agent.

Household bleach is a 5 percent aqueous solution of sodium hypochlorite (NaOCl). Sodium hypochlorite is an effective oxidizing agent. It is reduced in redox reactions according to the following equation:

$$NaOCl + H_2O + 2\ e^- \longrightarrow NaCl + 2\ OH^-$$

Bleach is normally used to eliminate stains and make dirty white clothes clean and white again. The molecules in stains possess electrons that can move around easily. These molecules are colored because they absorb visible light energy that hits them and energizes their electrons. When a dirty

Sodium hypochlorite decolorizes cloth.

article of clothing is bleached, the sodium hypochlorite in the bleach grabs the less tightly bonded electrons from the molecules in the stains. When these electrons are removed, the molecules no longer absorb visible light. Consequently, the color disappears, and the cloth is once again white. In effect, bleach does not remove stains; it merely decolorizes them. Because sodium hypochlorite can also oxidize hair and stained dentures, it is found in commercial products used for these purposes.

The Early Atmosphere

"Mathematics and dynamics fail us when we contemplate the earth, fitted for life but lifeless, and try to imagine the commencement of life upon it. This certainly did not take place by any action of chemistry, or electricity, or crystalline grouping of molecules under the influence of force, or by any possible kind of fortuitous concourse of atmosphere. We must pause, face to face with the mystery and miracle of the creation of living things."

—Lord Kelvin, quoted in *Battle for Creation*

Scripture reveals that during the six-day Creation period, the world was fully prepared for man's use. The atmosphere was formed on the second day in preparation for the plants that appeared on the third day. Animals and man were then created on the fifth and sixth days. At the end of the sixth day, God viewed everything He had made and pronounced it "very good," implying a state of completeness—all that was needed for life was present. One can then deduce that the earth's earliest atmosphere contained oxygen. However, evolutionists have a different theory about the early atmosphere.

In 1929 A. I. Oparin, from Russia, and John Haldane, from Britain, both biochemists, suggested that the earth's atmosphere was at one time without oxygen. They proposed that the earth's original atmosphere consisted primarily of methane, ammonia, hydrogen, and water vapor— completely devoid of oxygen. This supposed early atmosphere is described as reducing; it would have supplied electrons to substances and chemically reduced them to other forms. Although Oparin and Haldane agreed that spontaneous generation was not possible in the present-day atmosphere, they proposed that in the reducing atmosphere the conditions would favor spontaneous generation of life.

American chemist Harold Urey (1893-1981) published an elaboration of the Oparin-Haldane theory in 1952. Working in Urey's laboratory as a graduate student, Stanley Miller (b. 1930) developed an experiment to test a hypothesis—could life have originated in the reduced atmosphere? In his experiment he simulated a reducing atmos-

Balancing Redox Reactions: The Half-Reaction Method

Balanced chemical equations accurately describe the quantities of reactants and products in chemical reactions. They serve as the basis of stoichiometry by showing how atoms and mass are conserved during reactions. Oxidation and reduction reactions need to be balanced just as all other reactions do, but they are often quite complicated. Balancing redox reactions involves balancing not only atoms but also positive and negative charges.

phere, and using electrodes to provide energy, Miller was able to produce some basic amino acids. Although the newspapers of the day claimed that he had "created life," those amino acids were no more living creatures than a pile of lumber, nails, and wire is a house!

Is there any evidence to support the idea of an early reducing atmosphere? Proponents of this theory point to geologic formations from the Precambrian era containing various sediments of reduced minerals. However, these same geologic formations that evolutionists use to support their claim for the reducing atmosphere contain large amounts of calcium carbonate ($CaCO_3$) and iron oxide (Fe_2O_3), but these compounds could have formed only if large amounts of oxygen were present.

Two other problems face the idea of a reducing atmosphere. The first is that ultraviolet radiation from the sun in a reducing atmosphere without the protective ozone layer would have decomposed ammonia long before the envisioned life-forming reactions could have taken place. The second problem is, where did the oxygen come from? Two general approaches have been used in an effort to explain the buildup of oxygen in the atmosphere. The first theory states that early plant life formed the oxygen by photosynthesis.

It is unrealistic to think that plants could have appeared before oxygen, but assuming they did, another difficulty would arise. Photosynthesis produces not only oxygen but also organic matter. But the decaying organic matter would have depleted the oxygen.

The second idea that has been proposed to explain the oxygen buildup is that steam from volcanoes was decomposed to hydrogen and oxygen by ultraviolet energy from the sun. The hydrogen then presumably escaped into space, leaving the oxygen behind. However, this scheme could not have worked unless the early sun had several hundred times its present ultraviolet output.

Over the years, evolutionists have contrived many theories of earth history and the origin of life. However, none of the theories are anything more than speculation. Since theories have been known to change, it is possible that the reducing atmosphere may not be a permanent feature of evolutionary theory. Efforts will be made to patch it up for a time, but if it is finally deemed hopeless, it will be replaced by something else. It is interesting to note that the odds of a single cell's originating in a primitive environment, given a 4.6 billion-year time frame, are one chance in $10^{40,000}$.

Balancing either the atoms or the charges is not hard, but the process becomes difficult when both must be balanced simultaneously. Chemists have developed a way to balance redox reactions in an organized and straightforward process.

Redox reactions that are difficult to balance can be managed more easily if the processes of oxidation and reduction are considered separately before they are considered together. In the **half-reaction method,** the redox reaction is split into two hypothetical parts called **half-reactions.** The oxidation half-reaction deals with all the substances that become oxidized and all the electrons they lose. The reduction half-reaction involves substances that become reduced and all the electrons they gain.

The half-reaction method consists of eight steps that help balance reactions in an organized fashion. This method can reduce the task of balancing even complicated reactions such as the one between nitric acid and copper (I) oxide to a series of manageable procedures.

$$HNO_3 + Cu_2O \longrightarrow Cu(NO_3)_2 + NO + H_2O$$

Step 1. Assign oxidation numbers to all atoms, and determine which atoms are oxidized (LEO) and which ones are reduced (GER). Since the oxidation numbers of hydrogen and oxygen remain unchanged in this particular reaction, ignore them for now and focus on the other atoms. As previously discussed, ion charges are written with the charge designation after the numeral (3-), and the oxidation numbers are written with the charge designation preceding the numeral (+5). This will help avoid confusion when dealing with reactions containing ions.

$$\overset{+5}{H}\overset{}{N}O_3 + \overset{+1}{Cu_2}O \longrightarrow \overset{+2}{Cu}(\overset{+5}{N}O_3)_2 + \overset{+2}{N}O + H_2O$$

In this reaction, the oxidation number of nitrogen changes from +5 in nitric acid to +2 in nitrogen (II) oxide; so nitrogen is reduced (GER: gains 3 electrons). The oxidation number of copper changes from +1 in copper (I) oxide to +2 in copper nitrate; so copper is oxidized (LEO: loses 1 electron).

Step 2. Write a half-reaction for the oxidation process and one for the reduction process. Labeling each half-reaction with either LEO or GER will help keep them separate as they are balanced in the following steps. The half-reactions include only those substances that were either oxidized or reduced. Since the oxidation number of copper went from +1 to +2, it lost an electron: thus it was oxidized (LEO). All the compounds with copper atoms in them should be included in the oxidation half-reaction.

$$LEO: Cu_2O \longrightarrow Cu(NO_3)_2$$

The reduction half-reaction includes only those substances containing the atoms that are reduced. The oxidation number of nitrogen went from +5 to +2 by gaining three electrons (GER). Notice that even though copper nitrate contains nitrogen, it is not included in the reduction half-reaction because the nitrogen in copper nitrate is not reduced (the oxidation number remains +5).

$$\text{GER: } HNO_3 \longrightarrow NO$$

Step 3. Balance the atoms that were reduced or oxidized in the two half-reactions by inspection. A coefficient of 2 was added to the right side of the oxidation half-reaction to balance the copper atoms. In the reduction half-reaction the nitrogen atoms are balanced already.

$$\text{LEO: } Cu_2 \longrightarrow \mathbf{2}\ Cu(NO_3)_2 \quad 2\ Cu = 2\ Cu$$
$$\text{GER: } HNO_3 \longrightarrow NO \quad\quad\quad 1\ N = 1\ N$$

Step 4. Balance all other atoms by inspection. Most redox reactions (at least all the ones that will be considered in this text) take place in acidic solutions where there is an unlimited supply of hydrogen ions and water molecules. When needed, these can be added to one side of the equation or the other. If additional oxygen is needed to balance the reaction, add water. If additional hydrogen is needed, add hydrogen ions. In the oxidation reaction (LEO), the addition of 4 HNO_3 to the left side is needed to balance the nitrogen atoms.

$$\text{LEO: } \mathbf{4\ HNO_3} + Cu_2O \longrightarrow 2\ Cu(NO_3)_2$$

The extra oxygen on the left side can be balanced by the addition of one water molecule to the right-hand side. The extra hydrogens can be balanced by the addition of two hydrogen ions also to the right-hand side.

$$\text{LEO: } 4\ HNO_3 + Cu_2O \longrightarrow 2\ Cu(NO_3)_2 + \mathbf{H_2O} + \mathbf{2\ H^+}$$

The balancing of the reduction (GER) half-reaction involves balancing the hydrogen and the oxygen atoms and is performed using water and hydrogen ions.

$$\text{GER: } HNO_3 + \mathbf{3\ H^+} \longrightarrow NO + \mathbf{2\ H_2O}$$

Step 5. Balance the charges in each half-reaction by adding electrons until the total charge is the same on both sides. *Note:* the total charge does not have to equal zero.

$$\text{LEO: } 4\ HNO_3 + Cu_2O \longrightarrow 2\ Cu(NO_3)_2 + H_2O + 2\ H^+ + \mathbf{2\ e^-}$$

In the above oxidation reaction, the total charge of the left side is zero, and the right side is +2 from the previously added hydrogen ions. Therefore, two electrons must be added to the right side to make its total charge zero.

$$\text{GER: } HNO_3 + 3\ H^+ + \mathbf{3}\ e^- \longrightarrow NO + 2\ H_2O$$

In the reduction reaction, the total charge of the right side is zero, and the left side is +3; therefore, three electrons must be added to the left side to equalize the charges. Now both half-reactions are balanced, and the total charges of each side are equal.

Step 6. Multiply each half-reaction by an appropriate whole number so that the number of electrons produced by the oxidation half-reaction equals the number used by the reduction half-reaction. In this case, multiply all the coefficients in the oxidation half-reaction by three.

$$\text{LEO: } \mathbf{3}(4\ HNO_3 + Cu_2O \longrightarrow NO + 2\ H_2O + 2\ e^-)$$

$$\mathbf{12}\ HNO_3 + \mathbf{3}\ Cu_2O \longrightarrow \mathbf{3}\ NO + \mathbf{6}\ H_2O + \mathbf{6}\ e^-$$

To make both half-reactions show the same number of electrons, multiply all the coefficients in the reduction half-reaction by two.

$$\text{GER: } \mathbf{2}(HNO_3 + 3\ H^+ + 3\ e^- \longrightarrow NO + 2\ H_2O)$$

$$\mathbf{2}\ HNO_3 + \mathbf{6}\ H^+ + \mathbf{6}\ e^- \longrightarrow \mathbf{2}\ NO + \mathbf{4}\ H_2O$$

Step 7. Add the oxidation and reduction half-reactions together. Add the reactants from both reactions together and put the result on the left-hand side. The products from both half-reactions are then totaled and placed on the right-hand side.

$$12\ HNO_3 + 3\ Cu_2O \longrightarrow 6\ Cu(NO_3)_2 + 3\ H_2O + 6\ H^+ + 6\ e^-$$
$$+ 2\ HNO_3 + 6\ H^+ + 6\ e^- \longrightarrow 2\ NO + 4\ H_2O$$
$$\overline{14\ HNO_3 + 3\ Cu_2O + 6\ H^+ + 6\ e^- \longrightarrow}$$
$$6\ Cu(NO_3)_2 + 2\ NO + 7\ H_2O + 6\ H^+ + 6\ e^-$$

Step 8. Cancel any quantities that appear on both sides of the overall reaction.

$$14\ HNO_3 + 3\ Cu_2O + \cancel{6\ H^+} + \cancel{6\ e^-} \longrightarrow$$
$$6\ Cu(NO_3)_2 + 2\ NO + 7\ H_2O + \cancel{6\ H^+} + \cancel{6\ e^-}$$

Table 16A-1 Steps to Balancing Redox Reactions

Step 1	Assign oxidation numbers.
Step 2	Write half-reactions.
Step 3	Balance reduced/oxidized atoms in each half-reaction.
Step 4	Balance all other atoms.
Step 5	Balance charges in each half-reaction.
Step 6	Equalize electrons in each half-reaction.
Step 7	Add half-reactions together.
Step 8	Cancel quantities.

Since there are six hydrogen ions and six electrons on both sides, they cancel out.

$$14 \, HNO_3 + 3 \, Cu_2O \longrightarrow 6 \, Cu(NO_3)_2 + 2 \, NO + 7 \, H_2O$$

The reaction is balanced. The result is the same as the one that would have been obtained by the old balance-by-inspection method, but the extra steps help to keep you on the right track throughout the entire balancing process. Balancing redox reactions may seem difficult at first, but do not despair; practice makes perfect.

Sample Problem Sulfite ions and permanganate ions can react to form sulfate ions and manganese ions. Balance this reaction, using the half-reaction method.

$$SO_3^{2-} + MnO_4^- \longrightarrow SO_4^{2-} + Mn^{2+}$$

Solution

Step 1. Assign oxidation numbers and determine which atoms are reduced and oxidized.

$$\overset{+4}{S}O_3^{2-} + \overset{+7}{M}nO_4^- \longrightarrow \overset{+6}{S}O_4^{2-} + \overset{+2}{M}n^{2+}$$

The sulfur lost 2 electrons (LEO), so it was oxidized; and the manganese gained 5 electrons (GER), so it was reduced.

Step 2. Identify half-reactions.

LEO: $\overset{+4}{S}O_3^{2-} \longrightarrow \overset{+6}{S}O_4^{2-}$

GER: $\overset{+7}{M}nO_4^- \longrightarrow \overset{+2}{M}n^{2+}$

Step 3. Balance the reduced/oxidized atoms in each of the half-reactions. In these half-reactions, the atoms are already balanced.

Step 4. Balance the remaining atoms in the half-reactions, using water and H⁺ where needed.

LEO: $SO_3^{2-} + H_2O \longrightarrow SO_4^{2-} + 2 \, H^+$

GER: $MnO_4^- + 8 \, H^+ \longrightarrow Mn^{2+} + 4 \, H_2O$

Step 5. Balance the charges in each half-reaction. There is a -2 charge on the left side of the oxidation reaction and zero on the right side (the +2 of the hydrogen ions will cancel out the −2 on the sulfate ion). Remember that the charges do not have to equal zero, but they do need to be equal. Therefore 2 electrons can be added to the right side to equalize the charges.

LEO: $SO_3^{2-} + H_2O \longrightarrow SO_4^{2-} + 2 H^+ + 2 e^-$

The right side of the reduction half-reaction has a +2 charge; the left side has a +7 charge—the sum of a −1 from the manganate ion and +8 from the 8 hydrogen ions. Therefore, to equalize the charge of +2 on the right side, 5 electrons must be added to the left side.

GER: $MnO_4^- + 8 H^+ + 5 e^- \longrightarrow Mn^{2+} + 4 H_2O$

Step 6. Multiply all of the coefficients in both half-reactions so that the electrons will cancel out. In this case the oxidation half-reaction should be multiplied by five, and the reduction half-reaction by two.

LEO: $5 SO_3^{2-} + 5 H_2O \longrightarrow 5 SO_4^{2-} + 10 H^+ + 10 e^-$

GER: $2 MnO_4^- + 16 H^+ + 10 e^- \longrightarrow 2 Mn^{2+} + 8 H_2O$

Step 7. Add the half-reactions.

$5 SO_3^{2-} + 5 H_2O \longrightarrow 5 SO_4^{2-} + 10 H^+ + 10 e^-$
$+ 2 MnO_4^- + 16 H^+ + 10 e^- \longrightarrow 2 Mn^{2+} + 8 H_2O$
$\overline{5 SO_3^{2-} + 2 MnO_4^- + 5 H_2O + 16 H^+ + 10 e^- \longrightarrow}$
$5 SO_4^{2-} + 2 Mn^{2+} + 8 H_2O + 10 H^+ + 10 e^-$

Step 8. Cancel. All electrons, ten hydrogen ions, and five water molecules will cancel from both sides, resulting in the final reaction as follows:

$5 SO_3^{2-} + 2 MnO_4^- + 6 H^+ \rightarrow 5 SO_4^{2-} + 2 Mn^+ + 3 H_2O$

Redox Titrations: Finding Unknown Concentrations

A titration is an experiment in which chemists react a solution of known concentration with a solution of unknown concentration in order to determine the unknown concentration. A redox titration is a titration in which one of the reacting substances is an oxidizing agent and the other is a reducing agent. As with acid-base titrations, the concentrations and volumes of the two reacting substances are related to each other by the equation

$$V_u N_u = V_k N_k$$

The concentrations are measured in normality (the number of equivalents per liter of solution). In the context of redox chemistry, an equivalent is defined as the number of moles of the substance that either loses or gains 1 mole of electrons in a balanced half-reaction. You can determine the number of electrons that are either lost or gained from the balanced half-reactions in the redox equation.

Sample Problem Calculate the number of equivalents (electrons lost or gained) by copper (I) oxide and nitric acid in the following reaction.

$$14\ HNO_3 + 3\ Cu_2O \longrightarrow 6\ Cu(NO_3)_2 + 2\ NO + 7\ H_2O$$

Solution
The balanced half-reactions for this redox reaction have been determined earlier in the chapter on page 426.

LEO: $3\ Cu_2O + 12\ HNO_3 \longrightarrow 6\ Cu(NO_3)_2 + 3\ H_2O + 6\ H^+ + 6\ e^-$

GER: $2\ HNO_3 + 6\ H^+ + 6\ e^- \longrightarrow 2\ NO + 4\ H_2O$

Thus, 3 moles of copper (I) oxide produce 6 moles of electrons (a 1:2 ratio). Therefore, there are 2 equivalents in 1 mole of copper (I) oxide in the oxidation reaction.

In the reduction half-reaction, 2 moles of nitric acid react with 6 moles of electrons—a 1:3 ratio. Thus, there are 3 equivalents in 1 mole of nitric acid.

The calculations involved in redox titrations follow the same patterns as those for acid-base titrations. The equation $V_u N_u = V_k N_k$ is used in both cases.

Sample Problem One liter of a solution contains 4.74 g of potassium permanganate ($KMnO_4$). If 0.030 L of this solution titrates 0.032 L of a sodium sulfite (Na_2SO_3) solution, what is the normality of the sodium sulfite solution? The simple ionic equation for the reaction is

$$5\ SO_3^{2-} + MnO_4^- + 6\ H^+ \longrightarrow 5\ SO_4^{2-} + 2\ Mn^{2+} + 3\ H_2O$$

Solution

Before you can use the relationship $V_u N_u = V_k N_k$ to find the normality of the sodium sulfite solution, you must calculate the normality of the potassium permanganate solution. To calculate the normality, first determine the number of equivalents for the permanganate ion.

$$2\ Mn^{7+} + 10\ e^- \longrightarrow 2\ Mn^{2+} \quad \text{a 5:1 ratio}$$

In this reaction the permanganate ion has 5 equivalents per mole. Using unit analysis, the normality of the potassium permanganate is

$$\frac{4.74\ \text{g KMnO}_4}{\text{L solution}} \left| \frac{1\ \text{mol KMnO}_4}{158.04\ \text{g KMnO}_4} \right| \frac{5\ \text{eq KMnO}_4}{1\ \text{mol KMnO}_4} = 0.150\ N$$

Now you can calculate the normality of sodium sulfite by substituting the given volumes and the normality of the potassium permanganate into the standard equation.

$$V_u = 0.032 \qquad N_u = \text{unknown}$$
$$V_k = 0.030 \qquad N_k = 0.150\ N$$

$$V_u N_u = V_k N_k$$

$$N_u = \frac{(V_k)(N_k)}{V_u} = \frac{(0.030\ L)(0.150\ N)}{0.032\ L} = 0.14\ N\ Na_2SO_3$$

Section Review Questions 16A

1. Balance the following equations. Identify the oxidizing and reducing agents.
   a. $SnCl_2 + O_2 + HCl \longrightarrow H_2SnCl_6$
   b. $H_2SO_4 + HI \longrightarrow H_2S + I_2$
   c. $Cr_2O_7^{2-} + C_2O_4^{2-} \longrightarrow Cr^{3+} + CO_2$
   d. $MnO_2 + HNO_2 \longrightarrow Mn^{2+} + NO_3^-$

2. A 25.0 mL sample of barium hydroxide solution, $Ba(OH)_2$, was titrated with 0.150 N HCl. The titration required 45.3 mL of HCl. What was the normality of the barium hydroxide solution?

3. Balance the following reaction equations.
   a. $H_2S + HNO_3 \longrightarrow H_2SO_4 + NO_2 + H_2O$
   b. $As + HNO_3 + H_2O \longrightarrow H_3AsO_4 + NO$
   c. $KOH + CrCl_3 + Cl_2 \longrightarrow K_2CrO_4 + KCl + H_2O$

4. Calculate the normality for 10.0 g of the following reagents dissolved in 50.0 mL of H_2O.
   a. H_2S \qquad b. As \qquad c. $CrCl_3$

16B Electrochemical Reactions

Electrochemistry deals with redox reactions that are manipulated either to produce or to consume electricity. It includes the operation of batteries, the electroplating of metal objects, and the liberation of useful elements from their stable compounds.

Electrochemical Cells: Wires, Electrodes, and Electrolytes

Metals can conduct electricity because metallic bonds allow electrons to move freely throughout a piece of metal. Some solutions can conduct electricity, but not for the same reason. Water can conduct electricity only when some ionic substance is dissolved in it. An **electrolyte** is any substance that, when dissolved in water, allows the resulting solution to conduct electricity. When an electrolyte is dissolved in water, anions (negative ions) and cations (positive ions) are formed. The ions move freely in the solution and therefore may carry a charge. Solutions of strong electrolytes conduct electricity well because they completely dissociate to put many ions into the solution. Most salts, strong acids, and strong bases fill the list of strong electrolytes (NaCl, HCl, H_2SO_4, HNO_3, and NaOH). Substances that do not ionize completely, such as weak acids, weak bases, and hard-to-dissolve salts, are classified as weak electrolytes. Solutions of weak electrolytes conduct electricity but not as well as solutions of strong electrolytes. Nonionic substances such as sugar, alcohol, and oxygen might dissolve in water, but they cannot conduct electricity. For this reason, such substances are called **nonelectrolytes.**

All electrochemical techniques rely on combinations of metals and solutions that conduct electricity. In order to be useful in a variety of ways, however, these substances must be put together in just the right way. The fundamental apparatus used in electrochemistry is the electrochemical cell. An **electrochemical cell** consists of two electrical contacts, called **electrodes,** immersed in an electrolyte solution with a wire joining the electrodes. The electrodes are nothing more than metal rods or wires. They are commonly made of metals such as zinc, platinum, or copper.

There are two types of electrochemical cells, galvanic (voltaic) and electrolytic. In **galvanic cells,** a chemical reaction spontaneously occurs to produce electrical energy. All batteries consist of one or more galvanic cells. In **electrolytic cells,** electrical energy is used to force a nonspontaneous chemical reaction to occur. In both kinds of cells, oxidation occurs at the anode and reduction occurs at the cathode.

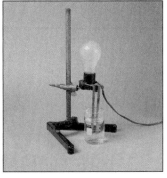

Solutions of a strong electrolyte (HCl), a weak electrolyte (acetic acid), and a nonelectrolyte (sugar) have different electrical conductivities.

Chapter 16B

Figure 16B-1

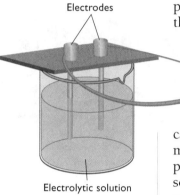

Galvanic cell

Electrodes

Electrolytic solution

Electrolytic Cells: Forcing reactions to occur

Electrolysis is the process of forcing an otherwise nonspontaneous redox reaction to occur with the aid of an electrical current in an electrochemical cell. For example, normally stable water molecules can be pulled apart by an electrical current to produce hydrogen and oxygen gases. Current can be passed through an electrolytic cell when a source of electricity is connected to two electrodes immersed in an electrolyte solution.

What happens when electrons flow through the cell? First, electrons flow from the source of the electrical energy into the electrode called the **cathode**, giving it a negative charge. The electrons in the cathode become available for reduction reactions. A redox reaction cannot occur unless oxidation and reduction reactions occur simultaneously. The oxidation occurs at the **anode**, which is the positively charged electrode. Electrons then flow back to the source of electrical energy. Since any anions in the solution are negatively charged, they migrate to the positively charged anode. Similarly, positively charged cations migrate to the negatively charged cathode. That is why positively charged ions are called cations, and negatively charged ions are called anions. If the electrical forces between the two electrodes are large enough, a redox reaction will occur.

Chemical engineers use electrolysis to purify active metals. Metals such as copper, tin, and iron, which are low on the activity series, can be freed from their natural compounds by chemical means. However, active metals such as sodium, lithium, and aluminum bond too strongly for those "mild" techniques to work, so engineers must separate the compounds in electrolytic cells. In nature, aluminum atoms are oxidized by oxygen atoms in an ore commonly known as bauxite. To get pure aluminum metal from the very stable aluminum oxide (Al_2O_3), the natural oxidation reaction must be reversed—electrons must be forced back into the aluminum ions.

The electrolysis process used to produce aluminum in industry is called the Hall-Héroult process. This process is named after Charles Hall and Paul Héroult, who developed the process independently in 1886. In the Hall-Héroult process, aluminum oxide is dissolved in molten cryolite and electrolyzed with carbon electrodes.

$$3\ C + 4\ Al^{3+} + 6\ O^{2-} \longrightarrow 4\ Al + 3\ CO_2$$

Sodium hydroxide and chlorine gas are produced by an electrolytic process in the chlor-alkali industry from a concentrated sodium

Figure 16B-2

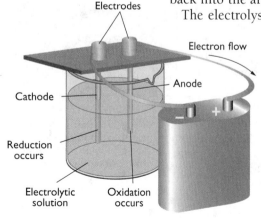

Electrolytic cell

Electrodes

Electron flow

Anode

Cathode

Reduction occurs

Electrolytic solution

Oxidation occurs

chloride salt solution called brine. Two different types of electrolytic cells are used in the chlor-alkali industry, as seen in Figure 16B-3a and b. The older type of cell uses a graphite anode and a mercury cathode. Chlorine gas is produced at the anode by reduction reaction, and a mercury-sodium alloy is produced at the cathode.

Anode: $2\ Cl^-\ (aq) \longrightarrow Cl_2\ (g) + 2\ e^-$
Cathode: $2\ Na^+\ (aq) + 2\ e^- \longrightarrow 2\ Na/Hg$

The mercury-sodium alloy (denoted by Na/Hg) is constantly removed from the cell and reacted with pure water to obtain sodium hydroxide.

$2\ Na/Hg + 2\ H_2O\ (l) \longrightarrow 2\ NaOH\ (aq) + H_2\ (g) + 2\ Hg\ (l)$

The second type of cell used in the chlor-alkali industry utilizes a graphite anode and a steel cathode placed in the brine separated by a membrane. Although the membrane will allow cations to pass through it, it prevents the mixing of the products that are produced at each electrode. The reaction at the anode is the same as in the previously discussed cell; however, the reaction at the cathode is

$2\ H_2O\ (l) + 2\ e^- \longrightarrow H_2\ (g) + 2\ OH^-\ (aq)$

The membrane prevents the OH⁻ ions in the cathode compartment from migrating to the anode compartment. The membrane cell was developed to avoid working with mercury, which can be toxic to humans and can cause ecological problems if released into the environment.

Where half-reactions occur

Anode—Oxidation
Both words begin with a vowel.

Cathode—Reduction
Both words begin with a consonant.

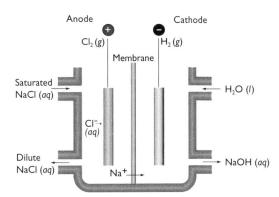

Figure 16B-3a

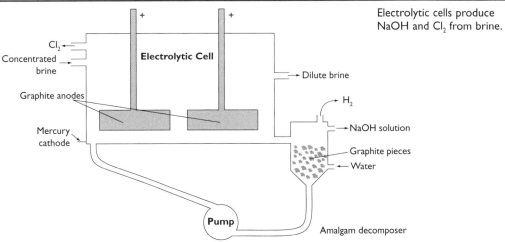

Figure 16B-3b

Electrolytic cells produce NaOH and Cl₂ from brine.

Galvanic Cells: Getting Electricity from Reactions

Electrolytic cells force nonspontaneous redox reactions forward by driving them with electricity. Galvanic or **voltaic cells**, however, do the opposite using spontaneous redox reactions to produce electricity. These cells are sometimes called galvanic cells in honor of Luigi Galvani (1737-98), who discovered the "galvanic effect" of two dissimilar metals, even though he incorrectly interpreted his experiments. Additionally, they are sometimes called voltaic cells, in honor of Allessandro Volta (1745-1827) who correctly interpreted the results of Galvani's experiments and later went on to invent the first battery in 1800. **Batteries** are collections of voltaic cells.

All voltaic cells have a negative electrode (anode) and a positive electrode (cathode). This designation appears to be opposite of the electrolytic cell convention where the positive electrode is the anode and the negative electrode is the cathode. However, in both cases, oxidation occurs at the anode and reduction at the cathode. This difference in sign convention will be explained later in this discussion. All voltaic cells contain an anode that loses electrons, a cathode that gains electrons, and an electrolyte between them.

In the voltaic cell diagrammed in Figure 16B-4, the reaction occurring at the anode is the oxidation of zinc.

$$\text{Anode: } Zn \longrightarrow Zn^{2+} + 2\ e^-$$

$$\text{Cathode: } Cu^{2+} + 2\ e^- \longrightarrow Cu$$

The electrons from the oxidation half-reaction travel along the wire to the cathode. At the cathode, copper (II) is reduced to metallic copper by the electrons coming from the anode. The steady flow of electrons from the anode to the cathode can be harnessed to make a light bulb glow or to power a small radio. As the reactions in each half-cell continue, the zinc solution builds up a positive charge from the accumulation of zinc ions. As the copper ions plate out as copper, that solution builds up a negative charge. The reactions will stop unless a mechanism is supplied to prevent the build up of charges in the solutions. This is accomplished by the use of a salt bridge. A **salt bridge** is a tube of electrolytic gel that connects the two half-cells of a voltaic cell. The salt bridge allows the flow of ions but prevents the mixing of the two different solutions.

A salt bridge is illustrated in Figure 16B-4. Current flows between the solutions in the form of migrating ions. As zinc is oxidized, excess zinc ions accumulate

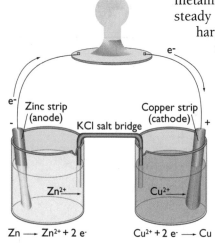

Figure 16B-4

in the solution around the anode. Chloride ions migrate from the salt bridge toward the concentration of positive charges and keep the solution close to neutral. As copper (II) is reduced to metallic copper at the cathode, positive charges are removed from the solution around the cathode. Potassium ions migrate from the salt bridge into the solution in order to keep the solution electrically neutral. If the salt bridge were removed, current would not flow.

The sign conventions of the anodes and cathodes are a common source of confusion. In one case, the anode has a positive (+) value, and in another, it has a negative (-) value. The confusion can be avoided if one remembers that the potential source electrode that emits the electrons is always negative. In voltaic cells, the potential source is the electrolytic solution. The zinc will tend to lose electrons more readily than copper. Thus, the zinc electrode is supplying the electrons (it is "electron rich") and is deemed negative. Since oxidation is occurring at this electrode, it is the anode. Recall that oxidation always occurs at the anode. In the electrolytic cell, the electrical device, not the solution, is the potential source (Figure 16B-5). The cathode is now electron rich and is the potential source electrode that emits electrons, so the cathode is negative. In the solution, the cations migrate to the cathode where a reduction reaction occurs. The anions in the solution migrate to the anode—now positively charged—where oxidation takes place.

Figure 16B-5

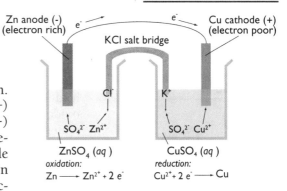

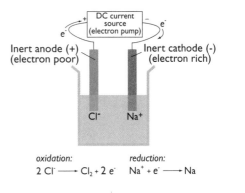

A common type of battery is the dry cell. The diagram of a dry cell shown in Figure 16B-6 on the next page shows that the cell consists of a zinc container filled with an electrolyte paste (made of MnO_2, $ZnCl_2$, NH_4Cl) and a binder that keeps it all together. The zinc can just inside the cardboard tube acts as the anode and loses electrons. Inserted into the electrolyte paste is a graphite (carbon) rod that acts as the cathode. The coated paper separator acts as the electrolyte layer as well as a means of preventing a short circuit.

Table 16B-1 Electrode Conventions

	Cathode	Anode
Ions attracted	Cations	Anions
Electron movement	Into cell	Out of cell
Half-reaction	Reduction	Oxidation
Sign		
electrolytic cell	Negative	Positive
galvanic cell	Positive	Negative

Although the reactions occurring at these electrodes are complicated, they can be summarized as follows:

anode reaction:
$$Zn \longrightarrow Zn^{2+} + 2\ e^-$$

cathode reaction:
$$2\ NH_4^+ + 2\ MnO_2 + 2\ e^- \longrightarrow Mn_2O_3 + H_2O + 2\ NH_3$$

Automobiles get their starting power from a series of six lead storage cells linked together so that the voltages add to each other. The cathode of a lead storage cell consists of a series of lead-antimony alloy plates permeated with lead (IV) oxide (PbO_2). The anode is a series of lead-antimony alloy plates filled with spongy lead. The cathode and anode are immersed in sulfuric acid.

The oxidation half-reaction is
$$Pb + SO_4^{2-} \longrightarrow PbSO_4 + 2\ e^-$$

The reduction half-reaction is
$$PbO_2 + 4\ H^+ + SO_4^{2-} + 2\ e^- \longrightarrow PbSO_4 + 2\ H_2O$$

As strange as it seems, lead atoms are oxidized on one plate and reduced on another. Lead is oxidized to Pb^{2+} at the anode, and Pb^{4+} is reduced to Pb^{2+} at the cathode.

The three major kinds of batteries most familiar to the consumer are dry cells, alkaline cells, and lead-acid batteries, as shown in Table 16B-2. All of these batteries are based on redox reactions.

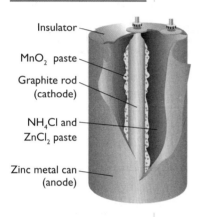

Figure 16B-6

- Insulator
- MnO_2 paste
- Graphite rod (cathode)
- NH_4Cl and $ZnCl_2$ paste
- Zinc metal can (anode)

Table 16B-2 Types of Batteries in Common Use

Type	Anode	Cathode	Electrolyte	Uses
Leclanché Dry Cell	*Zn/Hg	MnO_4	NH_4^+ and Zn^{2+} chlorides	Low-cost use
Alkaline Cells:				
Nickel-Cadmium	Cd	Nickel oxides	NaOH or KOH	Rechargeable batteries
Silver-Zinc	*Zn/Hg	Ag_2O	KOH	Military applications
Silver-Cadmium	CdO	Ag	KOH	Satellite uses
Silver-Iron	Fe	Ag	KOH and LiOH	Emergency power uses
Mercuric oxide-Zinc	*Zn/Hg	HgO	KOH	Transistorized equipment
Lead-acid	Pb	PbO_2	H_2SO_4	Automotive and industrial uses

*Zn/Hg is amalgamated zinc

The Battle Against Corrosion

Corrosion is a general term applied to the chemical destruction of a metal by its immediate surroundings. In order for iron to rust, it must be in contact with both air and moisture. Dry air alone will not corrode iron, nor will pure water that is free of dissolved oxygen. Rusting is also aided by impurities in the iron (such as carbon), impurities in the water (such as acids or other electrolytes), heat, physical strains in the metal, and the presence of a metal that is less active than iron. As a general rule, when two metals are in contact with each other, the more active metal undergoes corrosion while the less active metal is protected.

In spite of man's advanced knowledge of chemistry, losses caused by corrosion in the United States alone run into the hundreds of billions of dollars annually. Iron may be safeguarded from corrosion by being connected electrically to a metal such as zinc or magnesium. The more active metal in such an arrangement is called the sacrificial anode. The Alaskan oil pipeline is a large-diameter iron pipe that is protected by heavy zinc wires. Ships having exposed metal surfaces underwater are similarly protected from the corrosive action of seawater by zinc anodes.

Another strategy used to discourage corrosion is the alloying of iron with other metals. Stainless steel, an alloy noted for its resistance to rust and tarnish, is a mixture of iron and chromium. Finally, there are several materials that can be used to coat iron to protect it against corrosion. The well-known "tin can" is actually an iron can coated with tin. Galvanized iron is iron coated with zinc. It is used for such items as trash cans and chain-link fences. Paints, lacquers, and varnishes are also used to protect the surfaces of iron and steel. Paints containing red lead or zinc chromate are especially effective for preventing corrosion.

Chemical alteration of the surface of iron can also form a protective coating. When red-hot iron is treated with steam, a thin coating of black iron oxide (Fe_3O_4) is formed. The black color of stovepipes results from this process. The iron oxide coating affords good protection to the metal even at high temperatures.

The battle against corrosion is a never-ending one. Corrosion is a relentless degenerative process. Many theologians think that it is a consequence of the curse that was placed on the earth after man first sinned. Others hold the view that corrosion is a necessary result of the way the laws of nature were established from Creation. Whichever the case, a study of corrosion should serve to illustrate the futility of putting faith in the material objects of this world. The sight of rust should remind men of the Lord's admonition in Matthew 6:19-20: "Lay not up for yourselves treasures upon earth, where moth and rust doth corrupt, and where thieves break through and steal: But lay up for yourselves treasures in heaven, where neither moth nor rust doth corrupt, and where thieves do not break through nor steal."

Figure 16B-7

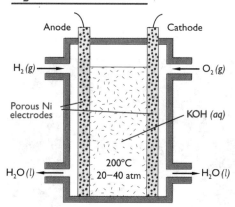

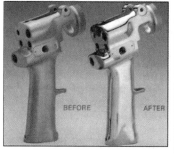

Magnesium electroplated with nickel and chrome.

Reversing the polarity of the electrodes recharges batteries. This reverses the redox reaction and regenerates the cell. Some cells cannot or should not be recharged. Some manufacturers put a label on their alkaline batteries to warn the customers against recharging the batteries. They do not mean that alkaline batteries *cannot* be recharged; they mean that they *should not* be recharged, because to do so would be dangerous. Even though the redox reaction in an alkaline cell is reversible, recharging produces some gases. Since some alkaline cells have no vents to release gases, they could explode if someone tried to recharge them.

Fuel cells are a special class of batteries that are very efficient. A fuel cell resembles a voltaic battery in that redox reactions occur, releasing electrons from one electrode to flow through a circuit to another electrode. However, there is one major difference. In the voltaic cell, the active ingredients are included within the cell, and are depleted as the redox reactions occur. In a fuel cell, a gas or liquid fuel is supplied to one electrode and oxygen or air to the other from an external source. Fuel cells have a long life and have been used extensively in space vehicles since the 1960s. Figure 16B-7 shows a schematic of an oxygen-hydrogen fuel cell.

Electroplating: A Cover-up

Sterling silver is at least 92.5 percent silver. Less expensive silverware is made of some common metal that is plated with a thin layer of silver. How is the thin silver layer deposited onto the inexpensive metal? One could pound, melt, or glue the silver onto the metal, but there is a much better way. Metallic ions in solution can be forced to cling to metal knives and forks. The electrochemical process of depositing one metal onto another is called **electroplating**.

Electroplating is a type of electrolysis. The cathode consists of the metal item, such as a fork, to be plated. The anode is made out of the metal that is to be plated onto the cathode. Silver plating is performed with a silver anode. The electrolyte solution contains silver ions. When the current in the cell is turned on, silver ions migrate to the cathode and are reduced to metallic silver. The silver plates onto the fork.

$$Ag^+ + e^- \longrightarrow Ag$$

Figure 16B-8

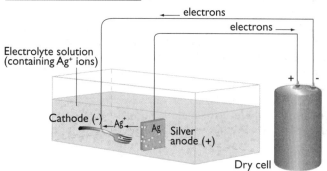

Zinc is sometimes electroplated onto steel to protect the steel against rust. Galvanized steel is steel that has been electroplated with zinc. Zinc corrodes, but the corrosion product does not flake off; it protects the zinc from further oxidation. Therefore, the layer of zinc protects the iron in the steel from being oxidized. Even if a small crack forms in the zinc plating, the iron is still protected because zinc is more easily oxidized than iron. Any oxidation that occurs will be the oxidation of zinc.

Nonmetals can also be electrodeposited on metal surfaces. The automotive industry uses an electrodeposition process to paint car bodies. Positively charged paints are deposited onto the sheet metal. This technique more effectively covers the metal and gives better corrosion-resistance than conventional spray painting.

Galvanizing steel helps to prevent corrosion.

Section Review Questions 16B

1. What is the major difference between a voltaic cell and an electrolytic cell?
2. Will pure water conduct electricity?
3. What reaction occurs at the anode of a voltaic cell? At the anode of an electrolytic cell?
4. What is the main difference between a voltaic cell and a fuel cell?

Chapter Review

Coming to Terms

oxidation-reduction reaction
redox reaction
oxidation
reduction
reducing agent
oxidizing agent
half-reaction method
half-reaction
electrolyte
nonelectrolyte
electrochemical cell

electrode
galvanic cell
electrolytic cell
electrolysis
cathode
anode
voltaic cell
battery
salt bridge
fuel cell
electroplating

Chapter 16 Review

Review Questions

1. Does the process of chemical oxidation require that oxygen atoms be present?
2. Tell whether each reaction is a redox reaction.
   a. $2 \text{ Fe (s)} + 3 \text{ Cl}_2 \text{ (g)} \longrightarrow 2 \text{ FeCl}_3 \text{ (s)}$
   b. $\text{CaO (s)} + 2 \text{ HCl (g)} \longrightarrow \text{CaCl}_2 \text{ (s)} + \text{H}_2\text{O (l)}$
   c. $2 \text{ C}_2\text{H}_6 \text{ (g)} + 7 \text{ O}_2 \text{ (g)} \longrightarrow 4 \text{ CO}_2 \text{ (g)} + 6 \text{ H}_2\text{O (l)}$
   d. $\text{Zn (s)} + \text{CuSO}_4 \text{ (aq)} \longrightarrow \text{Cu (s)} + \text{ZnSO}_4 \text{ (aq)}$
   e. $\text{Ba (s)} + 2 \text{ H}_2\text{O (l)} \longrightarrow \text{Ba(OH)}_2 \text{ (s)} + \text{H}_2 \text{ (g)}$
   f. $\text{Pb (s)} + \text{H}_2\text{SO}_4 \text{ (aq)} \longrightarrow \text{PbSO}_4 \text{ (s)} + \text{H}_2 \text{ (g)}$
   g. $\text{AgNO}_3 \text{ (aq)} + \text{HCl (aq)} \longrightarrow \text{AgCl (s)} + \text{HNO}_3 \text{ (aq)}$

Questions 3-7 refer to the following unbalanced redox reactions.
   a. $\text{SO}_4^{2-} + \text{Zn} \longrightarrow \text{Zn}^{2+} + \text{SO}_2$
   b. $\text{S}_2\text{O}_3^{2-} + \text{OCl}^- \longrightarrow \text{Cl}^- + \text{S}_4\text{O}_6^{2-}$
   c. $\text{I}^- + \text{SO}_4^{2-} \longrightarrow \text{I}_2 + \text{H}_2\text{S}$
   d. $\text{H}_2\text{S} + \text{CrO}_4^{2-} \longrightarrow \text{S} + \text{Cr}^{3+}$
   e. $\text{SO}_2 + \text{MnO}_4^- \longrightarrow \text{SO}_4^{2-} + \text{Mn}^{2+}$
   f. $\text{MnO}_4^- + \text{Fe}^{2+} \longrightarrow \text{Mn}^{2+} + \text{Fe}^{3+}$
   g. $\text{SO}_4^{2-} + \text{C} \longrightarrow \text{CO}_2 + \text{SO}_2$

3. Identify the substance being oxidized in each reaction.
4. Identify the substance being reduced in each reaction.
5. Identify the oxidizing agent in each reaction.
6. Identify the reducing agent in each reaction.
7. Balance each equation for the reaction in an acid solution.
8. A chemist titrates a solution of sodium oxalate ($\text{Na}_2\text{C}_2\text{O}_4$) with a solution of KMnO_4. The balanced half-reactions occurring during the titration are

$$5 \text{ Na}_2\text{C}_2\text{O}_4 \longrightarrow 10 \text{ CO}_2 + 10 \text{ Na}^+ + 10 \text{ e}^-$$
and
$$2 \text{ KMnO}_4 + 16 \text{ HCl} + 10 \text{ e}^- \longrightarrow 2 \text{ MnCl}_2 + 8 \text{ H}_2\text{O} + 2 \text{ KCl} + 10 \text{ Cl}^-$$

The gram-formula mass of $\text{Na}_2\text{C}_2\text{O}_4$ is 134.00 g/mol, and that of KMnO_4 is 158.04 g/mol.
   a. If 3.580 g of KMnO_4 are dissolved in 1.000 L of solution, what is the normality of the KMnO_4?
   b. If 87.58 mL of 0.1000 N KMnO_4 are required to titrate 63.87 mL of $\text{Na}_2\text{C}_2\text{O}_4$ solution, what is the normality of the $\text{Na}_2\text{C}_2\text{O}_4$ solution?

9. A bologna manufacturer uses $NaNO_2$ as a preservative in his product. To determine how much $NaNO_2$ is in a solution being used in the process, an official from the Food and Drug Administration (FDA) titrates the solution with a $K_2Cr_2O_7$ solution. The balanced half-reactions that occur during the titration are

$$3\ NaNO_2 + 3\ H_2O \longrightarrow 3\ NaNO_3 + 6\ H^+ + 6\ e^-$$

and

$$K_2Cr_2O_7 + 14\ HCl + 6\ e^- \longrightarrow 2\ CrCl_3 + 7\ H_2O + 2\ KCl + 6\ Cl^-$$

The gram-formula mass of $NaNO_2$ is 69.00 g/mol, and that of $K_2Cr_2O_7$ is 294.20 g/mol.

a. If 5.000 g of $K_2Cr_2O_7$ are dissolved in 1.000 L of solution, what is the normality of the $K_2Cr_2O_7$?

b. If 38.73 mL of 0.3270 N $K_2Cr_2O_7$ are required to titrate 45.00 mL of the $NaNO_2$ solution, what is the normality of the $NaNO_2$?

10. In a voltaic cell, identify the electrical charge on the anode, the type of ion that migrates toward the anode, and the process (oxidation or reduction) that occurs at the surface of the anode.

11. In a voltaic cell, identify the electrical charge on the cathode, the type of ion that migrates toward the cathode, and the process (oxidation or reduction) that occurs at the surface of the cathode.

12. What is the difference between a voltaic cell and an electrolytic cell?

13. Describe how the Hall-Héroult process frees aluminum atoms from bauxite.

14. Why did a special process have to be invented for the purification of active metals such as aluminum and sodium when other metals were being purified from their ores by heating and chemical reactions?

15. Why is a salt bridge used in voltaic cells?

16. Why are some batteries considered non-rechargeable even though they are based on the same reversible redox reactions as those in rechargeable batteries?

17. Zinc plating will protect steel from corroding even if the plating is cracked. Give an explanation for this fact.

17A Building an Organic Compound page 443
17B Hydrocarbons page 446
17C Substituted Hydrocarbons page 456
17D Organic Reactions page 466
FACETS: Fullerenes and the Buckyball page 464

Organic Chemistry 17

Spotlight on Carbon

Grapefruit, margarine, wood, plastic wrap, rubbing alchohol, aspirin, lipstick, this textbook, and skin have at least one thing in common. They all contain carbon. Because they contain carbon, they belong to the most extensive branch of chemistry: organic chemistry.

17A Building an Organic Compound

Organic chemists determine the structure of organic molecules, study the way in which those molecules react, and develop ways to synthesize new organic compounds. Organic chemistry has had a profound effect on our society. Many important industrial chemicals are organic compounds. Some of these compounds are listed in Table 17A-1.

Structural Formulas: Small Sketches for Big Molecules

Organic chemists study more than the types of atoms in molecules. They also study how the molecules are arranged. If an analytical chemist announced that he had isolated a compound with the molecular formula C_2H_6O, his colleagues would not know what compound he was talking about. C_2H_6O could be ethanol or it could be dimethyl ether. The arrangement of the atoms in the molecule makes the difference. For this reason, structural formulas are often used in organic chemistry. Many times, to make things simpler, the hydrogen atoms are left out of the drawing.

Table 17A-1 Important Organic Industrial Chemicals

Name	Structure	Family	Uses
1. Ethylene (Ethene)	$H_2C=CH_2$	alkene	fruit ripening, plastic wrap
2. Urea (carbonyl diamide)	$H_2N-C(=O)-NH_2$	amide	fertilizer, animal feed
3. Benzene (cyclohexatriene)	(benzene ring)	aromatic hydrocarbon	chemical intermediate
4. Methanol (wood alcohol)	H_3C-OH	alcohol	gasoline additive
5. Acetone	$CH_3-C(=O)-CH_3$	ketone	chemical intermediate, solvent
6. Formaldehyde (methanal)	$H-C(=O)-H$	aldehyde	embalming fluid
7. Phenol	(benzene ring with OH)	alcohol	aspirin, disinfectant
8. Ethylene glycol (1,2-ethanediol)	$H_2C(OH)-CH_2(OH)$	alcohol	antifreeze
9. Cyclohexane	(cyclic CH_2 ring)	cyclic alkane	nylon, solvent
10. Isopropyl alcohol	$CH_3-CH(OH)-CH_3$	alcohol	rubbing alcohol, after-shave lotion

Table 17A-2 Structural Comparisons of Ethanol and Dimethyl Ether

Name	Molecular Formula	Structural Formula		
Ethanol	C_2H_6O	H:C:C:O:H (with H's)	or H-C-C-OH (with H's)	or -C-C-OH
Dimethyl ether	C_2H_6O	H:C:O:C:H (with H's)	or H-C-O-C-H (with H's)	or -C-O-C-

These formulas quickly show that each carbon atom has four bonds, each oxygen atom has two bonds, and each hydrogen atom has one bond (Table 17A-2).

The Unique Carbon Atom: Multitudes of Bonds

The elements other than carbon can combine to form several hundred thousand compounds. Contrast that number with the four million compounds that incorporate the carbon atom. It seems amazing that the number of compounds that contain carbon is at least ten times greater than the number of compounds of all the other elements. What makes the carbon atom so versatile?

Carbon atoms form so many compounds because they have unique bonding abilities. There are three important properties of carbon that enable it to form large, stable molecules.

- Carbon has four valence electrons; therefore, it must form four bonds to obtain an octet.
- Carbon has the ability to form strong chemical bonds to other carbon atoms, which allows for almost infinite chains of carbon.
- There is a lack of reactivity between the carbon-hydrogen bond due to its nonpolar nature. C–H bonds are nonpolar because the electronegativity difference between carbon (2.5) and hydrogen (2.1) is small.

Carbon atoms can bond to a wide variety of atoms. Carbon atoms bonded together can form chains of various lengths. Chains of carbon atoms may be straight or branched. Carbon atoms can even form rings. In addition to this collection of possibilities, numerous others can result from double and triple bonds, making the possibilities seem endless.

Figure 17A-1

Variations in carbon-carbon bonding

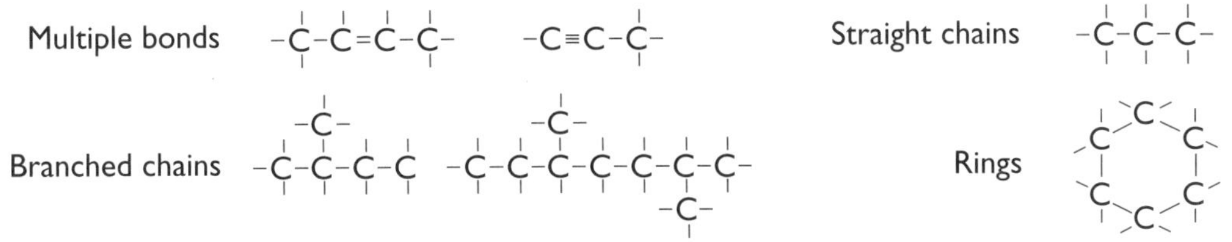

Multiple bonds: $-\overset{|}{C}-\overset{|}{C}=\overset{|}{C}-\overset{|}{C}-$ $-C\equiv C-\overset{|}{C}-$

Branched chains: (structures shown)

Straight chains: $-\overset{|}{C}-\overset{|}{C}-\overset{|}{C}-$

Rings: (hexagonal ring of carbons shown)

Carbon-carbon bonds in the "backbones" of the molecules account for only two of the four bonds. The remaining bonds are supplied by seemingly endless combinations of hydrogen, phosphorus, oxygen, nitrogen, sulfur, halogens, and other atoms.

Classification: A Map Through a Jungle of Compounds

Friedrich Wöhler

Organic chemistry nowadays almost drives me mad. To me it appears like a primeval forest full of the most remarkable things, a dreadful endless jungle into which one dare not enter for there seems to be no way out.

—Friedrich Wöhler, 1835

If Wöhler were alive today, he would be amazed to see how much larger the "jungle" has grown. Approximately seventy-five thousand new compounds are synthesized for the first time each year. If a person wishes to find his way through this ever-growing jungle, he must use some guidelines. A classification scheme that organizes compounds into easily identifiable groups serves as the map through the jungle.

Organic compounds can be divided into two large groups: aliphatic compounds and aromatic compounds. **Aliphatic compounds** include straight-chain compounds and those rings that could be formed by the bending and closing of the straight chains. Cyclohexane is an example of an aliphatic compound. **Aromatic compounds,** like benzene, have ringed shapes, but they are unlike aliphatic rings in one important aspect. Their electrons are not held down to specific bonds; instead, they can migrate in circular clouds above and below the carbon atoms.

Section Review Questions 17A

1. Write a job description of an organic chemist.
2. What characteristics of carbon enable it to be found in millions of compounds?
3. Carbon can bond to a wide variety of other atoms. Name several examples.
4. List and define the two large groups into which organic compounds are divided.

17B Hydrocarbons

Petroleum plant

This chapter will first survey the simplest aliphatic and aromatic compounds: hydrocarbons. As their name implies, **hydrocarbons** contain only hydrogen and carbon. The primary sources of hydrocarbons are natural gas, coal, and petroleum. This chapter will survey some of the ways in which hydrocarbons can be modified by other elements.

Hydrocarbons

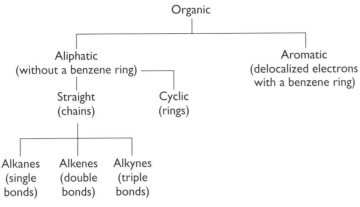

Figure 17B-1

Classification of hydrocarbons

Coal is one of the three primary sources of hydrocarbons.

The classification scheme of Figure 17B-1 was determined from the structures of compounds. All compounds in a class have a common structural feature. Usually, compounds in the same class also have similar physical and chemical properties.

Alkanes: Chains with Single Bonds

Alkanes are the fuels on which society relies. The natural gas used to heat homes contains methane (CH_4) as its principal component. Portable gas barbecue grills use bottles of pressurized propane, and automobiles burn a mixture of alkanes called gasoline. All of these compounds contain only carbon and hydrogen. They have structural formulas that resemble open chains, and they have only single bonds. According to the terms on the classification chart, **alkanes** are aliphatic, open-chained hydrocarbons that contain only single bonds.

Methane (CH_4) is the simplest alkane. Its one carbon atom is bonded to four hydrogen atoms. Other alkanes are formed as additional carbons lengthen the chain. Ethane (C_2H_6) has two carbons, propane (C_3H_8) has three, and butane (C_4H_{10}) has four. Each carbon atom in an alkane is surrounded by four other atoms—the maximum number possible. For this reason, the molecules of alkanes are said to be **saturated.**

A group of chemists belonging to IUPAC have devised a system that accurately names organic compounds. The IUPAC system works for alkanes and all the other types of organic molecules that will be studied in this chapter. The system relies on a series of prefixes that indicate the number of carbon atoms present.

These prefixes help in naming many classes of compounds in addition to the alkanes. The prefix gives the number of carbons in the longest chain. The suffix identifies the

Figure 17B-2

Straight-chain alkane molecules actually have a zig-zag shape because of the carbon atoms' tetrahedral bonding angles.

Table 17B-1 Numerical Prefixes	
C_1	meth-
C_2	eth-
C_3	prop-
C_4	but-
C_5	pent-
C_6	hex-
C_7	hept-
C_8	oct-
C_9	non-
C_{10}	dec-

type of compound. For example, a particular branched alkane has the following structure:

$$\begin{array}{c} -\overset{|}{\underset{|}{C}}- \\ -\overset{|}{\underset{|}{C}}-\overset{|}{\underset{|}{C}}-\overset{|}{\underset{|}{C}}- \end{array}$$

There are three carbons in this alkane's longest chain. Therefore, according to the IUPAC system it is named a propane—methylpropane to be exact. Methyl is the name of the alkyl group CH_3, which is attached to the second carbon on the carbon chain. The **alkyl groups** are named with the prefixes in Table 17B-1. Methyl is a type of substituent. A **substituent** is an atom or a group of atoms that can substitute for a single hydrogen.

The names, structural formulas, and some properties of the first ten alkanes are listed in Table 17B-2.

Are these molecules polar or nonpolar? Bonds between two carbon atoms are not polar because both atoms have the same electronegativity. The bonds between carbon and hydrogen are arranged symmetrically, so whatever polarity they have is canceled. Alkanes dissolve well in nonpolar solvents such as carbon tetrachloride. In polar substances such as water, however, the liquid alkanes form a slimy oil slick.

Crude oil straight from the well contains several types of alkanes. Chains from four carbons up to twenty carbons are put together into one seemingly inseparable mixture. Fractions of the mixture can be separated so that they can be used as lubricating oils, gasoline, and kerosene. How do chemical engineers separate the individual compounds? Petroleum engineers use the fact that the boiling points of alkanes rise about 20-30°C for each additional carbon that is added to the chain.

First, a batch of crude oil is heated and dumped into the bottom of a fractional distillation tower. The mixture is heated even further until a large percentage of the compounds boil into vapor. The vapors then rise through the cooling tower. Alkanes with high boiling points (the larger molecules) condense near the bottom of the tower. Smaller alkanes rise farther into the tower before they condense into liquids. By regulating the temperatures of the different portions of the tower, chemical engineers can collect distinct fractions of alkanes. A diagram of a fractional tower is shown in Chapter 10.

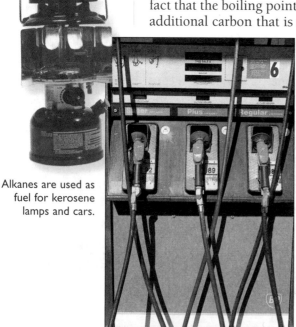

Alkanes are used as fuel for kerosene lamps and cars.

Table 17B-2 Straight-Chained Alkanes

Name	Structural Formula	Melting Point (°C)	Boiling Point (°C)
Methane	H–C(H)(H)–H	-183	-162
Ethane	H–C(H)(H)–C(H)(H)–H	-172	-88.5
Propane	CH₃–CH₂–CH₃	-187	-42
Butane	CH₃–(CH₂)₂–CH₃	-138	0
Pentane	CH₃–(CH₂)₃–CH₃	-130	36
Hexane	CH₃–(CH₂)₄–CH₃	-95	69
Heptane	CH₃–(CH₂)₅–CH₃	-90.5	98
Octane	CH₃–(CH₂)₆–CH₃	-57	126
Nonane	CH₃–(CH₂)₇–CH₃	-54	151
Decane	CH₃–(CH₂)₈–CH₃	-30	174

Structural Isomers: Variations on the Alkane Theme

The carbon chains of alkanes do not have to be completely straight. Some alkanes have branched-chain structures. The molecular formulas of the two substances listed below, butane and isobutane, are identical, but the molecules are obviously different. Compounds that have the same molecular formulas but different structural formulas are called **structural isomers**. Normal butane (n-butane, the straight-chained form) and its structural isomer, iso-butane, would be expected to have slightly different physical properties, and indeed they do.

Table 17B-3 Comparison of Butane and Its Isomer

Isomer	Structural Formula	Melting Point (°C)	Boiling Point (°C)	Density	Solubility
Butane	-C-C-C-C-	-138	0	0.579 g/mL	1813 mL/100 mL ethanol
Isobutane	-C-C(-C-)-C-	-159	-12	0.549 g/mL	1320 mL/100 mL ethanol

Sample Problem Draw structural formulas for all the structural isomers of pentane (C_5H_{12}).

Solution

Dimethyl propane   2-methyl butane   n-pentane

The "2" before methyl butane means that the alkyl group is attached to the second carbon from the shortest end on the carbon chain. There is no 2,2 before dimethyl propane because there is no other place that the alkyl groups could be attached and still yield propane.

Alkenes: Chains with Double Bonds

Carbon does not always form single bonds. Double bonds appear quite frequently. The simplest hydrocarbon that contains a double bond is ethene. No doubt you have heard the expression that one bad apple ruins the whole barrel. Rotting apples give off ethene, which speeds up the ripening of other fruits. As early as 1910 it was reported that ethene coming from oranges could ripen bananas. Today ethene is used to ripen green tomatoes, walnuts, and grapes. The rotten-apple principle applies to the friends that surround you as well. I Corinthians 15:33 says, "Be not deceived: evil communications corrupt good manners."

Hydrocarbons that contain double bonds are called **alkenes.** Because double bonds reduce the number of hydrogen atoms in the molecules, alkenes are said to be **unsaturated.** As in the alkane series, new compounds result when more singly bonded carbons are added to the chain. These molecules are named with the appropriate prefix and the ending *-ene*, which means that the compound has a double bond. Ethene contains only two carbons. When the carbon chain is longer than three carbons, the double bond could be in several locations. An alkene's name pinpoints the location of the double bond by giving the number of the first carbon that is doubly bonded.

Table 17B-4 **Alkenes**

Name	Structural Formula	Boiling Point (°C)	Melting Point (°C)
Ethene	−C=C−	−104	−169
Propene	−C=C−C−	−47	−185
1-butene	−C=C−C−C−	−6.3	−185
2-butene	−C−C=C−C−	4	−139.3

The physical properties of alkenes are very much like those of alkanes. The first few are gases at room temperature. Pentene and larger compounds are liquids at room temperature because of greater intermolecular attractions. Alkenes are relatively nonpolar.

Alkynes: Chains with Triple Bonds

A triple bond between two carbon atoms identifies a member of the **alkyne** family. The most common alkyne is also the simplest. Ethyne, commonly called acetylene, consists of two carbons joined by a triple bond. This compound is often used as a fuel for welding torches and as an ingredient for polymers.

$$H\!:\!C\!:\!:\!:\!C\!:\!H \quad \text{or} \quad H-C\equiv C-H$$

Larger alkynes have additional carbons. Names of alkynes are formed from a prefix that tells how many carbons are in the molecule. The suffix *-yne* signifies that a triple bond is present. When necessary, a number is used to tell where the triple bond occurs.

Physically, alkynes are similar to other hydrocarbons. They are practically nonpolar, so they are insoluble in water and very soluble in nonpolar solvents. Their boiling points rise as the carbon chains get longer. Table 17B-5 is a list of alkynes.

Table 17B-5 Alkynes

Name	Structural Formula	Boiling Point (°C)	Melting Point (°C)
Ethyne (acetylene)	$-C\equiv C-$	-75	-82
Propyne	$-C\equiv C-C-$	-23	-101.5
1-butyne	$-C\equiv C-C-C-$	9	-122
2-butyne	$-C-C\equiv C-C-$	27	-24
1-pentyne	$-C\equiv C-C-C-C-$	40	-98
2-pentyne	$-C-C\equiv C-C-C-$	55	-101
1-hexyne	$-C\equiv C-C-C-C-C-$	72	-124
2-hexyne	$-C-C\equiv C-C-C-C-$	84	-92
3-hexyne	$-C-C-C\equiv C-C-C-$	81	-103
1-heptyne	$-C\equiv C-C-C-C-C-C-$	100	-80
1-octyne	$-C\equiv C-C-C-C-C-C-C-$	126	-70
1-nonyne	$-C\equiv C-C-C-C-C-C-C-C-$	151	-65
1-decyne	$-C\equiv C-C-C-C-C-C-C-C-C-$	182	-36

Cyclic Aliphatic Compounds: Chains in Rings

Petroleum from California is unique. For some unknown reason it contains an unusually large number of carbon compounds whose chains have been bonded into rings. Such compounds are called **cyclic aliphatic compounds.**

A great variety of rings is possible, but five-carbon and six-carbon rings with single bonds are the most abundant. Simple alkenes and alkenes with more than one double bond multiply the number of possible structures. A few structures and their names are shown below.

Cyclohexane   Cyclopentene   1,3-cyclohexadine

Some very unusual structures are possible when several rings are combined.

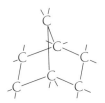

Bicyclo [2.2.1] heptane   Basketane

The chemical activity of cyclic compounds is about the same as that of other members of their parent families. Cycloalkanes act like alkanes; cycloalkenes act like alkenes. Cyclic compounds have several unique uses. Cyclopropane is used as an anesthetic. Cyclopentane works well as a cleaner in the fuel sytem of a car. Added to the gasoline, it dissolves deposits in the intake system, carburetor, cylinders, and top piston valves. Cyclohexane is used in soap making, dry cleaning, insecticides, and germicides.

Aromatic Substances: Roaming Electrons

Aromatic substances played an important role in early societies: the wealthy used them instead of soap and water. Since clean water was scarce in many areas, frequent baths were not possible. The poor learned to accept the odor, but the wealthy fought their body odors with layers of perfumes. Since many of these perfumes contained benzene compounds, any compound that contained a form of benzene was soon classified as aromatic, or good-smelling. Eventually, however, this rule failed. Chemists found several benzene compounds that had odors far from pleasant—some were actually quite distasteful. Chemists also found other substances that smelled good even though they did not contain benzene. Nevertheless, by this time the term *aromatic* had been associated with benzene compounds for so long that the name remained.

Benzene was isolated in 1825 by Michael Faraday. An analysis of the elements in benzene and a determination of its molecular weight showed that the molecular formula was C_6H_6. This compound is the simplest aromatic compound known. Today it is one of industry's most important compounds.

The structural formula of benzene puzzled scientists for thirty years after Faraday discovered the compound. Many clues were gathered, but they did not seem to fit together. The molecular formula C_6H_6 led chemists to believe that the molecule must have several double or triple bonds. Yet the chemical reactions of benzene did not support this idea. It behaved like an alkane, not an alkene or alkyne. When scientists determined the bond lengths between the carbon atoms, they found that the distances were not those of single or double bonds: they were in between. It was as if benzene used one and one-half bonds. Furthermore, it became known that the carbons were arranged in a ring and that all the carbon atoms had identical bonds.

In 1865 August Kekule proposed a structure that could account for most of the observations. He described benzene in terms of two symmetrical ring structures.

Each benzene molecule was thought to switch rapidly back and forth between the two forms. The bonds were mobile, not tied down to one location. The development of the quantum model of atoms modified this concept. Chemists realized that electrons existed in orbitals and that in benzene these orbitals overlapped and became one big, doughnut-shaped area of electron concentration. The electrons in these clouds are free to roam throughout the entire "doughnut." For this reason, they are called **delocalized electrons.**

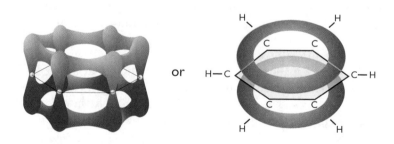

Because the electrons are not bound between any two carbon atoms, chemists now draw the structure of benzene as

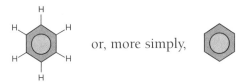 or, more simply,

The modern definition of aromatic compounds has nothing to do with their smell. Rather, it deals with the nature of the bonds between atoms. All aromatic compounds have cyclic clouds of delocalized electrons. Benzene is the parent compound for a huge number of substances, many of which are used commercially. Figure 17B-3 shows some of these commercial uses. Many other compounds contain multiple benzene rings that are fused together. Figure 17B-4 shows three examples.

Figure 17B-4

Compound containing multiple benzene rings

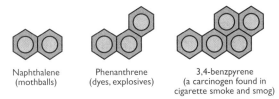

Finally, many ointments, perfumes, and oils contain aromatic compounds. One of the ointments used instead of the daily bath in the Middle Ages was made from crushed vanilla beans. The good-smelling aromatic compound is named vanillin.

Section Review Questions 17B

1. What do hydrocarbons contain and where are they obtained?
2. Explain the differences between saturated and unsaturated hydrocarbons.
3. What is a substituent?
4. What is a structural isomer?

Figure 17B-3

Commercial uses for benzene

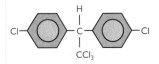

Dichlorodiphenyl trichloroethane (DDT, insecticide)

Trinitrotoluene (TNT explosive)

Chlorobenzene

Aniline (used in dyes)

Phenol (disinfectant)

Acetyl salicylic acid (aspirin)

5. Why are organic compounds mainly insoluable in water?
6. Compare and contrast alkanes, alkenes, and alkynes.
7. Define each molecular formula as an alkane, alkene, or alkyne.
    a. $C_{10}H_{22}$
    b. C_2H_4
    c. C_3H_{10}
    d. C_4H_6
    e. C_2H_2
    f. C_5H_8
8. What are carbon compounds whose chains have been bonded into rings?
9. Describe the nature of the bonds between atoms in aromatic compounds.

Chemists modify hydrocarbons for use in common household products.

17C Substituted Hydrocarbons

Functional groups are the hot spots of chemical activity on an organic molecule. They greatly modify the behavior of hydrocarbons. In alkenes and alkynes the reactive sites are the multiple bonds. The electrons in the bonds are available for a variety of chemical reactions. The multiple bonds give the molecules characteristic structures and chemical properties. There are many kinds of functional groups. Some contain sulfur; some contain halogens. Most functional groups in organic chemistry contain arrangements of oxygen or nitrogen atoms. In each case the functional group greatly influences the molecule's properties.

Halides: Hydrocarbons Plus Halogens

When a halogen and an alkyl group are combined, an alkyl halide is formed. An **alkyl halide** is a combination of an alkyl group and a fluorine, chlorine, bromine, or iodine atom. When a halogen is attached to an **aryl group** (an aromatic group that lacks a hydrogen atom), the resulting molecule is called an **aryl halide.** Several halides are used by industries today.

Table 17C-1 Industrial Uses of Halides

Name	Structural Formula	Application
Trichloromethane (chloroform)	H—C(Cl)(Cl)—Cl	early anesthetic
Tetrachloromethane (carbon tetrachloride)	Cl—C(Cl)(Cl)—Cl	nonpolar solvent (dry-cleaning agent)
Triiodomethane (iodoform)	H—C(I)(I)—I	veterinary antiseptic
Dichlorodifluoromethane (freon -12)	F—C(Cl)(F)—Cl	compressor gas used in refrigerators

Alkyl halides are ingredients in polymer plastics.

Many other halides serve as intermediates in the synthesis of other compounds.

Table 17C-2 Halides as Intermediates in the Synthesis of Compounds

Name	Structural Formula	Applications
Chloroethene (vinyl chloride)	$H_2C=CHCl$	building block for polyvinyl chloride (credit cards, plastic pipes)
1,1-dichloroethene (vinylidene chloride)	$Cl_2C=CH_2$	building block for plastic wrap
Tetrafluoroethene	$F_2C=CF_2$	building block for Teflon®

Alcohols: Molecules with OH Groups

To most people the word *alcohol* evokes images of society's most-abused drug. Many people do not realize that this misused substance is just one member of a very useful family of organic compounds.

Compounds that have a covalently bonded OH group attached to an alkyl group are classified as **alcohols.** The general formula for the whole family is R–OH, where R represents an alkyl group. The simplest alcohol is methanol. In this case the R group is the smallest one possible, CH_3.

$$H-\underset{\underset{H}{|}}{\overset{\overset{H}{|}}{C}}-OH$$

Alcohol names consist of the standard prefixes that tell how long the carbon chain is plus an *-ol* ending. If the OH group is attached to a carbon other than the end carbon, its position is indicated by a numerical prefix. For instance, what is commonly called *rubbing alcohol* is a three-carbon chain with the OH group attached to the middle carbon. The IUPAC name is 2-propanol.

The physical properties of alcohols are a result of the two following factors:

1. a polar OH group.
2. a nonpolar alkyl group.

The combination of these two opposites determines the behavior of each specific molecule. If the hydrocarbon chain is relatively short, the OH group dominates the molecule. As a result, it behaves as a polar molecule. If the hydrocarbon chain is very long, the chain dominates and imparts nonpolar characteristics to the molecule. Small alcohols, under the influence of the OH groups, form hydrogen bonds, have high boiling points, and are soluble in water. Large alcohols, under the influence of their hydrocarbon chains, are insoluble in water and soluble in nonpolar solvents.

Table 17C-3 Alcohols

Name	Structural Formula	Applications
Methanol (wood alcohol)	—C—OH	solvent, fuel
Ethanol (grain alcohol)	—C—C—OH	nonpolar solvent (dry-cleaning agent)
Propanol	—C—C—C—OH	solvent, antifreeze, after-shave lotion
2-propanol (rubbing alcohol) (isopropyl alcohol)	—C—C(OH)—C—	solvent, antifreeze, after-shave lotion
Butanol	—C—C—C—C—OH	solvent, shellac, varnish

Many common household products contain alcohols.

Alcohols that contain more than one OH group are called **polyhydroxy alcohols.** One common polyhydroxy alcohol is 1,2-ethanediol (also called ethylene glycol), which is used as an antifreeze in car radiators. Another common alcohol—1,2,3-propanetriol (glycerol)—is put into moisturizers in cosmetics.

OH OH
| |
—C—C—
| |

Ethylene glycol

OH OH OH
| | |
—C—C—C—
| | |

Glycerol

Industries use alcohols as solvents, paint thinners, antifreezes, and ingredients in after-shave lotions. When ethanol is required, product manufacturers face the prospect of paying the same stiff taxes that are imposed on liquor. To avoid these extra costs, they denature the ethanol, or make it totally unfit to drink. Poisons and foul-tasting substances such as gasoline, methanol, or 2-propanol are added to the ethanol. It may surprise some people to know that other alcohols are much more toxic than ethanol. Ingestion of rubbing alcohol causes headaches, nausea, comas, and possibly death. Methanol typically brings the same symptoms with one significant addition. It can cause chronic eye problems and even blindness because it can dissolve the fatty sheath around the optic nerve.

Ethers: Molecules with −O− Links

Compounds that have the general formula R_1-O-R_2 are called **ethers.** R_2 stands for a second alkyl group. Ethers are distinguished by the oxygen bridge between two carbon chains. The name of an ether includes the names of the alkyl groups on each side of the oxygen (smaller one first) and the word *ether* on the end.

$$-\overset{|}{\underset{|}{C}}-\overset{|}{\underset{|}{C}}-O-\overset{|}{\underset{|}{C}}-\overset{|}{\underset{|}{C}}- \qquad -\overset{|}{\underset{|}{C}}-O-\overset{|}{\underset{|}{C}}- \qquad -\overset{|}{\underset{|}{C}}-O-\overset{|}{\underset{|}{C}}-\overset{|}{\underset{|}{C}}-$$

      Diethyl ether   Dimethyl ether   Methyl ethyl ether

More complicated ethers have been synthesized, but diethyl ether is by far the most common. When people say "ether," they are usually referring to this compound. A Georgian doctor named Crawford Long made this ether famous when he painlessly removed a tumor from a patient's neck in 1842. The reason that the operation was painless was that ether is an anesthetic: it puts a person "to sleep." Ether served the medical profession for a long time until other anesthetics that had fewer side effects were developed. Ethers are now used as solvents for perfumes, primers for engines, and reagents in the synthesis of organic materials.

Working with ethers in the laboratory requires great caution. Ethers vaporize quickly, and the vapors are very flammable. Since the vapors are more dense than air, they sink to a tabletop or floor and then spread out over a great area. For this reason, there should be no open flames in the laboratory when ethers are present.

Aldehydes: Molecules with C=O Groups on the End

Aldehydes are organic compounds that contain a double-bonded oxygen on the end carbon. Their general structure is

$$R-\overset{\overset{O}{\|}}{C}-H$$

All aldehydes contain the C=O group, which is called the **carbonyl group**. According to the IUPAC rules, the name of an aldehyde is formed with an *-al* ending on the name of the corresponding alkane.

Propanal   2-methylpentanal

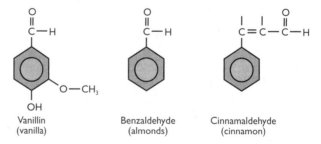

Biological materials used to be preserved with formalin. Propylene glycol is a safer preservative.

The simplest aldehyde (and one of the most important ones in industry) is known by the common name formaldehyde. Its IUPAC name is methanal. This substance is a colorless gas with a piercing odor. Since the gas is difficult to handle, it is often dissolved in water to make a 37 percent solution called formalin. This is the solution that has been used in the past to preserve frogs, fetal pigs, and other creatures to be dissected in biology labs. However, because of health concerns about its possible link to cancer and allergic reactions, the use of formalin has greatly declined. Safer methods of preservation are used today. Formalin is also used as an embalming fluid. It reacts with proteins and starches to form insoluble compounds. Other important aldehydes give unique flavors.

Vanillin (vanilla)   Benzaldehyde (almonds)   Cinnamaldehyde (cinnamon)

Ketones:
Molecules with C=O Groups in the Middle

If a double-bonded oxygen is attached to a carbon that is not on the end of the chain, the compound is called a **ketone**. The general formula of ketones is

$$R_1-\overset{\overset{O}{\|}}{C}-R_2$$

The simplest ketone is commonly called acetone. According to IUPAC rules, its name is propanone. The name of a ketone is derived from the name of the alkane that has the same number of carbon atoms; the *-e* ending of the alkane is changed to *-one*. For chains over four carbons long, the location of the carbonyl group is shown by a numerical prefix.

Acetone (propanone)

Butanone (methyl ethyl ketone)

Acetone is an excellent solvent. While the molecule's methyl groups are nonpolar, the carbonyl group is polar. As a result, acetone dissolves most organic compounds but still mixes well with water. Acetone is widely used as a solvent for lacquers, paint removers, explosives, plastics, and disinfectants. It is also the active ingredient in some nail-polish removers.

Since both ketones and aldehydes have the C=O group, it is not surprising that they have similar physical and chemical properties. As a rule, though, aldehydes are slightly more chemically reactive because their functional group is exposed on the end of the carbon chain.

Products containing acetone

Carboxylic Acids:
Ant Bites, Bee Stings, and Vinegar

The Ant

The ant has made himself illustrious
Through constant industry industrious.
So what?
Would you be calm and placid
If you were filled with formic acid?

—Ogden Nash

A carboxylic acid puts the "fire" in fire ants and the "ouch" in bee stings. Because of their abundance in nature, carboxylic acids were among the first organic compounds to be studied in detail. Since the IUPAC rules were still decades away from being formulated, the acids acquired common names from their most familiar sources. The Latin word for vinegar was *acelum,* so the acid in vinegar was called acetic acid. Butyric acid (from the Latin *butyrum*) gives rancid butter, aged cheese, and human perspiration their nasty odors. The Latin word for ant is *formica.* Consequently, the stinging acid of ants was called formic acid.

All **carboxylic acids** have the general formula

$$R-C\begin{matrix}\\\nearrow O\\\searrow OH\end{matrix}$$

The COOH group is called the **carboxyl group.** According to IUPAC nomenclature, the *-e* ending of the corresponding alkane is changed to *-oic,* and the word *acid* is added to form the name of a carboxylic acid.

HCOOH	CH₃COOH	CH₃(CH₂)₂COOH
Formic acid (methanoic acid)	Acetic acid (ethanoic acid)	Butyric acid (butanoic acid)

Esters are responsible for a flower's distinct smell and for a fruit's natural flavoring.

The smaller carboxylic acids are liquids at room temperature and have sharp or unpleasant odors. The acids with longer carbon chains are usually waxy solids. When the carbon chains are between twelve and twenty carbon atoms long, the compounds are often called **fatty acids.** Carboxylic acids are quite polar. They can form hydrogen bonds between themselves and other molecules. This explains why even the smallest members of the family are liquids at room temperature. Since they form hydrogen bonds with water, carboxylic acids are soluble in water unless their hydrocarbon chains are so long that they dominate the molecules.

To be an acid, a compound should have an ionizable hydrogen. The hydrogen that gets ionized in a carboxylic acid is the one in the functional group. Since only a small fraction of carboxylic acid molecules dissociate, these acids are weak. Only four out of one hundred acetic acid molecules ionize. The dissociation constant is only 1.754×10^{-5}. Although they are weak acids, carboxylic acids react quickly with strong bases to form salts. Many of these salts are used commercially. Most soaps are sodium or potassium salts of long-chain (fatty) acids.

Esters: Sweet-Smelling Chemicals

If the hydrogen of a carboxyl group is replaced with an alkyl group, an **ester** forms.

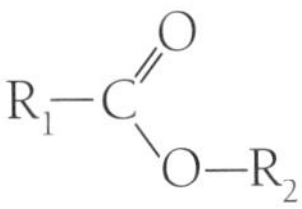

$$R_1-C\begin{matrix}\nearrow O\\ \searrow O-R_2\end{matrix}$$

Unlike their cousins the carboxylic acids, esters generally have appealing smells. These compounds are responsible for the flavors of many fruits and the fragrances of many flowers. In the naming of an ester, the R_2 group is indicated with its alkyl name, and the carboxylic acid part is given an *-oate* ending. Meat fat consists largely of solid esters, and oils are liquid esters.

Table 17C-4 Esters

Name	Structural Formula	Flavor or Odor
2-methylpropyl methanoate	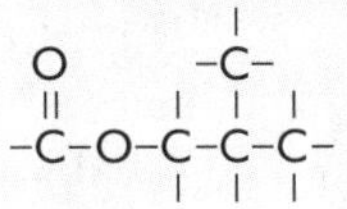 -C-O-C-C-C- with =O on first C and -C- branch on middle C	raspberry
Pentyl ethanoate	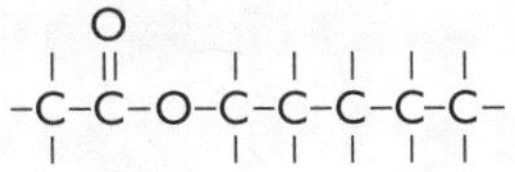 -C-C-O-C-C-C-C-C- with =O on second C	banana
Octyl ethanoate	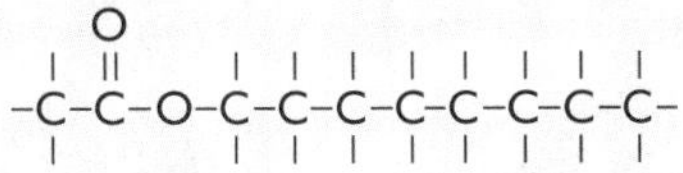 -C-C-O-C-C-C-C-C-C-C-C- with =O on second C	orange
Pentyl propanoate	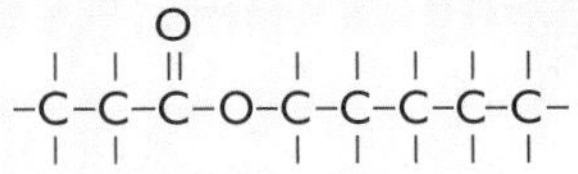 -C-C-C-O-C-C-C-C-C- with =O on third C	apricot
Ethyl butanoate	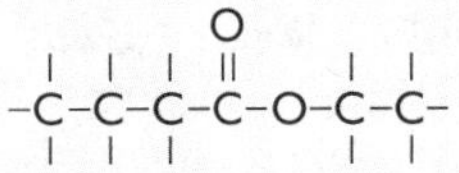 -C-C-C-C-O-C-C- with =O on fourth C	pineapple
Ethyl heptanoate	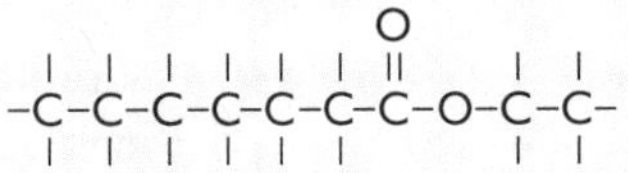 -C-C-C-C-C-C-C-O-C-C- with =O on seventh C	grape

Amines and Amides: Nitrogen Compounds

Nitrogen can bond into organic molecules in several different ways. The **amines** are a family of organic compounds that have ammonia as their parent. Derivatives are formed when the hydrogen atoms are replaced with other atoms or groups of atoms. One, two, or even three hydrogens may be replaced. The names of the compounds that result commonly have the word *amine* as a suffix after the names of the alkyl groups.

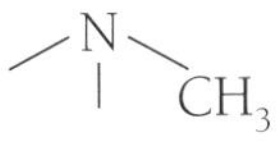

Methylamine 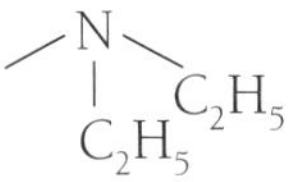   Diethylamine 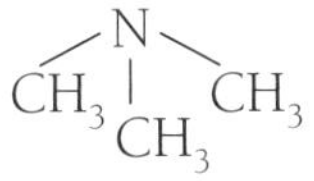   Trimethylamine

Another family of compounds that contain nitrogen is called **amides.** Amides result when an amine group takes the place of an −OH group in a carboxylic acid. All the members of this group have the following structure in common.

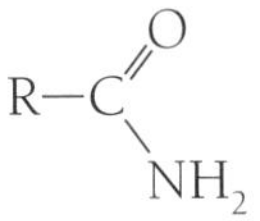

$$R-C\begin{matrix}\nearrow O\\ \searrow NH_2\end{matrix}$$

This structure is especially important because it holds the amino acids in proteins together.

Section Review Questions 17C

1. What is the main difference between an alkyl halide and an aryl halide?
2. The physical properties of alcohol are a result of what two factors?
3. What influence does the length of the hydrocarbon chain have on an alcohol?

Facets of Chemistry: Fullerenes and the Buckyball

Until 1985 there were only two known natural forms of pure carbon—graphite and diamonds. In that year American chemist R. E. Smalley, British chemist H. F. Kroto, and graduate students working under their direction made a significant discovery while studying the nature of interstellar matter. While vaporizing carbon with a laser, they created sixty-carbon molecules in the shape of soccer balls. They called these molecules *buckminsterfullerenes*, or *buckyballs* (C_{60}). Buckminsterfullerenes are named in honor of R. Buckminster Fuller, who pioneered the use of light, strong, unusually shaped domes in architecture. Spaceship Earth at Epcot, Walt Disney World, is the most notable of these dome structures.

When the buckyball was discovered, scientists already knew that long chains of carbon were present in space. These chains were identified using readings gathered from radio telescopes. The carbon chains exhibited characteristic readings, much like a fingerprint, on the radio telescope that could be compared to fingerprints of known molecules on earth. This allowed the previously unknown carbon chains to be characterized. R. E. Smalley generated long-chain carbon molecules in order to measure their spectroscopic fingerprints. In the apparatus Smalley created, a laser was aimed at a rotating graphite disk in a helium-filled vacuum chamber. The laser delivered a short, high-energy burst that converted light energy to chemical energy. This rapid, intense heating of the graphite surface caused many carbon bonds in the graphite to break. As a result, carbon atoms were released from the graphite surface and collided in the helium vacuum. New bonding arrangements of carbon were produced, including the buckyball.

Buckyballs are part of a class of carbon molecules called fullerenes. Fullerenes are carbon molecules that are arranged in the form of a closed, hollow sphere, cylinder, or the like. Fullerenes have the ability to trap other atoms in the spaces between their carbon atoms. Structures have been created in many different shapes, such as fullerene "diamonds," monster fullerene balls, and honeycomb-shaped tubules. Buckyballs are the only fullerenes to form a hollow sphere. They are not aromatic hydrocarbons as one might think, but are described as electron-deficient alkenes.

It has been estimated that buckyballs can contain between 30 and 980 carbon atoms in their stuctures. The C_{60} buckyball, however, is the most stable form and the largest possible symmetrical molecule. Because of symmetry, the C_{60} buckyball is resistant to high-speed collisions. Although spinning at one hundred million times per second,

4. Which compounds are distinguished by an oxygen bridge between two carbon chains?
5. What is the main difference between an aldehyde and a ketone?
6. What is the main difference between an amine and an amide?
7. Identify the organic compounds that have the following IUPAC endings to their names:
   a. -e ending changed to -one
   b. ends in -ol
   c. ends in -al
   d. ends in -oate
   e. ends in -oic acid

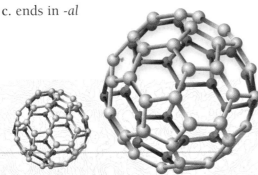

C_{60} buckyballs can withstand a collision into a stainless steel plate at 15,000 mph.

Buckyballs are very useful fullerenes since they form stable crystals that are nonreactive. They are nonreactive because all of the carbons are bonded to other carbons. When compressed to 70 percent of its original size, a pure mass of buckyballs becomes more than twice as hard as diamonds. Once buckyballs could be produced in large amounts, a solid form, fullerite, was produced. Fullerite is a transparent yellow solid whose molecules are stacked together in a close-packed arrangement like a pile of cannon balls. Tubular versions of fullerenes have also been produced in solid forms.

Worldwide, chemists have been studying the behavior and potential uses for fullerenes. New compounds and polymers with strange configurations and properties have been produced. Fullerenes have already been used experimentally as superconductors and to create small diamonds. Since buckyballs can trap any molecule within themselves, they are extremely practical for medical research. The administration of medicine molecularly through a buckyball has great possibilities. Treatment for cancer could become more localized, unlike chemotherapy. Buckyballs could also be used for batteries, lubricants, rocket fuel, plastics, and carbon fibers.

The History of Fullerenes

1985 Buckyballs were discovered.
1990 Researchers discovered an efficient way to manufacture buckyballs. C_{70} and C_{84} molecules were produced.
1991 Buckminsterfullerene was named "molecule of the year" by *Science* magazine.
1992 Chemists found fullerenes in a meteorite and in a natural black rock called shungite. Shungite is a rare, carbon-rich rock. Fullerenes have since been found in a glassy rock known as fulgerite. Fulgerite forms when lightning strikes the ground.
1993 Researchers formed a fullerene lubricant.
1994 Scientists discovered that buckyballs were impervious (or impenetrable) to laser beams. This makes buckyballs of great significance to the military.
1996 The discoverers of buckyballs were awarded the Nobel Prize for chemistry.

   IBM scientists in Zurich reported that they had built an abacus using buckyballs.

   Buckyballs containing helium isotopes in a ratio observed only in meteorites were discovered.
1999 Fullerenes were found within a meteorite that had hit the earth.
2000 Scientists speculated that fullerenes are abundant in the universe, particularly near red giant stars.

17D Organic Reactions

The number of organic compounds is continually growing. These compounds can participate in a multitude of reactions. Most biological processes rely on chemical reactions between organic molecules. These reactions are responsible for the movement of muscles, the digestion of food, the transmission of nerve impulses, and the sensing of light upon the retina. Industrial chemists manipulate molecules to form plastics, fuels, synthetic fabrics, and a host of other products. Some of these reactions are quite complicated. This text will survey a few of the basic reactions that characterize organic compounds.

Oxidation-Reduction. These two types of reactions were discussed in Chapter 16. Oxidation is the loss of electrons, and reduction is the gain of electrons. Whenever oxygen and carbon atoms bond, the carbon is oxidized. Since oxygen atoms have high electronegativities, they pull shared electrons away from carbon. This means that the addition of oxygen always makes the oxidation number of carbon more positive. Adding hydrogen atoms reduces carbon atoms.

Combustion oxidizes all the carbon atoms in an organic molecule to carbon dioxide. All hydrocarbons burn in oxygen to form carbon dioxide, water, and heat. The energy from these combustion reactions is used in engines to move cars, in furnaces to heat homes, and in gas lanterns to light up dark campsites.

Fatty acid chains in food are oxidized by the human body in much the same way that other hydrocarbons are burned. Providentially, the body controls the oxidation precisely so that only small amounts of energy are released at any instant. This regulation keeps the temperatures during oxidation tolerable and allows the body to use most of the released energy.

Substitution. **Substitution reactions** replace one part of a molecule with another part. Typically, most of the reactions of unreactive compounds are substitution reactions. Alkanes are not very reactive, but this is not surprising. Every carbon is already saturated with four single bonds. Furthermore, each of these bonds is very stable. When alkanes are heated to high

Rapid oxidation of fatty acids in a potato chip

Figure 17D-1

Substitution of Cl atoms for H atoms in CH_4

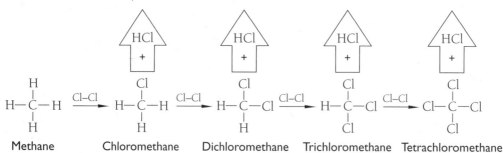

Methane — Chloromethane — Dichloromethane — Trichloromethane (chloroform) — Tetrachloromethane (carbon tetrachloride)

Table 170-1 Families of Organic Compounds

Hydrocarbons

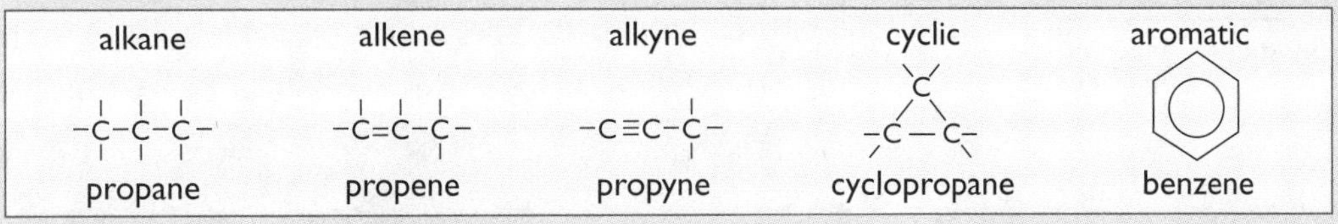

alkane	alkene	alkyne	cyclic	aromatic
–C–C–C–	–C=C–C–	–C≡C–C–	C with –C–C– (triangle)	benzene ring
propane	propene	propyne	cyclopropane	benzene

Substituted Hydrocarbons

Family	Structural Formula	Suffix or Group Name	Typical Compound	Name of Compound
halide	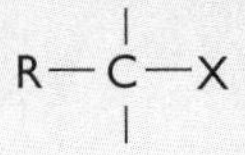 R–C–X	chloro- bromo- fluoro- iodo-	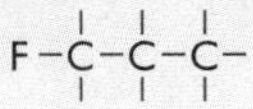 F–C–C–C–	fluoropropane
alcohol	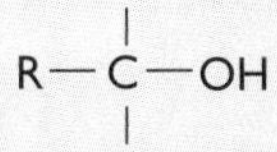 R–C–OH	-ol	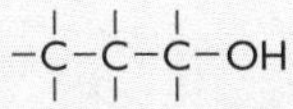 –C–C–C–OH	1-propanol
ether	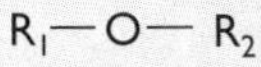 R_1–O–R_2	ether	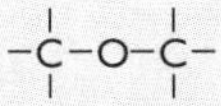 –C–O–C–	dimethyl ether
aldehyde	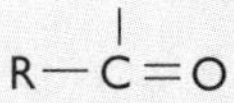 R–C=O	-al	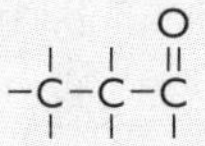 –C–C–C(=O)	propanal
ketone	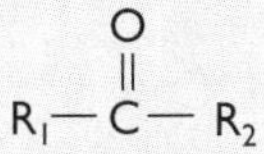 R_1–C(=O)–R_2	-one	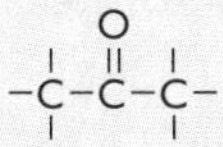 –C–C(=O)–C–	propanone
carboxylic acid	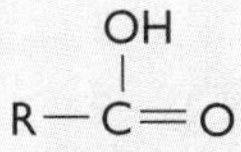 R–C(OH)=O	-oic acid	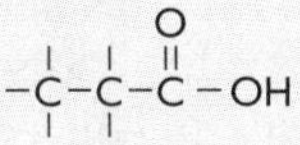 –C–C–C(=O)–OH	propanoic acid
ester	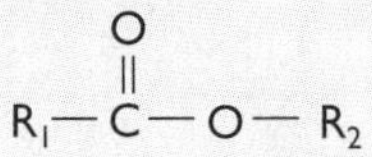 R_1–C(=O)–O–R_2	-oate	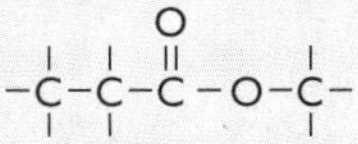 –C–C–C(=O)–O–C–	methyl propanoate
amine	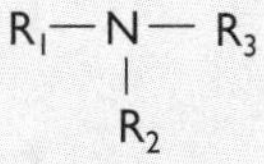 R_1–N(R_2)–R_3	-amine	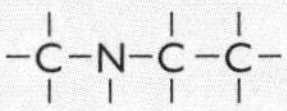 –C–N–C–C–	methylethylamine
amide	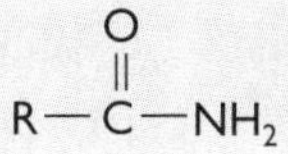 R–C(=O)–NH_2	-amide	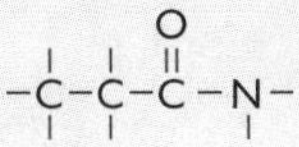 –C–C–C(=O)–N–	propanamide

temperatures or exposed to energetic ultraviolet light, hydrogen atoms can be replaced by other atoms. For example, methane and chlorine can react to form a variety of substitution products.

Figure 17D-2

Substitution reactions of benzene

Various groups can replace one or more of the hydrogens of aromatic compounds. Benzene can be nitrated, halogenated, or even alkylated.

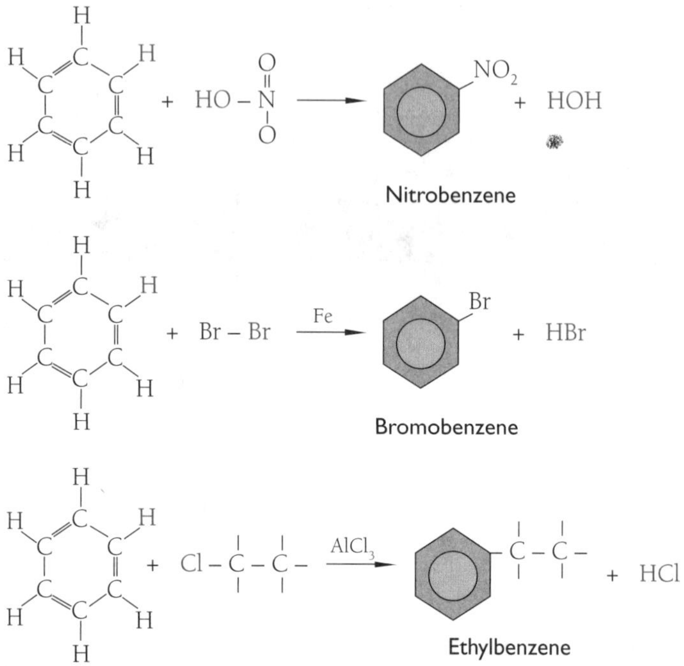

C₆H₆ + HO–NO₂ ⟶ C₆H₅NO₂ (Nitrobenzene) + HOH

C₆H₆ + Br–Br →[Fe] C₆H₅Br (Bromobenzene) + HBr

C₆H₆ + Cl–CH₂–CH₂–H →[AlCl₃] C₆H₅–CH₂–CH₂– (Ethylbenzene) + HCl

Addition. Compared to carbon-carbon single bonds, double bonds and triple bonds are very reactive. Consequently, the double and triple bonds in a molecule react first and determine the molecule's behavior. An **addition reaction** is a reaction in which a multiple bond is broken and two atoms or groups of atoms are added. This is a characteristic reaction of unsaturated molecules.

Figure 17D-3

Addition reactions of propene

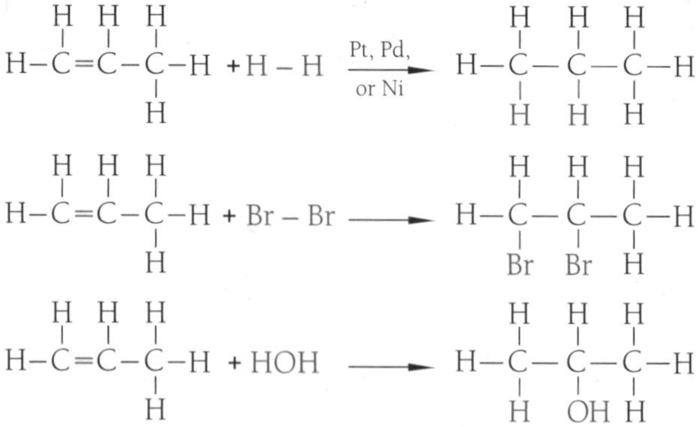

$H-C=C-C-H + H-H \xrightarrow{Pt, Pd, \text{ or } Ni} H-C-C-C-H$ (with appropriate H's)

$H-C=C-C-H + Br-Br \longrightarrow H-C-C-C-H$ (with Br, Br on adjacent carbons)

$H-C=C-C-H + HOH \longrightarrow H-C-C-C-H$ (with H, OH added)

Alkanes and alkenes look alike, but they can be distinguished by their chemical reaction with bromine. When bromine in carbon tetrachloride is first added to a hydrocarbon, the resulting mixture is reddish brown. If the reactive double bonds of alkenes are present, the bromine will be added to the molecule, and the color will disappear. If only carbon-carbon single bonds are present, the reddish brown color will remain because alkanes are unreactive under normal conditions.

Condensation and Hydrolysis. Like oxidation and reduction reactions, these reactions are opposites. Reactions in which molecules combine with each other and lose a water molecule are called **condensation reactions.** Two identical alcohol molecules can be made to join together to form an ether under special reaction conditions.

$$R-OH + R-OH \xrightarrow{H_2SO_4} R-O-R + HOH$$

Esters form when carboxylic acids and alcohols go through a condensation reaction. This condensation reaction is called **esterification.** An artificial banana flavoring can be made when ethanoic (acetic) acid and pentanol are mixed.

Ethanoic acid (acetic acid)   Pentanol

Pentyl ethanoate   Water

A bromine solution is added to an alkane and an alkene (top). The alkene reacts with the bromine and the color disappears (bottom).

A condensation reaction is responsible for much of the clothing that people wear. **Polymers** are substances that consist of huge molecules that have repeating structural units. Dacron® polyester, one of the more common polymers, forms when ethylene glycol and terephthalic acid condense.

Dacron® polyester

Table 17D-2, on the following page, lists some common polymers, the materials from which they are made, and some uses.

Hydrolysis is the reverse process of condensation. A water molecule works its way into the functional group of a large molecule and splits it. The OH group of the water attaches to one of the newly formed molecules, and the hydrogen attaches to the other

Table 17D-2 Common Polymers

Starting Materials	Polymer	Uses
1. Ethylene glycol and terephthalic acid	Dacron®	fabric
2. Vinyl chloride	Polyvinylchloride (PVC)	flooring, raincoats, pipes
3. Tetrafluoroethylene	Teflon®	bakeware coating
4. Ethylene	Polyethylene	plastic bags, toys
5. Propylene	Polypropylene	plastic milk containers
6. Sodium bisphenol A and phosgene	Polycarbonate	lenses, baby bottles
7. Acrylonitrite	Polyacrylonitrite	acrilon fibers, rugs
8. Styrene	Styrofoam® (polystyrene)	molded objects, coffee cups

molecule. The artificial banana flavoring made by a condensation reaction exists in equilibrium with a hydrolysis reaction.

$$R_1-\underset{OH}{\overset{O}{\overset{\|}{C}}} + R_2-OH \underset{\text{Condensation}}{\overset{\text{Hydrolysis}}{\rightleftharpoons}} R_1-\underset{O-R_2}{\overset{O}{\overset{\|}{C}}} + HOH$$

Carboxylic acid   Alcohol   Ester   Water

Cooking involves the partial breakdown of carbohydrates and proteins by heat and hydrolysis. Soap making, or **saponification**, also illustrates hydrolysis. The large molecule to be split is a fat molecule with three ester linkages. Each ester link is susceptible to hydrolysis. When steam hits the fats, water molecules split the large fat molecules.

$$\begin{array}{c} H \\ | \\ H-C-O-\overset{O}{\overset{\|}{C}}-(CH_2)_{16}CH_3 \\ | \\ H-C-O-\overset{O}{\overset{\|}{C}}-(CH_2)_{16}CH_3 + 3\,H_2O \\ | \\ H-C-O-\overset{O}{\overset{\|}{C}}-(CH_2)_{16}CH_3 \\ | \\ H \end{array} \longrightarrow \begin{array}{c} H \\ | \\ H-C-OH \\ | \\ H-C-OH \\ | \\ H-C-OH \\ | \\ H \end{array} + \begin{array}{c} O \\ \| \\ HO-C-(CH_2)_{16}CH_3 \\ O \\ \| \\ HO-C-(CH_2)_{16}CH_3 \\ O \\ \| \\ HO-C-(CH_2)_{16}CH_3 \end{array}$$

Fat   Glycerol   Fatty acids

Once the fatty acids have been separated, a strong base can be added easily. The result is the salt of a fatty acid, otherwise known as soap. In early American days fats and oils were boiled with lye (NaOH) for many hours in large kettles. The hydrolysis and the reaction with the sodium hydroxide base occurred at the same time. Figure 17D-4 shows the structural formula for a soap molecule.

Figure 17D-4

Structural formula of the soap molecule, sodium stearate

$$-\underset{|}{\overset{|}{C}}-\underset{|}{\overset{|}{C}}-\underset{|}{\overset{|}{C}}-\underset{|}{\overset{|}{C}}-\underset{|}{\overset{|}{C}}-\underset{|}{\overset{|}{C}}-\underset{|}{\overset{|}{C}}-\underset{|}{\overset{|}{C}}-\underset{|}{\overset{|}{C}}-\underset{|}{\overset{|}{C}}-\underset{|}{\overset{|}{C}}-\underset{|}{\overset{|}{C}}-\underset{|}{\overset{|}{C}}-\underset{|}{\overset{|}{C}}-\underset{|}{\overset{|}{C}}-\underset{|}{\overset{|}{C}}-\underset{|}{\overset{|}{C}}-\underset{\text{O}^- \text{Na}^+}{\overset{\text{O}}{\underset{\|}{C}}}$$

Section Review Questions 17D

Determine which organic reaction is taking place in the following examples:
1. Hydrogen atoms are replaced by chlorine atoms when an alkane is heated.
2. Unsaturated molecules react.
3. Carbon and oxygen atoms bond.
4. The functional group of a large molecule is split by a water molecule.
5. A condensation reaction forms esters.
6. Molecules combine and a water molecule is lost.
7. Fatty acids are separated from a fat molecule, and the salts of the fatty acids are obtained.

Chapter Review

Coming to Terms

aliphatic compound
aromatic compound
hydrocarbon
alkane
saturated
alkyl group
substituent
structural isomer
alkene
unsaturated
alkyne
cyclic aliphatic compound

delocalized electrons
functional group
alkyl halide
aryl group
aryl halide
alcohol
polyhydroxy alcohol
ether
aldehyde
carbonyl group
ketone
carboxylic acid

carboxyl group
fatty acid
ester
amine
amide
substitution reaction
addition reaction
condensation reaction
esterification
polymer
hydrolysis
saponification

Chapter 17 Review

Review Questions

1. Modify the structural formula of the alkane 2-methylbutane to create the various kinds of compounds requested. Remember that carbon atoms always have four bonds. Add or delete hydrogens when necessary.

$$-\underset{|}{\overset{|}{C}}-\underset{|}{\overset{|}{C}}-\underset{|}{\overset{|}{C}}-\underset{|}{\overset{|}{C}}-$$
$$-\underset{|}{\overset{|}{C}}-$$

   a. an alkene
   b. an alkyne
   c. an alkyl halide containing one iodine atom
   d. an alcohol
   e. an aldehyde
   f. a ketone
   g. a carboxylic acid
   h. an ester (Use an ethyl group to modify the carboxylic acid drawn in (g).)
   i. an amine
   j. an amide (Modify the carboxylic acid drawn in (g).)

2. Classify each of the following compounds according to its general family.

   a. $-\overset{|}{C}-\overset{|}{C}-Br$

   b. $-\overset{|}{C}-\overset{|}{C}-\overset{O}{\overset{\|}{C}}-\overset{|}{C}-$

   c. $-\overset{|}{C}-\overset{|}{C}-\overset{O}{\overset{\diagup}{C}}$
      $-\overset{|}{C}-OH$

   d. $-\overset{|}{C}-\overset{|}{\underset{|}{C}}-\overset{|}{C}-\overset{|}{C}-\overset{O}{\overset{\diagup}{C}}$
      $\diagdown NH_2$

   e. $-\overset{|}{C}-C\equiv C-\overset{|}{\underset{|}{C}}-\overset{|}{C}-$

   f. $-\overset{|}{C}-\overset{|}{C}-\overset{|}{C}-NH_2$

   g. $-\overset{|}{C}-\overset{|}{C}-\overset{|}{\underset{|}{C}}-\overset{|}{C}-\overset{|}{C}-$

   h. $-\overset{|}{C}-\overset{O}{\overset{\diagup}{C}}$
      $\diagdown O-\overset{|}{C}-$

   i. $-\overset{|}{C}-\overset{|}{\underset{|}{C}}-\overset{|}{C}-\overset{|}{C}-\overset{|}{C}-OH$

   j. $-\overset{|}{C}-\overset{|}{C}=\overset{|}{C}-\overset{|}{C}-$

   k. (benzene ring with Cl)

   l. $-\overset{|}{C}-\overset{|}{C}-O-\overset{|}{C}-\overset{|}{C}-$

3. Draw the structural formula for each of the following compounds. Assume that the carbon chain is straight in each case.
   a. hexane
   b. 1-heptene
   c. 2-octyne
   d. 1-pentanol
   e. butanol
   f. 1-chloropropane
   g. ethyl butanoate
   h. hexanoic acid
   i. 3-octanone
   j. methyl ethyl ether
   k. butylamine

4. Name each of the following compounds.

   a. 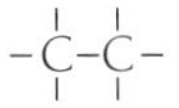 −C−C−

   b. −C=C−C−C−

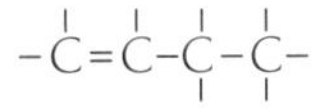

   c. −C−C−OH

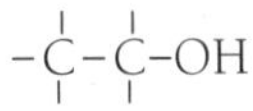

   d. 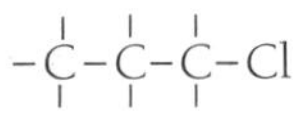 −C−C−C−Cl

   e. 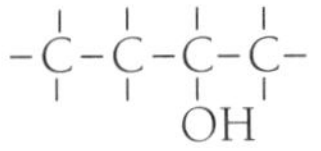 −C−C−C−C−
              |
              OH

   f. 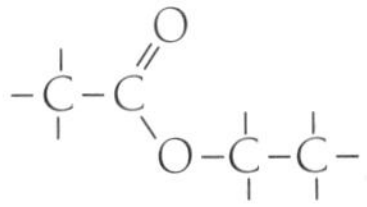 −C−C(=O)−O−C−C−

   g. 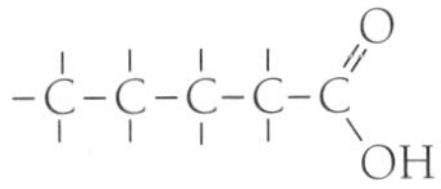 −C−C−C−C−C(=O)−OH

   h.  ⟋N⟍CH₃

   i.  −C−C≡C−C−

   j. 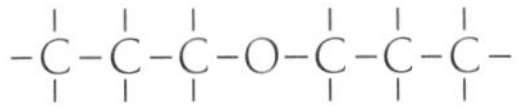 −C−C−C−O−C−C−C−

   k. 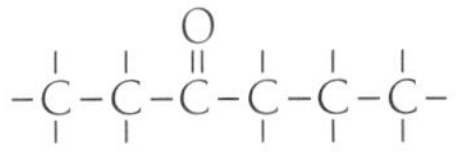 −C−C−C−C(=O)−C−C−

5. Name a class of organic compounds that contains a substance that can be used as
   a. a fuel for welding.
   b. a fuel in automobiles.
   c. an antifreeze.
   d. a refrigerator coolant.
   e. an ingredient in soap.
   f. a paint remover.

6. Why do straight-chained alkanes have higher boiling points than alkanes with branched chains?

7. How can you tell when two compounds are structural isomers?

8. Draw an electron-dot structure of propyne. Would you expect this molecule to be polar or nonpolar? Why?

9. What is unique about the carbon-carbon bonds in benzene and other aromatic compounds?

10. Methane (CH_4) is a gas at room temperature, whereas methanol (CH_3OH) is a liquid. Aside from the difference in molecular masses, suggest an explanation for their different melting and boiling points.

11. Methanol is soluble in water, but larger alcohols like octanol are not. Why is this?

12. Alcohols and metallic hydroxides both have OH groups in their structural formulas. Why are metallic hydroxides such as NaOH much more caustic than alcohols?

13. How are HCl and CH_3COOH similar? How are they different?

14. Give one way in which NaCl and soap are similar and one way in which they are different.

15. Draw a structural formula of acetic acid (CH_3COOH), and identify the hydrogen atom that is ionized.

18A Carbohydrates page 476
18B Proteins page 481
18C Lipids page 484
18D Cellular Processes page 487
FACETS: The Human Genome Project page 496

Biochemistry 18

The Miraculous Chemistry of Life

Biochemistry is a springboard; no one who studies it in depth can stay secluded in the field of chemistry for long. As its name implies, the subject of biochemistry leads students from the study of traditional chemistry into a field where many sciences blend together. The distinctions between biology, chemistry, and mathematics quickly become blurred. The marvelous complexities of life cannot be studied within the confines of a single subject area.

Biochemistry leads from strictly scientific matters into the realm of philosophical and spiritual issues. As scientists unravel the mysteries of DNA, they begin to see some sobering capabilities of genetic engineering. Can genetic material be altered? Can new forms of life be produced from old forms? Can humans be cloned? These questions naturally lead to the question, "Should these things be done?" Researchers need Bible-based ethical and moral guidelines that stem from true spiritual wisdom to direct their work.

Finally, biochemistry diverts attention from man to God. The infinite wisdom of the Creator is on display in the precise architecture of a protein chain, the complexity of a metabolic pathway, and the efficiency of an enzyme. Every molecule bears witness to the power, engineering ingenuity, and foresight of God. Man has been granted the privilege of making many discoveries in the field of biochemistry. These discoveries are exciting, but they are also humbling, for they reveal an omniscient, omnipotent Creator.

18A Carbohydrates

Carbohydrates are the most abundant biological compounds. Sugars and starches make up a large part of the human diet. Each year photosynthetic processes in plants convert water and carbon dioxide into one hundred billion tons of carbohydrates. The exoskeletons of all the insects, crabs, and lobsters on this planet contribute even more carbohydrates to this already impressive amount.

What Carbohydrates Are and What They Do

Carbohydrate literally means "water of carbon." The name goes back to the days when these compounds were thought to be hydrates of carbon molecules, and the empirical formula of all carbohydrates was CH_2O. However, as larger and larger molecules were discovered, this formula did not hold true for all carbohydrate molecules. All carbohydrates have several of the OH (hydroxy) groups that are common to alcohols. They also have the C=O (carbonyl) group of aldehydes and ketones. Concisely stated, **carbohydrates** are polyhydroxy aldehydes or ketones.

Carbohydrates have three primary functions: energy storage, energy source for cellular functions, and structural elements in plants and animals. The carbohydrates cellulose and chitin form the structural support for plants and animals. Other carbohydrates, such as glucose, supply energy for cell activities.

Carbohydrates can be classified into three groups: monosaccharides, disaccharides, and polysaccharides. *Monosaccharides* are simple sugars that contain one sugar unit. *Disaccharides* contain two simple sugar units, and *polysaccharides* contain many sugar units.

Both trees and crustaceans have carbohydrate structural elements.

Monosaccharides

Monosaccharides, or simple sugars, have one polyhydroxy aldehyde or ketone. All known monosaccharides—and there are many of them—are colorless, crystalline solids that dissolve in water. Most have a sweet taste. They rarely occur in nature as free molecules but are usually bonded to a protein, a fat, or another carbohydrate. Two significant exceptions existing as free molecules are glucose and fructose.

Glucose:
$$\begin{array}{c} H \\ | \\ C=O \\ | \\ H-C-OH \\ | \\ HO-C-H \\ | \\ H-C-OH \\ | \\ H-C-OH \\ | \\ CH_2OH \end{array}$$

Fructose:
$$\begin{array}{c} CH_2OH \\ | \\ C=O \\ | \\ HO-C-H \\ | \\ H-C-OH \\ | \\ H-C-OH \\ | \\ CH_2OH \end{array}$$

Glucose is the most abundant sugar in nature. Ripe berries, grapes, and oranges contain 20 to 30 percent glucose. The human body maintains a reasonably constant level of 80 to 120 mg of glucose per 100 mL of blood. Glucose is also the fundamental building block of the most common long-chain carbohydrates. Fructose may be found in ripe fruits and honey.

Monosaccharides are not always straight, chainlike molecules. They usually exist in rings that form when atoms near the end of the chain bond to atoms near the beginning of the chain. Glucose in an aqueous solution exists in an equilibrium between the two ring forms and the straight-chain form. The equilibrium lies strongly in favor of the ring forms.

Figure 18A-1

Formation of the ring structure of glucose

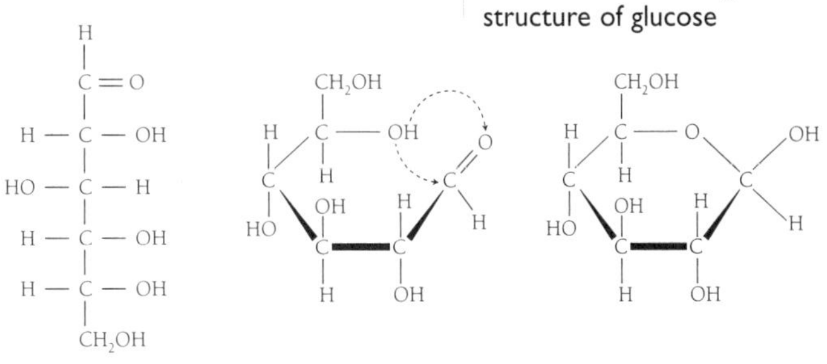

Straight form — Bent form — Ring form

Figure 18A-2

Glucose molecules exist in one of three forms.

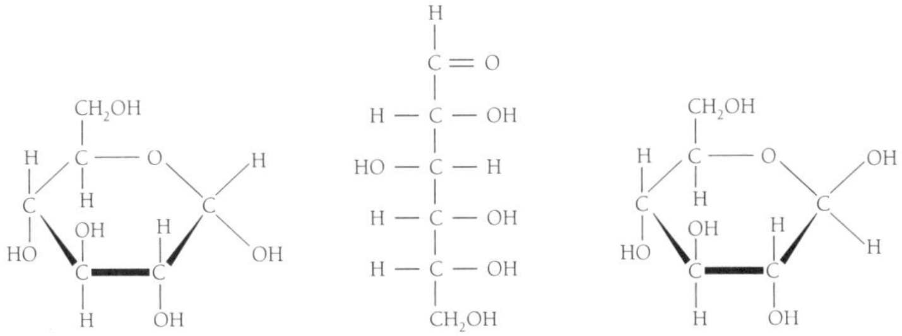

36% — less than 0.1% — 63%

Fructose also forms ring structures when it is in solution. Since the carbonyl group is not at the end of the carbon chain, the ring that forms has only five members.

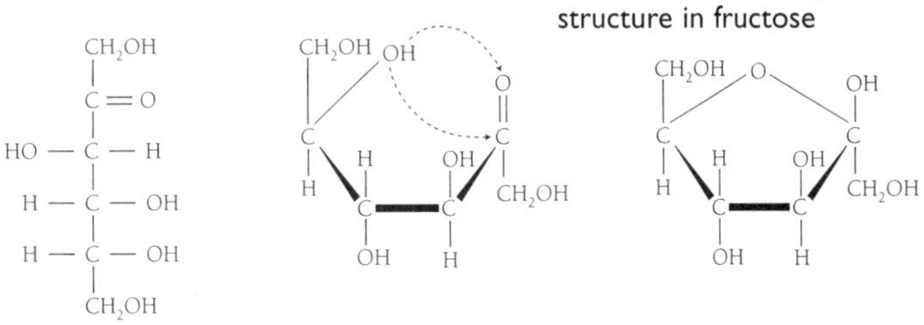

Figure 18A-3

Formation of the ring structure in fructose

Straight form — Bent form — Ring form

Disaccharides: Table Sugar and Other Goodies

As their name implies, **disaccharides** contain two monosaccharide units. An oxygen bridge between the two monosaccharides holds the two units together. While a large number of monosaccharides exist, an even larger number of disaccharides can be formed from combinations of monosaccharides. Of the numerous possibilities, three disaccharides play an important part in the human diet: maltose, lactose, and sucrose.

Figure 18A-4

Formation of a disaccharide from two monosaccharides. The bond that joins the two units may point in one of two directions.

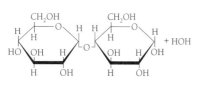

Maltose

Lactose

Sucrose

Maltose is the name given to two glucose molecules that are bonded together. The bond forms between the first carbon of one molecule and the fourth carbon of the other. Maltose is found in germinating grain and is produced during the digestion of starches.

Lactose is a disaccharide found only in milk. It consists of an isomer of glucose joined to a glucose molecule. The sugar is not very sweet, but it is an important ingredient in milk. If milk has become sour, it is because the lactose has broken down into lactic acid.

Sucrose—common household sugar—consists of glucose and fructose molecules. A bond links the first carbon of glucose to the second carbon of fructose.

Sucrose occurs abundantly in nature; fruits, sugar cane, sugar beets, and nectar are the major sources. This disaccharide can be split by a hydrolysis reaction if it is boiled in the presence of an acid or is acted upon by an enzyme (biological catalyst). People who make jellies and jams at home frequently use this reaction. A small amount of lemon juice is commonly added to preserves before they are cooked. Traditionally, people say that this action serves to make the jam sweeter. Lemon juice is sour because it contains citric acid. The citric acid hydrolyzes the sucrose in the fruit. The result of this reaction is two sugar molecules (one glucose and one fructose) instead of one sucrose molecule. The extra sugar molecules make the jam sweeter than it would have been had the lemon juice been left out. Honeybees cause the same type of reaction to occur when they make honey. They catalyze the reaction by secreting a special enzyme that splits sucrose molecules at hive temperatures.

Polysaccharides: Starch, Glycogen, and Cellulose

Polysaccharides are molecules that contain many sugar units. This is evident from their large molecular masses (up to several million amu). Multitudes of sugar units are bonded into long chains. Polysaccharides may be built from several different monosaccharides, but this text will concentrate on the polymers of glucose.

Figure 18A-5
Several units in a straight-chained starch molecule

Plants such as rice, potatoes, wheat, and oats store food in polysaccharide deposits called starch. **Starch** is a mixture of two glucose polymers: a straight chain and a branched chain. Together these polysaccharides supply nearly three-fourths of the world's food energy. The straight chain is an extended pattern of maltose disaccharide units. The bonds between the maltose units twist the long chain into a long spiral, or helix.

Figure 18A-6
Straight-chained starch molecules coil into a helix.

The branched polymer has the same backbone of glucose molecules as the straight chain. Branches off this main chain occur when the sixth carbon of a glucose in the chain bonds to the first carbon of another glucose.

Animals store food in the form of glycogen, not starches. **Glycogen** is a branched polymer of glucose in which the branches occur more frequently. In both starch and glycogen, thousands of glucose molecules are condensed into a single polymer. Why make these huge molecules just to be broken down before the cell can utilize them? Recall that osmotic pressure is regulated by the number of particles contained in the cell, not the size of the particles. Thousands of glucose molecules would cause the cell to burst due to the increased osmotic pressure, yet one molecule of glycogen (composed of thousands of glucose molecules) has minimal effect on the osmotic pressure.

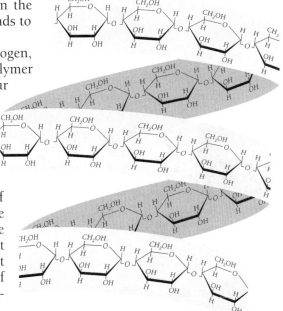

Horses, cows, and sheep eat grass. Why do humans not eat it? Aside from the taste consideration, the reason that humans do not eat such plants lies in the structure of the cellulose polymer. Cellulose, the major structural material of plants, is actually a glucose polysaccharide that is similar to starch and glycogen. In the polysaccharides that humans eat, the bonds between glucose units point downward; however, the bonds in cellulose point

Figure 18A-7

Several units in a branched starch molecule

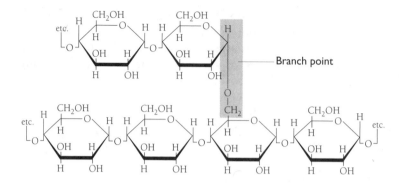

Figure 18A-8

Branched starch molecules form complex networks.

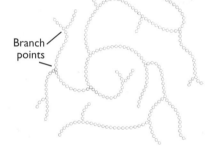

upward. Thus the difference between the starch in a tasty potato and the cellulose fibers in grass or a splintery piece of wood is the orientation of the bonds.

Human digestive tracts contain enzymes that can hydrolyze bonds that point downward. However, they have no such enzymes for the bonds that point upward. Cows, horses, and sheep are blessed with colonies of special bacteria that supply the needed enzymes in their intestines so that the cellulose can be digested.

The digestive system of a cow can hydrolyze the bonds between the glucose units in cellulose.

Figure 18A-9

Like starch, cellulose contains glucose molecules, but the bonds between the glucose units point in a different direction.

Section Review Questions 18A

1. Define *carbohydrate*.
2. List the three groups of carbohydrates.
3. Why is glucose not stored as separate molecules?
4. What products result from the hydrolysis of sucrose?
5. List the functions of carbohydrates.

18B Proteins

The Greek word *protos* means "first," and the French word *protéine* means "primary substance." It is not surprising, then, that the English word *protein* refers to one of the most important substances in the bodies of man and animals. Proteins serve as the building blocks for muscles, hair, blood cells, skin, spider webs, silk, enzymes, insulin, and even snake venom. They are essential nutrients, and they have many life-sustaining functions. Protein chains exhibit an intricate architecture that is marvelous to behold.

Figure 18B-1

General formula for amino acids

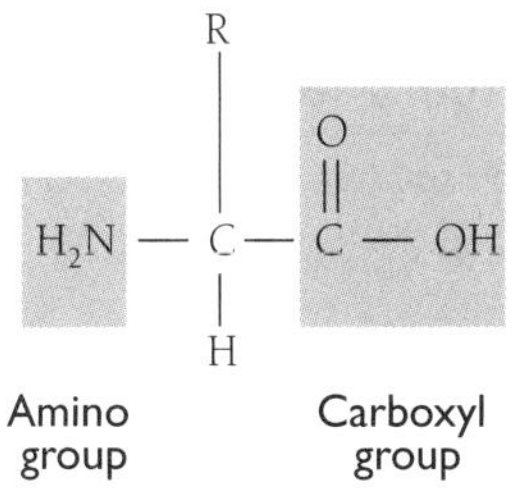

$$H_2N - \underset{\underset{H}{|}}{\overset{\overset{R}{|}}{C}} - \overset{\overset{O}{\|}}{C} - OH$$

Amino group — Carboxyl group

Amino Acids: The Building Blocks

Amino acids are the building blocks of proteins. As their name suggests, they are molecules that contain an amine (NH_2) group and the carboxyl (COOH) group of carboxylic acids. Various side chains (R groups) attached to the carbon adjacent to the nitrogen atom result in twenty different common amino acids. It is the side chains that determine the properties of the protein.

Our bodies can synthesize 12 of the 20 amino acids. The other eight must come from proper diet. Those amino acids that cannot be manufactured by the body are called **essential amino acids.** Table 18B-1 lists the essential amino acids.

Table 18B-1 Essential Amino Acids

Name	R group	Adult Requirements per Day
Isoleucine	$-\underset{\underset{CH_3}{|}}{CHCH_2CH_3}$	1.4 g
Leucine	$-CH_2CH(CH_3)_2$	2.2 g
Lysine	$-CH_2CH_2CH_2CH_2NH_2$	1.6 g
Methionine	$-CH_2CH_2SCH_3$	2.2 g
Phenylalanine	$-CH_2-\bigcirc$ 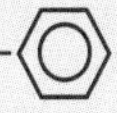	2.2 g
Threonine	$-\underset{\underset{CH_3}{|}}{CHOH}$	1.0 g
Tryptophan	$-CH_2-\text{(indole)}$ 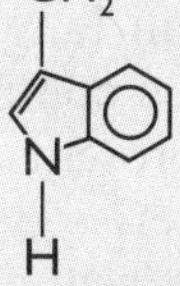	0.5 g
Valine	$-CH(CH_3)_2$	1.6 g

481

Polypeptide Chains

Amino acids join together when the amine group of one amino acid reacts with the carboxyl group of another. The bond that links the two amino acids is called a **peptide bond.** Molecules that contain two amino acids are called **dipeptides.** Aspartame (aspartyl phenylalanine), a popular artificial sweetener, is a dipeptide. **Polypeptides** are polymers of many amino acids.

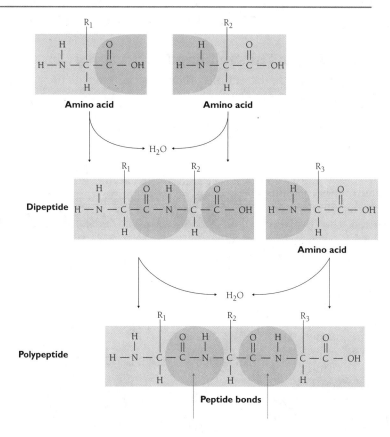

Figure 18B-2
Formation of peptide bonds

A **protein** is a polymer of one or more polypeptide chains. Some proteins contain only several hundred amino acids; others have many thousands of these building blocks. A protein's molecular weight can range from 5000 to millions of atomic mass units. Even with a comparatively short chain, the possible number of different proteins is staggering. From the twenty common amino acids, a series of fifty amino acids could be arranged in 3×10^{64} different sequences, with each sequence forming a different protein. When one considers that most proteins are much longer, the complexity and variety of proteins in God's creation become apparent.

Like polysaccharides, polypeptides tend to form regular shapes. The exact conformation of the chain depends on the amino acids that are present and the order in which they appear. The shapes and electrical natures of amino acids cause many proteins to coil themselves into helixes. This tight spiral of amino acids is held in place by hydrogen bonds between certain atoms in the amino acid backbone of the protein. A hydrogen in one amino acid forms a hydrogen bond with an oxygen of the amino acid four positions away. The side-chains of each amino acid extend like spokes from the surface of the helix. It is amazing that these intricate coils get their form from ordinary, but carefully arranged, hydrogen bonds. The structure truly is a masterpiece of art and engineering. Helixes known as alpha helixes impart toughness to hair, wool, fingernails, claws, antlers, hooves, and turtle shells. In most of these instances, alpha-helix coils wind around each other to form larger strands and fibers.

Another preferred shape of polypeptide chains is that of a pleated sheet. The chains stretch out in an almost flat, but folded, pattern. Hydrogen bonds, this time between amino acids in different chains, hold the structure together. Pleated sheets of proteins are largely responsible for the resilience of spider webs, silk, and feathers.

Each protein's distinctive shape (determined by the side chain) equips it to perform a unique function. For example, hemoglobin, the protein that carries oxygen in the bloodstream, consists of four interwoven chains. The shape of the protein provides perfectly shaped crevices for oxygen molecules to nestle into. Table 18B-2 lists the various protein classes and functions.

Figure 18B-3

Helical arrangement of amino acids around an axis with the side chains extending outward

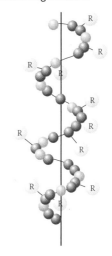

Table 18B-2 Function of Proteins

Class	Function
Enzymes	Catalyze biochemical reactions
Transport	1. Bind to various molecules/ions for transport via bloodstream, e.g., hemoglobin 2. Transport substances across membranes
Structural	Twisted helixes form fibers that bind skin, hair, blood clots, and tendons together
Motion	Produce movement by contraction, as muscle fibers do
Hormones	Regulate cellular activities, such as metabolism and growth
Antibodies	Provide protection against disease and foreign bodies

Section Review Questions 18B

1. Define *protein*.
2. Which proteins are essential? Why are they considered essential?
3. What is a polypeptide?
4. List the classes of proteins and their functions.

18C Lipids

Although water is sometimes called the universal solvent, it does not mix with all the compounds in the human body. An entire class of compounds called **lipids** cannot dissolve in water. Fatty acids, fats, oils, waxes, and steroids are examples of vital lipids.

Fats and Oils: Unsaturated or Saturated?

Fats and oils are esters of glycerol and fatty acids. **Glycerol** is a three-carbon molecule that has three OH groups on separate carbons. **Fatty acids** are carboxylic acids with long hydrocarbon tails. Ester linkages join the COOH groups of acids to the OH groups in the glycerol to construct a fat or oil molecule and water.

$$\begin{array}{c}\text{H}\\|\\\text{H}-\text{C}-\text{O}-\text{H}\\|\\\text{H}-\text{C}-\text{O}-\text{H}\\|\\\text{H}-\text{C}-\text{O}-\text{H}\\|\\\text{H}\end{array} + \begin{array}{c}\text{O}\\\|\\\text{H}-\text{O}-\text{C}-\text{R}\\\text{O}\\\|\\\text{H}-\text{O}-\text{C}-\text{R}\\\text{O}\\\|\\\text{H}-\text{O}-\text{C}-\text{R}\end{array} \longrightarrow \begin{array}{c}\text{H}\\|\\\text{H}-\text{C}-\text{O}-\text{C}-\text{R}\\|\\\text{H}-\text{C}-\text{O}-\text{C}-\text{R}\\|\\\text{H}-\text{C}-\text{O}-\text{C}-\text{R}\\|\\\text{H}\end{array} + 3\,\text{H}_2\text{O}$$

Glycerol + Three fatty acids → Fat (triglyceride) + Three molecules of water

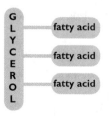

The acids that bond to the glycerol usually contain a large hydrocarbon chain (12 to 18 carbons are common), and they usually contain an even number of carbons (the chains are made in two-carbon units). Some chains contain only single carbon-carbon bonds, and some contain a few double bonds. This difference leads to the distinction between fats and oils.

Fats exist as solids at room temperature. Because most of their fatty acid components have a large degree of saturation (single bonds), the chains tend to extend in straight lines from the glycerol backbone. As a result, they fit together well and have effective dispersion forces. Recall that intermolecular forces can have a great effect in determining the physical properties of matter. It is these intermolecular attractions that cause fats such as lard to be a solid at room temperature.

Oils, on the other hand, are liquids at room temperature. Generally, their fatty acid components contain more double bonds than fats. Because of these double bonds, oils are considered to be unsaturated (recall that double bonds lessen the number of hydrogen atoms that are bonded). The double bonds in oils introduce bends in the chains. Because these irregularly shaped molecules do not fit together as well as the molecules in fats do, their dispersion forces are less effective. As a result, the oil molecules do not experience strong intermolecular forces—resulting in liquidity at room temperature.

Like carbohydrates and protein, dietary fat is a vital nutrient to help promote a healthful lifestyle, especially as a source of energy for the body. It is the most concentrated source of energy in the diet, providing nine calories per gram compared to four calories per gram from either carbohydrates or protein. Dietary fat is needed to transport fat-soluble vitamins, and it aids in their adsorption in the small intestine. It is also important in maintaining healthy skin and in regulation of cholesterol metabolism. Fat is an important ingredient in food because it enhances the taste, aroma, and texture of the food. Since it is digested more slowly than protein or carbohydrates, it helps provide a sense of fullness, or satiety, after eating.

All fats have a tendency to break down when they are exposed to air, especially the unsaturated fats. To prevent this breakdown, a process termed *hydrogenation* was developed in the early 1900s. Hydrogenation adds hydrogen molecules directly to an unsaturated fatty acid such as vegetable oil to convert it into a semisolid such as margarine. These hydrogenated oils are often substituted for saturated fats. Since a diet high in saturated fats tends to raise the level of cholesterol in the blood and increase the risk of heart disease, researchers thought

Figure 18C-1

Structural formula and space-filling model of a molecule of fat

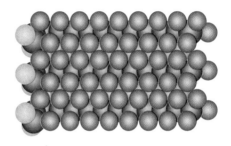

Double bonds make the difference between solid and liquid.

hydrogenated oils were safer because they contained lesser amounts of saturated fats. However, recent research has revealed that when oils are hydrogenated, trans fats (fats whose molecular shape is *transformed* by the hydrogenation process) are formed, and they can be just as bad as a saturated fat with respect to cardiac health. The majority of experts agree that calories from fats should not exceed 30 percent of the total calorie intake.

Figure 18C-2

Structural formula and space-filling model of linseed oil

Steroids: Cholesterol and Company

Steroids are lipids, but their structures do not resemble triglyceride fats and oils in any way. They are the nonester lipids. The basic structure of a steroid is a combination of three six-membered rings and one five-membered ring. Functional groups attached to various points on the rings result in a wide variety of steroids. Cholesterol, vitamin D, cortisone, testosterone, and estrogen all use this common "chicken wire" frame.

The study of steroids ranks as one of the most active areas of chemical research. One steroid that has been studied much in recent years is cholesterol—a compound that, although vital to good health, can increase the risk of heart disease if excessive amounts are present in the blood. Contrary to popular belief, most of the cholesterol found in the body is manufactured by the body itself. The average total blood cholesterol in Americans is 203 milligrams per deciliter (mg/dL). The National Institutes for Health consider less than 200 mg/dL desirable and greater than 240 mg/dL high.

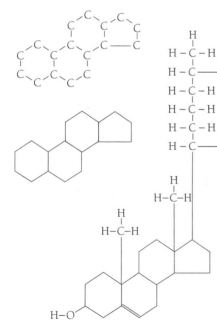

Figure 18C-3

General structure of steroids (left). Structural formula and space-filling model of cholesterol (right).

Researchers have identified two types of cholesterol, termed high-density lipoprotein (HDL) and low density lipoprotein (LDL). The LDL is referred to as "bad cholesterol," because it contains most of the cholesterol in the blood and is associated with the cholesterol deposits on artery walls. It is these deposits that cause clogged arteries and can lead to heart attacks. The HDL, or "good cholesterol," is believed to carry cholesterol out of the blood and to the liver for breakdown and excretion. It appears that monounsaturated fats tend to lower LDL while raising HDL. Olive oil and canola oils are among the highest in monounsaturated fat.

In addition to diet, blood cholesterol is influenced by a wide variety of factors, including heredity, smoking, age, race, gender, high blood pressure, obesity, and activity level. Researchers are also seeking to isolate a gene that may be used to identify individuals at risk and apply early treatment.

Section Review Questions 18c

1. Explain the difference between saturated and unsaturated fats.
2. How does the difference in saturation affect physical properties?
3. What is the basic structure of a steroid?
4. List the two types of cholesterol that have been identified, and explain their relationships to heart disease.

18D Cellular Processes: Carbohydrates, Proteins, and Lipids Work Together

An organism that contains carbohydrates, proteins, and lipids may be complete structurally, but it still does not have all the compounds it needs to carry on its cellular processes. Specialized proteins called *enzymes* facilitate almost every reaction that takes place in living organisms. *Vitamins* are small but indispensable nutrients that organisms cannot produce by themselves. *Hormones* serve as chemical messengers from one part of an organism to another. Finally, *nucleic acids* guide the construction of proteins and carry genetic information to offspring. These, too, are the molecules of life.

Enzymes: Organic Catalysts

Enzymes are protein molecules that act as catalysts. For example, an equilibrium between carbon dioxide, water, and carbonic acid functions as the red blood cells pick up carbon dioxide from muscles, transport it to the lungs, and then release it during respiration. Once carbon dioxide is released, red blood cells immediately pick up oxygen. An enzyme named carbonic anhydrase catalyzes both of these reactions. Without this enzyme, one molecule of hemoglobin would react approximately every 100 seconds. With the enzyme, however, 100,000 molecules can react every second. In this reaction, the rate is multiplied by a factor of 10^7. Without that increase, there would be no way for the lungs to exchange enough gases to support the respiratory process needed to maintain life.

$$CO_2 + H_2O \xrightleftharpoons{\text{carbonic anhydrase}} H_2CO_3$$

Enzymes act with amazing precision. God, in His omnipotence, designed the enzymes so that they would select just the right types of molecules on which to act. An enzyme called trypsin catalyzes a reaction that breaks down protein chains. What is remarkable is that the enzyme splits protein chains only on a certain side of two amino acids; it ignores all other sites on the protein chain.

How can enzymes work so fast, in so many applications, and with such precision? Researchers have proposed several models to account for what has been observed. One widely accepted model is the **lock-and-key model.** It holds that enzymes catalyze reactions by positioning reactants in ideal positions for the reactions to occur. When an enzyme and the substance on which it works combine, collisions with other reactants are more effective than without the enzyme.

A complete understanding of how enzymes work will probably involve a thorough knowledge of the shapes and sizes of the biological molecules—not an easy task, considering how complex large proteins can be. Like the study of steroids, enzyme research is currently one of the most exciting and active areas in biochemistry.

Figure 18D-1

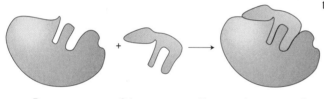

Enzyme   Substrate   Enzyme-substrate complex

Vitamins: Micronutrients

As far as biochemistry is concerned, sailors had it rough in the old days. For British seamen long trips almost certainly meant bleeding gums, loose teeth, cuts that did not heal, and weight loss. Japanese navigators faced stiff lower limbs and possible paralysis. The causes of these plights had nothing to do with the sea, but rather with the unbalanced diets of the seamen. These men had the vitamin deficiency diseases now known as scurvy and beriberi.

Vitamins are organic substances that are essential for normal nutrition, but they are not carbohydrates, lipids, proteins, or fats. The British sailors lacked ascorbic acid, or vitamin C, in their on-board diets. A daily ration of lemon or lime juice provided the necessary nutrient to help control the symptoms of scurvy and earned the sailors the nickname "limeys." The Japanese navy fought the disease of beriberi in the late 1800s by introducing portions of wheat and barley and unpolished rice into their diets. The hulls of these grains contain vitamin B_1, which relieved the sailors' suffering. The doctor who first isolated the organic substance that cured beriberi found that it contained an amine (NH_2) group; he called it a "vital amine," or vitamin.

Vitamins are vital to the proper functioning of enzymes. Vitamin C is connected with the formation of the intercellular "glue" in bones, connective tissue, and cartilage. The B vitamins help enzymes break down carbohydrates. Without them, pyruvic acid, a product of incomplete carbohydrate breakdown, accumulates in the muscles and causes the pain of beriberi. Today approximately twenty-one vitamins are known, but researchers suspect that several others exist. The known vitamins can be classified as either water soluble or fat soluble.

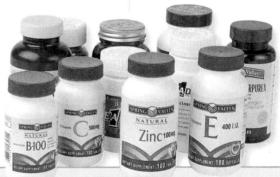

Table 18D-1 Fat-Soluble Vitamins

Vitamin	Functions	Deficiency Symptoms	Source
A	contributes to visual pigments in the eye	inflammation of eyes, night blindness, drying of mucous membranes	eggs, butter, cheese, liver, dark green and deep orange vegetables
D	aids in absorption and deposition of calcium	rickets (defective bone growth)	fish-liver oil, fortified milk
E	protects blood cells, unsaturated fatty acids, and vitamin A from oxidation	anemia, bursting of red blood cells	wheat germ, whole grain cereals, liver, margarine, vegetable oil, leafy green vegetables, egg yolk
K	aids clotting of blood	increased clotting time of blood	liver, cabbage, potatoes, peas, leafy green vegetables

Tables 18D-1 and 18D-2 list the common vitamins, their functions, and the possible consequences of deficient intake.

Water-soluble vitamins can be excreted from the kidneys; therefore, they need to be replenished constantly. Unlike water-soluble vitamins, excessive amounts of fat-soluble vitamins can accumulate in the fatty tissue. Although vitamins are essential to good health, megadoses of some vitamins can be harmful. Excessive amounts of vitamin A can result in brain and kidney damage. Too much vitamin C has been linked to kidney stones.

Table 18D-2 Water-Soluble Vitamins

Vitamin	Functions	Deficiency Symptoms	Source
B_1 (thiamine)	aids in carbohydrate metabolism	fatigue, beriberi, accumulation of body fluids	pasta, bread, lima beans, wheat germ, nuts, milk, liver, peas, pork
Niacin	aids in energy utilization	inflammation of nerves and mucous membranes, dermatitis	meat, whole grains, poultry, fish, peanuts
B_2 (riboflavin)	aids in protein metabolism	dermatitis, inflammation of the tongue, anemia	milk, meat, eggs, mushrooms, dark green vegetables, pasta, bread, beans, peas
B_6 (pyridoxine)	aids in amino acid metabolism	convulsions in infants, inflammation of the tongue, increased susceptibility to infections	muscle meats, liver, whole grains, poultry, fish
B_{12}	aids in the formation of nucleic acids	retarded growth, spinal cord degeneration	meat, fish, eggs, milk, kidneys, liver
C (ascorbic acid)	builds strong connective tissues in bones, cartilage, and blood vessels	slow wound healing, scurvy, anemia	citrus fruits, melons, tomatoes, green peppers, berries, leafy green vegetables
Pantothenic acid	aids in respiration	gastrointestinal disturbances, depression, mental confusion	whole grain cereals, bread
Folic acid	aids in formation of heme groups and nucleotides	various types of anemia	kidneys, liver, leafy green vegetables, wheat germ, peas, beans

Hormones: Chemical Messengers

Complex organisms have many specialized body parts that must constantly communicate with each other. The brain must know when the eyes see something as significant as a charging rhinoceros, the adrenal medulla gland must know when it is time to secrete adrenaline, and the muscles must know when to respond to the danger with increased activity. Nerves, which carry electrical messages, handle many of the body's communications, but they are not alone in this task. **Hormones** are chemical messengers that are produced by the endocrine glands and are transported by the bloodstream to various areas of the body.

Hormones are a chemically diverse lot. They can be steroids, polypeptide chains, or proteins. These compounds, when released into the bloodstream or other body fluids, travel throughout the body. Despite the fact that they come into contact with many cells, they act only on their target cells. Research findings suggest that the "target" cells contain receptor molecules that recognize specific hormones by their shapes.

Hormones are also found in insects and plants. Biochemical researchers use insect sex attractant hormones called pheromones to trap and disorient insect pests. The pheromone for the gypsy moth, for example, is effective at a level of 1×10^{-13} gram. Some herbicides are modeled after plant hormones in order to be very selective in the plants that they kill. Continued biochemical research will enable us to replace broad-spectrum pesticides and herbicides with compounds that will act against specific targets.

Nucleic Acids: Chemical Blueprints

When cells reproduce, they pass genetic information to one another in long-chain molecules called **chromosomes.** Human body cells contain forty-six of these long molecules in their nuclei. Segments of chromosomes, called **genes,** carry the coded information that directs the production of specific polypeptide chains. Genes are constructed of many **nucleotides,** which are the building blocks of **nucleic acids.**

Each nucleotide consists of three units: a pentose sugar, a phosphate group, and a ring-shaped nitrogenous base. The pentose sugar and the phosphate groups alternate to form a long chain that supports the nitrogenous bases. In RNA (ribonucleic acid), the pentose sugar is ribose,

Figure 18D-2

The nucleus (a) contains chromosomes (b), which are made of DNA strands (c), which are made of nucleotides (d).

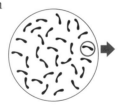

a. Cell nucleus

b. One chromosome

c. One DNA strand

d. One nucleotide

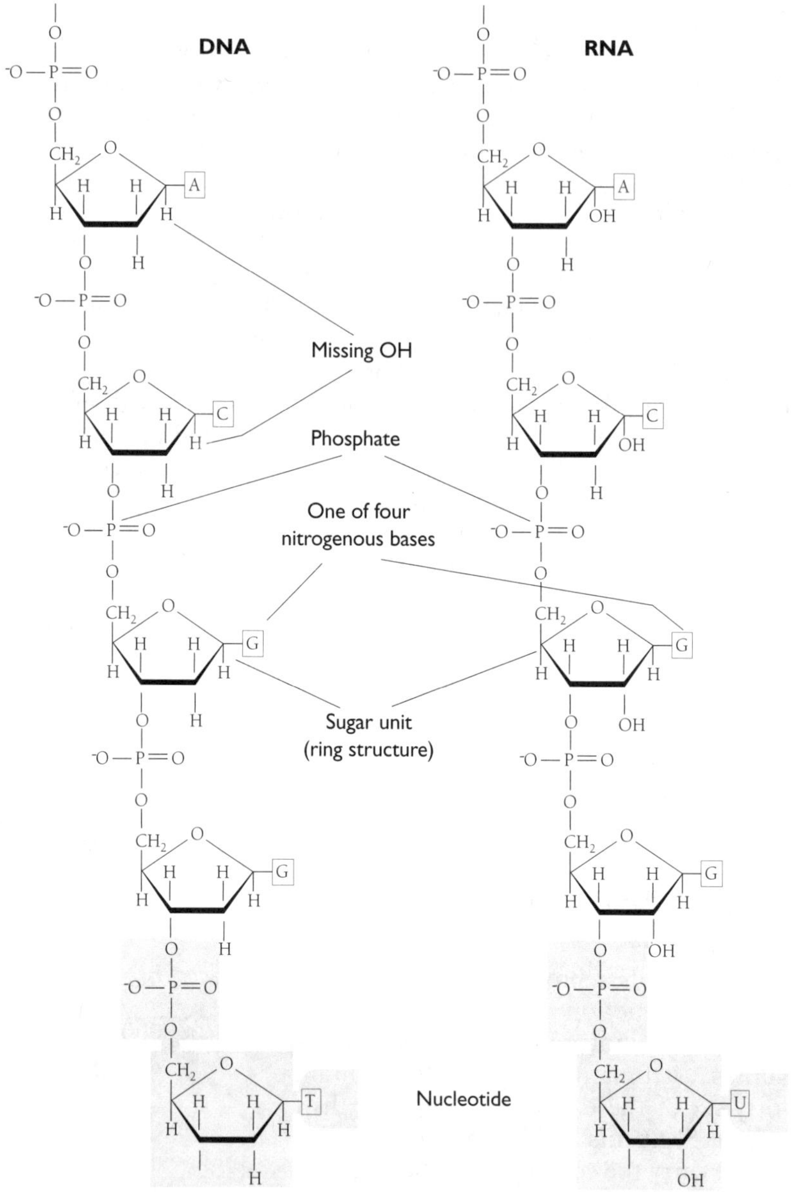

Figure 18D-3

Single strands of DNA and RNA

while in DNA (deoxyribonucleic acid) it is deoxyribose, which is a ribose with one less OH group.

The nitrogenous bases attached to the sugar may be one of five possibilities: adenine, cytosine, guanine, thymine, or uracil (abbreviated A, C, G, T, and U). DNA strands contain only A, C, G, and T bases, and RNA strands have A, C, G, and U. Of these bases, C, T, and U are small, single-ringed structures, and A and G are larger, double-ringed structures.

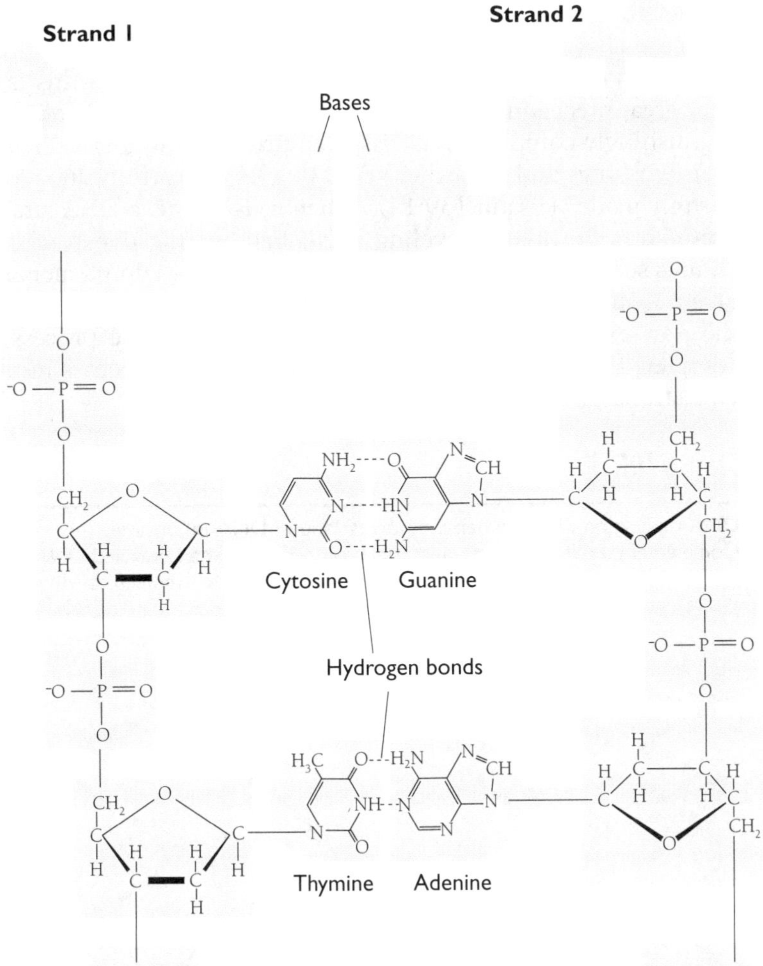

Figure 18D-4

A two-nucleotide sequence of double-stranded DNA

DNA has been designed so that two strands coil around each other in a double helix. It looks somewhat like a ladder that has been twisted several times along its long axis. The backbones circle around the outside, and the bases mesh together inside the coil. Large, double-ringed adenine bases pair with smaller thymine bases, while large guanine bases fit neatly next to smaller cytosine bases. Hydrogen bonds hold the complementing bases snugly in position.

Nucleic acids function in a marvelous, almost miraculous manner. They are responsible for the faithful duplication of an organism's distinctive characteristics, and they perform this task with great precision and accuracy. In this regard God's creation is astonishingly complex. Man has been privileged to unravel some of its workings and to further verify that he is "fearfully and wonderfully made" (Psalm 139:14). When cells divide, a DNA strand reproduces by first unraveling itself. Each of the two resulting strands serves as a template, or pattern, for a new complementary chain. Complementing bases on the newly forming strand match the now-exposed bases of the old strand. When the process is completed, two identical double-stranded DNA molecules result. One strand goes to each half of the dividing cell.

Figure 18D-5

DNA replication occurs when a double strand of DNA (a) unravels (b). Complementary strands form and attach onto each of the unraveled DNA strands (c), producing two identical double-stranded DNA molecules (d).

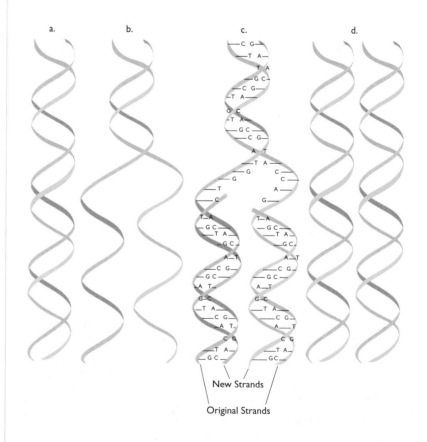

Figure 18D-6

Transfer and cloning of the human insulin gene

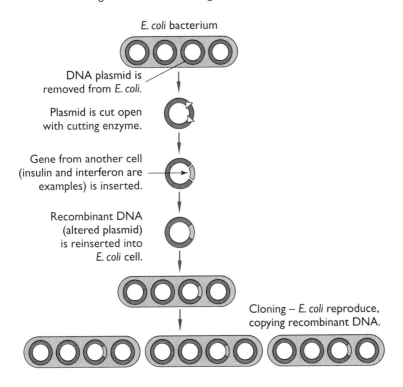

DNA and RNA molecules provide sets of "blueprints" instructing cells on how to make proteins. Amazingly, researchers have discovered how to translate the genetic code in these blueprints. The order of nitrogenous bases along a DNA strand will dictate which amino acids are to follow each other in a protein. For instance, the sequence guanine-guanine-adenine along a DNA strand calls for the amino acid glycine. Other sequences of three bases call for other amino acids until entire polypeptide chains have been constructed.

If DNA and RNA are the "blueprints," can they be used to build anything? Yes. The biotechnological industry based on the synthesis and cloning of recombinant DNA molecules has produced many complex human hormones as well as plants with improved disease and pest resistance. Since the 1980s, the pharmaceutical industry has used these techniques to produce human insulin for the treatment of diabetes. Although the basic process is painstaking, it can be discussed in terms of cutting and pasting. In the case of insulin, the gene sequence for the

The Human Genome Project

The Human Genome Project (HGP), formally begun in 1990, is a multiyear project of international researchers whose goal is to gain a basic understanding of the entire genetic blueprint of the human being. It grew out of a series of scientific conferences from 1985 to 1987. In the United States, the HGP is funded by the National Institutes of Health and the Department of Energy. One of the first directors was American biochemist James Watson, who shared the 1962 Nobel Prize in physiology with British biophysicist Francis Crick for the discovery of the double helical structure of DNA. Many nations are participating in the HGP including Germany, France, the United Kingdom, and Japan. Additionally, several private companies worked on their own to sequence the human genome.

A genome is the complete collection of a particular organism's genetic code. In a human genome, there are between 50,000 and 100,000 genes located on 23 pairs of chromosomes. A chromosome contains more than 250 million DNA base pairs, and researchers have estimated that the entire human genome may consist of as many as 3 billion base pairs. The specific sequencing and pairing of these 3 billion base pairs determine all of the various characteristics of an individual. At completion, the HGP had prepared high-resolution maps of the genes of the human chromosomes.

Why spend so much time and effort on a task as daunting as this? Knowledge about the effects of DNA differences between individuals can lead to better methods of diagnosis and treatment of diseases. It is well known that many diseases are associated with heredity, such as cystic fibrosis, Huntington's disease, and muscular dystrophy. In fact, the genes associated with these diseases have already been identified. Once the map is completed, researchers say that better genetic screening tests, new drugs, and genetic therapies can be developed to predict, prevent, and treat genetic diseases.

However, with increased knowledge comes increased accountability—the greater the technological power, the greater the potential for evil. Complete knowledge of the human genome has implications for the born and the unborn, for the well and the ill, that raise many social, legal, and moral questions. If you are healthy, but genetic testing reveals that you will develop a disease, when would you want to know? Do you want to know? If you know that you carry a deficient trait that could be passed on to your children, should you have children? Can someone tell you not to have children? Will health care be denied to those who may carry a "defective" gene because of the increased cost of medical treatment? If an unborn child has been determined to have a "defective" trait, what should the parents do? Who determines what is a "defective" trait? Will you be able to predetermine your children's traits? Will "high-grade" embryos be "manufactured" *in vitro* so that parents can select their child's traits? As unsettling and cold as these questions may seem, they must be contemplated.

If you think that no one would ever consider a human life in such a cold-hearted way, recall the Holocaust suffered by the Jews at the hands of Nazi Germany. More recently, there was the rampant genocide that was carried out in parts of Africa, Bosnia, and Serbia during the late 1990s. Though it may seem hard to believe, thousands of people are still held in bondage as slaves in many parts of the world. How are we to act toward our fellow man? Jesus Christ is certainly the ultimate example, and He gave the definitive answer when the religious leaders tried to trick Him by asking which was the greatest commandment. Christ's response can be found in Matthew 22:37-40: "Thou shalt love the Lord thy God with all thy heart, and with all thy soul, and with all thy mind. This is the first and great commandment. And the second is like unto it, Thou shalt love thy neighbor as thyself. On these two commandments hang all the law and the prophets." Because of man's sinful state, what may be used for good can always be used for evil. As technology progresses, answers for questions such as these will need to be found. We as Christians, using the Bible as our guide, must be the salt of the earth and not shrink from our responsibilities toward our neighbors.

production of insulin in humans is "cut out and pasted" into the DNA of the common bacterium *E. coli*. Then, as the bacteria reproduce, billions of copies of the gene used to produce human insulin are made, and each bacterium becomes a microscopic insulin factory. Figure 18D-6 is a simplified illustration of the very complex process of producing recombinant DNA for human insulin production.

Section Review Questions 18D

1. What is an enzyme?
2. Explain the lock-and-key model of enzyme function.
3. Define *vitamin*.
4. What are the two classes of vitamins?
5. Why would megadoses of vitamin A not be recommended?
6. Define *hormone*.
7. What differentiates RNA from DNA on the sugar base?

Chapter Review

Coming to Terms

carbohydrate
monosaccharide
disaccharide
polysaccharide
starch
glycogen
amino acid
essential amino acid
peptide bond
dipeptide
polypeptide
protein
lipid
glycerol
fatty acid
fat
oil
steroid
enzyme
lock-and-key model
vitamin
hormone
chromosome
gene
nucleotide
nucleic acid

Review Questions

1. List three major functions of carbohydrates.
2. The melting point of glucose ($C_6H_{12}O_6$) is 146°C. This seems quite high compared to the 6.5°C melting point of cyclohexane. What functional groups on the glucose are responsible for the difference? What do these functional groups do to raise the melting point?
3. Describe what is meant by each term, and give one example of each.
   a. monosaccharide   b. disaccharide   c. polysaccharide
4. How are cellulose and glycogen similar? How are they different? How does the difference between cellulose and glycogen affect their nutritional value to man?
5. Postulate a reason that termites can eat wood but people cannot.
6. Why are amino acids named *amino acids*? Draw the structural formula of an amino acid, and identify the functional groups.
7. Polysaccharides and polypeptides are both polymers. What is the difference between them?
8. Why is the three-dimensional shape of a protein important?
9. Can a protein consist of more than one polypeptide? Explain.
10. List three functions of proteins.
11. What is the difference between a fat molecule and an oil molecule?
12. What distinguishes steroids from other lipids?
13. What do enzymes do? In terms of energy, how do they do this? In terms of chemical structure, how do they do this?
14. What are the two general classes of vitamins?
15. Some people think that massive doses of vitamins can be beneficial. They routinely take large doses of vitamin C, yet these people do not take large doses of vitamin A. Why is this?
16. What are the functions of hormones in the human body?
17. What is a nucleotide? What are nucleotides composed of? How do the nucleotides of DNA and RNA differ?
18. What is the general shape of a DNA molecule? What is the function of DNA?

19. Match each of the following structural formulas to the general class of compound that it represents.

amino acid   monosaccharide   fat   polysaccharide

disaccharide   polypeptide   oil   steroid

a.

b.

c.

d.

e.

f.

g.

h.

19A Natural Radioactivity page 501
19B Induced Reactions page 513
FACETS: Radioactive Age-Dating Methods page 510

Nuclear Chemistry 19

Getting to the Heart of Matter

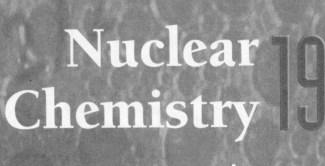

On August 6, 1945, the United States dropped the first nuclear weapon in the history of warfare on the Japanese city of Hiroshima. No one doubts that nuclear power can be destructive. However, this same energy source can offer numerous peaceful uses. Radioactive tracers, for instance, are used medically to diagnose diseases. Properly constructed nuclear power plants offer a clean, reliable source of electricity.

What causes radiation? How are nuclear reactions different from chemical reactions? Where does the awesome power come from? What makes a nuclear weapon different from a nuclear power plant? The answers to all these questions lie in the remarkable nature of the atomic nucleus.

19A Natural Radioactivity

In the chemical activities studied in the previous chapters, the atom's valence electrons play the crucial role. But this is not so in nuclear chemistry. The reactions that occur in the nuclei of atoms have nothing to do with the electrons in the atoms' energy levels. The nuclei have a chemistry all their own.

The Discovery of Radiation: Becquerel's Mysterious Rays

Soon after the discovery of X rays, a French physicist named Henri Becquerel (beh KREL) (1852-1908) set out to determine whether some "glow-in-the-dark" crystals he possessed would emit X rays. Becquerel gathered crystals of potassium uranyl sulfate that had been exposed to bright sunlight. These crystals have the strange ability to glow in the dark after they have been exposed to sunlight. To determine whether light from the crystals contained X rays, he covered a photographic plate with black paper and exposed it to glowing crystals. He reasoned that only X rays could pass through the paper to expose the film. As expected, the glowing crystals left marks on the plate. Becquerel correctly concluded that potassium uranyl sulfate emits some kind of penetrating rays. Yet Becquerel soon made an unexpected discovery that caused him to wonder just what kind of rays were coming out of the crystals.

Toward the end of his study, Becquerel determined that the crystals left marks on photographic plates even though they had not been exposed to the sun. Further investigations showed that other uranium compounds also darkened photographic plates. The crystals constantly gave off energetic rays despite being melted, dissolved, and recrystallized. No one knew where the energetic rays came from.

In 1898 two scientists in France, Marie Sklodowska Curie and her husband, Pierre, discovered another element that seemed to defy the law of energy conservation. They coined the word **radioactivity** to describe the spontaneous emission of the penetrating rays. The Curies correctly concluded that radioactivity is an atomic property that does not depend on how the atoms are chemically bonded. Becquerel's experimental results were then understood. The uranium atoms, not the potassium uranyl sulfate crystals, were responsible for the radiation. While Becquerel had actually discovered radioactivity, the Curies correctly identified its source.

In time, scientists analyzed the rays from radioactive substances. They passed streams of radiation through powerful electrical and magnetic fields. By charting the deflections of the rays, they could determine the electrical charges of the different types of radiation. They found that radiation had three components: one with a positive charge, one with a negative charge, and one with no charge at all.

Henri Becquerel

Figure 19A-1

Experiment for separating and analyzing the components of radiation

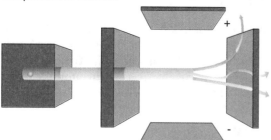

Alpha Particles: Helium Nuclei

A stream of positively charged particles was deflected slightly toward the negative electrical plate. Because the particles swerved only slightly, scientists deduced that they must be more massive than the other types of radiation. They later found that these **alpha particles** contained two protons and two neutrons. Interestingly enough, these were the same particles that Rutherford had used in his famous experiments with gold foil that were studied in Chapter 4. Isotopic notation shows that alpha particles are actually $^{4}_{2}\text{He}$ nuclei. The superscripted number shows that the mass of the particles is approximately 4 amu, and the subscripted number gives the atomic number of the element. Another name for an alpha particle is the He^{2+} ion.

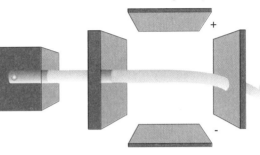

Figure 19A-2

Alpha particles bend slightly toward the negative plate.

While alpha radiation will penetrate matter, it can be stopped easily. The alpha particles will interact quickly with the matter they strike to produce helium atoms. For instance, alpha particles cannot go through this page. If He^{2+} ions hit the page, they would immediately grab two electrons from the paper and turn into ordinary helium atoms. Although alpha particles cannot pass through the skin, they can be ingested or inhaled. Once inside the body, they can cause biological damage as they ionize molecules that are vital to biological processes. The damage caused by the alpha particles may take the form of burns, sickness, or even death. Some of the superficial burns that resulted from the nuclear weapons dropped on Hiroshima and Nagasaki were caused by the enormous amounts of alpha particles released by the bombs. A nuclear bomb, however, is not the only source of alpha particles—they could be in your own house. Radon gas can seep into basements of houses and emit alpha particles as it decays:

$$^{222}_{86}\text{Rn} \longrightarrow {}^{4}_{2}\text{He} + {}^{218}_{84}\text{Po}$$

Radon detectors are used to determine whether toxic levels of radon gas exist in a house.

It originates in the earth as a decay product of uranium. Homeowners should consider radon testing and corrective actions such as venting of the basement to the outside air to prevent the buildup of radon in the house.

Beta Particles: Electrons from the Nucleus

Like alpha radiation, the second type of radiation is also composed of particles. **Beta particles** carry a negative charge and have very little mass. Scientists noted that beta particles acted much like electrons when they swerved sharply toward a positively charged plate. The direction of the swerve identified the particle's charge, and the amount of the swerve indicated that the particle had negligible mass. Further studies gave a surprisingly simple identity to beta particles: they are electrons—electrons

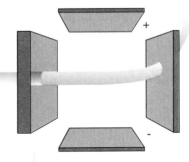

Figure 19A-3

Beta particles bend sharply toward the positive plate.

503

Figure 19A-4

A Geiger counter can detect radiation passing through its chamber because gases ionize and allow an electrical current to flow. The current is then amplified and converted to sound.

Figure 19A-5

Gamma rays are not deflected by electrical charges.

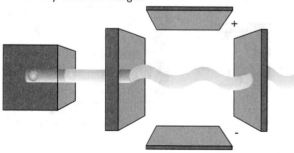

that have been formed in and emitted from the nucleus. Beta particles are represented as $_{-1}^{0}e$ in isotopic notation. This convention shows both the negligible mass and the -1 charge of the particle.

Beta radiation penetrates into substances deeper than alpha radiation. Beta particles can zip through sheets of paper, but wood or metal can stop them. Like alpha particles, they ionize atoms and molecules in the matter they hit, but not as readily as alpha particles do. Because they do not ionize as readily, beta particles penetrate deeper and are more dangerous than alpha particles to the inside of the body.

Gamma Rays: High-Energy Waves

Unlike alpha and beta radiation, gamma radiation is not composed of particles, and its emission does not require the nucleus to be unstable. It consists of electromagnetic waves similar to those of visible and ultraviolet light, but with more energy. **Gamma rays** are uncharged and are undeflected in an electrical field, as can be seen in Figure 19A-5.

If no particles are formed by gamma emission, why are gamma rays produced? Gamma rays are emitted by nuclei in excited states. An excited nucleus is nothing more than a normal nucleus with extra energy. Nuclei in excited states are sometimes produced as the result of alpha and beta emission. An excited nucleus returns to its ground state by releasing energy in the form of gamma rays.

Of the three types of radiation, gamma rays are the most harmful. Partly because they have no electrical charge and partly because they are not particles, gamma rays have a lower ionizing ability than alpha or beta particles. As a result, gamma rays penetrate more deeply than either alpha or beta particles; therefore, they do more damage. They are not stopped by ionization reactions near the surface of a body. They can, however, be stopped by several feet of concrete or a sheet of lead.

Measurement of nuclear radiation involves two types of units, depending on what is being measured—the physical radiation itself, or the biologic effect caused by the radiation. Physical radiation units measure the activity of the radiation source, using an SI unit called the **becquerel** (Bq). A radiation

Table 19A-1 Three Types of Radiation

Name	Symbol	Identity	Charge	Mass	Penetration
Alpha	^4_2He, α	helium nucleus	+2	4 amu	low
Beta	$^0_{-1}e$, β	electron	-1	$\frac{1}{1836}$ amu	medium
Gamma	$^0_0\gamma$	electromagnetic radiation	0	0	high

source with an activity of one becquerel has one disintegration per second. Another unit, which is older but still used, is the curie (Ci). One **curie** is the number of disintegrations per second in a one-gram sample of radium; one curie is 3.7×10^{10} Bq.

Biological radiation units measure the nuclear radiation's effect on living tissue. The **gray** is the SI unit of biological radiation effect. A gray is equivalent to the transfer of one joule of energy to one kilogram of living tissue. The **rad** (**r**adiation **a**bsorbed **d**ose) is an older term, though it is still used, and is equal to 0.01 gray. The **roentgen** (R) (RENT gin) was originally used to measure gamma and X rays. One roentgen measures the quantity of radiation, and is equal to 93.3×10^{-7} joules of energy per gram of tissue. One roentgen is 0.0096 gray—almost one rad.

Recall that the different types of radiation have different energies and penetration depths. Therefore, a 1 R dose of alpha radiation will have a different effect than a 1 R dose of gamma radiation. To overcome these differences, a unit was devised to measure the effects of different types of radiation on man. The **rem** (**r**adiation **e**quivalent in **m**an) is the dose of any type of radiation that has the same health effect as one roentgen of X ray or gamma radiation. Table 19A-2 shows the dose-related health effects of radiation measured in rems.

Nuclear Equations: Describing Nuclear Reactions

Chemical equations describe the reactants and products in a chemical reaction. **Nuclear equations** describe what occurs when nuclei split, fuse, or release radiation. These equations identify the nuclei that react and the products of the reaction. Where appropriate, they show whether alpha particles, beta particles, or gamma rays leave nuclei. For example, U-238 emits alpha particles. Because two protons in the alpha particle depart, the atomic number decreases by two, and the nucleus becomes a thorium nucleus (atomic number = 90). Because the departed alpha particle had a

Table 19A-2 Health Effects of Radiation

Dose	Health Effects
0–25 rem	No detectable clinical effect in humans.
25–100 rem	Slight short-term reduction in number of some types of red blood cells; severe sickness uncommon.
100–200 rem	Fatigue/nausea; vomiting if >150 rem; longer-term decrease in some types of blood cells.
200–300 rem	Nausea/vomiting first day of exposure; two-week latent period followed by malaise, appetite loss, sore throat, diarrhea. Recovery usually in 3 months unless complicated by infection.
300–600 rem	Nausea/vomiting/diarrhea in first few hours, with one-week latent period followed by fever, loss of appetite, malaise in the second week, followed by inflammation of the mouth and throat, diarrhea, emaciation. Some deaths in 2 to 6 weeks. Death likely for 50% if exposure was above 450 rem; others will recover in about 6 months.
600+ rem	Nausea/vomiting/diarrhea in first few hours, followed by rapid emaciation and death possibly within 2 weeks. Death probable for 100%.

mass of 4 amu, the mass of the remaining nucleus is 234 instead of the original 238. A nuclear equation describes this process.

$$^{238}_{92}U \longrightarrow\ ^{234}_{90}Th + ^{4}_{2}He$$

Remember that $^{4}_{2}He$ is an alpha particle. Two things can be noted about the nuclear equation for an alpha-emitting process. First, the sum of the atomic numbers of the reactants equals the sum of the atomic numbers of the products (92 = 90 + 2). Second, the sum of the mass numbers on one side of the equation equals the sum of the mass numbers on the other (238 = 234 + 4).

Sample Problem Write the nuclear equation that describes the alpha decay of U-234.

Solution
Alpha decay implies that an alpha particle $\left(^{4}_{2}He\right)$ is one of the products. U-234 is the reactant.

$$^{234}_{92}U \longrightarrow\ ^{4}_{2}He + ^{a}_{z}X$$

where X is the unknown element, a is the mass number, and z is the atomic number.

The sum of the mass numbers and the atomic numbers of the

products must equal the mass number and the atomic number of the reactant. From the equation above, $a + 4 = 234$ and $z + 2 = 92$. Therefore, the product nucleus must be thorium (atomic number = 90) with a mass of 230, or $^{230}_{90}Th$. Thus, the final equation is:

$$^{234}_{92}U \longrightarrow {}^{4}_{2}He + {}^{230}_{90}Th$$

Scientists have concluded that beta particles emitted from nuclei are actually electrons. How can electrons be in the nucleus? Scientists theorize that neutrons produce beta radiation by breaking apart, forming a proton and an electron. The proton stays in the nucleus, and the electron is emitted as a beta particle. A carbon nucleus with six protons and eight neutrons is prone to this kind of reaction.

$$^{14}_{6}C \longrightarrow {}^{14}_{7}N + {}^{0}_{-1}e$$

The mass numbers of the reactant (C-14) and the product (N-14) are the same. The reacting nucleus does not lose nuclear particles when a neutron changes into a proton. The atomic number of the product nucleus is one greater than that of the reactant nucleus because of the extra proton.

Sample Problem Write the nuclear equation for the beta decay of Al-28.

Solution
Since beta decay implies the emission of an electron $\left({}^{0}_{-1}e \right)$ from the nucleus, the equation can be set up as follows:

$$^{28}_{13}Al \longrightarrow {}^{0}_{-1}e + {}^{a}_{z}X$$

$a + 0 = 28$ and $z + (-1) = 13$, therefore, $z = 14$. The unknown element is $^{28}_{14}Si$. The final equation is:

$$^{28}_{13}Al \longrightarrow {}^{0}_{-1}e + {}^{28}_{14}Si$$

Gamma emission is not accompanied by any changes in mass number or atomic number. The only change that occurs in gamma emission is a change in energy. For instance, an excited technetium-99 nucleus decays to form an unexcited technetium-99 nucleus and a gamma ray.

$$\text{Excited } {}^{99}_{43}Tc \longrightarrow {}^{99}_{43}Tc + \text{gamma ray}$$

Radioactive Decay Series: The Long Road to Stability

Unstable nuclei attain stable states by emitting various types of radioactivity. Some nuclei release alpha particles and some release beta particles. Some highly radioactive nuclei, such as U-238, do not stop decaying after just one alpha emission. The Th-234 that is produced by alpha emission is also an unstable nucleus, and it decays by beta emission.

$$^{234}_{90}\text{Th} \longrightarrow ^{234}_{91}\text{Pa} + ^{0}_{-1}e$$

The protactinium-234 produced by this reaction is unstable, and it decays by beta emission.

$$^{234}_{91}\text{Pa} \longrightarrow ^{234}_{92}\text{U} + ^{0}_{-1}e$$

U-234 is also unstable, and it too decays by a radioactive decay process. Do these decay reactions ever stop? Yes, but not until a stable nucleus is formed. When the product of a nuclear decay is stable, the radiations cease. Sequential alpha and beta emissions often form long series of nuclear reactions called **radioactive decay series**. When U-238 decays, nuclear decay reactions proceed until Pb-206 is formed. Figure 19A-6 shows the complete radioactive decay series of U-238.

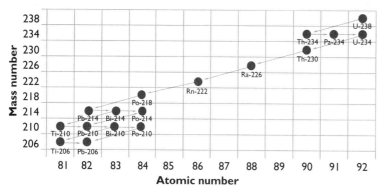

Figure 19A-6

The radioactive decay series of U-238. On the graph, alpha decays angle down and to the left, and beta decays progress horizontally to the right.

All natural nuclei with more than eighty-three protons (bismuth) exhibit radioactivity. They all decay according to one of three natural radioactive decay series: the uranium series, the thorium series, or the actinium series. The uranium series begins with U-238 and ends with Pb-206. The actinium series begins with U-235 and ends with Pb-207. The thorium series begins with Th-232 and ends with Pb-208.

Half-Life: Two Do Not Make a Whole

How quickly do nuclear reactions occur? It might seem that individual decay steps occur instantaneously. But this is not necessarily so. While some radioactive nuclei release particles and waves quickly, others release them only occasionally.

Scientists use the half-life concept to describe the decay rate of a particular isotope. The **half-life** of a radioactive

Table 19A-3 Half-Lives

Nucleus	Half-Life
O-13	0.0087 second
Br-80	17.6 minutes
Mg-28	21 hours
Rn-222	3.8 days
Th-234	24.1 days
H-3	12.26 years
C-14	5730 years
U-238	4,510,000,000 years

element is the amount of time that elapses while half of the nuclei in a sample of an element decay to form another element. Short half-lives mean that the nuclei decay quickly into other kinds of nuclei. Long half-lives show that radioactive decay proceeds more slowly.

The half-life of Th-234 is 24.1 days. Of 40 grams of Th-234, 20 grams will decay into other elements within 24.1 days. How much Th-234 will be left after two half-lives? It might seem logical to say "none"; however, that is not the correct answer. During each half-life, half of the *remaining* mass decays. Twenty grams remain when the second half-life begins, and 10 grams remain after 24.1 days. After another 24.1 days, half of the 10 grams will decay, leaving 5 grams. Where is the mass going? It is not disappearing, but rather is being converted into another element whose identity depends on the type of decay (α, β, or γ) that is occurring.

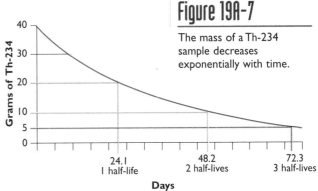

Figure 19A-7
The mass of a Th-234 sample decreases exponentially with time.

Sample Problem A 192-gram sample of Th-234 decays for 96.4 days. At the end of this time, how much Th-234 is left?

Solution
Since the half-life is 24.1 days, you know that four half-lives have passed.

$$\frac{96.4 \text{ days}}{1} \cdot \frac{1 \text{ half-life}}{24.1 \text{ days}} = 4.00 \text{ half-lives}$$

During each half-life the mass of the Th-234 in the remaining sample is halved. Since there are 4 "halvings," $\frac{1}{2}$ is multiplied by itself four times, or raised to the fourth power:

$$\left(\frac{1}{2}\right)\left(\frac{1}{2}\right)\left(\frac{1}{2}\right)\left(\frac{1}{2}\right) = \left(\frac{1}{2}\right)^4 = \frac{1}{16}$$

The amount of Th-234 left after 96.4 days (4 half-lives) is:

$$\frac{1}{16}(192 \text{ g Th-234}) = 12.0 \text{ g Th-234}$$

Spent fuel rods from nuclear power plants contain plutonium-239 with a half-life of 24,000 years. This raises the question of nuclear waste disposal. Special long-term nuclear waste facilities have to be constructed so that the waste is well contained. Often, where to store the waste is not a scientific problem, but a political problem.

Radioactive Age-Dating Methods

Newspaper headline: "Three-Million-Year-Old Human Fossil Found in Kenya." Evolutionists often use radioactive age-dating methods to support their claims that fossils and rocks are millions or billions of years old. But these radioactive "clocks" usually do not tell time accurately.

A clock is a device or system in which some component changes at a constant rate of speed. To be useful, the change must be observable so that men can distinguish one moment in time from another. The clock must also be "set," or calibrated, with another clock; a clock that has not been set properly is worthless for telling time, even though it runs at a constant rate.

Radioactive dating techniques attempt to use radioactive reactions as clocks. A parent radioactive substance decays and forms a daughter product presumably at a constant rate. The quantities of each substance can be measured and compared. Problems arise because no one has calibrated the system. The amount of each material present at the beginning of the process cannot be known. In the absence of these needed measurements, the investigator is forced to guess at the original quantities. This guesswork is not based on observation and is therefore not science. Many of the "scientific" age-dating methods are actually very unscientific. The following two examples illustrate unscientific methods; the third example has scientific merit when properly used.

The uranium-lead method measures the radioactive decay of uranium into lead. It has been applied to both earth rocks and moon rocks that contain these elements. The greater the amount of lead present when compared to the amount of uranium in a sample, the older it is assumed to be. The major problem, of course, is that much of the lead could already have been present in the original rock. A revealing fact about this method is that widely differing results are often obtained on different portions of the same sample. Yet it is impossible for one rock to have several different ages.

The potassium-argon method has recently become popular because it produces large numbers—larger than those from the uranium-lead method. In this procedure the parent substance is potassium, and the daughter substance is argon. The major problem of this method is similar to that of the uranium-lead method. Much of the observed argon could have been present in the original rock. There is strong experimental evidence against this method. When used on 170-year-old Hawaiian lava, potassium-argon dating gave results ranging from 22 million to 3 billion years. Yet the entire lava flow had been observed flowing and hardening in the years 1800 and 1801.

The radiocarbon method, based on carbon-14, differs from those discussed above because a calibration curve has been set up according to measurements of the radioactivity in several samples of known age. Since it is calibrated against known dates, it has scientific legitimacy if it is used properly. The calibration curve extends back only 5000 years, since that is the age of the oldest carbon-containing sample of independently known age: an Egyptian mummy.

Carbon-14 is a radioactive isotope of carbon produced when cosmic rays from the sun strike nitrogen in the upper atmosphere. The isotope has a half-life of 5730 years. Carbon-14 mixes with the other carbon in the world's supply and eventually finds its way into all living organisms. When an organism dies, no new carbon-14 is brought in from the outside. That which is present decays into nitrogen, and radioactivity continually decreases. The older the sample is, the less radioactivity there will be. Once the radioactivity level of a new sample has been determined, it can be compared to the calibrated curve. The radiocarbon method works acceptably when it is kept within the range of its calibration curve: the past 5000 years. Going beyond that involves an extrapolation (projecting a trend into an uncharted region of space or time), and the results are unreliable at best.

Nuclear Stability: Why Nuclei Do Not Always Fly Apart

One obvious question that has been ignored so far is "What keeps nuclei together?" Nuclei contain many protons that are packed together. Since protons are positively charged, they should repel each other and immediately fly apart. Yet some nuclei are so stable that they exist for centuries without changing. Other nuclei are so unstable that their half-lives must be measured in units of seconds. Why are some nuclei stable (nonradioactive) and others unstable (radioactive)?

Scientists explain these facts by saying that **strong nuclear forces** hold the protons together. Nuclear forces are not well understood. Scientists postulate that these forces work well only over small distances (less than 10^{-13} cm) to account for the fact that protons repel each other when they are not bound in nuclei. Scientists say that stable nuclei have effective nuclear forces, while unstable nuclei have insufficient forces to keep them together.

While unsaved scientists wonder what strong nuclear forces are, Christians have an authoritative source of information about this seeming mystery. Colossians 1:17 reveals that Jesus Christ "is before all things, and by him all things consist." God is ultimately responsible for holding all the protons in all the atoms of this world together.

The protons and neutrons in a nucleus seem to have energy levels similar to the energy levels in the electron shells. The **nuclear shell model** is a nuclear model in

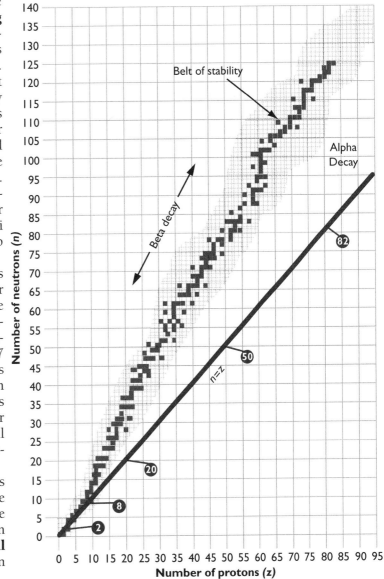

Figure 19A-8
Stable Nuclei

which protons and neutrons exist in levels, or shells, that are analogous to the energy levels that exist for electrons. Just as "full" electron shells are more stable than those that are less full, there are "full" nuclear shells that appear to be more stable. The number of protons in a full shell is called the **magic number** for that shell. For protons, the magic numbers are 2, 8, 20, 28, 50, and 82. Neutrons have these same numbers, with the addition of 126.

A brief analysis of the stable nuclei of the elements reveals an interesting trend. Figure 19A-8 represents the set of nuclei that researchers have observed to be stable. When plotting the number of protons (x-axis) and the number of neutrons (y-axis), if the intersection point on the graph that corresponds to a particular nucleus is shaded in, that nucleus is stable and not radioactive. Empty spots on the graph represent nuclei that are unstable. The curved band of points in this figure is called the **belt of stability** because it identifies the nuclei that do not undergo radioactive decay.

Sample Problem Are $^{110}_{50}$Sn atoms radioactive or stable?

Solution

$^{110}_{50}$Sn atoms have fifty protons and sixty neutrons. Referring to Figure 19A-8, the point on the graph corresponding to the intersection of fifty protons and sixty neutrons is not filled in. This represents an unstable, radioactive nucleus.

The straight line in Figure 19A-8 represents nuclei that contain equal numbers of protons and neutrons. Note that only small nuclei in this group are stable. Larger nuclei need extra neutrons to be stable. In the absence of neutrons, even nuclear forces cannot keep the many protons from repelling each other. The more protons a nucleus contains, the more neutrons it needs to remain stable.

The graph of stable nuclei can also be used to predict how unstable nuclei will decay. Nuclei with atomic numbers less than eighty-three that fall above the belt of stability have more neutrons than their stable isotopes. It is reasonable, then, that they will transform one of these extra neutrons into a proton and will emit a beta particle. For instance, Tl-208 (81 protons) decays by beta emission to form Pb-208 (82 protons), which has a stable nuclear structure. Nuclei with eighty-three or more protons frequently demonstrate alpha decay. $^{222}_{86}$Rn decays by alpha emission to form $^{218}_{84}$Po.

Sample Problem What type of radioactive decay (if any) are the following nuclei most likely to undergo?

a. $^{23}_{11}Na$ b. $^{230}_{90}Th$ c. $^{29}_{13}Al$

Solution

a. This element has 11 protons and 12 neutrons. When plotted on Figure 19A-8, the intersection of this nucleus is on the line of stability; therefore, no radioactive decay occurs.
b. Alpha emission is most likely because the atomic number is greater than eighty-two.
c. Beta emission is most likely because the nucleus has less than eighty-two protons, and when plotted on Figure 19A-8, it falls above the line of stability.

Section Review Questions 19A

1. List the types of natural radioactivity. Which are particles?
2. Fill in the missing isotope in each of the following reactions.
   a. $^{150}_{64}Gd \longrightarrow {}^{4}_{2}He + \underline{}$
   b. $^{245}_{96}Cm \longrightarrow {}^{4}_{2}He + \underline{}$
   c. $\underline{} \longrightarrow {}^{255}_{100}Fm + {}^{4}_{2}He$
   d. $^{141}_{56}Ba \longrightarrow {}^{0}_{-1}e + \underline{}$
   e. $\underline{} \longrightarrow {}^{0}_{-1}e + {}^{32}_{16}S$
   f. $^{239}_{94}Pu \longrightarrow {}^{4}_{2}He + \gamma + \underline{}$
   g. $^{40}_{19}K \longrightarrow {}^{0}_{-1}e + \gamma + \underline{}$
3. Barium-122 has a half-life of 2 minutes. A fresh sample weighing 80 g was obtained. If it takes 10 minutes to prepare the experiment, how much Ba-122 remains when the experiment begins?
4. Iodine-131 is used in the treatment of a hyperactive thyroid gland. The half-life of I-131 is 8 days. If a hospital receives a 200 g shipment of I-131, how much will be left after 32 days?

Particle accelerators are used to induce nuclear reactions.

19B Induced Reactions

The alchemists dreamed of the day that they could transform base metals into gold. Of course, they worked on that project before modern chemistry proved that transformations like this are impossible. Or are they? The transformation of lead into gold requires the removal of three protons and eight neutrons from a Pb-208 nucleus. Ordinary chemical reactions cannot alter the nucleus, but particle accelerators can. Lead can be changed into gold, but this nonspontaneous reaction can be forced to occur only with many millions of dollars' worth of equipment and massive inputs of energy—in other words, it would cost more to make the gold than the gold itself would be worth. Non-spontaneous nuclear reactions that are forced to occur are called induced nuclear reactions.

Fission: Splitting Nuclei

Extra neutrons induce some nuclei to split apart into smaller nuclei in a process called **nuclear fission.** For instance, U-235 can be bombarded with neutrons until the nucleus splits and releases various fragments, some liberated neutrons, and a tremendous amount of energy. There may be more than one way for the nucleus to split. Two of these ways are shown for U-235.

$$^{235}_{92}U + ^{1}_{0}n \longrightarrow ^{139}_{56}Ba + ^{94}_{36}Kr + 3\,^{1}_{0}n + \text{Heat}$$

$$^{235}_{92}U + ^{1}_{0}n \longrightarrow ^{144}_{55}Cs + ^{90}_{37}Rb + 2\,^{1}_{0}n + \text{Heat}$$

The energy released by nuclear fission is the result of the transformation of a small amount of nuclear mass into energy. According to Albert Einstein's theory of relativity, mass and energy are related by the equation $E = mc^2$, where E is energy, m is mass, and c is the speed of light. Since the speed of light is a fantastic 3×10^{10} centimeters per second, the energy released by the conversion of even a small amount of mass into energy is huge. A careful examination of the products and reactants of fission reveals that some matter is missing.

Figure 19B-1

Fission of a U-235 nucleus

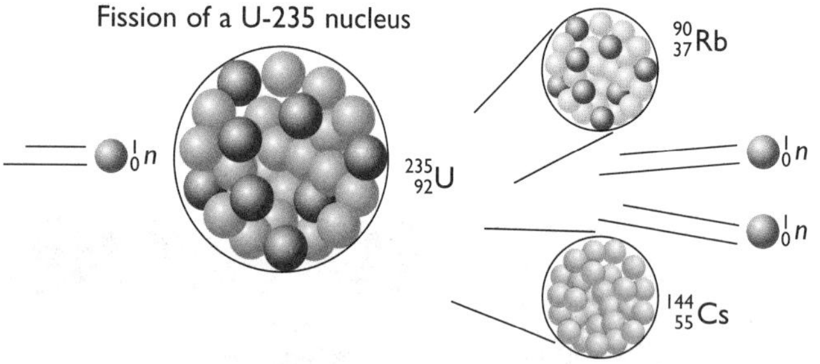

Chain Reactions: One Thing Leads to Another

The fission of U-235 produces, among other things, two neutrons. What happens to these neutrons? They can either escape from the sample of uranium or they can collide with other U-235 atoms. If the two neutrons hit and fuse into two other U-235 nuclei, the resulting nuclei will immediately undergo fission.

The neutrons produced by these two reactions can, in turn, initiate more nuclear fission reactions, and so on. This ongoing, self-sustaining fission process is called a **chain reaction.** For this to occur, the sample of uranium must be large enough to intercept many of the released neutrons. If the mass of uranium is too small, many neutrons will escape without initiating more nuclear fission reactions, and a chain reaction does not occur. The smallest mass of a fissionable substance that can sustain a chain reaction is called the **critical mass** of the substance.

Chain reactions release large amounts of energy because many nuclei are split apart. Each split converts a minute amount of matter into energy. If technicians control the reaction in nuclear reactors, the heat may be harnessed and transformed into useful

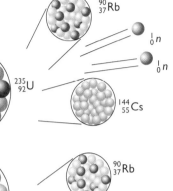

Figure 19B-2

A chain reaction fission of U-235

forms of energy (such as electricity). If the mass of fissionable material is larger than the critical mass, many neutrons are produced. The chain reaction proceeds quickly, even explosively, if it is not controlled. The amount of fissionable material that can support an explosion is called the **supercritical mass.** An atomic bomb uses a supercritical mass of a fissionable substance to achieve extraordinarily large energy releases and great destruction.

Fusion: Bringing Nuclei Together

Nuclear fusion is the transmutation process of directly combining several light (low atomic number) nuclei to form different, heavier elements. As with fission, there is a discrepancy of mass between the products of fusion and the reactants. The nuclei that form the larger nucleus have more mass than the product of the reaction. The nuclear mass of helium is 4.0015 amu. However, the combined masses of the two protons (2×1.0073 amu) and the two neutrons (2×1.0087 amu) equal 4.0320 amu—a 0.0305 amu difference. Just as in fission, this lost mass has been converted into an enormous amount of energy.

Although the concept of fusing two atoms may appear simple, before nuclear fusion can occur, light nuclei must collide with enough kinetic energy to overcome the natural repulsion between nuclei. Temperatures close to 100 million °C are required before nuclei have enough energy for fusion to proceed spontaneously. Needless to say, large amounts of energy must be expended to heat atoms to this temperature. In fact, in the first fusion reactions (hydrogen bombs) atomic bombs had to be used to supply the vast amount of kinetic energy required to initiate the fusion reaction.

Nuclear chemists and physicists hope to control nuclear fusion reactions and to convert their energy into electricity. The formation of helium by the fusion of two hydrogen isotopes $\left(^{2}_{1}H \text{ and } ^{3}_{1}H\right)$ offers particular promise.

$$^{2}_{1}H + ^{3}_{1}H \longrightarrow ^{4}_{2}He + ^{1}_{0}n$$

However, using fusion to produce energy is a formidable problem for researchers because once the 100 million °C temperatures are reached, they must be contained. Currently there are no materials that can withstand such enormous temperatures. Researchers have devised two methods to confine the high temperatures. One approach is called inertial confinement in which a small pellet of frozen hydrogen is compressed and heated by a laser so quickly that fusion occurs before the atoms can fly apart. Another approach is magnetic confinement, in which a strong magnetic field holds the ionized atoms together in a "magnetic bottle" while they are heated by microwaves, X rays, or other energy sources. Most nuclear scientists agree that the magnetic bottle is the more viable method for commercial energy production. Researchers at the Sandia National Laboratory in Albuquerque, New Mexico, using the magnetic field process and high-energy X rays, have caused temperatures of 1.8 million K (April 1998), and they predict that they will be able to reach 3 million K soon. This is the critical temperature that researchers feel must be attained to fuse deuterium and tritium (isotopes of hydrogen) for energy production. If the research proceeds at its current pace, small-scale energy production by fusion could soon become a reality.

Why all the interest in fusion? If scientists are successful in harnessing fusion, it could provide an energy source that is almost inexhaustible. The fuel for fusion would be ordinary water. In every 6500 atoms of water there is one deuterium $\left(^{2}_{1}H\right)$ atom, giving a gallon of water the energy content of 300 gallons of gasoline. There are no hydrocarbon by-products from fusion. The products of fusion (helium and a neutron) are not radioactive, and compared with the current fission reactors, there is minimal nuclear waste. All of these make fusion an attractive energy source.

Synthetic Elements: Bigger, Yes, But Are They Better?

Before the atomic age began, scientists knew of no elements with atomic numbers greater than that of uranium. The elements following uranium on the periodic table were synthesized with induced nuclear reactions. Scientists call elements with atomic numbers higher than ninety-two **transuranium elements.**

Transmutation is any process that converts one element into another. Ernest Rutherford, who discovered the nucleus, discovered transmutation in 1919. He bombarded nitrogen with alpha particles and found that protons and an isotope of oxygen were produced.

$$^{14}_{7}N + ^{4}_{2}He \longrightarrow ^{17}_{8}O + ^{1}_{1}H$$

Transmutation can be performed on other light (low atomic number) elements, but (as with Rutherford's experiments) no new elements will be produced. If, however, heavier elements are bombarded, it is possible to produce an element with an atomic number greater than ninety-two.

Soon after discovering that beta decay increased the atomic number of an atom, scientists reasoned that this process could be the key to producing new elements. If uranium could be forced to emit an electron from one of its neutrons, the remaining part of the neutron would become a proton. The added proton would turn the uranium nucleus into element number ninety-three. To make a uranium atom emit a beta particle, scientists bombarded a sample of uranium with neutrons to produce U-239.

$$^{238}_{92}U + ^{1}_{0}n \longrightarrow ^{239}_{92}U$$

Neptunium was first synthesized in 1940 when U-239 atoms released beta particles.

$$^{239}_{92}U \longrightarrow ^{239}_{93}Np + ^{0}_{-1}e$$

A year later the same technique produced element number ninety-four. Larger elements resulted when alpha particles were used as projectiles. Einsteinium and fermium emerged from the fireball of an experimental hydrogen bomb in 1952. More sophisticated techniques and larger projectiles have since produced additional new elements. Scientists produced element 105 in 1970 when they bombarded californium-249 with nitrogen-15 nuclei, and ununbium ($^{227}_{112}Uub$) was produced in a German laboratory in 1996 by bombarding lead atoms with zinc atoms. In 1999 at a laboratory in Berkeley, California, ununoctium ($^{293}_{118}Uuo$) was fabricated by accelerating a beam of Kr-86 ions to an energy of 449 million electron volts (one electron volt = 1.602×10^{-19} joules) onto a target of Pb-208. After eleven days three atoms of ununoctium were produced. These atoms lasted less than one millisecond before they decayed by alpha emissions.

Mass Defect and Nuclear Binding Energy: Transforming Matter into Energy

All nuclei have slightly less mass than their components. The idea that energy and mass are interconvertible explains this otherwise inexplicable physical phenomenon. The difference between the mass of an atom and the total mass of all its components is called the **mass defect** of the nucleus.

While scientists do not fully understand the forces that hold a nucleus together, they do know that energy is required to separate all the protons and neutrons in the nucleus. This energy is called the **nuclear binding energy** of the nucleus. For the purpose of comparing the elements, researchers have determined the nuclear binding energies of various nuclei. They then divided this value by the total number of particles in the nuclei to obtain the binding energy per nuclear particle. Figure 19B-3 shows the binding energy per nuclear particle plotted as a function of mass number. It shows that nuclei with atomic masses near 50 amu have the most binding energy per nuclear particle.

Figure 19B-3

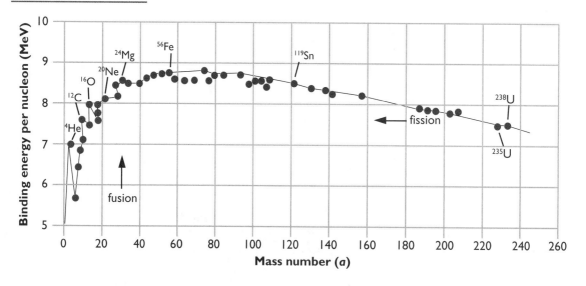

Nuclei with high binding energies per nuclear particle are more stable than nuclei with lower binding energies per nuclear particle. As expected, the observed mass defects for the various elements match the observed binding energies. Nuclei with masses near 50 amu have the most mass defect per nuclear particle. It is reasonable to assume that the mass that has "disappeared" (transformed into energy) serves as the nuclear binding energy.

Sample Problem Which of the following nuclei is the most stable: H-2, As-75, or U-235?

Solution
From Figure 19B-3 you can see that of the three nuclei mentioned, As-75 has the highest binding energy per nuclear particle. Therefore, it is the most stable (it is closest to 56 amu).

Both fission and fusion will proceed in the direction that produces stable nuclei as end products. Figure 19B-3 helps us to understand this idea. The y-axis displays the binding energy per nucleon, and the x-axis shows the mass number of the elements. The graph peaks at a mass number of 56, demonstrating that the greater the binding energy released per nucleon, the more stable the nucleus. Any element with the mass number of 56 will achieve the greatest stability possible. Thus, elements to the right of 56 when bombarded by nuclear particles split (fission) into smaller elements as they attempt to achieve more stability. Elements to the left of 56 will fuse when bombarded in order to form more stable elements. During both of these processes, vast amounts of energy are produced.

Destruction in Nagasaki after an atomic bomb was dropped.

Responsible Citizenship

The awesome power of the atom brings real risks. Nuclear weapons could spread large quantities of radioactive substances throughout the atmosphere. Nuclear fallout, as these airborne radioactive by-products are called, would spread throughout the earth. Water and food supplies would be contaminated. Since some of the radioactive substances have long half-lives, health problems from the fallout (birth defects and cancer) would continue for years.

Similar difficulties could result if a massive accident occurred at a nuclear power plant. Such an accident did occur in 1986 at a nuclear reactor at Chernobyl in the Ukraine. Although there was no "meltdown" of the nuclear core and no radioactive decay products made their way into the ground water, there was a massive cloud of radioactive material released into the atmosphere and carried across the earth by the upper level winds. It was estimated that the fallout effects from this accident exposed each person in the world to the same radiation dosage as that obtained from the increased cosmic radiation one would receive in one trans-Atlantic flight. A nuclear explosion at a nuclear power plant is unlikely because reliable methods are used to keep the masses of fissionable materials below the supercritical level.

Nuclear accidents such as the one at Three Mile Island in Pennsylvania have caused many people to be concerned about the safety of nuclear energy production.

Chapter 19B

We must be responsible in using the power God has allowed us to possess.

Technical considerations about nuclear plant safety and political concerns regarding the disposal of spent fuel rods have made nuclear energy very costly in the United States. Any new nuclear power plant will need to be smaller, simpler, standardized in design, and built with interchangeable components to decrease the cost of initial construction and continued maintenance.

The Bible makes it clear that life on earth will not be accidentally wiped out because of a nuclear accident or war (Rev. 20:7-9). God, not man, controls the destiny of the human race. While God will not allow man to completely destroy the earth, He has given man the responsibility to care for it (Gen. 1:28). Ignoring this responsibility and misusing nuclear power could result in the deaths of many people and long-term damage to the environment.

Although man has released radiation into the environment, human activity is not the only source of radiation. Background radiation, or natural radiation as it is sometimes called, is always present in the environment. Cosmic rays coming to earth from the sun and other stars consist of alpha and beta radiation. Once cosmic rays reach the earth's atmosphere, they induce reactions in the atmosphere that release gamma radiation. These processes create a significant amount of radiation in the environment.

Science alone cannot supply the answers to the problems associated with nuclear power. Men need common sense, a code of ethics, and wisdom in order to apply science correctly. The best source of guidance is God's Word, the Bible. In it, Christians can find the truth and insight necessary to make difficult decisions.

Section Review Questions 19B

1. Differentiate between nuclear fission and fusion.
2. Of the following pairs, which will be more stable? (Use Figure 19B-3.)
   a. O-16, Sn-119
   b. Mg-24, C-12
   c. U-238, Ne-20
   d. Sn-119, He-4
3. What are transuranium elements?
4. True or False
   a. Fusion processes convert energy into matter.
   b. The greater the binding energy, the less stable the nucleus.
   c. Light nuclei tend to undergo fission in order to become more stable.
   d. The mass needed to sustain an explosion is known as the critical mass.
   e. The high temperatures of fusion can be confined within magnetic fields.

Chapter Review

Coming to Terms

radioactivity
alpha particle
beta particle
gamma ray
becquerel
curie
gray
rad
roentgen
rem
nuclear equation
radioactive decay series
half-life
strong nuclear force
nuclear shell model
magic number
belt of stability
nuclear fission
chain reaction
critical mass
supercritical mass
nuclear fusion
transuranium elements
transmutation
mass defect
nuclear binding energy

Review Questions

1. Which types of radiation consist of particles and which type consists of waves?
2. What are the charges on the three types of radiation?
3. Rank the three types of radiation according to ionizing power and then according to penetrating power.
4. Explain why the different types of radiation have different ionizing capabilities.
5. Explain the relationship between ionizing energy and depth of penetration.
6. Gamma radiation, visible light, and radio waves are all forms of electromagnetic radiation. Why are gamma rays much more dangerous than visible light and radio waves?
7. Write the nuclear equation for the following:
    a. alpha decay of $^{217}_{86}Rn$
    b. alpha decay of $^{238}_{94}Pu$
    c. beta decay of $^{214}_{83}Bi$
    d. beta decay of $^{234}_{91}Pa$
    e. gamma decay of $^{152}_{63}Eu$
    f. gamma decay of $^{165}_{66}Dy$

8. Complete these nuclear reactions.
   a. $^{242}_{95}\text{Am} \longrightarrow {}^{242}_{96}\text{Cm} + \underline{\quad}$
   b. $^{220}_{87}\text{Fr} \longrightarrow {}^{216}_{85}\text{At} + \underline{\quad}$
   c. $\underline{\quad} \longrightarrow {}^{126}_{54}\text{Xe} + {}^{0}_{-1}e$
   d. $\underline{\quad} \longrightarrow {}^{216}_{85}\text{At} + {}^{4}_{2}\text{He}$

9. Identify each of the reactions in the previous question as alpha, beta, or gamma decay.

10. Why do the decay processes in a radioactive decay series not continue indefinitely?

11. Referring to Table 19A-3, which nucleus decays more quickly, Br-80 or Mg-28?

12. The half-life of H-3 is 12.26 years. If 35.8 g of H-3 are allowed to decay for 36.78 years, what mass of H-3 will be left?

13. Scientific theories are based on observation. What observation led to the concept of strong nuclear forces?

14. Refer to Figure 19A-8, and tell whether the following nuclei are stable or unstable. If they are unstable, predict whether the nucleus will undergo alpha or beta emission.
   a. $^{154}_{64}\text{Gd}$
   b. $^{59}_{26}\text{Fe}$
   c. $^{232}_{94}\text{Pu}$
   d. $^{200}_{80}\text{Hg}$
   e. $^{106}_{44}\text{Ru}$

15. Give two ways in which nuclear fission and nuclear fusion are similar and one way in which they are dissimilar.

16. Why do nuclear fission and nuclear fusion both release energy?

17. Why do some nuclei undergo fission but not fusion?

18. Determine which nucleus is the more stable by referring to Figure 19B-3.
   a. Li-6, Mg-24
   b. U-235, Fe-56
   c. He-4, U-235

Appendixes

Glossary

Index

Photograph Credits

Appendix A - Physical Constants

Quantity	Symbol	Traditional Units	SI Units
Atomic Mass Unit ($\frac{1}{12}$ the mass of C-12 atom)	amu	1.66×10^{-24} g	1.66×10^{-27} kg
Avogadro's number	N_A	6.022×10^{23} particles/mol	6.022×10^{23} particles/mol
Charge-to-mass ratio of electron	e/m	1.76×10^{8} coulomb/g	1.76×10^{11} C/kg
Electronic charge	e	1.60×10^{-19} coulomb	1.60×10^{-19} C
Electron rest mass	m_e	9.11×10^{-28} g 0.000549 amu	9.11×10^{-31} kg
Gas constant	R	0.08206 L·atm/mol·K 62.36 L·torr/mol·K	8.3145 J/mol·K
Molar volume (STP)	V_m	22.4 L/mol	2.24×10^{-2} m³/mol
Neutron rest mass	m_n	1.67495×10^{-24} g 1.008669 amu	1.67495×10^{-27} kg
Proton rest mass	m_p	1.6726×10^{-24} g 1.007277 amu	1.6726×10^{-27} kg
Velocity of light	c	2.9979×10^{10} cm/s (186,281 mi/s)	2.9979×10^{8} m/s

Appendix B - Unit Conversion Factors

The number that is used to convert one unit to another may be found in the box in which the columns containing the two units cross. The conversion factor can be used in unit analysis.

Example: The table shows that 1.0 centimeter = 6.214×10^{-6} mile. The conversion factor 1.0 cm/6.214×10^{-6} mile can be constructed from this equality.

Length	cm	m	km	in.	ft	mi
1 centimeter =	1	10^{-2}	10^{-5}	0.3937	3.281×10^{-2}	6.214×10^{-6}
1 meter =	100	1	10^{-3}	39.37	3.281	6.214×10^{-4}
1 kilometer =	10^5	1000	1	3.937×10^4	3281	0.6214
1 inch =	2.540	2.54×10^{-2}	2.540×10^{-5}	1	8.333×10^{-2}	1.578×10^{-5}
1 foot =	30.48	0.3048	3.048×10^{-4}	12	1	1.894×10^{-4}
1 mile =	1.609×10^5	1609	1.609	6.336×10^4	5280	1

1 angstrom (Å) = 1.0×10^{-10} m

1 light-year = 9.4600×10^{12} km

Time	s	min	hr	d	yr
1 second =	1	1.667×10^{-2}	2.778×10^{-4}	1.157×10^{-5}	3.169×10^{-8}
1 minute =	60	1	1.667×10^{-2}	6.944×10^{-4}	1.901×10^{-6}
1 hour =	3600	60	1	4.167×10^{-2}	1.141×10^{-4}
1 day =	8.640×10^{4}	1440	24	1	2.738×10^{-3}
1 year =	3.156×10^{7}	5.259×10^{5}	8.766×10^{3}	365.2	1

Density	g/cm^3	kg/m^3	$lb/in.^3$	lb/ft^3
1 gram per cubic centimeter =	1	1000	3.613×10^{-2}	62.43
1 kilogram per cubic meter =	0.001	1	3.613×10^{-5}	6.243×10^{-2}
1 pound per cubic inch =	27.68	2.768×10^{4}	1	1728
1 pound per cubic foot =	1.602×10^{-2}	16.02	5.787×10^{-4}	1

Volume	cm^3	m^3	L	$in.^3$	ft^3
1 cubic centimeter =	1	1.0×10^{-6}	1.000×10^{-3}	6.102×10^{-2}	3.531×10^{-5}
1 cubic meter =	10^{6}	1	1000	6.102×10^{4}	35.31
1 liter =	1000	1.000×10^{-3}	1	61.02	3.531×10^{-2}
1 cubic inch =	16.39	1.639×10^{-5}	1.639×10^{-2}	1	5.787×10^{-4}
1 cubic foot =	2.832×10^{4}	2.832×10^{-2}	28.32	1728	1

1 U.S. fl. gal. = 4 U.S. fl. qt. = 8 U.S. pt. = 128 U.S. fl. oz. = 231 $in.^3$

1 British imperial gal. = 277.4 $in.^3$

Pressure	atm	in. of water	torr	Pa	$lb/in.^2$
1 atmosphere =	1	406.8	760	1.013×10^{5}	14.70
1 inch of water at 4°C =	2.458×10^{-3}	1	1.868	249.1	3.613×10^{-2}
1 torr at 0°C =	1.316×10^{-3}	0.5353	1	133.3	0.01934
1 pascal =	9.869×10^{-6}	4.015×10^{-3}	7.501×10^{-3}	1	1.450×10^{-4}
1 pound per square inch =	6.805×10^{-2}	27.68	51.71	6.895×10^{3}	1

Mass	g	kg	amu	oz	lb	t
1 gram =	1	0.001	6.024×10^{23}	3.527×10^{-2}	2.205×10^{-3}	1.102×10^{-6}
1 kilogram =	1000	1	6.024×10^{26}	35.27	2.205	1.102×10^{-3}
1 atomic mass unit =	1.660×10^{-24}	1.660×10^{-27}	1	5.855×10^{-26}	3.660×10^{-27}	1.829×10^{-30}
1 ounce =	28.35	2.835×10^{-2}	1.708×10^{25}	1	6.250×10^{-2}	3.125×10^{-5}
1 pound =	453.6	0.4536	2.732×10^{26}	16	1	5.0×10^{-4}
1 ton =	9.072×10^{5}	907.2	5.465×10^{29}	3.2×10^{4}	2000	1

Energy, heat	BTU	J	kcal	cal
1 British thermal unit =	1	1055	0.252	252.0
1 joule =	9.481×10^{-4}	1	2.389×10^{-4}	0.2389
1 kilocalorie =	3.9868	4.186×10^{3}	1	1000
1 calorie =	3.98×10^{-3}	4.186	0.001	1

Appendix C ▪ Electron Configurations of the Elements

Element	Atomic Number	1s	2s	2p	3s	3p	3d	4s	4p	4d	4f	5s
H	1	1										
He	2	2										
Li	3	2	1									
Be	4	2	2									
B	5	2	2	1								
C	6	2	2	2								
N	7	2	2	3								
O	8	2	2	4								
F	9	2	2	5								
Ne	10	2	2	6								
Na	11	\[Neon core\]			1							
Mg	12				2							
Al	13				2	1						
Si	14				2	2						
P	15				2	3						
S	16				2	4						
Cl	17				2	5						
Ar	18	2	2	6	2	6						
K	19	\[Argon core\]						1				
Ca	20							2				
Sc	21						1	2				
Ti	22						2	2				
V	23						3	2				
Cr	24						5	1				
Mn	25						5	2				
Fe	26						6	2				
Co	27						7	2				
Ni	28						8	2				
Cu	29						10	1				
Zn	30						10	2				
Ga	31						10	2	1			
Ge	32						10	2	2			
As	33						10	2	3			
Se	34						10	2	4			
Br	35						10	2	5			
Kr	36	2	2	6	2	6	10	2	6			
Rb	37	\[Krypton core\]										1
Sr	38											2
Y	39									1		2
Zr	40									2		2
Nb	41									4		1
Mo	42									5		1
Tc	43									6		1
Ru	44									7		1
Rh	45									8		1
Pd	46									10		
Ag	47									10		1
Cd	48									10		2

Element	Atomic Number		4d	4f	5s	5p	5d	5f	6s	6p	6d	7s
In	49		10		2	1						
Sn	50		10		2	2						
Sb	51		10		2	3						
Te	52		10		2	4						
I	53		10		2	5						
Xe	54		10		2	6						
Cs	55		10		2	6			1			
Ba	56		10		2	6			2			
La	57		10		2	6	1		2			
Ce	58		10	2	2	6			2			
Pr	59		10	3	2	6			2			
Nd	60		10	4	2	6			2			
Pm	61		10	5	2	6			2			
Sm	62		10	6	2	6			2			
Eu	63		10	7	2	6			2			
Gd	64		10	7	2	6	1		2			
Tb	65		10	9	2	6			2			
Dy	66		10	10	2	6			2			
Ho	67		10	11	2	6			2			
Er	68		10	12	2	6			2			
Tm	69		10	13	2	6			2			
Yb	70		10	14	2	6			2			
Lu	71		10	14	2	6	1		2			
Hf	72		10	14	2	6	2		2			
Ta	73		10	14	2	6	3		2			
W	74		10	14	2	6	4		2			
Re	75	Krypton core	10	14	2	6	5		2			
Os	76		10	14	2	6	6		2			
Ir	77		10	14	2	6	9					
Pt	78		10	14	2	6	9		1			
Au	79		10	14	2	6	10		1			
Hg	80		10	14	2	6	10		2			
Tl	81		10	14	2	6	10		2	1		
Pb	82		10	14	2	6	10		2	2		
Bi	83		10	14	2	6	10		2	3		
Po	84		10	14	2	6	10		2	4		
At	85		10	14	2	6	10		2	5		
Rn	86		10	14	2	6	10		2	6		
Fr	87		10	14	2	6	10		2	6		1
Ra	88		10	14	2	6	10		2	6		2
Ac	89		10	14	2	6	10		2	6	1	2
Th	90		10	14	2	6	10		2	6	2	2
Pa	91		10	14	2	6	10	2	2	6	1	2
U	92		10	14	2	6	10	3	2	6	1	2
Np	93		10	14	2	6	10	5	2	6		2
Pu	94		10	14	2	6	10	6	2	6		2
Am	95		10	14	2	6	10	7	2	6		2
Cm	96		10	14	2	6	10	7	2	6	1	2
Bk	97		10	14	2	6	10	9	2	6		2
Cf	98		10	14	2	6	10	10	2	6		2
Es	99		10	14	2	6	10	11	2	6		2
Fm	100		10	14	2	6	10	12	2	6		2
Md	101		10	14	2	6	10	13	2	6		2

GLOSSARY

A

absolute zero 0 K; theoretically, the temperature at which all molecular motion would cease; the coldest temperature possible.

accuracy How close a measurement is to an accepted reference or theoretical value.

acid dissociation constant The special name given to the constant that describes the equilibrium between an acid and its conjugate base.

acid rain The lowering of the pH of rainwater because of the airborne concentration of SO_2, SO_3, and NO_2.

acidic anhydride A nonmetal oxide that forms an acid when added to water. Example: SO_3.

acidic salt A salt formed by the reaction of a weak base with a strong acid.

acidic solution A solution with more H_3O^+ ions than OH^- ions, resulting in a pH less than 7.

actinide series A portion of the seventh series of the periodic table that includes the inner transitional metals from Actinium to Lawrencium.

activated complex An unstable, energetic group of reactants that forms as a transitional structure during a chemical reaction.

activation energy The minimum amount of kinetic energy that must be possessed by reactants before they can have an effective collision.

activity series A table of metals or nonmetals arranged in order of descending activities.

addition reaction An organic reaction in which one reactant joins another reactant at the site of a double or triple bond.

adsorption The attachment of charged particles to the particles in a colloid.

alchemy The ancient study of transmutations between base metals and gold, sickness and health, age and youth, or even earthly and supernatural existence.

alcohol An organic compound of the general form R–OH, having a covalently bonded OH^- functional group attached to a nonaromatic group.

aldehyde Any organic compound of the general form
$$R-\overset{\overset{O}{\|}}{C}-H,$$ having an aldehyde group in its structure.

aliphatic compound An open-chain compound and any cyclic compound whose bonds resemble those of an open-chain compound.

alkali metal A family IA metal, which has only one valence electron.

alkaline earth metal A family IIA metal, which has two valence electrons.

alkane An open-chain, aliphatic hydrocarbon that contains only single bonds.

alkene An open-chain, aliphatic hydrocarbon that contains at least one carbon-carbon double bond.

alkyl group (-R group) A group of bonded atoms that can be thought of as an alkane with one hydrogen atom missing.

alkyl halide An organic compound that contains an alkyl group and a halogen as a functional group.

alkyne An open-chain, aliphatic hydrocarbon containing at least one triple bond.

allotrope One of two or more forms of a polymorphic element that exist in the same physical state. Example: O_2 and O_3 are allotropic forms of oxygen.

allotropic A term describing elements that can exist in more than one form.

alpha particle (α) The nucleus of the helium atom (two protons, two neutrons); represented by He^{2+} or $He\text{-}4^{2+}$.

amalgam An alloy of mercury with some other metal.

amide An organic compound of the general form
$$R-\overset{\overset{O}{\|}}{C}-NH_2,$$ in which an amine is substituted for an OH group in a carboxylic acid.

amine An organic compound which can be thought of as an ammonia molecule whose hydrogen atoms have been replaced by other atoms or groups of atoms.

amino acid The building block of a protein; a carboxylic acid containing the amino group.

amorphous solid A solid without any definite shape or crystalline structure.

amphiprotic A term describing a substance that can act as either an acid or a base by releasing or accepting a proton.

analytical chemistry The techniques by which chemists devise equipment and methods to discover what is in a sample of material and to quantify its constituents.

angstrom (Å) A unit of length equal to 10^{-10} m, or 10^{-8} cm.

anhydride A compound that reacts with water in a composition reaction.

anhydrous The form of a hydrate compound that has lost the H_2O from its crystalline structure.

anion A negative ion.

anode The electrode at which oxidation occurs during an electrochemical reaction; the electrode that attracts anions because of its positive charge.

apothecary A person who prepares and sells medicines (a person similar to a modern pharmacist).

aqueous Dissolved in water.

aromatic compound Benzene and any compound that has a structure resembling benzene's characteristic ring structure; aromatic compounds are all cyclic compounds with clouds of delocalized electrons.

Arrhenius acid A substance that releases hydrogen ions (H^+) into aqueous solutions.

Arrhenius base A substance that releases hydroxide ions (OH^-) into aqueous solutions.

aryl group (-Ar group) A group of bonded atoms that can be thought of as an aromatic compound with one hydrogen atom missing.

aryl halide An organic compound that contains an aryl group and a halogen as a functional group.

atmosphere A standard unit of pressure equal to the pressure exerted by a column of mercury exactly 760 mm high at 0° C; 760 mm Hg = 760 torr = 1 atm = 101,325 Pa.

atom A neutral particle with a centrally located nucleus consisting of protons and neutrons with electrons around it; the smallest representative unit in an element.

atomic mass The average mass of the isotopes of an element expressed in atomic mass units (amu).

atomic mass unit (amu) One-twelfth the mass of one atom of carbon-12; 1.6606×10^{-24} g (which is very close to the mass of one proton or one neutron).

atomic number The number of protons in the nucleus of an atom.

Aufbau principle The principle that states that the electron configuration of an atom may be obtained by building on the electron configuration of an atom of lower atomic number; the electrons fill the sublevels in the order given by the diagonal rule.

autoionization See **autoprotolysis constant of water.**

autoprotolysis A process in an acid-base reaction in which one molecule donates a proton to another molecule of the same substance.

autoprotolysis constant of water (K_w) The special name given to the dissociation constant of water molecules at 25° C; $K_w = 1 \times 10^{-14}$.

Avogadro's number (N_A) 6.022×10^{23}; the number of atoms in exactly 12.0 g of carbon-12.

Avogadro's principle The principle that equal volumes of gases at the same temperature and pressure contain equal numbers of molecules.

azeotrope A homogeneous mixture of liquids that forms a constant boiling point and cannot be separated by normal distillation techniques.

B

balanced chemical equation A chemical equation in which coefficients are arranged to show conservation of mass in a reaction.

barometer An apparatus that measures atmospheric pressure by allowing it to support a column of liquid.

basic anhydride A metal oxide that forms a base when added to water. Example: Na_2O.

basic salt A salt produced by the reaction of a strong base with a weak acid.

basic solution A solution with fewer H_3O^+ ions than OH^- ions, resulting in a pH greater than 7.

battery One or more galvanic cells arranged to produce electricity.

becquerel (Bq) The SI unit of measure describing the radioactivity of a substance; 1 Bq = 1 disintegration per second.

belt of stability The group of stable nuclei represented on a graph of atomic numbers versus number of neutrons.

bent A structural arrangement in which two particles are bonded to a central particle nonlinearly.

beta particle (β) An electron that has been emitted from a nucleus.

binary acid A two-element compound in which hydrogen is the first element. Example: HCl.

binary compound A compound that contains only two kinds of elements.

biochemistry The study of the chemical processes in living things.

boiling A physical change from the liquid state to the gaseous state that occurs when the vapor pressure of a liquid equals the prevailing atmospheric pressure.

boiling point The temperature at which the vapor pressure of a liquid equals the applied atmospheric pressure.

boiling point elevation A raising of the boiling point of a solvent caused by the presence of solute particles.

Bose-Einstein condensate The theoretical fifth state of matter that will exist at absolute zero.

Boyle's law A law of gas behavior that states that the pressure of a gas is inversely proportional to its volume if the temperature is held constant ($PV = k$).

bridge element An element that has properties of another family. Example: boron.

British thermal unit (BTU) The quantity of heat energy required to raise one pound of water by 1° F.

Brönsted-Lowry acid A substance that donates protons.

Brönsted-Lowry base A substance that accepts protons.

Brownian movement The random, chaotic movements of microscopic particles in a colloidal dispersion.

buckminsterfullerene The most common and stable of the fullerene molecules consisting of 60 carbon atoms, arranged in the shape of a soccer ball; named after R. Buckminster Fuller because of their resemblance to his geodesic dome structures.

buckyball See **buckminsterfullerene.**

buffer A solution that can receive moderate amounts of either acid or base without a significant change in pH.

C

calorie (cal) The amount of energy required to raise the temperature of 1 g of water 1° C.

Calorie (Cal; kcal) One kilocalorie.

capillary rise The movement of a liquid up a narrow tube caused by the attraction of the molecules in the walls of the tube for the molecules of the liquid.

carbohydrate A polyhydroxy aldehyde or ketone; a compound that can be hydrolyzed to form a polyhydroxy aldehyde or ketone.

carbonyl group A carbon atom with a doubly bonded oxygen atom attached to it.

carboxyl group The $-\overset{\overset{\displaystyle O}{\|}}{C}-OH$ group.

carboxylic acid An organic compound of the general form $R-\overset{\overset{\displaystyle O}{\|}}{C}-OH$, having a carboxylic acid functional group in its structure.

catalyst A substance that changes a reaction rate without being permanently changed by the reaction.

cathode The electrode at which reduction occurs during an electrochemical reaction; the electrode that attracts cations because of its negative charge.

cathode rays The stream of electrons emitted from the cathode in a cathode-ray tube (CRT).

cation A positive ion.

Celsius scale A temperature scale, proposed by Anders Celsius, that divides the range from the freezing point of water (0° C) to the boiling point of water (100° C) into 100 increments and labels absolute zero as -273° C.

chain reaction A self-sustaining fission process in which neutrons produced from fission reactions cause more fission reactions.

characteristic The power to which 10 is raised in scientific notation expressions; the number that indicates how many places the decimal point must be moved to return to conventional decimal format. Example: 3 is the characteristic in 1.200×10^3.

Charles's law A law of gas behavior that states that the volume and absolute temperature of a gas are directly proportional to one another when the pressure is held constant ($V/T = k$).

chemical bond A force of attraction that holds atoms together in compounds; an attraction produced by the transfer or sharing of electrons.

chemical change See chemical reaction.

chemical equation An expression that represents a chemical reaction by using chemical formulas, chemical symbols, and coefficients.

chemical equilibrium The state of balance attained in a reversible reaction in which the forward and reverse reactions proceed at the same rate.

chemical property A property of matter that describes how one substance reacts in the presence of other substances.

chemical reaction A change in which a substance loses its characteristics and becomes one or more new substances.

chemistry The study of the composition and properties of matter and the energy transformations accompanying changes in the fundamental structure of matter.

chlor-alkali industry The industrial processes responsible for the production of much of the chlorine gas used by industry.

chromosome A strand of DNA combined with proteins; it is usually formed within the nucleus of the cell.

coagulation The process whereby the dispersed phase of a colloid is made to aggregate and separate into a separate phase.

coefficient A number that appears in front of a chemical formula and indicates how many units of that substance are present. Examples: 2 O_2—two oxygen molecules; 4 K_2SO_4—four potassium sulfate molecules.

colligative property A property of solutions that depends only on the number of particles present, without regard to type.

collision theory The theory that states that molecules and atoms must undergo forceful, properly oriented collisions before they can react.

colloid A mixture of fine particles (between 1 and 1000 nanometers in size) that do not settle out of the mixture.

combination reaction A chemical reaction of the general form $A + B \longrightarrow AB$, in which two or more reactants combine into a single product.

combined gas law A law of gas behavior that combines Boyle's law and Charles's law ($P_1V_1/T_1 = P_2V_2/T_2$).

common ion effect An equilibrium phenomenon in which two or more substances dissolve and release a common ion, thereby decreasing the ionization of the weaker electrolyte.

compound A substance that consists of atoms of different elements chemically bonded together.

compressibility The property of a substance that allows its particles to be squeezed into smaller volumes.

concentrated solution A solution whose ratio of solute to solvent is relatively high.

condensation A physical change from the gaseous state to the liquid state.

condensation reaction An organic reaction in which two compounds combine with each other by losing water; the opposite of hydrolysis.

conductivity A physical property of matter indicating the ability to transfer heat or electrons through a substance.

conjugate acid The structure formed when a base is protonated.

conjugate base The structure formed when an acid has donated a proton.

conjugate pair Two particles that differ from each other by only a hydrogen ion.

continuous spectrum A spectrum that has segments that blend into each other without distinct boundaries.

conversion factor A ratio that is constructed from the relationship between two units and is equal to 1. Example: 1 kg/1000 g.

corrosion The chemical destruction of a metal by its immediate surroundings.

covalent bond Two atoms held together in an ion or molecule by their attraction for the same pair of shared electrons.

critical mass The smallest mass of a fissionable substance that can sustain a chain reaction.

critical pressure The pressure needed to liquefy a gas at its critical temperature.

critical temperature The temperature above which no amount of pressure will liquefy a gas.

crystal A solid in which the particles occur in a regular, repeating pattern.

crystal lattice A three-dimensional structure of points or objects that represents the regular alternating pattern of positive and negative ions.

crystalline solid See **crystal**.

curie (Ci) A measure of radioactivity based on the nuclear decay of radium; 1 Ci = 1 disintegration / sec of 1 gram of radium.

cyclic aliphatic compound An aliphatic, organic compound whose carbon chains are bonded in ring shapes. Examples: cycloalkanes, cycloalkenes.

D

Dalton's law of partial pressures The total pressure of a mixture of gases equals the sum of the partial pressures of the constituent gases.

decomposition reaction A chemical reaction of the general form $AB \longrightarrow A + B$, in which a single reactant breaks down into two or more products.

deductive reasoning A process that intends to prove that the conclusions must be true, and that the argument is always logical and factual.

deliquescence The action of a compound that absorbs enough water from the air to dissolve itself and form a solution.

delocalized electrons Electrons that can move between several different bonds; in benzene, delocalized electrons can move throughout circular spaces above and below the plane of bonded nuclei.

density A measure of the concentration of matter; it is expressed as a ratio of the object's mass to its volume.

deprotonation The process of losing a proton. Example: $H_2SO_4 \longrightarrow H^+ + HSO_4^-$.

descriptive chemistry The study of elements and compounds that stresses identification of properties rather than theoretical calculations.

desiccator An airtight container with a hygroscopic substance that removes moisture to protect the compound being stored.

diagonal rule A mnemonic device that gives the energy levels and sublevels in their order of filling.

diatomic element An element whose atoms bond into two-atom units. Examples: N_2, O_2, H_2, Cl_2, F_2, Br_2.

diffusion Spontaneous mixing caused by particle motion.

dilute solution A solution whose ratio of solute to solvent is relatively low.

dimensional analysis See **unit analysis**.

dipeptide Two amino acids joined by a peptide bond.

dipole moment The product of the distance between charges and the strength of the charges; a measure of the polarity of a molecule.

dipole-dipole interaction The attraction of the positive end of one polar molecule to the negative end of another polar molecule.

diprotic acid An acid that can donate two protons.

disaccharide A carbohydrate composed of two monosaccharide units. Example: sucrose.

dispersion force (London force) An electrostatic attraction that arises between atoms or molecules because of the presence of instantaneous and induced dipoles.

dissociation The process in which a solvent disrupts the attractive forces in a solute and pulls the solute apart.

distillation A laboratory technique by which chemists separate a mixture by evaporating its components at their boiling points and then condensing and collecting the vapors.

double covalent bond A covalent bond in which two atoms share two pairs of electrons. Example: SO.

double replacement reaction A chemical reaction of the general form $AX + BZ \longrightarrow AZ + BX$, in which the cation of one compound combines with the anion of another compound, and vice versa.

ductility A physical property of matter describing its ability to be drawn into a wire.

dynamic equilibrium An equilibrium (chemical) in which there is a continuation of two or more opposing events occurring at the same rate but resulting in no net change.

E

efflorescence The action of a hydrate that loses part or all of its water of hydration when exposed to air.

electrochemical cell An apparatus consisting of two electrical contacts (electrodes) that are immersed in an electrolyte solution and joined by a wire.

electrode A conductor that allows an electrical current to enter or leave an electrolytic cell (or other apparatus).

electrolysis A process that uses electricity to force an otherwise nonspontaneous chemical reaction to occur; the separation of a compound into simpler substances by an electrical current.

electrolyte A substance that releases ions and conducts electricity when it dissolves in water.

electrolytic cell An electrochemical cell used to split compounds by electrolysis.

electron A particle with a -1 charge and a mass of 0.00055 amu found orbiting the nucleus in an atom.

electron affinity The amount of energy released when an electron is added to an atom to form a negative ion.

electron configuration A representation of how electrons are positioned in an atom: a number indicates the principal energy level, a letter indicates the sublevel, and a superscript denotes the number of electrons contained within the sublevel.

electron-dot structure A representation of the electronic structure of atoms and compounds: chemical symbols represent nuclei, and dots represent valence electrons. Examples: Na·, Na:Cl: .

electronegativity A measure of the tendency of bonded atoms to attract electrons.

electron-sea theory A theory that offers an explanation of how metals bond; the valence electrons of atoms are said to be freely shared among all atoms.

electroplating The deposition of a metal on a surface by means of an electrical current.

electrostatic attraction Forces between particles caused by their opposite electrical charges.

element A substance that cannot be broken down by ordinary chemical means into anything that is both stable and simpler and whose atoms all have the same atomic number.

empirical formula A formula that tells the types of atoms that are present in a compound and the simplest whole number ratio between the atoms. Example: The empirical formula of C_2H_4 is CH_2.

emulsion A suspension of substances, in a dispersed form, that will not mix. Example: mayonnaise.

end point The point in a titration reaction at which the reaction is observed to be complete.

endothermic A term describing a process that absorbs heat energy.

energy The ability to do work.

enthalpy (H) The energy content of a system at constant pressure.

enthalpy of bond formation (bond enthalpy) The enthalpy change that occurs when 1 mole of bonds in a gaseous compound are broken.

enthalpy of formation (ΔH_f°) The change in enthalpy that occurs when 1 mole of a compound is formed from its elements; abbreviated .

entropy The measure of randomness or disorder in a specified portion of the universe; a measure of the increasing unavailability of energy to perform useful work.

enzyme A protein molecule that acts as a catalyst.

equilibrium constant A mathematical expression of the ratio between the concentrations of the products and reactants at equilibrium: each concentration in the expression is raised to the power that matches the substance's coefficient in the balanced chemical reaction.

equivalence point A point in a titration reaction at which an equivalent amount of titrant has been added; the number of H_3O^+ ions equals the number of OH^- ions.

equivalent For redox reactions, one equivalent is defined as the amount of substance that gains or loses a mole of electrons; for an acid-base reaction, it is the amount of substance that gains, loses, or neutralizes a mole of hydrogen ions.

essential amino acid An amino acid required for life that cannot be manufactured by the body and must be obtained by diet.

ester An organic compound of the general formula
$$R_1-\overset{O}{\underset{\|}{C}}-O-R_2$$, having a carbonyl-oxygen-carbon system.

esterification The formation of an ester through the reaction of an acid with an alcohol.

ether An organic compound of the general form R_1-O-R_2, in which an oxygen atom links alkyl groups. Ethers differ from esters in that the carbons next to the oxygen link are not bonded to another oxygen atom.

evaporation A physical change from the liquid state to the gaseous state that occurs at the surface of a liquid.

exact number A number that has infinite significance or precision.

exothermic A term describing a process that releases heat energy.

expansibility The property of a substance that allows its particles to spread out.

exponential notation See **scientific notation**.

F

Fahrenheit scale (F) A temperature scale proposed by Gabriel Fahrenheit in which there are 180 divisions between the freezing point of water (32° F) and the boiling point of water (212° F) at 1 atm.

family A vertical column of elements in the periodic table with similar physical and chemical properties.

fat A molecule formed from glycerol and three carboxylic acids with mostly saturated carbon chains; a fat differs from an oil in that it is solid at room temperature, while an oil is liquid.

fatty acid A carboxylic acid with a long, aliphatic chain that can be obtained by the hydrolysis of animal fat or vegetable oils.

first law of thermodynamics The physical law that states that energy cannot be created nor destroyed but can be converted from one form into another (also called the law of conservation of mass/energy).

flocculating agent A substance whose ions can block the repulsive forces between colloidal particles, causing them to aggregate.

flow chart A diagram showing the progress of work through a sequence of operations.

formula A combination of subscripts and chemical symbols that indicate the number and kinds of atoms that are present in a compound.

formula unit The simplest ratio between the different kinds of atoms in an ionic compound.

free energy (ΔG) (Gibbs free energy) A term that includes the enthalpy and entropy of a substance.

free energy change The driving force of a reaction and the indicator of spontaneity; $\Delta G = \Delta H - T\Delta S$. If ΔG is negative for a process, the process is energetically favorable.

freezing A physical change from the liquid state to the solid state.

freezing point depression A lowering of the freezing point of a solvent caused by the presence of solute particles.

fuel cell A class of battery that operates with a continuous supply of reactants for fuel from an external source and the products are continually removed.

fullerene A class of carbon molecule in which the carbon atoms are arranged in the form of a hollow sphere or cylinder.

functional group An atom or group of atoms that is common to the members of a family of compounds and imparts characteristic chemical properties to that family.

G

galvanic cell An electrochemical cell in which a spontaneous redox reaction produces electricity.

gamma ray (γ) Electromagnetic waves of very high frequency and short wavelength.

gas A state of matter in which the particles have enough energy to overcome the attractive forces: a gas has no definite size or shape.

Gay-Lussac's law A law of gas behavior that states that the pressure of a confined gas is proportional to its absolute temperature, provided its volume is held constant ($P/T = k$).

gene A segment of DNA capable of producing a specific polypeptide that is responsible for a particular characteristic (hair color, for example).

glycerol A three-carbon molecule with three hydroxyl groups.

$$-\underset{|}{\overset{|}{C}}-\underset{|}{\overset{|}{C}}-\underset{|}{\overset{|}{C}}-$$
$$\ \ \ OH\ \ OH\ \ OH$$

glycogen A branched polymer of glucose that serves to store energy for animals.

gram-atomic mass The mass in grams of a mole of atoms; a value numerically equivalent to the average atomic mass expressed in grams.

gram-equivalent mass The mass of one equivalent expressed in grams.

gram-formula mass The mass in grams of a mole of formula units; a value numerically equivalent to the average formula mass expressed in grams.

gram-molecular mass The mass in grams of a mole of molecular units; a value numerically equivalent to the average molecular mass expressed in grams.

gray (Gy) The SI unit of biologic radiation effect.

Greek prefix system A system of prefixes used to indicate the number of atoms in a binary covalent compound or the number of water molecules in a hydrate. Examples: dinitrogen pentoxide, sodium carbonate monohydrate.

group See **family**.

H

Haber process The industrial preparation of ammonia from nitrogen and hydrogen gas that uses high temperatures, high pressures, and catalysts.

half-life ($t\frac{1}{2}$) The amount of time required for one-half of the nuclei in a radioactive sample to decay into another kind of nucleus.

half-reaction A hypothetical portion of a redox reaction that consists of either the substances involved in oxidation or the substances involved in reduction.

half-reaction method A method of balancing redox equations using half-reactions.

halogen A family VIIA element, which has seven valence electrons.

heat Thermal energy in transit from one object to another.

heat of condensation The amount of heat that must be removed from a vapor at its boiling point to condense it to a liquid at the same temperature.

heat of fusion The amount of heat required to change 1 g of a substance at its melting point from a solid to a liquid; also known as latent heat of fusion.

heat of reaction (ΔH) The change in enthalpy that occurs during a reaction.

heat of vaporization The amount of heat required to change a liquid at its boiling point to a gas at the same temperature.

Heisenberg uncertainty principle The principle that states that it is impossible to know the energy (velocity) and exact position of an object at the same time.

Henry's law The law that states that the solubility of gases increases with pressure.

Hess's law The law that states that the enthalpy change of a reaction equals the sum of the enthalpy changes for each step of the reaction.

heterogeneous catalyst A catalyst that is in a separate phase from the reactants.

heterogeneous mixture A mixture composed of two or more distinctly separate phases that have their own properties. Example: a suspension.

homogeneous catalyst A catalyst that is in the same phase as a reactant, or in solution with the reactant.

homogeneous mixture A mixture existing in only one distinctly separate region with its own properties. Example: a solution.

hormone A steroid, polypeptide chain, or protein that serves as a chemical messenger and is transported by the blood stream.

Hund's rule The rule that states that as electrons fill a sublevel, all orbitals receive one electron before any receive two.

hybridization The process of forming new kinds of orbitals with equal energies from a combination of orbitals of different energies.

hydrate A compound that has water molecules in its crystalline structure. Example: $CaCO_3 \cdot H_2O$.

hydration The process by which water molecules surround and interact with solute particles (the type of solvation in which water molecules act as the solvent).

hydrocarbon An organic compound containing only hydrogen and carbon atoms.

hydrogen bond The electrostatic attraction between an unshared pair of electrons in a highly electronegative atom and a hydrogen atom that is bonded to a different highly electronegative atom.

hydrolysis A reaction in which one molecule is split by the addition of water.

hygroscopic compound Any substance that absorbs water from the air.

hypothesis An educated guess about the solution to a problem; when supported by scientific facts, it may become a theory.

I

ideal gas A hypothetical gas that behaves exactly according to the ideal gas law.

ideal gas law A law of gas behavior that relates pressure, volume, temperature, and amount for an ideal gas ($PV = nRT$).

immiscible A term describing two liquids that are not soluble in each other.

indicator A substance that changes color when the pH of a solution changes.

inductive reasoning The process of beginning with known facts and proceeding to an unknown conclusion.

inert A substance that will not readily react with other substances.

inhibitor A substance used to slow a reaction or reduce the effect of a catalyst.

inner transition metal A member of the lanthanide and actinide series.

inorganic chemistry The study of all elements and compounds other than covalent compounds containing carbon-carbon bonds.

intermolecular force An electrostatic attraction between molecules; it is much weaker than the bonds that form within molecules.

ion An atom or group of atoms that has acquired an electrical charge by either losing or gaining one or more electrons.

ion product constant of water See **autoprotolysis constant of water**.

ionic bond The electrostatic attraction between two oppositely charged ions in a solid.

ionic compound A type of compound that consists of positive and negative ions whose electrical charges hold them together while neutralizing each other.

ionic equation An equation that represents all the substances present during a reaction, including the spectator ions, non-ionic products, and insoluble precipitates.

ionization energy The minimum amount of energy needed to remove an electron from an atom.

irreversible reaction A reaction that proceeds in only one direction: the reactants change into products but not vice versa.

isotope One of two or more atoms of the same element with the same number of protons (atomic number) but with different numbers of neutrons.

isotopic notation A convention that includes the symbol, atomic number, and atomic mass of an element: it specifies the exact composition of an atom.

IUPAC The International Union of Pure and Applied Chemistry; the body responsible for the standardization of chemical nomenclature and usage.

J

joule (J) The SI unit of energy equal to one $kg \cdot m^2/s^2$.

K

Kelvin scale A temperature scale that divides the range from the freezing point of water (273 K) to the boiling point of water (373 K) into 100 increments and labels absolute zero (0 K) as its zero point; the absolute scale.

ketone Any organic compound of the general form
$$R_1 - \overset{\overset{\displaystyle O}{\|}}{C} - R_2$$
in which an interior carbon double bonds to an oxygen atom.

kinetic energy Energy due to motion.

kinetic theory A theory that states that the particles of matter are in constant motion and that the properties of matter are consequences of that motion; usually used in reference to gases and their properties but sometimes applied to solids and liquids.

kinetics In chemistry, the study of reaction rates and the mechanisms by which reactions occur.

L

lanthanide series A portion of the sixth series of the periodic table that includes the inner transitional metals from lanthanum to lutetium.

latent heat The heat energy that results in a phase change of a substance with the temperature remaining constant.

lattice energy The energy released when gaseous particles form a crystal.

law A description of the behavior of matter and energy based on the results of many experiments.

law of combining volumes A law of gas behavior that states that gases at the same temperature and pressure react with one another in volume ratios of small whole numbers.

law of definite composition A law that states that every compound has a definite composition by mass.

law of mass/energy conservation A corollary to the first law of thermodynamics: during ordinary physical and chemical processes, mass is neither created nor destroyed, only converted from one form to another.

Le Châtelier's principle The principle that when a system at equilibrium is subjected to stress, the equilibrium is shifted in the direction that relieves the stress.

Lewis acid Any substance that can accept a pair of electrons.

Lewis base Any substance that can donate a pair of electrons.

line spectrum A spectrum showing only certain colors or wavelengths of light.

linear A structural arrangement in which particles are positioned in a straight line.

lipid A member of the large class of biological molecules that are not soluble in water. Examples: fat, oil, wax, steroid, fatty acid.

liquid A state of matter in which the particles have enough energy to partially overcome the attractive forces; the liquid will conform to the shape of its container.

lock-and-key model A theory that maintains that enzymes catalyze reactions by positioning reactants in ideal positions for reactions to occur.

M

magic number The number of protons or neutrons in a completed nuclear shell, according to the nuclear shell model.

main group One of the eight A groups in the periodic table that contain elements whose outermost electrons are in s or p sublevels.

malleability A physical property of matter indicating the capability of matter to be shaped by pounding.

mantissa The part of the scientific notation expression that is the actual measurement, with a decimal point following the first significant figure. Example: 1.200 is the mantissa in the expression 1.200×10^3.

mass A measure of the amount of matter in a given substance.

mass defect The difference between the mass of a nucleus and the sum of the masses of the particles from which it was formed.

mass number The sum of the number of protons and neutrons in the nucleus of an atom.

matter Anything that occupies space and has mass.

measurement A number that indicates quantity and is followed by a unit.

melting A physical change from the solid to the liquid state.

meniscus The curved upper surface of a column of liquid.

metal An element located to the left of, but not touching, the heavy, stair-step line in the periodic table; an element that is typically malleable, ductile, lustrous, and conductive of electricity and that forms positive ions when it gives away its few valence electrons.

metallic bond A communal sharing of electrons between metal atoms. See **electron-sea theory**.

metallic hydride A compound in which the oxidation number of hydrogen is -1; formed when hydrogen reacts with an active metal. Examples: LiH, NaH, MgH_2.

metalloid An element whose properties lie between those of metals and nonmetals; a compound found along the heavy, stair-step line in the periodic table.

metallurgy The process of extracting metals from their ores and adapting them for commercial use.

millimeters of mercury (mm Hg) A standard unit of pressure derived from the fact that normal atmospheric pressure can support 760 mm Hg in a column; 1 mm Hg = 1 torr.

miscible A term describing two liquids that are completely soluble in each other.

mixture Two or more pure substances physically combined with no definite proportions.

model A working representation of experimental facts.

molal boiling point elevation constant (K_{bp}) A number that relates the change in boiling point of a particular solvent to the concentration of solute particles.

molal freezing point depression constant (K_{fp}) A number that relates the change in freezing point of a particular solvent to the concentration of solute particles.

molality (m) A quantitative measure of concentration equal to the number of moles of solute per kilogram of solvent; m = moles solute/kg solvent.

molar volume of a gas The volume that a mole of gas occupies if it is at standard temperature and pressure: 22.4 L.

molarity (M) A quantitative measure of concentration equal to the number of moles of solute per liter of solution; M = moles solute/L solution.

mole The amount of substance contained in 6.022×10^{23} units.

molecular compound A compound made of separate, distinct, independent units (molecules).

molecular formula A formula that shows the types of atoms involved and the exact composition of each molecule. Example: C_2H_4.

molecular orbital theory A bonding theory that holds that atomic orbitals are replaced with molecular orbitals when bonding occurs. These orbitals, both bonding and non-bonding, are linear combinations of the atomic orbitals.

molecule Two or more covalently bonded atoms found as a separate, distinct, independent unit.

monatomic element An element whose atoms exist independently. Example: noble gases.

monoprotic acid An acid that can donate only one proton.

monosaccharide A three- to six-carbon carbohydrate with attached hydroxyl groups and either an aldehyde or ketone group that cannot be hydrolyzed into simpler compounds. Example: glucose.

N

net ionic equation An equation that shows only the substances actually involved in a reaction and excludes spectator ions.

neutral salt A salt that causes no change in pH when dissolved in water.

neutral solution A solution with equal numbers of H_3O^+ and OH^- ions, resulting in a pH of 7.

neutralization reaction The reaction of an acid and a base to produce a neutral (pH = 7) solution of water and a salt.

neutron A neutral particle with a mass of 1.0087 amu found in the nucleus of an atom.

noble gas A family VIIIA element, which has a full outer energy level.

nomenclature A system or set of names used by a branch of learning, such as the system of names for compounds in chemistry.

non-electrolyte A substance that will not conduct electricity when melted or dissolved, because it does not release ions.

nonmetal An element located to the right of, but not touching, the heavy, stair-step line in the periodic table; an element that is nonductile, nonmalleable, and nonconducting and that usually forms negative ions because it has a strong attraction for its numerous valence electrons.

normal boiling point The boiling point temperature of a liquid at STP.

normality (N) A quantitative measure of concentration equal to the number of equivalents of solute per L of solution; $N = eq/L$.

nuclear binding energy The energy required to separate all the protons and neutrons in a specific nucleus from each other; the energy equivalent of the nucleus's mass defect.

nuclear chemistry The study of radioactivity, the nucleus, and the changes the nucleus undergoes.

nuclear equation An equation that identifies the nuclei that react and are produced when nuclei split, fuse, or release radiation.

nuclear fission The process of splitting a massive nucleus, usually with the release of great amounts of energy and two large fragments of comparable mass.

nuclear fusion The process of combining two or more smaller nuclei into one larger nucleus, releasing great amounts of energy.

nuclear shell model A model that states that the protons and neutrons in an atomic nucleus exist in different shells within the nucleus.

nucleic acid A large molecule that stores and translates genetic information in living cells and consists of sugar units, nitrogenous bases, and phosphate groups.

nucleotide The "building block" of a DNA or RNA molecule: each block is made of a sugar unit, a phosphate group, and a nitrogenous base (adenine, guanine, cytosine, thymine, or uracil).

nucleus The dense central part of an atom made up of protons and neutrons; the nucleus contains virtually all of the atom's mass but only a small portion of the atomic volume.

O

octet rule The rule that states that an atom tends to gain, lose, or share electrons until its outer level s and p orbitals are filled with eight electrons: this gives the element the electron configuration of a noble gas.

oil A molecule formed from glycerol and three carboxylic acids in which the carbon chains have a high degree of unsaturation; it differs from a fat in that it is liquid at room temperature, while a fat is solid.

orbital A four-dimensional region of space in which as many as two electrons may exist; sections of the sublevels.

orbital notation A diagrammatic representation that uses dashes and arrows to show the principal energy levels and sublevels for all the electrons in an atom.

organic chemistry The study of carbon-containing compounds, their structures, and the reactions they undergo.

osmosis Diffusion of pure solvent molecules, such as water, through a membrane.

osmotic pressure The pressure required to prevent a solution from gaining water by osmosis.

oxidation A chemical process in which electrons are lost and an oxidation number increases.

oxidation number (oxidation state) A number that reflects the charge that an atom in a compound would have if all the bonding electrons were arbitrarily assigned to the most electronegative element.

oxidation-reduction reaction (redox reaction) Any chemical reaction in which electrons transfer or shift; the shift of electrons is shown by changes in oxidation numbers.

oxide A binary compound in which the oxidation number of oxygen is -2. Example: Li_2O.

oxidizing agent The atom or ion that receives electrons during a redox reaction; the substance that causes other substances to be oxidized; the substance that is reduced.

oxyanion An anion composed of oxygen and one other element. Example SO_4^{2-}.

P

paramagnetism A weak attraction of a substance by a magnetic field, usually a result of unpaired electrons.

particle molality The molality of a solution based on the number of solute particles in the solution. Example: A 1 molal solution of $AlCl_3$ has a particle molality of 4 (three Cl^- ions and one Al^+ ion).

pascal (Pa) The SI unit of pressure; See Appendix A.

Pauli Exclusion Principle The rule that mandates that an orbital can hold only two electrons with opposite spin. Therefore, no two electrons in the same atom can have the same four quantum numbers.

peptide bond The bond between an amino group of one amino acid and a carbonyl group of another.

peptidization The process whereby coagulated colloidal particles are made to disperse and return to a colloidal state.

percent by mass A quantitative measure of concentration in which the mass of the solute is compared to the mass of the solution; per composition = mass solute/mass solution $\times$ 100%.

percent composition A percent that gives the relative amount (based on mass) of each element present. Example: There is 52.9% Al and 47.1% O by mass in Al_2O_3.

period A horizontal row of elements in the periodic table; also called a series.

periodic law The law that states that the properties of elements are a periodic function of their atomic numbers.

periodic table A table in which elements are arranged in order of increasing atomic numbers so that the elements with similar properties fall into the same vertical columns, or families.

permeability The property of a substance that allows other substance particles to spread or flow throughout it.

peroxide A compound that contains an oxygen-oxygen bond; the name of the O_2^{-2} ion.

pH A measure of the hydronium ion (H_3O^+) concentration; the negative logarithm of the H_3O^+ ion concentration: pH = -log [H_3O^+].

pharmacology The science of developing, using, and studying the effects of drugs.

phase A homogeneous region of a heterogeneous mixture.

phase diagram A graphic representation of the function of pressure versus temperature; a diagram showing conditions under which a pure substance exists as a solid, liquid, or gas.

pheromones A chemical messenger secreted by one animal and affecting the behavior of another member of the same species.

phlogiston A non-existent substance once thought to be the volatile component of combustible material.

physical change A change that alters the physical properties of a substance (state, size) but that does not change its identity.

physical chemistry The branch of chemistry that examines the structure and interactions of matter and the energy changes that accompany those interactions.

physical property A property of matter that results from the position and characteristics of its particles and that can be measured without causing a change in the identity of the material.

pi bond A bond in which there is a side-by-side overlap of the orbitals.

plasma The most abundant form of matter, consisting of a gaseous sea of high-velocity electrons, ions, and neutral atoms.

pOH A measure of OH^- ion concentration; the negative logarithm of the OH^- ion concentration: pOH = -log [OH^-].

polar Having unequally distributed electrical charges.

polar covalent bond A chemical bond that has partially positive and partially negative ends because of unevenly shared electrons. All bonds between non-identical atoms are polar.

polyatomic compound A compound that contains at least three different elements. Examples: $NaNO_3$ and NH_4OH.

polyatomic element An element whose atoms bond in multi-atom units. Example: S_8.

polyatomic ion A group of atoms that maintains a constant electrical charge while existing as a unit in a wide variety of chemical reactions. Example: SO_4^{-2}.

polymer A substance consisting of huge molecules that have repeating structural units.

polymorphous A term describing substances (either elements or compounds) that form more than one crystalline form.

polypeptide A series of many amino acids joined by peptide bonds.

polyprotic acid An acid that can donate more than one proton.

polysaccharide A carbohydrate composed of many monosaccharide units.

post-transition metal A metal found in families IIIA, IVA, or VA in the periodic table.

pounds per square inch (psi) A unit of pressure equal to 6.895×10^3 Pa or 51.71 torr.

precipitate A solid that separates from a solution.

precipitation The separation of a solid from a solution.

precision The agreement between two or more measurements.

pressure Average force per unit area.

principal energy level A region around the nucleus containing a specified group of electrons in sublevels and orbitals.

product An element or compound that is produced from a chemical change and is usually written to the right of the arrow in a chemical equation.

protein A complex structure of many amino acids that is joined by peptide bonds and has a molecular mass greater than 10,000 amu.

proton A particle with a +1 charge and a mass of 1.0073 amu in the nucleus of an atom.

protonation The process of gaining a proton. Example: $H_2O + H^+ \longrightarrow H_3O^+$.

pure substance A substance that is made up of only one kind of particle and has uniform composition.

pyramidal A structural arrangement in which three particles and an unshared electron pair surrounding a central particle are oriented toward the corners of a four-sided pyramid.

Q

quantized A term describing something that has separate, discrete values.

quantum (pl. *quanta*) A discrete amount of energy.

quantum numbers Four numbers that describe the location of an electron in an atom: the first number identifies the relative size of the principal energy level, the second describes the type of sublevel, the third indicates the direction of the orbital in space, and the fourth describes the spin of the electron.

R

rad (radiation absorbed dose) A unit of absorbed radiation corresponding to 0.01 gray.

radioactive decay series A series of sequential reactions of alpha and beta emissions that change a larger, unstable nucleus to a smaller, stable nucleus.

radioactivity The spontaneous emission of penetrating rays from nuclei.

Raoult's law The law that states that the lowering of a solvent's vapor pressure is directly proportional to the concentration of solute particles.

rate law A mathematical equation that describes how fast a reaction occurs.

reactant An element or compound that undergoes chemical change and is usually written to the left of the arrow in a chemical equation.

reaction mechanism The series of steps that make up a reaction.

reaction rate The speed at which reactants disappear or products appear in a chemical reaction.

redox reaction Short name for an oxidation-reduction reaction.

reducing agent The atom or ion that supplies electrons during a redox reaction; the substance that causes other substances to be reduced; the substance that is oxidized in a redox reaction.

reduction A chemical process in which electrons are gained and an oxidation number decreases.

rem (radiation equivalent in man) A measure of any type of radiation that has the same adverse health effect as 1 roentgen of X ray or gamma radiation.

reversible reaction A reaction in which the products can change back into the original reactants so that an equilibrium is reached.

roentgen (R) A measure of dosage for radioactivity; 1 R = 0.0096 gray.

Roman numeral system See **Stock system**.

S

salt A compound formed from the positive ions of a base and the negative ions of an acid.

saponification The reaction by which soaps are made from an ester and an aqueous hydroxide: an alcohol and the salt of a large carboxylic acid result.

saturated A term describing a solution that contains the maximum amount of solute possible at a given set of conditions; a term describing an organic compound that contains the maximum possible number of hydrogens; the compound has no double or triple bonds.

science A systematic study of nature based on observations.

scientific method A logical method of problem-solving that starts with observations based on inductive reasoning.

scientific notation A convenient way of expressing very large and very small numbers as a mantissa between 1 and 10, which is multiplied by 10 raised by some power. Example: 1,200 is 1.200×10^3.

second law of thermodynamics The physical law that states that during any energy transformation, some energy goes to an unusable form.

self-ionization See **autoprotolysis**.

semiconductor A substance with an electrical conductivity intermediate between a conductor and an insulator.

semipermeable membrane A barrier that allows small particles (ions and small molecules) to pass through but that will stop large particles.

semiconductor A substance with an electrical conductivity intermediate between a conductor and an insulator.

sensible heat The heat energy that when applied to a substance results in a change in temperature of that substance.

series See **period**.

SI See **Système Internationale**.

sigma bond A hybrid bond in which an s and p orbital overlap end-to-end.

significant digits The digits in the numerical value of a measurement that indicate its precision.

single replacement reaction A chemical reaction of the general form $A + BZ \rightarrow B + AZ$, in which an active element replaces a less active element in a compound.

solid A state of matter in which the particles have relatively little energy and cannot overcome the attractive forces: the particles of a solid remain in fixed positions with set distances between them.

solubility product constant The equilibrium constant for the dissolving of a slightly soluble salt.

solute One of the least abundant substances in a solution; the substance that is dissolved.

solution A homogeneous mixture of two or more substances.

solvation The process in which solvent particles surround and interact with solutes.

solvent The most abundant substance in a solution; the substance that does the dissolving.

specific heat The amount of heat required to raise the temperature of 1 g of a substance 1° C.

spectator ion An ion present on both sides of an ionic equation: it does not actually participate in the reaction.

sp hybrid One of two orbitals arising from the hybridization of one s and one p orbital, and arranged in a linear configuration.

sp^2 hybrid One of three orbitals arising from the hybridization of one s and two p orbitals, and arranged in a trigonal planar configuration.

sp^3 hybrid One of four orbitals arising from the hybridization of one s and three p orbitals, and arranged in a tetrahedral configuration.

standard molar enthalpy of formation (ΔH_f°) The enthalpy change for the reaction that produces 1 mole of a compound in its standard state from its elements in their standard states.

standard state An accepted set of conditions for a substance that specifies temperature, pressure, and concentration; for thermodynamics, the standard state is usually 298 K and 1 atm; for gas laws, it is usually 273 K and 1 atm.

standard temperature and pressure (STP) A set of agreed-upon conditions; for gas laws, standard temperature and pressure are 0° C (or 273 K) and 1 atm (or 760 torr).

starch A mixture of straight and branched polymers of glucose that serves to store energy for plants.

steroid A member of the class of lipids that contains a characteristic set of three six-membered rings and one five-membered ring. Example: cholesterol.

Stock system A convention used to show the oxidation state of a metal ion. Example: The Stock system nomenclature for $CuCl_2$ is copper (II) chloride.

stoichiometry The measurement and calculation of the mass and molar relationships between reactants and products in chemical reactions.

strong acid A substance that gives up protons very easily and ionizes to a great extent in water.

strong base A substance that readily accepts protons.

strong nuclear forces A term used to describe the forces that hold a nucleus together.

structural formula A formula that shows the types of atoms involved, the exact composition of a molecule, and the location of chemical bonds. Example: The structural formula of C_2H_4 is $-\overset{|}{C}=\overset{|}{C}-$.

structural isomers Compounds that have the same molecular formula but different structural formulas.

sublevel A portion of a principal energy level made up of one or more orbitals. Examples: *s, p, d,* and *f* sublevels.

sublimation A physical change directly between the solid and gaseous states; usually refers to the change from solid to gas. Example: solid CO_2 (dry ice) changing to gaseous CO_2.

subscript A number written at the lower right of a chemical symbol in a formula to indicate the number of components immediately preceding it.

substituent An atom or group of atoms that can substitute for a single hydrogen atom.

substitution reaction A reaction in which one atom or group replaces another atom or group in a molecule.

sulfide A binary compound in which the oxidation number of sulfur is -2. Example: H_2S.

supercritical mass The amount of fissionable material that can support an explosion.

supersaturated A situation where more solute is dissolved in the solvent than occurs at equilibrium.

surface tension The apparent "skin" effect on the surface of a liquid caused by unbalanced forces on the surface particles.

surfactant A substance added to a liquid that acts to reduce the surface tension of that liquid.

symbol A one- or two-letter representation for an element.

Système Internationale An accepted system of units for physical measurements; the system is based on seven units: the meter, the kilogram, the second, the kelvin, the mole, the ampere (electrical current), and the candela (light intensity).

T

temperature A measure of the average kinetic energy of the atoms, molecules, or ions in matter (measured in degrees).

ternary acid An acid molecule that contains three different elements—hydrogen, oxygen, and another nonmetal; an oxyacid. Example: $HClO_4$.

tetrahedral A structural arrangement in which four particles surrounding a central particle are oriented toward the corners of a four-sided pyramid.

theory A tested explanation of scientific observations.

thermal energy The measure of the total kinetic energy of the molecules or ions in matter; usually measured in calories.

thermodynamics The study of energy transformations in chemical and physical processes.

titration The procedure for measuring the capacity of a solution of unknown concentration to react with one of known concentration.

titration curve A graph that plots the pH change of a solution versus the volume of added acids or bases.

torr See **millimeters of mercury.**

transition metal One of the B families of metals in the periodic table.

transmutation Any process that converts one element into another. Example: the changing of lead into gold.

transuranium element An element with an atomic number higher than ninety-two.

trigonal planar A structural arrangement in which three particles surrounding a central particle are oriented toward the corners of a flat triangle.

triple covalent bond A covalent bond in which two atoms share three pairs of electrons. Example: N_2.

triple point The temperature and pressure at which the solid, liquid, and gaseous states of a substance exist in equilibrium.

triprotic acid An acid that can donate three protons.

Tyndall effect The scattering of light by particles in a colloidal dispersion.

U

unit A label, such as "inches" or "meters," used to specify the terms in which a measurement is being reported.

unit analysis A problem-solving tool in which the units of numbers are changed by being multiplied by a conversion factor equal to 1.

unit cell The simplest unit of repetition in a crystal lattice.

universal gas constant (*R*) The constant *R* in the ideal gas law ($PV = nRT$) whose value and units depend on the units used for *P, V, n,* and *T*. Examples: 8.314 Pa·m^3/mole·K = 8.31 J/mole·K = 0.0821 L·atm/mole·K.

unsaturated A term describing a solution that contains less than the maximum amount of solute at a given set of conditions; a term describing organic compounds that contain less than the maximum possible number of hydrogen atoms because they have at least one double or triple bond between the carbon atoms.

V

valence bond theory The idea that covalent bonds are formed when orbitals of different atoms overlap.

valence electron The most loosely bound electron, which is usually found in the outermost energy level.

Valence Shell Electron Pair Repulsion theory (VSEPR) A theory that states that because of electron-electron repulsion, the electron orbitals in molecules are arranged so that they are as far apart as possible.

van der Waals forces A general term for intermolecular forces including dipole-dipole and London forces.

vapor pressure The pressure exerted by a vapor in equilibrium with its solid or liquid state at a specified temperature.

viscosity The ability of a liquid to resist flowing; the amount of internal resistance.

vitamin A micronutrient that is required by the body for normal metabolism, growth, and development and is acquired through the diet.

voltaic cell See **galvanic cell.**

W

water of hydration The water molecules held in some definite molar ratio to the rest of the substance; if the hydrate is a crystal, the water is often called the water of crystallization.

weak acid A substance that does not give up protons easily and does not ionize very much in water.

weak base A substance that is a poor proton acceptor.

weight A unit measure of the gravitational force exerted on a substance.

INDEX

A

absolute zero, 34
accuracy, 44
acetaminophen, 197
acetic acid, 370, 386, 399, 462
acetone
   boiling point, 324
acetylene, 451
acid-base reactions
   equivalents, 319
   See also **titration**,
   neutralization.
acid dissociation constant, 400-404
acidic anhydrides, 302
acid rain, 392
acids, 389-415
   carboxylic, 461-62
   definitions of, 389-92
   diprotic, 403
   monoprotic, 403
   nomenclature, 174, 178-79
   polyprotic, 403
   properties of, 392-93
   strengths of, 401
   strong, 399
   triprotic, 404
   weak, 399
acid salt, 407
actinide series, 107
activated complex, 353-54
activation energy, 353
activity series, 210
   halogens, 210
   metals, 210
addition reaction, 466-67
adenine, 492
adsorption, 330
air, 240, 311
   density of, 246
air pollution, 395
Alaskan pipeline, 437
alchemy, 9
alcohol, 457-59
aldehydes, 460
aliphatic compounds, 446
alkali metals, 118
alkaline earth metals, 119
alkane, 447-50
alkene, 450-51
alkyl group, 448
alkyl halide, 456-57
alkyne, 451-52
allotrope, 268-69
alloy, 310, 437
alpha decay, 508

alpha particles, 502-3
aluminum, 123
   in soil, 395
   production, 432
   uses, 124
aluminum sulfate, 187, 189-90
amalgam, 436
amides, 463
amines, 463
amino acids, 481
   evolution, 422-23
ammonia, 392, 402
   equilibrium, 369, 373-74, 378-79
   formation of, 378-79
   molecular shape of, 152
   pH of, 397
   reaction, 346
ammonium chloride, 350
ammonium ion, 392
   electron-dot structure, 146
amorphous solids, 263
amphiprotic substances, 402-3
Anabaena cylindrica, 299
analytical chemistry, 13
anesthetic, 459
anhydrides, 302
   metal oxides, 302
   nonmetal oxides, 302
anhydrous compound, 177
anion, 94, 432
   in salts, 406
anode, 432-36
antacid, 406
antilog, 398
apothecaries, 7
apple juice
   pH of, 397
applied science, 15
aqueous solution, 202
argon, 130, 232
aromatic compounds, 446, 453-55
Arrhenius, Svante, 389
arsenic, 126
aryl halide, 456
asbestos, 116
ascorbic acid, 436
aspirin, 197
atmosphere, 299
   early, 422-23
atmospheric pressure, 229
   effect on boiling, 276
atom, 25
atomic bomb, 501
atomic mass, 91
atomic mass unit (amu), 89

atomic models, 67-88
atomic number, 89
atomic radii, 110-12
atomic theory, 68
Aufbau principle, 83
autoclave, 277
autoprotolysis constant
   of water, 394
autoprotolysis of water, 394
Avogadro, Amedeo, 181
Avogadro's number, 181
Avogadro's principle, 244
azeotrope, 310

B

Bacon, Francis, 10, 375
baking soda, 385
balancing chemical
   equations, 202-6
   half-reaction method, 423-28
   limitations, 206
barium sulfate, 387
barometer, 229
bases
   definitions of, 389-92
   properties of, 392-93
   strengths of, 399-401
   strong, 400
   weak, 400
basic anhydrides, 302
basic salt, 407
batteries, 434
bauxite, 432
Becquerel, Henri, 501-2
becquerel (Bq), 504-5
bee stings, 462
benzene, 454-55
   formation of, 340
beriberi, 489
Berzelius, Jons, 26
beta particles, 503-4
binary acids, 178
binary compounds, 173-75
biochemistry, 13, 475-97
bleach, 421
blood
   pH of, 397
Bohr, Niels, 73
   model of atom, 74
boiling, 276
boiling point, 277
boiling point elevation, 323
Bonaparte, Napoleon, 180
bonds
   covalent, 141
   double, 142

542

enthalpy, 343-44
ionic, 138
metallic, 146
polar covalent, 154
properties of, 261
quantum model of, 109-10
stability of, 343-44
triple covalent, 142
types of, 137
boric acid, 393, 400
boron, 124
Bose-Einstein condensate, 38
Bose, Satyendra, 38
Boyle, Robert, 234, 375
Boyle's law, 234
brass, 310
bridge element, 119, 124
bright-line spectra, 73-74
British Thermal Unit (BTU), 34
bromine, 128-29
bond enthalpy, 343
properties of, 129
test for unsaturation, 468-69
Brönsted, J. N., 391
bronze, 7
Brown, Robert, 330
Brownian movement, 330
buckminsterfullerene, 464
buckyballs, 464-65
buffers, 410-11
Bunsen, Robert, 77
butane, 450

C

calcium, 119-20
calcium chloride, 326
formation of, 138-39
freezing point of solution, 326-27
calorie (cal), 34
calorimeter, 340
capillary rise, 273
carbohydrates, 476-80
carbon
bonding, 445
charcoal, 125
forms of, 125
hybridization of, 150
physical properties of, 125
uses, 125
carbon-14
radioactive dating, 510
carbonated drink, 208, 314-15
carbon dioxide, 232
empirical formula, 192
solutions of, 315
carbonic acid
in blood, 411, 488
carbonic anhydrase, 488

carbon monoxide, 233
carbon tetrachloride, 263
nonpolar bonds of, 156
carbonyl group, 460
carboxyl group, 462
carboxylic acid, 461-62
catalyst, 202, 356
See also **enzyme.**
cathode, 432-36
cathode rays, 70
cation, 94, 431-39
in salts, 406-7
Cavendish, Henry, 117
cellulose, 479-80
Celsius scale, 34
Chadwick, James, 73
chain reactions, 514-15
charcoal, 125
Charles's law, 235-36
chemical
bonds, 135-58
changes, 23
equations, 200
properties, 22
reactions, 207-12
symbols, 25
chemistry
history of, 6-13
major branches of, 13
study of, 14-16
vocational application, 14-15
Chernobyl, 395, 519
chitin, 476
chlor-alkali industry, 433
chlorine, 128-30, 233
covalent bonding of, 141
in redox reactions, 418, 420
production of, 433
chloroacetophenone, 233
chloroform, 456
cholesterol, 485, 486-87
chromosomes, 491
citric acid, 392, 478
classification
of matter, 21-24
of organic compounds, 447
clone, 495
coagulation, 330
coal, 125
dust, 354
coefficients, 28
colligative properties, 322
boiling point elevation, 323
decreased vapor pressure, 322
freezing point depression, 325
osmotic pressure, 327
Raoult's law, 323

collisions
in gases, 352
collision theory
of reactions, 352
colloids, 328-30
colorimetry, 116
combination reaction, 207
combined gas laws, 238
combustion, 23
spontaneous, 354
common ion effect, 383
compounds, 27
compression
of gases, 278-79
concentration effect
on equilibrium, 372-73
effect on reaction rate, 355-56
condensation, 39, 275
condensation reactions, 469
conductivity, 22
conjugate acid, 391
conjugate base, 391
conservation, 31, 338-39
See also **first law of thermodynamics.**
continuous spectrum, 73
conversion factor, 56-57
copper sulfate, 319-20
Cornell, Eric, 38
corrosion, 437
covalent bond, 141
of carbon, 445
Creation, 338, 437
criss-cross method, 166-67
critical mass, 515
critical pressure, 279
critical temperature, 278
critical thought, 8
cryogenics, 281
cryosurgery, 281
crystal, 139
crystal lattice, 266
crystalline structures, 266-69
Curie, Marie and Pierre, 220, 502
curie (Ci), 505
cyclic aliphatic compounds, 452-53
cytosine, 492

D

Dalton, John, 26, 68-69
Dalton's law of partial pressures, 239-42
DDT, 193, 455
Dead Sea, 316
decomposition reactions, 208
deductive reasoning, 2

deliquescence, 305
delocalized electrons, 147, 454
Democritus, 68
denature, 459
density, 22, 60
    of gases, 228, 246
    of water, 294
deprotonation, 391
derivative unit, 45
descriptive chemistry, 117
desiccator, 305
detergent, 273
diamond, 125, 270-71
    oxidation of, 351
diatomic elements, 25
diffusion, 228, 312
dimensional analysis, 56
dimethyl ether, 444
dipeptide, 482
dipole-dipole interactions, 260
dipole moment, 155
disaccharides, 477-78
dispersion forces, 260
displacement reactions, 209
dissociation, 311
dissolving
    and equilibria, 366
    mechanism, 311-12
distillation, 277-78
    fractional, 278
DNA, 491-97
Döbereiner, Johann, 101
double bonds, 142
    in fats, 485
double replacement
    reaction, 211
dry cell, 435-36
dry ice, 266
ductility, 22

E

E. coli, 495, 497
ecology, 395, 519-20
efflorescence, 305
Egypt, 7
Einstein, Albert, 31
electrochemical cell, 431-36
electrode, 300, 431
electrolysis, 300, 432-33
electrolyte, 301, 431
electron affinity, 113
electron configuration, 84
    from the periodic table, 108
electron-dot symbol, 94
    of polyatomic ions, 142-43, 146
    steps for drawing, 142-44

electronegativity, 114
    in bonding, 137
    hydrogen bonds, 260
    of carbon, 445
    periodic trends, 114
electrons, 71
    as waves, 79
    electrostatic attraction, 111
    in quantum model, 84
    octet rule, 136
    oxidation numbers, 164-69
electron-sea theory, 147
electron spin, 84
electroplating, 438-39
element, 25
empirical formulas, 187
emulsion, 329-30
endpoint, 409
endothermic, 29
energy, 30
    in bonds, 337-39
    transformations, 31
energy diagrams, 351-52
enthalpy, 339-44
    of formation, 340-41
entropy, 32, 345-46
enzyme, 357, 483, 487-88
    in honey, 478
Epsom salts, 120, 171
equations
    ionic, 211
    special symbols, 202
equilibria, 365
equilibrium, 365-84
    autoprotolysis, 394
    forms of monosaccharides, 477
    vaporization, 275
equilibrium constant, 368-70
equivalence point, 409
equivalents, 319
essential amino acids, 481
ester, 463
    formation of, 469
esterification, 469
ethane, 447-49
    combustion of, 204-5
ethanol, 444, 458
ether, 459
ethics, 281, 496
evaporation, 274
    equilibria, 336
evolution, 375, 422-23
ex nihilo, 21
exothermic, 29
exponential factor, 52
explosives, 377

F

Fahrenheit scale, 35
faith
    of scientists, 375
Faraday, Michael, 454
fats, 485
fatty acid, 462, 484
ferrous sulfate, 197
fertilizer, 377-79
film, 418-19
first law of thermodynamics, 31, 338-39
fission, 514-15
flavoring, 377, 463
flocculating agent, 330
fluoridation, 303
fluorine, 129
    dipole moment, 158
    reaction with water, 301
fluorosis, 303
fog, 329
formaldehyde, 188, 460
formic acid, 462
formula, 27
    empirical, 187
    molecular, 187
    percent composition, 188-89
    structural, 187
formula units, 140
Forshufvud, Sten, 180
fractional distillation, 278
    of gasoline, 279
fragrance, 463
free-energy, 346
free-energy change, 346-50
freezing, 39
    of cells, 281
freezing point depression, 325
freon, 456
fructose, 476
fuel cell, 436
fullerenes, 464
Fuller, R. Buckminster, 464
functional groups, 456
fusion, 515-17

G

Galileo, 1
Galvani, Luigi, 431
galvanic cell, 431
galvanization, 437, 439
gamma emission, 504
gamma rays, 504-5
gases, 37, 227-53
    and pressure, 230
    and temperature, 230
    and volume, 230

collection, 240-41
critical values of, 278-79
densities of, 228, 246
from batteries, 438
ideal, 249-53
kinetic description of, 227-28
mixtures of, 239-42
molar volume, 244
physical properties of, 228-31
stoichiometric conversions, 248
sublimation, 265
gas laws, 227-53
gastric juice
pH of, 397
Gay-Lussac, Joseph, 236
Gay-Lussac's law, 236
Geiger counter, 504
genes, 491
genetic engineering, 475
Gibbs, J. Willard, 346
glass, 263
glucose, 476
glycerol, 458
glycogen, 479
gold, 121
gram-atomic mass, 186
gram-equivalent mass, 319
gram-formula mass, 186
gram-molecular mass, 186
of gases, 246-53
graphite, 125
gray (Gy), 505
Greek prefix system, 172
Greeks, 7, 8
groups, 106
alkali metals (IA), 118
alkaline earth metals (IIA), 119
halogens (VIIA), 128
hydrogen, 117
nitrogen and phosphorus (VA), 125-26
noble gases (VIIIA), 130
oxygen and sulfur (VIA), 126-28
post-transition metals, 123
transition metals 121
guanine, 492
gunpowder, 338

H

Haber, Fritz, 378
Haber process, 378-79
Haldane, John, 422-23
half-life, 508-9
half-reaction, 424
balancing equations, 423-28
Hall-Héroult process, 432

halogens, 128
activity series, 210
diatomic molecules, 129, 145
oxidation numbers, 165
reactions, 210, 301
heartburn, 406
heat, 33
heat of fusion, 264
heat of reaction, 334
See enthalpy.
heat of vaporization, 274-75
Heisenberg uncertainty principle, 78
Heisenberg, Werner, 78
helium, 232, 503
See also alpha particles.
helix
DNA, 493
in starch, 479
protein, 483
hemoglobin, 121
Henry's law, 315
Hess's law, 342
heterogeneous mixtures, 24
hexane, 312
Hiroshima, 501
homogeneous catalyst, 357
homogeneous mixtures, 24
honey, 477
hormones, 483, 490-91
Human Genome Project, 496
Hund's rule, 85
hybridization, 150
hydrates,, 177, 304-5
hydration, 311
hydrocarbons, 446-47
hydrofluoric acid, 400
hydrogen, 117
bond, 260
density of, 117
formation of, 433
liquid, 278-79
molecular velocity of, 230
oxidation number, 165
reactions, 117-18
hydrogen bomb, 516
hydrogen bond, 260, 493
See hydrogen.
hydrogen carbonate, 411
hydrogen cyanide, 233
hydrogen sulfide, 233
hydrogenation, 485-86
hydrolysis reaction, 469-70
hydronium ion, 391-411
hydroxide ions
in bases, 391-411
hydroxides, 208
hygroscopic compounds, 305
hypothesis, 3

I

ideal gases, 249-50
ideal gas law, 250
immiscible, 310
indicators, 404-5
inductive reasoning, 2
inert, 5
gases, 130-31
inhibitor, 358
inner transition metals, 122
inorganic chemistry, 13
insecticide, 491
intermolecular forces
dipole-dipole interactions, 260
dispersion forces, 260
in alkenes, 451
in fats, 485
van der Waals, 259
International Union of Pure and Applied Chemistry (IUPAC), 27
iodine, 128-30
reactions, 355
sublimation of, 266
iodine chloride, 355
ionic bonds, 138
ionic compounds
nomenclature, 173-74
ionic equation, 211
ionic radii, 110-12
ionization energy, 112-13
ions
nomenclature, 94
spectator, 211
iron, 121, 437
isomers
structural, 449-50
isotopes, 90
isotopic notation, 90

J

Jabir ibn-Hayyan (Geber), 9
jelly, 478
Joliot, Irene and Frederic, 220
Joule, James Prescott, 34
joule (J), 34

K

Kekule, August, 454
Kelvin, Lord, 375
Kelvin scale, 34
ketone, 461
kilocalorie (Cal; kcal), 34
kinetic energy, 33
kinetic theory, 37
kinetics, 337, 351-60
Kirchhoff, Gustav, 76

L

lactose, 477-78
lanthanide series, 107
laser, 516
latent heat, 264
lattice energy, 139, 269
lattice structures, 266-67
laughing gas, 232
  See also **nitrous oxide**.
Lavoisier, Antoine, 11, 220
Lavoisier, Marie Anne, 11, 220
law of combining volumes, 243
law of definite composition, 68
law of energy conservation, 31
law of mass conservation, 31
law, 5
lead, 124, 264, 514
  oxidation number, 168
lead chromate, 211
Le Châtelier, Henri, 371
Le Châtelier's principle, 371-76
lemon juice
  pH of, 397
Lewis dot structures, 94
Lewis, Gilbert N., 392
light, 514
limestone, 120, 395
line spectrum, 73
linseed oil, 486
lipids, 484-87
lipoprotein
  high density (HDL), 487
  low density (LDL), 487
liquefaction, 38
  See also **condensation**.
liquid, 37
  distillation, 277-78
  kinetic description, 272
lithium, 118-19
litmus, 393, 405
lock-and-key model, 488
logarithm, 396-99
London Forces, 260
Long, Crawford, 459
Lowry, T. M., 391

M

magic number, 512
magnesium, 119-20
  reaction with acid, 209
  reaction with oxygen, 207
malleability, 22
maltose, 478
Mars Climate Orbiter, 46
mass, 46
mass composition, 191-92
mass defect, 518
matter, 21
  classification of, 21-24
  energy in, 29-34
  measurement, 43-47
  states of, 36-38
mayonnaise, 329-30
measurement, 44
measure of concentrations
  equivalents, 319
  molality, 321-22
  molarity, 318-19
  normality, 319-21
melting, 39, 264-65
melting points, 264-65
Mendeleev, Dmitri, 102
meniscus, 273
mercury
  in barometer, 229
  reaction with gold, 310
metal carbonates, 208
metal chlorates, 208
metal hydroxides, 208
  production of, 301
metallic bonding, 146
metallic hydrides, 118
metallurgy, 7
metal oxides, 207
metalloids, 106, 123
metals, 106
  activity series of, 210
  in ionic bonds, 137
  in metallic bonding, 146
  purification of, 432
methane, 447
  combustion, 342-43, 351
  molecular shape, 151
methanol, 312, 457
milk of magnesia, 397
Miller, Stanley, 422-23
miracles, 338
miscible, 310
mixtures, 24
model, 67
molal boiling point
  elevation constant, 323-25
molal freezing point
  depression constant, 325-27
molality, 321-22
  See also **colligative properties**.
molarity, 318-19
molar volume of a gas, 244
mole, 181-86
  conversion factors, 184
  stoichiometric
    conversions, 214-19
molecular formula, 187
molecular orbital theory, 157
molecular shapes, 150-54
molecule, 28
monatomic elements, 25
monosaccharides, 476-77
Mosley, Henry, 103
mothballs, 38
muscles, 483

N

Nagasaki, 503, 519
neon, 130-31
neptunium, 517
nerve gas, 233
net ionic equations, 211
neutralization reactions
  acids and bases, 392-93
  See also **titration**.
neutral salt, 407
neutron, 73
Newlands, John, 102
nitrogen, 232
  liquid, 281, 237
nitrogen oxide
  reaction with oxygen, 202-3
nitrous oxide, 232
noble gases, 130-31
Noddack, Ida and Walter, 220
nomenclature
  flow chart of, 172
  of binary acids, 178
  of binary compounds, 173
  of hydrates, 177
  of organic compounds,
    449, 451-52, 467
  of polyatomic ions, 175
  of ternary acids, 178
nonelectrolyte, 431
nonmetal oxides
  acid anhydrides, 302
nonmetals, 106
  covalent bonding, 141-45
  ionic bonds, 138-41
  multiple oxidation states, 168
nonpolar molecules
  intermolecular forces, 260
normal boiling point, 277
normality, 319-21
nuclear binding energy, 518
nuclear chemistry, 13, 501-20
nuclear equations, 505-7
nuclear fallout, 519
nuclear fission, 514-15
nuclear fusion, 515-17
nuclear power
  safety, 519-20
nuclear shell model, 511-12
nuclear weapons, 519
nucleic acids, 491
nucleotides, 491
nucleus, 72

O

oceans, 295
octet rule, 136
oils, 485
Oparin, A. I., 422-23
orbital notation, 84
orbitals, 78, 148
organic chemistry, 12, 443-71
osmosis, 327-28
osmotic pressure, 327
oxidation, 418
oxidation number, 163
   metallic hydrides, 170
   multiple, 167
   peroxides, 170
   polyatomic ions, 168
   rules, 164-69
oxidation-reduction reactions, 417-30
oxides, 127
oxidizing agent, 421
oxyacids, 208
oxyanions, 175
oxygen, 127, 232
   discovery of, 10
   in redox reactions, 418
   oxidation number of, 165
   test for, 355-56
ozone, 92

P

palladium, 4
Paracelsus, Philippus, 10
paraffin, 261
paramagnetism, 122, 157
partial charges, 154
partial pressures, 239
particle accelerators, 511-12
particle molality, 323
Pascal, Blaise, 229
pascal (Pa), 229
Pauli exclusion principle, 84
Pauling, Linus, 114
pentane
   isomers, 450
peptide bond, 482
peptidization, 330
perbromic acid, 179
percent by mass, 317-18
percent composition, 188-89
perfumes, 453
periodic table
   chart, 104-5, inside back cover
   European Convention Periodic Table, 107
   history of, 101-3
   IUPAC Periodic Table, 108
   North American Convention Periodic Table, 107
   parts of, 106
peroidic trends
   atomic radii, 110-11
   electron affinity, 113
   electronegativity, 114
   ionic radii, 110-11
   ionization energy, 112
periods, 106
permeability, 327-28
peroxides, 170
petroleum, 466
pharmacology, 10
phase, 24
phase diagram, 292-93
pheromones, 491
Philistines, 8
philosophy, 8
pH meter, 397
pH scale, 396-99
Phoenicians, 7
phosphate
   in DNA, 491-92
   ion, 411
phosphoric acid, 212-19, 390
phosphorus, 126
photography, 418-19
photosynthesis, 339
physical changes, 23
physical chemistry, 13
physical properties, 22
pi bond, 149
plasma, 37
plastic, 457
pleated sheet, 483
plum-pudding model, 71
pOH scale, 396-97
polar bonds, 154
polar molecules, 154-58
   intermolecular forces, 259-60
polarity, 155-56
polonium, 220
polyatomic compounds, 175
polyatomic elements, 25
polyatomic ions, 145, 175
polyester, 469
polyhydroxyl alcohols, 458
polymers, 469
polymorphs, 268
polypeptide, 482
polysaccharides, 479-80
post-transition metals, 123
potassium, 119
   reaction with water, 301
potassium permanganate, 429-30
potassium uranyl sulfate, 502
precipitate, 201
precision, 44
pressure
   atmospheric, 229
   Boyle's law, 234
   effect on equilibrium, 373-74
   effect on solubility, 314-15
   Gay-Lussac's law, 236
   law of partial pressures, 239-40
   measurement of, 229
   vapor, 241
Priestley, Joseph, 10
principal energy levels, 74
   capacity of, 81-83
   possible sublevels, 80-81
problem solving, 58-59
products, 201
propane, 447-49
proteins, 481-84
proton, 72
   in acids, 389-90
protonation, 391
pure number, 44
pure science, 15
pure substances, 24
pyramidal, 152

Q

qualitative analysis, 116
quantitative analysis, 116
quantized, 76
quantum model, 70, 80-88
   bonding, 148-59
   valence bond theory, 148
   valence shell electron pair repulsion theory, 150-51
quantum numbers, 86-88

R

R (universal gas constant), 251
rad, 505
radioactive age-dating
   carbon-14, 510
   potassium-argon, 510
   uranium-lead, 510
radioactive decay series, 508
radioactivity, 502
radium, 120, 220
radon, 503
Raleigh, Lord, 130
Raoult's law, 323
rate law, 359
rates of reactions, 355-60
reactants, 201
reaction mechanisms, 358

reactions
  addition, 466-67
  combination, 207
  condensation, 469
  decomposition, 208
  double replacement, 211
  oxidation-reduction, 417-30
  rates, 355-60
  single replacement, 209
  spontaneous, 337, 353-355
  substitution, 466-68
  synthesis, 207
  See also **equilibrium**.
reasoning
  deductive, 2
  inductive, 2
red blood cells, 488
redox reactions, 417-30
reducing agent, 420
reduction, 418
refining, 278
rem, 505
rhenium, 220
rhodium, 5
rickets, 489
RNA, 491-92
Roentgen, Wilhelm, 71
roentgen (R), 505
Roman numeral system, 176
rubbing alcohol, 457
Rush, Benjamin, 12
Rutherford, Daniel, 126
Rutherford, Ernest, 71
  model of atom, 72

S

salt, 119, 207
  formation of, 406
  solubility product constants, 315
  types of, 407
salt bridge, 434-35
Sandia National Laboratory, 516
saponification, 470-71
sarin, 233
saturated, 313, 447
scanning tunneling microscopy (STM), 67
science, 4
scientific law, 5
scientific method, 4
scientific notation, 52
scrubbers, 395
scurvy, 489
seawater, 397
second law of thermodynamics, 32, 344
semiconductors, 123
sensible heat, 264

SI, Système International, 45
sigma bond, 149
significant digits, 48
silicon, 125
silver, 121, 418-19
silver chloride, 383
silver nitrate, 383
single replacement reactions, 209
soap, 470-71
soda ash, 171
sodium, 118-19
  reactions, 209
sodium carbonate, 171
sodium chloride
  crystal structure, 138-41
  formation of, 138
  hydration of, 312
  unit cell, 267
sodium hydroxide, 471, 305, 397
solidification, 39
solids, 37
  amorphous, 263
  crystalline, 263
  freezing of, 264
  kinetic description of, 262-63
  melting of, 265
  properties of, 262-63
  sublimation of, 265
solubility
  effect of pressure, 315
  effect of temperature, 315
  factors that affect, 314-15
  graph, 315
  Henry's law, 35
  of salts, 379-84
solutes, 309
  effects on solutions, 322-28
  polar, 312-13
solutions, 24, 309-28
  concentrated, 317
  dilute, 317
  dissociation of, 311-12
  electrolytic, 431
  equilibria, 313-14
  measuring concentration of, 317-22
  molality, 321-22
  molarity, 318-19
  normality, 319-21
  percent by mass, 317-18
  rates of, 314
  saturated, 313
  solvation, 311
  supersaturated, 313-14
  types of, 309-11
  unsaturated, 313
  See also **acids** and **bases**.

solvation, 311
solvent, 309
  polar, 312
Sorensen, S. P. L., 396
specific heat, 279
spectator ions, 211
spectroscopy, 76-77
spontaneous combustion, 354
standard conditions
  gas laws, 234
  thermodynamics, 340-41
standard molar enthalpy of formation, 340
starches, 479-80
sterling silver, 438
steroids, 486-87
stewardship, 395, 520
Stock system, 176
  flow chart, 177
stoichiometry, 212-19
STP, standard conditions, 234
strong nuclear forces, 511
structural formula, 187
structural isomers, 449-50
sublevels, 80
  energies of, 82-83
  in periodic table, 109
  types of, 80-81
sublimation, 39, 265-66
substituent, 448
substituted hydrocarbons, 456-65
substitution reaction, 209, 466-68
sucrose, 478
sugar, 417-24
  hydration of, 311
sulfate ion
  nomenclature, 174
  oxidation number of, 169, 174
sulfides, 128
sulfur, 127
  allotropic forms, 268-70
  oxides, 127
sulfur dioxide, 233
sulfuric acid, 394, 401
  in batteries, 393
Sumerians, 7
supercritical mass, 515
supersaturation, 313-14
surface area
  effect on reaction rate, 356
surface tension, 272
surfactant, 273
symbols, 25
  history of, 26
synthesis reactions, 207
Système International (SI), 45

T

tear gas, 233
technetium, 220
Teflon®, 130
temperature, 33
  calorie, 34
  Charles's law, 235-36
  critical, 278
  effect on equilibrium, 374
  effect on reaction rate, 356
  effect on solubility, 314-15
  Gay-Lussac's law, 236
  Kelvin scale, 34
  specific heat, 279
ternary acids, 178
tetrahedral, 151
theory, 4
theory of relativity, 514
thermal energy, 33
thermodynamics, 31, 337-51
Thomson, J. J.
  model of atom, 69-71
thymine, 492
tin can, 437
titration
  acid-base, 408-10
  redox, 428-30
titration curve, 408
TNT, 193, 455
torr, 229
Torricelli, Evangelista, 229
transition metals, 121
transmutation, 517
transuranium elements, 122, 517
trigonal planar, 153
triple covalent bonds, 142
triple point
  of water, 293
Tyndall, John, 329
Tyndall effect, 329

U

ultraviolet radiation, 423
unit analysis, 56
unit cell, 267
unit of measurement, 44
universal gas constant (R), 251
unsaturated, 313, 451
uracil, 492
uranium, 220, 502
Urey, Harold, 422

V

valence bond theory, 148
valence electrons, 93
  in bonding, 135-36
valence shell electron repulsion theory (VSEPR), 150
van der Waals forces, 259
vanillin, 455
vapor pressure, 241, 275-76
  of water, 241
vinegar, 397
viscosity, 274
vitamin C, 489
vitamins, 488-91
  fat soluble, 490
  water soluble, 491
volcanoes, 243
Volta, Alessandro, 434
voltaic cell, 434-36
volume
  Boyle's law, 234
  Charles's law, 235-36
  water, 294

W

warming curve, 264-65
  specific heat, 280-82
water
  acid-base reactions, 299, 406-7
  amphiprotic, 403
  anhydrides, 302
  autoprotolysis, 394
  boiling points of, 291-92
  combustion, 298
  covalent bonding of, 141
  crystal structure of ice, 294-95
  decomposition of, 299-301
  electrolysis, 300
  enthalpy of formation, 339-40
  formation of, 298-99
  hydrates, 304-5
  hydration, 304
  hydrogen bonding, 290
  in compounds, 304-5
  melting point, 291-92
  molecular shape, 290-91
  percent composition, 188-89
  properties of, 291
  purification, 296-97
  reactions, 301-2
  stability of, 298
  surface tension, 294
  triple point, 293
  vapor, 232
  warming curve, 264, 281
water of hydration, 304
weight, 46
wetness, 273
wheat, 479
Wieman, Carl, 38
Wöhler, Friedrich, 446
Wollaston, William, 4
wood, 479-80
  See also **cellulose**.

X

X rays, 71

Z

zinc, 439
  galvanization, 437

Photograph Credits

The following agencies and individuals have furnished materials to meet the photographic needs of this textbook. We wish to express our gratitude to them for their important contribution.

Cover
Digital Stock (rock formation); PhotoDisc. Inc. (all others)

Front Matter
BJU Press Files i, ii-iii; Digital Stock iii(left, middle); PhotoDisc, Inc. iii(right)

Chapter 1
PhotoDisc, Inc. x-1(both), 14, 15(all except architect), 16(both); Unusual Films 2(vessels, coin), 15(architect); Egyptian Tourist Authority 2(King Tutankhamen); www.arttoday.com 2(bottom left), 3(Marie Curie), 11; Astronomy Charted 2(bottom right); Edgar Fahs Smith Collection, University of Pennsylvania Library 3(Mendeleev, Bohr); Parke Davis, division of Warner Lambert 7, 9; NPS, photo by Richard Frear 10; Eastman Chemicals 12; George R. Collins 18

Chapter 2
Unusual Films 20-21(background), 23(foreground), 24; PhotoDisc, Inc. 20-21(foreground), 22, 23(top, bottom background), 31, 37(left, right); Eastman Chemicals Division 28; NASA 37(middle)

Chapter 3
PhotoDisc, Inc. 42-43(background), 52, 53; Stem Labs, Inc. 43(inset); OHAUS Corporation 49(a); Unusual Films 49(b); Corning Glassworks 49(c)

Chapter 4
IBM Corporation, Research Division, Almaden Research Center 66-67; Fisher Scientific Company 68; Unusual Films 70(top); Edgar Fahs Smith Collection, University of Pennsylvania Library 70(bottom); www.arttoday.com 72, 76, 77(top left), 77(bottom); Princeton University, courtesy of AIP Emilio Segre Visual Archives 73, 77(top right)

Chapter 5
PhotoDisc, Inc. 100-101(both), 116(both), 120(c), 121(b), 125(all), 128(top); NASA 118(both); Unusual Films 119, 121(d), 124(bottom), 126(both), 129, 130(all); BJU Press Files 120(a); NPS, photo by Fred Mang, Jr. 120(b); Encore Medical Corp., Austin, Texas 121(a); Clemson University EM Lab 121(c); Ward's Natural Science Establishment, Inc. 124(top); Westfälisches Amt für Denkmalpflege 128(bottom both); Goodyear Tire and Rubber Co. 131(top); www.arttoday.com 131(bottom)

Chapter 6
PhotoDisc, Inc. 134(background); Eastman Chemicals Division 134(foreground); George R. Collins 140; Photo courtesy of NC State University Physics Demonstrations 157

Chapter 7
PhotoDisc, Inc. 162-63; Unusual Films 170, 171, 178, 182; Eastman Chemicals Division 190

Chapter 8
PhotoDisc, Inc. 198-99(all); Unusual Films 207, 209(both), 211; Edgar Fahs Smith Collection, University of Pennsylvania Library 220(top); New York Academy of Medicine Library 220(bottom)

Chapter 9
PhotoDisc, Inc. 226-27, 227, 250(top); Unusual Films 228(top-both, bottom background), 237, 240, 241, 250(bottom); www.arttoday.com 228(bottom foreground); NOAA 243, 246

Chapter 10
Brian D. Johnson 258-59; PhotoDisc, Inc. 259, 270(both), 273(top), 281(left); Ward's Natural Science Establishment, Inc. 263(top), 268(bottom); Unusual Films 263(bottom), 265, 266(all), 268(top-all), 269, 272(both), 273(middle, bottom), 274, 277, 281(right-both)

Chapter 11
PhotoDisc, Inc. 288-89, 303; Unusual Films 291, 295(right), 301(both), 302(all), 304, 305(all); Naval Photographic 295(left); Eastman Chemicals Division 297(left); USDA 297(center); George R. Collins 297(top right); Du Pont Co. 297(bottom right); BMW 298

Chapter 12
PhotoDisc, Inc. 308-9, 319, 328, 329; Stephen Christopher 310; Unusual Films 311, 312, 313, 315, 316(bottom), 324(both); Dead Sea Periclase, Ltd. 316(top); Digital Stock 326

Chapter 13
NOAA 336-37; PhotoDisc, Inc. 337, 339, 345(a), 354(bottom right); Unusual Films 338, 345(b), 351(both), 355(both); NETZSCH Instruments, Inc. 340; Edgar Fahs Smith Collection, University of Pennsylvania Library 347; Corel Corporation 354(left, top right)

Chapter 14
PhotoDisc, Inc. 364-65; Unusual Films 366, 372, 377, 383; Eastman Chemicals Division 371(top); Edgar Fahs Smith Collection, University of Pennsylvania Library 371(bottom); NASA 375(top right); www.arttoday.com 375(top left, bottom left); Fisher Scientific Co. 375(left middle); USDA 379

Chapter 15
Unusual Films 388-89, 392(bottom), 397, 404(bottom), 404(bottom), 405; PhotoDisc, Inc. 389, 395(left, top right); Edgar Fahs Smith Collection, University of Pennsylvania Library 390, 391(both), 392(top), 396; Brian D. Johnson 395(bottom right); E.M. Laboratories 404(top)

Chapter 16
PhotoDisc, Inc. 416-17, 417, 420, 421; Unusual Films 418, 431(all); George R. Collins 422; www.arttoday.com 437; American Electroplating Company 438; ALCOA 439

Chapter 17
PhotoDisc, Inc. 442-43, 446(bottom); Unusual Films 444, 455, 456(top), 458, 460, 461, 466, 469(both), 471; Edgar Fahs Smith Collection, University of Pennsylvania Library 446(top); Eastman Chemicals Division 447, 456(bottom); Corel Corporation 448(left), 462(bottom); Suzanne R. Altizer 448(right); Kenneth Frederick 462(top)

Chapter 18
PhotoDisc, Inc. 474-75, 475; NPS, photo by Fred Mang, Jr. 476(top); Corel Corporation 476(bottom); USDA 480; Unusual Films 485, 489

Chapter 19
PhotoDisc, Inc. 500-501, 501, 513, 520; Burndy Library, courtesy AIP Emilio Segre Visual Archives 502; Unusual Films 503, 504; U.S. Energy Department and Resource Administration 519(top); National Archives 519(bottom)

Back Matter
Digital Stock 523(left, middle); PhotoDisc, Inc. 523(right)

Chapter Opener Header, Sample Problem Box Background
PhotoDisc, Inc. (all chapters)